COMPREHENSIVE CHEMICAL KINETICS

COMPREHENSIVE

Section 1. THE PRACTICE AND THEORY OF KINETICS (3 volumes)

Section 2. HOMOGENEOUS DECOMPOSITION AND ISOMERISATION REACTIONS (2 volumes)

Section 3. INORGANIC REACTIONS (2 volumes)

Section 4. ORGANIC REACTIONS (6 volumes)

Section 5. POLYMERISATION REACTIONS (3 volumes)

Section 6. OXIDATION AND COMBUSTION REACTIONS (2 volumes)

Section 7. SELECTED ELEMENTARY REACTIONS (1 volume)

Section 8. HETEROGENEOUS REACTIONS (4 volumes)

Section 9. KINETICS AND CHEMICAL TECHNOLOGY (1 volume)

Section 10. MODERN METHODS, THEORY, AND DATA

CHEMICAL KINETICS

EDITED BY

R.G. COMPTON

M.A., D.Phil. (Oxon.)
University Lecturer in Physical Chemistry
and Fellow, St. John's College, Oxford

VOLUME 27

ELECTRODE KINETICS: REACTIONS

ELSEVIER
AMSTERDAM–OXFORD–NEW YORK–TOKYO
1987

ELSEVIER SCIENCE PUBLISHERS B.V.
Sara Burgerhartstraat 25,
P.O. Box 211, 1000 AE Amsterdam, The Netherlands

Distributors for the United States and Canada

ELSEVIER SCIENCE PUBLISHING COMPANY INC.
52 Vanderbilt Avenue
New York, NY 10017

ISBN 0-444-41631-5 (Series)
ISBN 0-444-42879-8 (Vol. 27)

with 140 illustrations and 8 tables

Printed in the Netherlands

COMPREHENSIVE CHEMICAL KINETICS

ADVISORY BOARD

Professor C.H. BAMFORD

Professor S.W. BENSON

Professor LORD DAINTON

Professor G. GEE

Professor G.S. HAMMOND

Professor W. JOST

Professor K.J. LAIDLER

Professor SIR HARRY MELVILLE

Professor S. OKAMURA

Professor N.N. SEMENOV

Professor Z.G. SZABO

Professor O. WICHTERLE

Volumes in the Series

	Section 1. THE PRACTICE AND THEORY OF KINETICS (3 volumes)
Volume 1	The Practice of Kinetics
Volume 2	The Theory of Kinetics
Volume 3	The Formation and Decay of Excited Species
	Section 2. HOMOGENEOUS DECOMPOSITION AND ISOMERISATION REACTIONS (2 volumes)
Volume 4	Decomposition of Inorganic and Organometallic Compounds
Volume 5	Decomposition and Isomerisation of Organic Compounds
	Section 3. INORGANIC REACTIONS (2 volumes)
Volume 6	Reactions of Non-metallic Inorganic Compounds
Volume 7	Reactions of Metallic Salts and Complexes, and Organometalic Compounds
	Section 4. ORGANIC REACTIONS (6 volumes)
Volume 8	Proton Transfer
Volume 9	Addition and Elimination Reactions of Aliphatic Compounds
Volume 10	Ester Formation and Hydrolysis and Related Reactions
Volume 12	Electrophilic Substitution at a Saturated Carbon Atom
Volume 13	Reactions of Aromatic Compounds
	Section 5. POLYMERISATION REACTIONS (3 volumes)
Volume 14	Degradation of Polymers
Volume 14A	Free-radical Polymerisation
Volume 15	Non-radical Polymerisation
	Section 6. OXIDATION AND COMBUSTION REACTIONS (2 volumes)
Volume 16	Liquid-phase Oxidation
Volume 17	Gas-phase Combustion
	Section 7. SELECTED ELEMENTARY REACTIONS (1 volume)
Volume 18	Selected Elementary Reactions
	Section 8. HETEROGENEOUS REACTIONS (4 volumes)
Volume 19	Simple Processes at the Gas–Solid Interface
Volume 20	Complex Catalytic Processes
Volume 21	Reactions of Solids with Gases
Volume 22	Reactions in the Solid State

Section 9. KINETICS AND CHEMICAL TECHNOLOGY (1 volume)

Volume 23 Kinetics and Chemical Technology

Section 10. MODERN METHODS, THEORY, AND DATA

Volume 24 Modern Methods in Kinetics
Volume 25 Diffusion-limited Reactions
Volume 26 Electrode Kinetics: Principles and Methodology
Volume 27 Electrode Kinetics: Reactions

Contributors to Volume 27

E.J. CALVO	Argentine Science and Technology Research Council (CONICET), Sector Electroquímica Aplicada, Instituto Nacional de Technologia Industrial (INTI), 1650 Buenos Aires, Argentina
A. HAMNETT	Inorganic Chemistry Laboratory, University of Oxford, Oxford OX1 3QR, Gt. Britain
E.J.M. O'SULLIVAN	IBM, T.J. Watson Research Center, P.O. Box 218 Yorktown Heights, NY 10598, U.S.A.
M.J. WEAVER	Department of Chemistry, Purdue University, West Lafayette, IN 47907-3699, U.S.A.

Preface

Volumes 26 and 27 are both concerned with reactions occurring at electrodes arising through the passage of current. They provide an introduction to the study of electrode kinetics. The basic ideas and experimental methodology are presented in Volume 26, whilst Volume 27 deals with reactions at particular types of electrode. Thus, Chapter 1 of the present volume deals with redox reactions at metal electrodes, Chapter 2 with semiconducting electrodes and Chapter 3 with reactions at metal oxide electrodes. Both theoretical aspects and experimental results are covered.

The editor thanks Mr. Geoffrey Stearn for his assistance in compiling the index.

Oxford
July 1987

R.G. Compton

Contents

Chapter 2 (A. Hamnett)

Chapter 3 (E.J.M. O'Sullivan and E.J. Calvo)

Chapter 1

Redox Reactions at Metal–Solution Interfaces

MICHAEL J. WEAVER

1. Introduction

The transfer of electrons between metals and solution species at electrode–electrolyte interfaces has long been the subject of experimental study. The kinetics of such electrode reactions have much in common with electron transfer and related processes in homogeneous solution. This is reflected in close similarities in the microscopic description of heterogeneous and homogeneous charge-transfer processes. Nevertheless, our understanding of the kinetics of electrode processes has developed in a distinctly different direction from that for other types of charge-transfer reaction. Research in the homogeneous redox area has emphasized the relationships between molecular structure and redox reactivity for an increasingly diverse range of systems (for example, for bioinorganic and intramolecular redox processes). Related theoretical work has chiefly been concerned with providing more detailed descriptions of the energetics of single electron-transfer steps, leading to an increasing interplay between theory and experiment [1]. However, studies of electrochemical kinetics have been relatively unaffected by these developments and have, until recently, remained largely on a phenomenological level. This can be traced not only to the apparently more complex methodologies employed to examine electrochemical reactions, but also to the distinctly different interests and outlook of electrochemists in comparison with researchers concerned with homogeneous redox processes in condensed phases.

The primary objective of the present chapter is to present a detailed description of mechanistically simple electrochemical reactions in relation to contemporary theoretical treatments of single electron-transfer steps. Although these treatments have been developed primarily in conjunction with homogeneous redox processes, with suitable modification they are equally applicable to electrochemical reactions. Besides emphasizing the close connections between heterogeneous and homogeneous redox processes, this approach serves to clarify the various structural factors expected to influence electrochemical kinetics. In addition, selected comparisons will be made with experiment in order to examine the strengths and limitations of the theoretical treatments and to suggest possible directions for future experimental and theoretical work. While a number of reviews dealing with the fundamentals of electron transfer have appeared recently [1], they are concerned almost exclusively with homogeneous-phase processes. It is hoped that this unified approach to electrochemical kinetics in relation to the more widely considered homogeneous redox processes will help to diminish the artificial gulf that persists between these two areas of research endeavor.

2. Basic characteristics. Reaction types

2.1 EXPERIMENTAL RATE PARAMETERS

We shall consider the simple generalized electrochemical reaction expressed as

$$\text{Ox} + n\,\text{e}^- \,(\text{electrode}, E) \rightleftharpoons \text{Red} \tag{1}$$

where Ox and Red are the oxidized and reduced forms of the redox couple, respectively, in bulk solution and E is the electrode potential of the metal–electrolyte interface measured versus a reference electrode. A description of phenomenological electrode kinetics and details of their measurement are provided in Vol. 26, Chap. 1 of this series. Nevertheless, in order to clarify the relationships between experimental kinetic parameters and the underlying theoretical models, a brief summary of the former will now be given, although methodological aspects are not considered here.

For solution redox couples uncomplicated by irreversible coupled chemical steps (e.g. protonation, ligand dissociation), a standard (or formal) potential, E^0, can be evaluated at which the electrochemical free-energy driving force for the overall electron-transfer reaction, $\Delta\bar{G}^0_{\text{rc}}$, is zero. At this potential, the electrochemical rate constants for the forward (cathodic) and backward (anodic) reactions k^{c} and k^{a} (cm s^{-1}), respectively, are equal to the so-called "standard" rate constant, k^{s}. The relationship between the cathodic rate constant and the electrode potential can be expressed as

$$k^{\text{c}} = k^{\text{s}} \exp\left[-\alpha F(E - E^0)/RT\right] \tag{2}$$

where α is the cathodic transfer coefficient. Since α is often predicted (and observed) to be close to 0.5 (vide infra), the measured rate constants are commonly very sensitive to the electrode potential, changing by tenfold every 100–150 mV. This emphasizes the unique feature of electrochemical processes in chemical kinetics in that their energetics can be drastically yet continuously modified by adjusting the free energy of the reactant (or product) electron(s) brought about by alterations in the independent electrical variable, the electrode potential.

In addition to rate constants measured as a function of potential at a given temperature, electrochemical activation parameters obtained from temperature-dependent rate data also yield useful information. Unfortunately, such measurements have seldom been made by electrochemists, probably due largely to confusion on the most appropriate way of controlling the electrical variable as the temperature is varied and the widespread (although incorrect) belief that greater interpretative difficulties are faced than with activation parameters for ordinary chemical processes.

Two types of electrochemical activation enthalpies can be distinguished [2]. The first, which have been termed "real" activation enthalpies, $\Delta H_r^{\ddagger}$, are obtained from the temperature dependence of standard rate constants or, more generally, from the temperature dependence of rate constants measured at a fixed standard overpotential $(E - E^0)$. The second, so-called "ideal" activation enthalpies, $\Delta H_i^{\ddagger}$, are obtained from the temperature dependence of the rate constant measured at a fixed metal–solution potential difference ϕ_m, the so-called "Galvani potential". Although it is not strictly possible to control ϕ_m exactly under these conditions, it can be achieved to a good approximation by holding the electrode potential constant using a non-isothermal cell with the reference electrode held at a fixed temperature [2]. The relation between these two types of activation enthalpies for one-electron reduction reactions at a given potential and temperature is [2, 3]

$$\Delta H_i^{\ddagger} = \Delta H_r^{\ddagger} + \alpha T \Delta S_{rc}^0 \tag{3}$$

where ΔS_{rc}^0 is the "reaction entropy", i.e. the difference in ionic entropy between Red and Ox [4]. Although $\Delta H_i^{\ddagger}$, rather than $\Delta H_r^{\ddagger}$, approximates the actual enthalpic barrier at the electrode potential at which it is determined, $\Delta H_r^{\ddagger}$ is nevertheless a valuable quantity since it embodies a correction for the entropic driving force, ΔS_{rc}^0. In particular, $\Delta H_r^{\ddagger}$ determined at the standard electrode potential approximates the intrinsic activation enthalpy, i.e. the activation enthalpy corresponding to the absence of an enthalpic driving force [2] ("thermoneutral condition", see Sect. 3.4). For "chemically irreversible" reactions, where E^0 is unknown, only $\Delta H_i^{\ddagger}$ can be obtained since k^s, and hence $\Delta H_r^{\ddagger}$, cannot be determined. Nevertheless, approximate estimates of $\Delta H_r^{\ddagger}$ can still be obtained in suitable cases by using eqn. (3) providing that ΔS_{rc}^0 can be estimated from experimental data for structurally similar redox couples.

Given that the reaction kinetics of the forward and backward reactions are first order in Ox and Red, respectively, measurements of k^s, k^c, or k^a, and α, $\Delta H_r^{\ddagger}$ and/or $\Delta H_i^{\ddagger}$ provide a detailed phenomenological description of the electrochemical kinetics for solution-phase reactants at a given electrode–electrolyte interface. It is also of fundamental interest, however, to evaluate rate parameters for adsorbed (or "surface attached") reactants or reaction intermediates (Sect. 2.3).

2.2 MECHANISTIC CLASSIFICATION. MULTIPLE-STEP ANALYSES

Before describing in detail the theoretical treatments appropriate to electrochemical reactions, it is desirable to outline the broad types of reaction mechanisms that are encountered.

Several distinctly different classifications can be established. The simplest of these concerns the spatial position of the reaction site in the interfacial region. So-called *outer-sphere* reaction pathways are defined as those where neither the reacting species, nor (where applicable) its coordinated ligands, penetrates the inner layer of solvent molecules immediately adjacent to the electrode surface in the transition state for electron transfer. By contrast, *inner-sphere* pathways involve the formation of a transition state where the inner layer is penetrated by the reacting species so that direct contact is made with the metal surface. These definitions are exactly analogous to those for homogeneous redox reactions [5] in that the inner-layer solvent can be regarded as the electrode's "coordination sphere", similar to the ligands surrounding metal ion reactants in homogeneous solution.

A related, yet distinct, classification of electrode reactions can be made on the basis of the degree of interaction between the reactant and the metal surface in the transition state for electron transfer. So-called *weak overlap* electron-transfer reactions are defined as those where the interaction between the two reacting centers (the reacting species and the electrode surface) are sufficiently weak and non-specific so that the activation free energy for the electron-transfer step is largely unaffected by their proximity, being determined instead by the properties of the isolated reactant. In contrast, *strong-overlap* reactions involve sufficiently strong interactions between the reacting centers so as to significantly diminish the activation energy [6].

The distinction between the energetics of weak- and strong-overlap

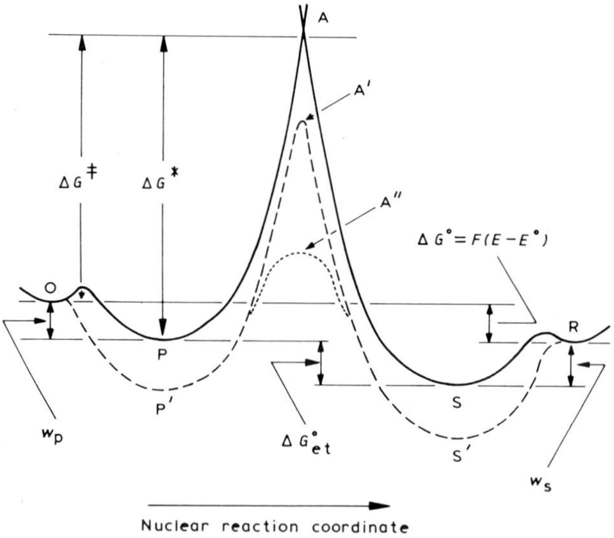

Fig. 1. Schematic free energy–reaction coordinate profiles for a single-electron electroreduction involving solution reactant O and product R at a given electrode potential E, occurring via three different reaction pathways, PAS, P′A′S′, and P′A″S″. Pathway PAS involves energetically favorable precursor and successor states (P and S) but with a weak-overlap transition state. Pathways P′A′S′ and P′A″S′ involve energetically similar precursor and successor states, but with the latter involving strong overlap in the transiton state.

processes for simple reactions is illustrated in Fig. 1. The solid curve represents a plot of the free energy of the reacting species against a reaction coordinate. The thermodynamic states O and R represent the bulk-phase oxidized and reduced forms, respectively, whereas P and S represent these reactant and product species (known as "precursor" and "successor" states) formed in the interfacial region that immediately precede and follow the transition state. The curve PAS represents the elementary free-energy barrier to electron transfer. For weak-overlap reactions, states O and P, and R and S will differ little in free energy; most importantly, the curve PAS will approximate that corresponding to large spatial separations between the electron donor and acceptor sites. This weak-overlap limit will also be approached for suitably large reactant–electrode separations, although the degree of electronic coupling between the donor and acceptor orbitals may be sufficiently weak as to yield only a very small probability of crossing between the OPA and ASR curves once the transition state A has been reached. (This circumstance is known as a "non-adiabatic" reaction pathway, vide infra.) The occurrence of weak-overlap pathways is unique to electron-transfer reactions and arises from the wave-like properties of electrons enabling reactions to occur even in the absence of chemical bond formation between the reacting centers.

The closer proximity of the reactant to the metal surface will not only increase this "crossing probability", κ_{el}, eventually to unity, yielding so-called adiabatic pathways, but can also result in increasing energy differences between O and P, and between R and S. Corresponding energy changes in A are also expected since the transition state, as well as the precursor and successor states, are also situated in the interfacial region. The broken curve OP′A′S′R schematically depicts the influence upon the free-energy profile caused by favorable reactant–electrode interactions. For strong-overlap reactions, which are anticipated when a chemical bond to the surface is formed, the shape of the elementary barrier can differ markedly from PAS. This is illustrated by the dotted curve P′A″S′, which reflects a stronger overlap pathway than the P′A′S′ profile. Depending on the extent of the reactant–surface interactions, the free energy of A will be decreased to a greater extent than that of P and S, yielding a smaller free-energy barrier ΔG^* (Fig. 1) (see Sects. 3.5.1 and 4.6).

Outer-sphere pathways are normally expected to be of the weak-overlap type since reactant–electrode bond formation is, by definition, absent. Nevertheless, substantial energy differences between O and P, and R and S may occur (so-called "double-layer" effects, Sect. 3.5.2), at least as a result of electrostatic reactant–surface interactions. Inner-sphere pathways may be either of the weak- or strong-overlap type, depending on the strength and specificity of the reactant–surface binding (Sect. 3.5.1).

The schematic free-energy profiles in Fig. 1 also illustrate the relationships between the various free-energy barriers of fundamental significance in electron-transfer kinetics. The activation free energy for the overall electroreduction reaction, $\Delta G^{\ddagger}$, is related to the activation free energy for the elementary step, ΔG^{*}, at a given electrode potential E by [7]

$$\Delta G^{\ddagger} = \Delta G^{*} + w_{\mathrm{p}} \tag{4a}$$

$$= \Delta G^{*} - RT \ln (K_{\mathrm{p}}/K_{\mathrm{o}}) \tag{4b}$$

where the "work term", w_{p}, (the free-energy difference between O and P) is equivalently expressed as an equilibrium constant, K_{p}, for forming the precursor state, together with a statistical term, K_{o}. The dependence of $\Delta G^{\ddagger}$ upon the electrochemical driving force $F(E - E^{0})$ may be expressed as [cf. eqn. (2)]

$$\Delta G^{\ddagger} = \Delta G^{\ddagger}_{E0} + \alpha F(E - E^{0}) \tag{5}$$

where $\Delta G^{\ddagger}_{E0}$ is the value of $\Delta G^{\ddagger}$ at the standard potential E^{0}.

A quantity of key theoretical significance is the so-called intrinsic barrier, $\Delta G^{*}_{\mathrm{int}}$; this equals ΔG^{*} for the particular case where the driving force for the elementary electron-transfer step, $\Delta G^{0}_{\mathrm{et}}$, equals zero. We can write

$$\Delta G^{*} = \Delta G^{*}_{\mathrm{int}} + \alpha_{\mathrm{et}} [F(E - E^{0}) + (w_{\mathrm{s}} - w_{\mathrm{p}})] \tag{6}$$

where α_{et} is the transfer coefficient (or symmetry factor) for the elementary step and w_{s} is the work of forming the successor state from the bulk product. (For comments on α_{et}, see Sect. 3.6.) The relation between $\Delta G^{\ddagger}_{E0}$ and $\Delta G^{*}_{\mathrm{int}}$ is obtained from eqns. (4)–(6).

$$\Delta G^{\ddagger}_{E0} = \Delta G^{*}_{\mathrm{int}} + [w_{\mathrm{p}} + \alpha_{\mathrm{et}} (w_{\mathrm{s}} - w_{\mathrm{p}})] \tag{7}$$

Equation (7) expresses an important distinction between the activation free energy for the overall electrochemical reaction in the absence of a *net* driving force, $\Delta G^{\ddagger}_{E0}$, and the intrinsic barrier for the electron-transfer step, $\Delta G^{*}_{\mathrm{int}}$. The former is most directly related to the experimental standard rate constant, whereas the latter is of more fundamental significance from a theoretical standpoint (vide infra). It is therefore desirable to provide reasonable estimates of w_{p} and w_{s} so that the experimental kinetics can be related directly to the energetics of the electron-transfer step.

For one-electron transfer reactions occurring via outer-sphere mechanisms, w_{p} and w_{s} can be estimated on the basis of electrostatic double-layer models. Thus, if the reaction site lies at the outer Helmholtz plane (o.H.p.), $w_{\mathrm{p}} = ZF\phi_{\mathrm{d}}$ and $w_{\mathrm{s}} = (Z - 1)F\phi_{\mathrm{d}}$, where Z is the charge number of the oxidized species and ϕ_{d} is the potential across the diffuse layer. Rewriting eqn. (7) in terms of rate constants rather than free energies yields the familiar Frumkin equation [8]

$$\ln k^{\mathrm{s}}_{\mathrm{ob}} = \ln k^{\mathrm{s}}_{\mathrm{corr}} + \frac{(Z - \alpha_{\mathrm{et}})F\phi_{\mathrm{d}}}{RT} \tag{8}$$

where k^s_{ob} and k^s_{corr} are the observed and work-corrected standard rate constants, respectively. The latter quantity is the estimated value of k^s_{ob} for the hypothetical case where the steps preceding and following the electron-transfer step (i.e. the work terms) are absent. This rate constant is clearly of fundamental interest because it is related directly to the intrinsic barrier for the single electron-transfer step (Sect. 3.6). The estimation of w_p and w_s for outer-sphere processes using more sophisticated treatments is noted in Sect. 3.5.2.

Such work-term corrections can be broadened to include rate constants at any electrode potential E. Thus, eqn. (7) can be generalized as

$$\Delta G^{\neq} = \Delta G^{*}_{corr} + [w_p + \alpha_{et}(w_s - w_p)] \tag{7a}$$

where ΔG^{*}_{corr} is the work-corrected free energy of activation (i.e. corresponding to $w_p = w_s = 0$) at the same electrode potential as ΔG^{*}; α_{et} can be related to the observed transfer coefficient α [eqns. (2) and (5)] by

$$\alpha = \alpha_{et} + \left[\alpha_{et}\left(\frac{dw_s}{FdE}\right) + (1 - \alpha_{et})\left(\frac{dw_p}{FdE}\right)\right] \tag{7b}$$

Equation (8) can also be expressed more generally as

$$\ln k_{ob} = \ln k_{corr} + \frac{(Z - \alpha_{et})F\phi_d}{RT} \tag{8a}$$

where k_{corr} is the value of k_{ob} that would be observed at a particular electrode potential in the absence of work terms and α_{et} is now related to α by

$$\alpha = \alpha_{et} + (Z - \alpha_{et})(d\phi_d/dE) \tag{8b}$$

Comparison of eqns. (4a) and (7a) shows that ΔG^{*} and ΔG^{*}_{corr} are related by

$$\Delta G^{*} = \Delta G^{*}_{corr} + \alpha_{et}(w_s - w_p) \tag{9}$$

This relation emphasizes that only part of the double-layer correction upon $\Delta G^{\neq}$ arises from the formation of the precursor state [eqn. (4a)]. Since the charges of the reactant and product generally differ, normally $w_p \neq w_s$ and so, from eqn. (9) the work-corrected activation energy, ΔG^{*}_{corr}, will differ from ΔG^{*}. [This arises because, according to transition-state theory, the influence of the double layer upon $\Delta G^{\neq}$ equals the work required to transport the *transition state*, rather than the reactant, from the bulk solution to the reaction site (see Sect. 3.5.2).] Equation (9) therefore expresses the effect of the double layer upon the elementary electron-transfer step, whereas eqn. (4a) accounts for the work of forming the precursor state from the bulk reactant. These two components of the double-layer correction are given together in eqn. (7a).

The above analysis emphasizes that all reactions involving bulk-solution species necessarily involve several elementary steps, each of which needs to be analyzed carefully in order to relate the observed kinetics to the electron-transfer barrier. Nevertheless, the influence of the steps preceding and succeeding electron transfer can be accounted for by using relationships such as eqn. (8a) if the reactant–electrode interactions are weak and non-specific. Such one-electron reactions can therefore be regarded as "single-step" processes. Examples are metal cation redox couples such as $Ru(NH_3)_6^{3+/2+}$ and $Cr(OH_2)_6^{3+/2+}$ and molecule–radical anion couples in aprotic media. Some processes involving strongly adsorbed reactants enable the kinetics of the electron-transfer step to be examined directly (Sect. 2.3) and are therefore true single-step reactions.

However, many electrochemical reactions involve other steps coupled to a rate-determining electron-transfer stage, involving other redox and/or chemical transformations for which the thermodynamics are unknown, thereby precluding estimation of E^0 and hence the standard rate constant, k_{corr}, for the desired electron-transfer step. Such processes can be regarded as occurring via "multiple-step" mechanisms. For example, $Co(NH_3)_6^{3+}$ reduction occurs via a one-electron rate-determining step followed by a rapid ligand dissociation step to form a solvated Co(II) species [5]. Since this latter step is rapid and irreversible, E^0 for the $Co(NH_3)_6^{3+/2+}$ couple is unknown so that k^s_{corr} for this couple cannot be determined even though the reaction may well occur via an outer-sphere mechanism. A similar difficulty is faced for many multiple electron-transfer reactions, such as Zn^{2+} reduction in aqueous solution. Although E^0 for the Zn^{2+}/Zn(Hg) couple can be determined, electron transfer probably occurs in two separate steps, the latter being coupled with ion (or atom) transfer between the solution and metal phases. Thus, even under conditions where the first step is rate-determining, the appropriate E^0 at which to determine k^s_{corr} is that for this *single* electron-transfer step, which is unknown.

Most electrochemical reactions involving non-metal inorganic or organic species, especially in protic media, also occur in several steps. Nevertheless, significant progress has been made recently in understanding the observed kinetics of some homogeneous reactions involving such species by isolating redox intermediates using pulse radiolysis, enabling the thermodynamics of the required single electron-transfer step to be extracted [9]. The majority of inner-sphere reactions occur via relatively complex multiple-step mechanisms; a well-studied example is that of oxygen reduction in aqueous media [10]. The electroreduction of solvated protons under conditions where proton discharge is rate-determining is often regarded by electrochemists as the archetypically simple electrode reaction. While certainly of much fundamental interest, our understanding of this process, even when proton transfer is rate-determining, is hindered by a lack of knowledge of w_s (i.e. the adsorption free energy of the hydrogen atom intermediate) on many metals, together with the thermodynamics of forming this species from molecular hydrogen. These complications commonly preclude estimating the thermodynamics of the charge-transfer step from the measured standard potential.

Therefore, an underlying problem faced in the fundamental interpretation of electrochemical kinetics, as with other chemical processes, is that it is often difficult to relate the measured rate constants to the intrinsic electron-transfer barrier and other parameters referring to the electron-transfer step since, often, only incomplete information is available on the thermodynamics of the individual steps comprising the overall reaction. For this reason, this chapter is concerned primarily with mechanistically simple electrochemical processes for which the observed kinetics can be related in a direct manner to parameters of contemporary theoretical significance. Nevertheless, useful information can be obtained by comparing rate parameters for structurally related reactions, where the reaction thermodynamics can be varied in a known fashion or held essentially constant. Much insight has also been obtained into the structural factors influencing electrochemical reactivity, even for quite complicated multiple-step reactions, by examining rate responses to systematic changes in system state under conditions where the unknown thermodynamic or other parameters can be arranged to cancel. Such "relative rate" comparisons are considered further in Sect. 4.1.

2.3 THE PRE-EQUILIBRIUM TREATMENT. UNIMOLECULAR REACTIVITIES

The foregoing discussion emphasizes the desirability of treating one-electron electrochemical reactions as involving an electron-transfer step occurring within a precursor state previously formed in the interfacial region. It is therefore useful to separate the overall observed rate constant, k_{ob}, into a precursor equilibrium constant, K_p (cm) [eqn. (4b)], and a unimolecular rate constant, k_{et} (s^{-1}), for the elementary electron-transfer step, related by [7]

$$k_{ob} = K_p k_{et} \tag{10}$$

This relation will be valid for both inner- and outer-sphere pathways provided that electron transfer, rather than precursor-state formation, is the rate-determining step. The precursor equilibrium constant can be expressed as

$$K_p = \frac{\Gamma_p}{C_r} \tag{10a}$$

where Γ_p (mol cm^{-2}) is the reactant surface concentration in the precursor state and C_r (mol cm^{-3}) is the corresponding reactant concentration in the bulk solution.

For processes involving strongly adsorbing reactants, Γ_p is often sufficiently large to be measured by surface electroanalytical methods (e.g. chronocoulometry, double-layer capacitance), enabling K_p to be evaluated. This allows k_{et} to be determined from k_{ob} using eqn. (10). Alternatively, k_{et} can often be evaluated directly in such circumstances by using electrochemical pulse techniques. An important class of such systems involves reactant molecules that are "covalently attached", or otherwise irreversibly adsorbed, to electrode surfaces [11]. The direct measurement of k_{et} for such systems is facilitated by the opportunity to generate high surface reactant concentrations in the presence of little or no bulk reactant. It is interesting to note that electron-transfer reactions involving such surface-attached (or specifically adsorbed) species are directly analogous to intramolecular reactions involving binuclear redox systems in homogeneous solution. Such heterogeneous reactions can therefore be perceived as "surface intramolecular" processes [12]. The experimental evaluation of k_{et} is normally limited to reactions proceeding via inner-sphere pathways since suitably large values of K_p usually require the formation of a reactant–surface bond. Nevertheless, k_{et} values may be obtained for some outer-sphere reactions that involve very strong electrostatic attraction of the reactant into the diffuse layer, such as that arising from multicharged cations at electrodes carrying large negative charges (see Sect. 2.4). This situation is exactly analogous to the measurement of k_{et} for outer-sphere homogeneous reactions between multicharged cationic and anionic complexes [1b].

The measurement of k_{et} for single electron-transfer reactions is of particular fundamental interest since it provides direct information on the energetics of the elementary electron-transfer step (Sect. 3.1). As for solution reactants, standard rate constants, k^{s}_{et}, can be defined as those measured at the standard potential, E^0_a, for the adsorbed redox couple. The free energy of activation, ΔG^*, at E^0_a is equal to the intrinsic barrier, ΔG^*_{int}, since no correction for work terms is required [contrast eqn. (7) for solution reactants] [3]. Similarly, activation parameters for surface-attached reactants are related directly to the enthalpic and entropic barriers for the elementary electron-transfer step [3].

2.4 THE LOCATION OF THE REACTION SITE. EXPERIMENTAL DISTINCTION BETWEEN OUTER- AND INNER-SPHERE PATHWAYS

In order to understand the manner in which the interfacial region influences the observed kinetics, especially in terms of the theoretical models discussed below, it is clearly important to gain detailed information on the spatial location of the reaction site as well as a knowledge of the mechanistic pathway. Information on the latter for multistep processes can often be obtained by the use of electrochemical perturbation techniques in order to detect reaction intermediates, especially adsorbed species [13]. Various in-situ spectroscopic techniques, especially those that can detect interfacial species such as infrared and Raman spectroscopies, are beginning to be used for this purpose and will undoubtedly contribute greatly to the elucidation of electrochemical reaction mechanisms in the future.

Even for mechanistically simple processes, determination of the nature of the reaction site can be far from straightforward. Since it is of direct relevance to the meaningful application of the electron-transfer models described here, this matter will now be briefly discussed.

It is usually assumed that all reactants which lack a functional group capable of binding to the metal surface will follow outer-sphere pathways (Sect. 2.2) where the reaction site lies essentially at the outer Helmholtz plane (o.H.p.). There are, however, several reasons to regard this conventional view as being seriously oversimplified. There is strong evidence that the inner-layer thickness depends upon the hydrated radius of the supporting electrolyte ion present predominantly in the diffuse layer at the relevant electrode charges [14]. The position of the reaction site relative to the o.H.p. should therefore be sensitive to the relative sizes of the reactant and supporting electrolyte species. This expectation is substantiated by experimental data, which indicates that the response of electrochemical reactivities to systematic changes in the double-layer structure can be altered substantially by variations in either the size of the reactant [15] or the supporting electrolyte cation [15b–17]. Especially large effects of varying the supporting electrolyte cation are seen for electrode kinetics in non-aqueous media [17]. The majority of these results can be rationalized in terms of variations in the electrostatic double-layer effects as determined by the average potential profile [15, 16]. Indeed, approximate estimates of the average reactant–electrode distance in the transition state can be extracted from such measurements [15a]. More complex effects, however, occur in non-aqueous media at large negative electrode charges, probably associated with the very high o.H.p. concentration of supporting electrolyte cations present under such conditions [17].

Most treatments of such double-layer effects assume that the microscopic solvation environment of the reacting species within the interfacial region is unaltered from that in the bulk solution. This seems oversimplified even for reaction sites in the vicinity of the o.H.p., especially since there is evidence that the perturbation of the local solvent structure by the metal surface [18] extends well beyond the inner layer of solvent molecules adjacent to the electrode [19]. Such solvent-structural changes can yield considerable influences upon the reactant solvation and hence in the observed kinetics via the work terms w_p and w_s in eqn. (7a) (Sect. 2.2). While the position of the reaction site for inner-sphere processes will be determined primarily by the stereochemistry of the reactant–electrode bond, such solvation factors can influence greatly the spatial location of the transition state for other processes.

In particular, the reactant may penetrate the inner layer and contact directly the metal surface even in the absence of bona fide chemical interactions as a result of stabilizing image or Van der Waals interactions. This is most likely to occur with relatively weakly solvated species. Reactions occurring via such transition states can, in a sense, be considered to be inner- rather than outer-sphere processes. In terms of the above reaction classification, they nevertheless may be of the weak-overlap type if the

References pp. 56–60

reactant–surface interactions are mild. The unambiguous occurrence of outer- rather than inner-sphere reaction pathways for inorganic redox processes in homogeneous solution is limited to reactions involving substitutionally inert, coordinatively saturated, complexes for which contact of the reacting centers via a common ligand is clearly eliminated. Since most metal surfaces are "substitutionally labile" in that the adsorption–desorption rates at electrode surfaces are rapid, either inner- or outer-sphere pathways may dominate the observed kinetics depending on which provides the most stable transition-state structure.

It is therefore by no means obvious whether true outer-sphere reactions are commonly followed, even for reactants lacking adsorbable functional groups. Unfortunately, this point is only infrequently recognized. Even when outer-sphere mchanisms prevail, the electrode surface may exert important influences upon the reaction kinetics via solvation and other factors. Nevertheless, there are a number of experimental approaches that, in favorable circumstances, can enable distinctions between alternative outer- and inner-sphere pathways to be made with reasonable confidence and especially can diagnose the presence of the latter mechanistic type. The most useful of these are now summarized.

(a) Direct detection of precursor intermediate. As noted in Sect. 2.3, the two-dimensional concentration of reactant in the precursor state, Γ_p, may be evaluated directly by electroanalytical methods provided that Γ_p is sufficiently large. Although such detection does not in itself identify a particular reaction site, such deductions can often be made on chemical grounds. Most obviously, ligand-bridged inner-sphere reaction pathways are clearly implicated by similarities in the adsorption thermodynamics to that for specific adsorption of the corresponding uncoordinated ligands [20]. The presence of outer-sphere precursor states can also be diagnosed in cases where sufficiently high Γ_p values can be generated by electrostatic attraction of the reactant to interfaces coated by a monolayer of anions or other charged or dipolar species that nevertheless exclude the reactant from the inner layer [21]. Surface Raman or infrared spectroscopies can also yield information on the extent of the adsorbate–surface interactions and hence imply the location of the reaction site from comparisons in the vibrational spectra of the interfacial and bulk-phase reactants [22].

For the majority of electrode reactions, however, for which the precursor states are insufficiently stable to allow analytical detection, resort must be made to less direct methods of determining the reaction site, such as the following.

(b) The response of the reaction rate to the addition of non-reacting adsorbates. Modification of the inner-layer composition by the addition of specifically adsorbed non-reactants, especially simple chemisorbing anions, can provide valuable mechanistic information since the nature and extent of the reactant–adsorbate interactions should be very sensitive to the location of the

reaction site. Broadly speaking, one can distinguish between the effects of low coverages of specifically adsorbed ions, where the observed rate responses arise from electrostatic interactions, and those for adsorbate coverages approaching monolayer levels, where the availability of surface binding sites provides a major influence.

The addition of specifically adsorbing anions can provide a useful means of distinguishing between inner- and outer-sphere pathways for reactions involving cationic transition-metal complexes under both the former [15a, 23] and latter [12] conditions. While large rate increases are both anticipated and commonly observed for cationic complexes upon the addition of specifically adsorbed anions due to the electrostatic attraction of reactant into the double layer, rate decreases are seen instead for complexes containing adsorbable anionic ligands, even at low anion coverages [15a, 23]. The latter provides strong evidence for a ligand-bridged transition state and can be most satisfactorily interpreted in terms of the effect of electrostatic repulsion between the bridging and adsorbed non-reacting anions within the inner layer outweighing the net electrostatic attraction of the cationic redox center to the adsorbed anions [23]. More striking rate decreases upon addition of adsorbing anions are seen at some solid metal surfaces for which anion monolayers can readily be formed (e.g. halides at platinum and silver surfaces [12]). These effects arise from the virtual elimination of inner-sphere pathways by surface site competition if the added anion is more strongly adsorbed than the reactant. Besides providing a means of mechanism diagnosis, such anion monolayers yield surfaces at which reactions are obliged to proceed via outer-sphere pathways in a manner similar to homogeneous metal complexes (e.g. $Ru(NH_3)_6^{2+}$, $Fe(CN)_6^{4-}$) that act almost exclusively as outer-sphere reagents in view of their coordinative saturation and inability to bind to incoming reactants [24].

This approach to mechanism diagnosis is much less straightforward with uncharged reactants, although in principle a distinction between inner- and outer-sphere transition states can still be achieved if high adsorbate coverages are employed.

It is important to note that, for substitutionally labile reactants (e.g. Cd^{2+}, Eu^{3+}, H_3O^+), the added anionic or neutral adsorbates can themselves provide catalytic ligand-bridged pathways. Such effects for cation reductions by adsorbed anions have been widely studied [25]; although it is usually difficult to distinguish true ligand-bridging from electrostatic double-layer effects, this can be achieved in some cases from an analysis of the rate dependence on the added adsorbate coverage [25d]. The incorporation of the ligand bridge into the reaction product [as for the formation of Cr(III) complexes from Cr(II)] can also diagnose the presence of inner-sphere pathways, although this evidence is not as unambiguous as for the corresponding reactions in homogeneous solution [23].

(c) The influence of the electrode material. In principle, inner-sphere pathways can be diagnosed from their much greater sensitivity to variations in the

electrode material since the transition-state stability should be influenced strongly by specific reactant–surface interactions. This approach can indeed be useful, although, as noted above, the kinetics of even bona fide outer-sphere reactions may depend greatly on the electrode material due to variations in the specific reactant solvation as well as to electrostatic double-layer effects (see Sect. 4.6).

(d) Relative rate comparisons between corresponding electrochemical and homogeneous reactions. The development of criteria for the distinction between inner- and outer-sphere pathways for inorganic redox reactions in homogeneous solution [26] and the identification of several redox reagents that can only react via the latter route opens up an interesting route to examining electrode reaction mechanisms by means of relative rate comparisons with corresponding homogeneous redox processes. Such comparisons are discussed in Sect. 4.7.

3. Electron-transfer models

3.1 RATE FORMALISMS

We shall now outline contemporary theoretical treatments of single electron-transfer steps and relate them to experimental electrochemical rate parameters in the same manner as for the more widely considered reactions between pairs of redox couples in homogeneous solution. The role of one of the reacting centers (electron acceptor or donor) is taken by the metal surface which provides a rigid two-dimensional environment where reaction occurs. In some respects, electrode reactions represent a particularly simple class of electron-transfer processes since only one redox center is required to undergo activation and the metal surface may yield only a weak non-specific influence on the activation energentics. Moreover, electrochemical processes provide a unique opportunity to examine the effects upon the rate parameters of continuous variations in the reaction thermodynamics, brought about by alterations in the electrode potential. The treatment presented here reflects the close similarities in the physical features of electron transfer in electrochemical and homogeneous reaction environments. More detailed descriptions of a number of the theoretical concepts considered here can be found in the plethora of recent reviews devoted to redox processes in homogeneous solution [1].

We shall consider the generalized single electron-transfer reaction [cf. eqn. (1)]

$$\text{Ox} + \text{e}^- \text{ [electrode } (E - E^0)] \rightleftharpoons \text{Red} \tag{11a}$$

where Ox and Red are stable solution species. This can be compared with the generalized homogeneous reaction

$$Ox_1 + Red_2 \rightleftharpoons Red_1 + Ox_2 \qquad (11b)$$

Both these processes can be considered to occur in several distinct stages as follows: (i) formation of precursor state where the reacting centers are geometrically positioned for electron transfer, (ii) activation of nuclear reaction coordinates to form the transition state, (iii) electron tunneling, (iv) nuclear deactivation to form a successor state, and (v) dissociation of successor state to form the eventual products. At least for weak-overlap reactions, step (iii) will occur sufficiently rapidly ($\lesssim 10^{-16}$ s) so that the nuclear coordinates remain essentially fixed. The "elementary electron-transfer step" associated with the unimolecular rate constant k_{et} [eqn. (10)] comprises stages (ii)–(iv).

This rate constant can be related to the corresponding barrier height ΔG^* (see Fig. 2) by [1a, 7]

$$k_{et} = \nu_n \kappa_{el} \Gamma_n \exp(-\Delta G^*/RT) \qquad (12)$$

where ν_n (s^{-1}) is the nuclear frequency factor, κ_{el} is the electronic transmission coefficient (i.e. the fractional electron-tunneling probability in the transition state), and Γ_n is the nuclear tunneling factor. (The last term constitutes a correction to the rate constant from molecules that react without fully surmounting the classical free-energy barrier, ΔG^* [27].) The theoretical model associated with eqn. (12) has been termed a "semi-classical" treatment in view of the inclusion of quantum-mechanical correction factors [27]. As detailed below, the magnitudes of both ν_n and ΔG^* are determined by a combination of the various reactant vibrational and solvent reorientation modes.

The conventional treatment of outer-sphere electrode reactions, as in homogeneous solution, envisages electron transfer taking place via collisions between the reacting centers, so that the pre-exponential factor is commonly assumed to contain a collision frequency. However, several authors have recently pointed out that it is preferable to describe the reaction as occurring via an intermediate precursor state as for the inner-sphere processes, but with K_p now describing the probability of forming an "encounter state" with the electron donor and acceptor sites in suitably close proximity [1a, 7, 28, 29]. A comparison has been made between this "encounter pre-equilibrium" model and the "collisional" treatment for electrochemical reactions [7]. Strictly speaking, the measured reaction rate will be an integral of local rates associated with a myriad of subtly distinct reaction sites, each with a different spatial position of the reactant and correspondingly different values of κ_{el} and ΔG^* (Sect. 3.5.2) [28]. As noted above, reacting species located far from the electrode will contribute little to the measured rate, even when they are suitably activated, since κ_{el} will be vanishingly small. It is therefore useful to conceive of a "reaction zone" close to the electrode surface within which the reactant has to be located in order to exchange electrons with sufficient efficiency to contribute significantly to the reaction rate (also see Sect. 3.5.2).

We can write, for outer-sphere electrode reactions [7]

$$K_p = K_0 \exp(-w_p/RT) \tag{13a}$$

$$= \delta r \exp(-w_p/RT) \tag{13b}$$

where w_p is the average work term as in eqn. (4a) and δr (= K_0) is the "reaction zone thickness". The combined quantity $\delta r \kappa_{el}^0$, where κ_{el}^0 is the electronic transmission coefficient at the distance of closest approach, can be regarded as the "effective electron-tunneling distance" since it would equal the reaction zone thickness if adiabaticity was maintained (i.e. $\kappa_{el} = 1$) throughout this region. This quantity appears in the combined expression from eqns. (10) and (11) [30].

$$k_{ob} = \delta r \kappa_{el}^0 \nu_n \Gamma_n \exp(-w_p/RT) \exp(-\Delta G^*/RT) \tag{14a}$$

where w_p and ΔG^* are average values within the reaction zone. This relation can also be expressed in terms of a work-corrected rate constant, k_{corr}, [cf. eqns. (7a) and (8a)]

$$k_{corr} = \delta r \kappa_{el}^0 \nu_n \Gamma_n \exp(-\Delta G_{corr}^*/RT) \tag{14b}$$

The magnitude of $\delta r \kappa_{el}^0$ clearly depends not only on the magnitude of κ_{el}^0 but also upon the dependence of κ_{el} on the electrode–reactant separation, r (Sect. 3.3.2). Strictly speaking, one should also consider the spatial variation of both ΔG^* and w_p. This matter is considered further in Sect. 3.5.2.

3.2 FREE ENERGIES OF ACTIVATION

A central preoccupation of electron-transfer models is to provide estimates of the activation free energy of a single electron-transfer step, ΔG^*, from the structural and thermodynamic properties of the system [31]. Two key features of these models should be noted at the outset. Firstly, as noted above, it is advantageous to separate ΔG^* into intrinsic and thermodynamic components according to [cf. eqn. (6)]

$$\Delta G^* = \Delta G_{int}^* + \alpha_{et} \Delta G_{et}^0 \tag{6a}$$

where ΔG_{et}^0 is the driving force for the elementary step given by

$$\Delta G_{et}^0 = F(E - E^0) + (w_s - w_p) \tag{6b}$$

Most theoretical models are concerned with descriptions of ΔG_{int}^* and α_{et}, E^0 being obtained experimentally, and w_p and w_s either measured or estimated. The magnitude of the symmetry factor α_{et} is predicted to depend on the relative shapes of the reactant and product free-energy profiles as well as on their displacement (i.e. upon the driving force). If these profiles are symmetrical (equal-shaped) parabolas, it follows that [32, 33]

$$\alpha_{et} = 0.5 + \frac{\Delta G_{et}^0}{16 \Delta G_{int}^*} \tag{15}$$

so that the slope of the activation free energy-potential dependence should decrease with increasing overpotential. Much interest has been shown in testing this relation for simple electrochemical reactions (see Sect. 4.3).

Secondly, the intrinsic barrier is usefully separated into so-called "inner-shell" and "outer-shell" components, ΔG^*_{is} and ΔG^*_{os}, respectively. The former is associated with reorganization of the internal reactant coordinates (e.g. changes in metal–ligand bond distances, reactant conformation), whereas the latter arises from alterations in the polarization of surrounding solvent molecules. The former can be estimated on the basis of a simple harmonic oscillator model from [1a]

$$\Delta G^*_{is} = 0.5\Sigma f_i (\Delta a/2)^2 \qquad (16)$$

where Δa is the change in distance of a given bond between the oxidized and reduced forms of the redox couple and f_i is the corresponding individual force constant. Values of Δa can be obtained in suitable cases from X-ray crystallography or from EXAFS measurements [35, 36]. Estimates of f_i can often be obtained from Raman or infrared spectroscopic data [36].

A complication arises when the force constants for a given bond differ in the oxidized and reduced forms. This is commonly expected to be the case since the addition or subtraction of electrons will generally yield significant and even substantial changes in the reactant bonding. For homogeneous self-exchange reactions, this difficulty is surmounted by employing so-called "reduced" (or mean) force constants, f_r, in eqn. (16), which are related to the corresponding individual force constants for the oxidized and reduced forms, f_{ox} and f_{red}, respectively by [1a]

$$f_r = \frac{2f_{ox}f_{red}}{f_{ox} + f_{red}} \qquad (17)$$

This relation arises because the structurally more favorable transition state, i.e. that having the minimum reorganization energy resulting from summing the *pairs* of free energy-bond coordinate profiles for the two reacting species, is structurally non-symmetrical, even for self-exchange reactions [1a]. A slightly different procedure applies for simple electrochemical reactions since only a *single* reacting species is involved. This nevertheless also yields a non-symmetrical transition-state structure for electrochemical exchange since the reactant and product wells will generally have a different shape.

Rather than employ eqn. (16), values of ΔG^*_{is} for electrochemical reactions can therefore be obtained more appropriately from the intersection point of the individual free-energy curves for the oxidized and reduced forms [37]. The individual curves can be defined conveniently in terms of "reorganization parameters", λ_f and λ_r, for the forward and reverse reactions [cf. eqn. (16)]

$$\lambda_i = 0.5\Sigma f_i (\Delta a)^2 \qquad (18)$$

The required intersection point, and hence ΔG^*_{is}, can be obtained from the solution for $\Delta G^0_{et} = 0$ of the simultaneous equations [37].

$$\Delta G^* = \lambda_f X^2 \tag{18a}$$

$$\lambda_f X^2 = \lambda_r (1 - X)^2 + \Delta G^0_{et} \tag{18b}$$

where X is a dimensionless parameter (equalling Marcus' m parameter [32]) which characterizes the nuclear reaction coordinate.

The outer-shell intrinsic reorganization energy, ΔG^*_{os}, is usually treated on the basis of the dielectric continuum treatment. For one-electron electrochemical reactions, this yields [32]

$$\Delta G^*_{os} = \frac{e^2}{8}\left(\frac{1}{a} - \frac{1}{R_e}\right)\left(\frac{1}{\varepsilon_{op}} - \frac{1}{\varepsilon_s}\right) \tag{19}$$

where e is the electronic charge, R_e is the distance from the reactant to its image in the metal surface (i.e. twice the reactant–surface distance), and ε_{op} and ε_s are the optical and static dielectric constants, respectively. Some doubts have been expressed concerning the validity of eqn. (19) and related relations for homogeneous processes [31d, 38]. Recent calculations using a molecular dynamical model yielded smaller free-energy barriers than those obtained by using the dielectric continuum approach [39]. This question is clearly an important one since outer-shell barriers in the range ca. 4–6 kcal mol^{-1} are commonly obtained from eqn. (19) using typical parameters, so that even a, say, 20% error in estimating ΔG^*_{os} can correspond to a tenfold uncertainty in the calculated rate constant.

Experimental tests of the dielectric continuum model based on variations of the solvent medium are, unfortunately, quite limited; these are summarized in Sect. 4.5. It is nonetheless useful to examine the physical basis of eqn. (19) in order to ascertain the likelihood of its applicability. An enlightening derivation of eqn. (19) involves treating the formation of the transition state in terms of a hypothetical two-step charging process [31b, 40]. First, the charge of the reactant Ox is slowly adjusted to an appropriate value so that the solvent molecules are polarized to an extent identical to that for the transition state (step 1). Second, the charge is readjusted to that of the reactant sufficiently rapidly so that the solvent orientation remains unaltered (step 2), thereby yielding the non-equilibrium solvent polarization appropriate to the transition state.

This process is illustrated schematically in Fig. 2 [40b]. Figure 2(a) traces the alterations in the free energy of the reacting species; the states O and R refer to the chemical free energies, G^0_{os}, of the oxidized and reduced forms of the redox couple, respectively, that are associated with solute–solvent interactions. Step 1 entails moving down the curve OSR; the bowed shape arises from the quadratic dependence of G^0_{os} on the ionic charge that is anticipated from the Born model. State S is that formed upon completion of step 1, so that the vertical line ST refers to step 2 yielding the transition state

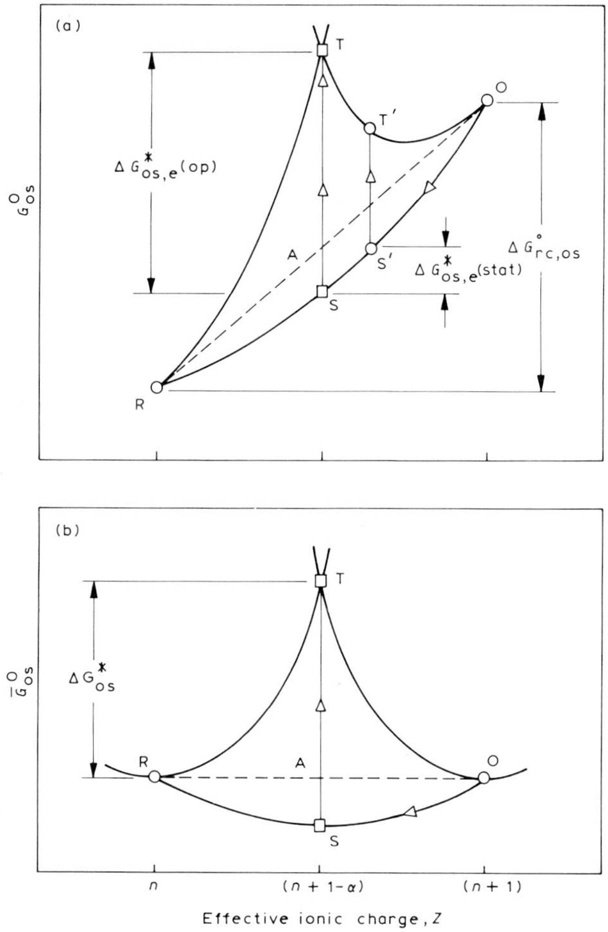

Fig. 2. Schematic plots outlining outer-shell free energy–reaction coordinate profiles for the redox couple O + e^- $\rightleftharpoons$ R on the basis of the hypothetical two-step charging process (Sect. 3.2) [40b]. The y axis is (a) the ionic free energy and (b) the electrochemical free energy (i.e. including free energy of reacting electron), such that the electrochemical driving force, $\Delta G^0 = F(E - E^0)$, equals zero. The arrowed pathways OT′S′ and OTS represent hypothetical charging processes by which the transition state, T, is formed from the reactant.

T. This state is at the intersection of the non-equilibrium polarization curves OT and RT corresponding to the oxidized and reduced forms, respectively. These curves can be constructed from this two-step charging process via a series of intermediate charged states such as the non-equilibrium polarization state T′ formed by the segments OS′ and S′T′ in Fig. 2(a).

Figure 2(b) is the corresponding plot for the *electrochemical* free energy, i.e. containing the free energy of the transferring electron, for the situation where $E = E^0$. At this potential, the change in solvational free energy between Ox and Red is balanced exactly by the electron contribution, so that O and R have the same electrochemical free energy. The symmetrical curves OT and RT represent the free energy-reaction coordinate profiles for the reactant and product species associated with solvent reorganization. In the dielectric continuum limit and ignoring reactant–electrode interactions, the energy change for step 1 [equal to AS in Fig. 2(b)] yields a "static" contribution to ΔG^*_{os}

$$\Delta G^*_{os}\ (\text{stat}) = -\frac{Ne^2}{8a\varepsilon_s} \tag{20}$$

whereas the energy change from step 2 (equal to AT) yields an "optical" contribution to ΔG^*_{os}

$$\Delta G^*_{os}\ (\text{op}) = \frac{Ne^2}{8a\varepsilon_{op}} \tag{21}$$

Combining these two contributions and accounting for reactant–electrode image interactions within the transition state [31b] yields eqn. (19).

The ε_s^{-1} and ε_{op}^{-1} terms in eqn. (19) can therefore be identified with steps 1 and 2, respectively. Since this relationship is derived from a similar analysis to the Born treatment of ionic solvation energies, it is often thought that the demonstrated severe deficiencies of the latter [41] are liable to be reflected in similarly large errors in estimating ΔG^*_{os} using the former. Fortunately, however, this supposition is liable to be incorrect. This is because the "static" Born changing component (step 1) typically makes only a small contribution to ΔG^*_{os}, since usually $\varepsilon_{op} \gg \varepsilon_s$. Thus, although the difference in solvation free energies between Ox and Red, ΔG^0_{os}, may be in severe and even qualitative disagreement with the Born model, the outer-shell intrinsic barrier necessarily refers to the condition where $\Delta G^0_{os} = 0$. The ε_s^{-1} term in eqn. (19) arises from the anticipated non-linear dependence of the solvation free energy on the ionic charge. The predominant "optical" contribution to ΔG^*_{os} (step 2) is much more likely to be described, at least approximately, by the dielectric continuum model since the relevant dielectric constant, ε_{op}, is determined solely by internal electronic polarization. This should be relatively insensitive to the microscopic solvent structure in the vicinity of the reacting species [40b].

The likely influence of specific reactant–solvent interactions upon ΔG^*_{os} can be estimated by evaluating the energetics of step 1 using experimental redox thermodynamic data in place of the Born model [40b]. This analysis shows that such effects may yield only small ($<$ 1–2 kcal mol^{-1}) contributions to ΔG^*_{os}, even when they provide dominant contributions to the redox thermodynamics [40b]. Such specific reactant–solvent interactions therefore appear to influence the electron-transfer barrier, ΔG^*, largely through the thermodynamic driving force component, $\alpha_{et}\Delta G^0_{et}$, rather than via the outer-shell intrinsic barrier [eqn. (6a), p. 16]. Nevertheless, the development of more molecular-based treatments of solvent reorganization seems overdue;

such approaches should yield further insight into the possible limitations of the dielectric continuum model of non-equilibrium solvent polarization.

It also is important to note that the aforementioned treatments of free-energy barriers refer only to weak-overlap reactions. This corresponds to the curve PAS in Fig. 1, where the transition-state energy is essentially unaffected (at least in a specific manner) by the proximity of the metal surface. When these reactant–electrode interactions become sufficiently strong and specific, marked decreases in both the inner- and outer-shell intrinsic barriers can be anticipated. This is discussed further in Sects. 3.5.1 and 4.6.

3.3 PRE-EXPONENTIAL FACTOR

Although preponderant attention has been devoted to theoretical descriptions of the barrier height, also of importance is the development of models for the dynamics of surmounting the electron-transfer barrier. As noted above, the pre-equilibrium treatment embodied in eqns. (10) and (12) is normally expected to be applicable to outer-, as well as inner-, sphere processes. We shall now consider theoretical models for each of the pre-exponential factors in the expression for k_{et} [eqn. (12)]: ν_n, κ_{el}, and Γ_n, in turn.

3.3.1 Nuclear frequency factor

For adiabatic reaction pathways (i.e. $\kappa_{el} = 1$) the nuclear frequency factor, ν_n (s^{-1}), represents the rate at which reacting species in the vicinity of the transition state is transformed into products. This frequency will be influenced by a combination of the various motions associated with the passage of the system over the barrier, approximately weighted according to their relative contributions to the activation energy. These motions usually involve bond vibrations and solvent motion, associated with the characteristic inner- and outer-shell frequencies, ν_{is} and ν_{os}, respectively. A simple formula for ν_n which has been employed recently is [1a, 7]

$$\nu_n = \left(\frac{\nu_{os}^2 \Delta G_{os}^* + \nu_{is}^2 \Delta G_{is}^*}{\Delta G_{os}^* + \Delta G_{is}^*}\right)^{1/2} \tag{22}$$

where ΔG_{os}^* and ΔG_{is}^* are now the outer- and inner-shell contributions to the free-energy barrier, ΔG_{et}^*.

This formula follows from the transition-state theory (TST) of unimolecular reactions [42]. Since it is commonly anticipated that $\nu_{is} \gg \nu_{os}$, eqn. (22) predicts that the effective frequency is dominated by the inner-shell frequency even when the outer-shell barrier provides a substantial contribution to ΔG_{et}^*. For metal–ligand, and other typical inner-shell vibrations, $\nu_{is} \approx 10^{13}\,s^{-1}$. Indeed, for the common circumstance where $\Delta G_{is}^* \gtrsim \Delta G_{os}^*$, we expect that $\nu_n \approx \nu_{is}$. This is intuitively reasonable since, on the basis of TST, we generally expect that the *fastest* motion along the reaction coordinate will control the frequency factor [28].

Recent theoretical treatments, however, suggest instead that the dynamics of solvent reorganization can play an important and even dominant role in determining ν_n, at least when the inner-shell barrier is relatively small [43–45]. The effective value of ν_{os} can often be determined by the so-called longitudinal (or "constant charge") solvent relaxation time, τ_L [43, 44]. This quantity is related to the experimental Debye relaxation time, τ_D, obtained from dielectric loss measurements using [43]

$$\tau_L = \frac{\varepsilon_\infty}{\varepsilon_s}\tau_D \tag{23}$$

where ε_∞ is the high- ("infinite-") frequency dielectric constant. [Note that $\varepsilon_\infty \neq \varepsilon_{op}$; normally for polar solvents $\varepsilon_\infty \sim$ 2–5 times ε_{op}.] The quantitative relation between τ_L^{-1} and the corresponding outer-shell frequency factor, ν_{os}^L, is dependent upon the shapes of the reactant and product free-energy profiles in the intersection region [43]. For exchange reactions where the electronic coupling is small ("weak overlap" limit) [43, 44]

$$\nu_{os}^L = \tau_L^{-1}\left(\frac{\Delta G_{os}^*}{4\pi k_B T}\right)^{1/2} \tag{24}$$

where k_B is the Boltzmann constant. Since τ_L commonly falls in the range ca. 10^{-11} to 2×10^{-13} s for polar solvents, for the typical value $\Delta G_{os}^* \sim 5$ kcal-mol^{-1}, ν_{os}^L lies in the range ca. 10^{11} to 3×10^{12} s^{-1}. For water, in this case $\nu_{os}^L \approx 3 \times 10^{-12}$ s^{-1} [46]. This frequency refers to "overdamped solvent relaxation", involving the concerted motion of large numbers of solvent molecules [43, 47]. For solvents where $\nu_{os}^L \lesssim 10^{-12}$ s, the rotational frequency of individual solvent molecules ("solvent inertial" effects), ν_{os}^R, can partly control ν_{os} [43].

It is important to recognize that, while ν_{os}^R represents solvent motion appropriate to the TST limit, and therefore is that relevant to eqn. (22), the situation where the solvent motion is predominantly overdamped (i.e. when $\nu_{os} \approx \nu_{os}^L$) refers to a breakdown of TST [47]. Physically, the latter corresponds to where the system is obliged to cross and recross the "intersection region", close (within k_BT) to the barrier top, a number of times before reaction occurs. This will yield a smaller rate than that corresponding to the TST limit since the latter refers to solvent reorientation enabling a single smooth transition of the system through the intersection region [47]. In the common circumstance where $\nu_{os}^L < \nu_{os}^R$, the former will dominate ν_{os}, corresponding to the "overdamped limit".

Although ν_{is} will still dominate ν_n if ΔG_{is}^* is sufficiently large, it is anticipated that $\nu_n \approx \nu_{os}^L$ for exchange reactions when the inequality

$$(\Delta G_{is}^*/\Delta G_{int}^*)^{1/2}\,\nu_{is}\exp(-\Delta G_{is}^*/k_BT) \gtrsim \tau_L^{-1} \tag{25}$$

is obeyed [48]. Note that, in contrast to eqn. (22), eqn. (25) predicts that ν_{os}^L will tend to dominate ν_n when $\nu_{os}^L \ll \nu_{is}$, i.e. the *slowest* activation mode controls the effective frequency factor. This is because eqn. (25) refers to systems for which the TST limit [associated with eqn. (22)], does not apply. Significantly, eqn. (25) predicts that the effective nuclear frequency factor can be dominated by solvent, rather than inner-shell, motion for significantly larger values of ΔG^*_{is} than are anticipated from eqn. (22) [45b]. Although the theoretical description of solvent motion for electron-transfer reactions is very incomplete as yet, these treatments demonstrate that solvent dynamics can exert an important influence upon rate constants for both electrochemical and homogeneous reactions via the frequency of crossing the free-energy barrier as well as the barrier height itself. Some comparisons with experimental data are noted in Sect. 4.4.

3.3.2 Electronic transmission coefficient

Relatively little attention has been given in the literature to the electronic transmission coefficient for electrochemical reactions. On the basis of the conventional collisional treatment of the pre-exponential factor for outer-sphere reactions, κ_{el} has commonly been assumed to equal unity, i.e. adiabatic reaction pathways are followed. Nevertheless, as noted above, the dependence of κ_{el} upon the spatial position of the transition state is of key significance in the "encounter pre-equilibrium" treatment embodied in eqns. (13) and (14). Thus, the manner in which κ_{el} varies with the reactant–electrode separation for outer-sphere reactions will influence the integral of reaction sites that effectively contribute to the overall measured rate constant and hence the effective electron-tunneling distance, $\delta r \kappa^0_{el}$, in eqn. (14).

The value of κ_{el} for a particular transition-state geometry is determined by the degree of overlap for the wave functions describing the donor and acceptor orbitals, as described by the electronic coupling matrix element H_{AB} [1a, d, 28, 49]. The degree of non-adiabaticity, as denoted by the extent to which $\kappa_{el} < 1$, depends upon the relative frequencies of electron tunneling, ν_{el}, and nuclear rearrangements, ν_n, in that $\kappa_{el} = \nu_{el}/\nu_n$. In other words, κ_{el} in eqn. (11) describes the extent to which electron tunneling within the transition state, rather than nuclear rearrangements, determines the effective frequency of crossing the free-energy barrier. Thus, when $\kappa_{el} \ll 1$, this frequency is actually determined entirely by ν_{el} rather than by ν_n.

The electron-tunneling frequency can be related to H_{AB} by [1a]

$$\nu_{el} = \frac{2H^2_{AB}}{h}\left(\frac{\pi^3}{4\Delta G^*_{int}RT}\right)^{1/2} \qquad (26)$$

where h is Planck's constant. To a first approximation, H_{AB} and hence ν_{el} are expected to depend exponentially on the donor–acceptor separation distance, r. Therefore, for non-adiabatic pathways ($\kappa_{el} < 1$), we can write

$$\kappa_{el}(r) = \kappa^0_{el} \exp\left[-\beta(r - \sigma)\right] \qquad (27)$$

where σ is the separation distance corresponding to the closest approach of the donor and acceptor sites and β is a constant. The last quantity is typically estimated to lie in the range ca. 1–2 Å [49] (see also Sect. 4.4).

Recent ab initio calculations of H_{AB} for $Fe(OH_2)_6^{3+/2+}$ self-exchange in homogeneous solution indicate that the electron transfer is non-adiabatic, even at the close contact distance, 6.5 Å, defined by the effective radii of the reactants [1d, 49a, b]. As a consequence, the optimum precursor complex geometry for electron transfer is predicted to involve interpenetration of the aquo ligands with a face-to-face configuration of the reactant octehedra so that $r = 5.5$ Å; even for this geometry, κ_{el} is calculated to be significantly below unity [49a]. Although so far limited to this reaction, these calculations are important because they suggest that outer-sphere reactions may utilize pathways which are marginally non-adiabatic, even for inorganic reactants with small saturated ligands.

Other theoretical activity has centered on the dependence of reaction non-adiabaticity upon the structure of the intervening medium as well as the donor–acceptor separation for intramolecular electron transfer [50], i.e. between donor and acceptor sites contained within a single species such as a binuclear complex. The electron-tunneling probability is predicted to be enhanced substantially by the presence of delocalized electron groups, such as aromatic ligands, between the reacting centers [50]. This is consistent with experimental studies of thermal and optically induced electron transfer within binuclear complexes [51].

Despite its importance, less attention has been devoted to reaction non-adiabaticity for electrochemical reactions. It has been claimed that electron transfer at metal electrodes should be adiabatic, even at large (10–20 Å) reactant–surface separations due to the continuum of surface electronic states [52]. Semi-empirical tunneling calculations performed for oxide-coated metal surfaces, however, suggest that the tunneling probability should decrease much more sharply with increasing reactant–surface separation [53]. A similar finding has been obtained from calculations for $Fe(OH_2)_6^{3+/2+}$ at platinum electrodes [54]. These latter assertions are borne out by recent experimental results (Sect. 4.4). Further theoretical work has distinguished between "elastic" and "inelastic" (or resonance) electron tunneling [53c]. The former refers to direct tunneling between isoenergetic states for the reactant and the metal conduction band and should normally predominate at bare metal surfaces or where the tunneling distances are relatively small ($\lesssim$ 10–15 Å). The latter involves tunneling via localized "impurity" and other states in the intervening medium (most plausibly in an oxide film or other "spacer" layer). Such inelastic tunneling mechanisms can involve different effective activation barriers and yield unusual potential-dependent behavior [53c].

3.3.3 Nuclear tunneling factor

The procedures for calculating the free-energy barrier outlined in Sect. 3.2 yield "classical" values of ΔG^*. In actuality, a certain fraction of molecules can react without fully surmounting this barrier, i.e. electron transfer can take place for nuclear configurations having energies below that for the classical transition state. This is referred to as "nuclear tunneling". The resulting correction to the rate constant, Γ_n, [eqn. (11)] is known as the nuclear-tunneling factor [1a, 27, 28]. Since this effect will always act to increase k_{et}, generally $\Gamma_n > 1$. The magnitude of Γ_n depends upon both the height and shape of the barrier as well as the temperature. Larger barriers and lower temperatures will yield greater values of Γ_n since the Boltzmann factor for surmounting the classical free-energy barrier will then be especially unfavorable. Larger values of Γ_n will also result from barriers associated with greater values of ν_n.

It should be recognized that, in contrast to the electron-tunneling probability κ_{el}, the nuclear tunneling factor is essentially a property of energy barriers associated with *individual* reactants rather than with *interactions between* the reacting centers. Consequently, the various detailed relationships for estimating Γ_n which have been outlined for homogeneous processes [1a] are applicable with only minor modification to electrochemical reactions. The magnitude of this nuclear-tunneling correction is commonly small, at least for outer-shell barriers and inner-shell barriers involving low-frequency (e.g. metal–ligand) modes, so that typically $\Gamma_n \sim 1$–3. This factor can nevertheless exert an important influence on activation parameters since Γ_n often increases sharply with decreasing temperature (Sect. 3.4).

3.3.4 *Numerical values of the pre-exponential factor*

It is of interest to estimate likely values of the combined pre-exponential factor in eqn. (14) for outer-sphere reactions

$$A_p = \delta r \kappa_{el}^0 \nu_n \Gamma_n \tag{28}$$

on the basis of the foregoing discussion, particularly in comparison with the more conventional collisional treatment. The numerical value of A can be very sensitive to the reactant structure as well as the reactant environment and mechanism, primarily through variations in the electron-tunneling and nuclear frequency factors. Nevertheless, for small inorganic reactants having moderate or large inner-shell barriers, $\nu_n \approx \nu_{is} \sim 1 \times 10^{13}\,\mathrm{s^{-1}}$ (Sect. 3.3.1). Some experimental evidence indicates that $\delta r \kappa_{el}^0 \sim 0.5$ Å for bona fide outer-sphere reactions, as anticipated for marginally non-adiabatic pathways yet with $\kappa_{el}^0 \to 1$ (Sect. 3.3.2) [30]. Inserting these values into eqn. (28) along with the assumption $\Gamma_n \sim 1$ yields $A_p = 5 \times 10^4\,\mathrm{cm\,s^{-1}}$.

This can be compared with the values of A determined from the collisional treatment whereby on the basis of the simple "gas-phase" model [32]

$$A_c = \left(\frac{k_B T}{2\pi m}\right)^{1/2} \tag{29}$$

where m is the reactant mass. Inserting the typical values $m = 200$, $T = 298\,\mathrm{K}$, yields $A_c = 5 \times 10^3\,\mathrm{cm\,s^{-1}}$. Although eqns. (28) and (29) are based on distinctly different physical models, the larger value of A_p compared with A_c can be attributed to the recognition of the pre-equilibrium model that molecules can react without having to collide with a particular reaction plane [7]. Similar numerical values of A_p and A_c can be predicted for certain conditions; for example, when $\nu_n \approx \nu_{os} \approx 10^{12}\,\mathrm{s^{-1}}$. Nevertheless, the pre-equilibrium model is generally to be preferred for estimating frequency factors since it provides the fundamentally more appropriate treatment for outer- as well as inner-sphere reactions under most experimental conditions [7].

3.4 ELECTROCHEMICAL ACTIVATION PARAMETERS

As for other types of chemical processes, examining the dependence of electrochemical rate constants upon temperature is of fundamental importance since, at least in principle, it provides a means of separating the pre-exponential and exponential factors contributing to electrochemical reactivity (Sect. 2.1). Equation (14b) can be written as

$$k_{corr} = A_p \exp(\Delta S^*_{corr}/R) \exp(-\Delta H^*_{corr}/RT) \tag{30}$$

where ΔS^*_{corr} and ΔH^*_{corr} are the work-corrected entropy and enthalpy of activation corresponding to ΔG^*_{corr}.

In order to extract estimates of A_p [eqn. (28)] from the temperature dependence of k_{corr}, it is clearly necessary to estimate both ΔS^*_{corr} and any temperature dependence of A_p. The activation entropies for electrochemical reactions are usually expected to be large since, in contrast to homogeneous redox processes, the net charge of the reactant and product species will inevitably differ, with consequent changes in the extent of solvent polarization attending electron transfer. Nevertheless, values of ΔS^*_{corr} can readily be estimated if the entropic driving force, ΔS^0_{rc}, ("reaction entropy") is known, using the relation [2c, 40a, 55]

$$\Delta S^*_{corr} = \alpha_{et} \Delta S^0_{rc} \tag{31}$$

providing that the solvation environment in the transition state (and hence the solvation entropy) is unaffected by its proximity to the electrode surface.

The influence of the entropic driving force can be circumvented automatically, however, by evaluating so-called "real" activation enthalpies, $\Delta H^{\ddagger}_r$ (Sect. 2.1), from the temperature dependence of the standard rate constant, k^s_{corr}, since eqn. (30) can be rewritten as [2b, c]

$$k^s_{corr} = A_p \exp(\Delta S^*_{int}/R) \exp(-\Delta H^*_{int}/RT) \tag{30a}$$

where ΔS^*_{int} and ΔH^*_{int} are the intrinsic entropic and enthalpic barriers (i.e. for which ΔS^0_{et} and ΔH^0_{et}, respectively, equal zero). Since ΔS^*_{int} is expected to be small or even negligible even when ΔS^0_{rc} is dominated by specific solute–solvent interactions [it is assumed to be zero in deriving eqn. (31)] then for $\Delta H_r^{\ddagger}$ determined at the standard electrode potential

$$\Delta H_{int}^{*} \approx \Delta H_r^{\ddagger} = -R\frac{d \ln k_{corr}^{s}}{d(1/T)} \qquad (32)$$

provided that the pre-exponential factor A_p is itself temperature-dependent.

However, either Γ_n and/or ν_n may vary significantly with temperature. The former circumstance will commonly be encountered for reactions with large inner-shell barriers, especially those associated with high ν_n values; the temperature dependence of Γ_n is conventionally treated formally in terms of a component of ΔS_{int}^{*} [27, 30]. The latter circumstance is anticipated when ν_n is dominated by outer-shell motion, especially as in eqn. (24) since τ_L can be strongly temperature-dependent [45]. Another complication is that the temperature dependence, as well as the absolute values, of the double-layer corrections may well be sufficiently uncertain as to yield large errors in estimating both ΔH_{int}^{*} and ΔS_{int}^{*} using this procedure. As noted in Sect. 2.1, values of $\Delta H_r^{\ddagger}$ cannot be obtained for chemically irreversible reactions. Estimates of A_p can nevertheless be obtained by evaluating "ideal" activation enthalpies, $\Delta H_i^{\ddagger}$, from the dependence of k_{corr} at a fixed non-isothermal electrode potential and obtaining corresponding values of $\Delta H_r^{\ddagger}$ from eqn. (3) provided that the entropic driving force term, ΔS_{rc}^{0}, can be estimated. Alternatively and equivalently, one can equate $\Delta H_i^{\ddagger}$ with ΔH_{corr}^{*} in eqn. (30) and extract A_p from this relation by inserting the value of ΔS_{corr}^{*} estimated from eqn. (31).

Similar analyses can be undertaken for surface-attached reactants (Sect. 2.3), with the advantage that a unimolecular frequency factor A_{et}, given by

$$A_{et} = \kappa_{el}\nu_n\Gamma_n \qquad (28a)$$

can be obtained after correction for the entropic driving force rather than the usual combined pre-exponential factor A_p [3, 7].

It should be emphasized that, contrary to widespread belief, the interpretation of electrochemical activation parameters involves no more complications and uncertainties than that for homogeneous redox or other chemical processes [2b, 3]. Thus, the estimation of the pre-exponential factor from activation parameters for homogeneous cross reactions, for example, requires a correction for the entropic driving force, ΔS_{rc}^{0}, in the same manner as for "ideal" electrochemical activation parameters (Sect. 2.1). This entropic correction is inherently contained in the "real" activation parameters determined for electrochemical exchange reactions (i.e. those measured at the standard electrode potential) as a consequence of variation in the electron free energy, $F\phi_m^0$, associated with the temperature dependence of the standard potential. [See eqn. (3); note that $F(dE_{ni}^0/dT) \approx F(d\phi_m^0/dT) = \Delta S_{rc}^0$, where E_{ni} is the non-isothermal standard cell potential [4]. For homogeneous self-exchange reactions, this entropic correction, although perhaps less self-evident, is provided by the co-reacting couple, since the entropic as well as enthalpic changes experienced by the redox couple undergoing reduction is exactly compensated by those for the reaction partner undergoing oxidation.

3.5 INFLUENCE OF REACTANT–SURFACE INTERACTIONS

3.5.1 Strong-overlap pathways

The foregoing theoretical treatment implicitly assumes that the interaction between the reacting species and the electrode is sufficiently weak and non-specific so that the energetics of the elementary electron-transfer step are determined by the properties of the isolated reactant and the surrounding solvent ("weak-overlap" pathway, Sect. 2.2). However, as noted in Sect. 2.2, the occurrence of inner-sphere pathways may not only alter the overall reaction energentics via stabilization of the precursor and successor states, but also via alterations in the shape of the electron-transfer barrier itself ("strong-overlap" pathway).

The circumstances that give rise to strong-, rather than weak-, overlap pathways can easily be understood in general terms with reference to corresponding arguments for inner-sphere reactions in homogeneous solution [1b, 51a]. Increasing spatial overlap between the donor and acceptor orbitals, even in the absence of chemical bonding between the reaction centers, can lead to significant "resonance splitting" of the potential-energy surfaces in the intersection region (transition state); this leads to an effective lowering of the free-energy barrier and is characterized by the electronic coupling matrix element H_{AB} [Sect. 3.3.2]. It has been estimated that H_{AB} values around 0.2–0.5 kcal mol^{-1} are sufficient to yield $\nu_{el} \gg \nu_n$ so that complete reaction adiabaticity is achieved [1d, 49a].

For some inner-sphere reactions, much larger H_{AB} values may occur if a strong chemical bond links the redox centers. Under these circumstances, the intrinsic free-energy barrier can be lowered substantially compared with that for the same reaction occurring via an outer-sphere or weak-overlap inner-sphere pathway [1b, 51a]. This is illustrated schematically in Fig. 1, where curve P′A″S′ would be typical for a strong-overlap pathway, and curves P′A′S and PAS for progressively weaker-overlap pathways. For electrochemical reactions, such a diminution of ΔG^*_{int} and hence ΔG^*_{et} can be regarded as "intrinsic catalysis" afforded by the inner-sphere pathway. Such catalysis can be distinguished from rate accelerations that arise purely from more favorable precursor stability, i.e. from greater cross-sectional reactant concentrations at the interface. The latter may be termed "thermodynamic catalysis" (see also Sect. 3.5.2).

It is important to note that systems that demonstrate strong "thermodynamic" catalysis do not necessarily give rise to significant "intrinsic" catalysis (even though the converse is usually expected to be true). This is because the former only requires that a surface bond is formed with a functional group residing on the reactant, perhaps distant from the redox center itself, whereas the latter requires that the electronic states of the donor and acceptor sites for electron transfer beocme strongly mixed. Extensive "thermodynamic" catalysis may then occur, even when the electron-transfer step itself is of the weak-overlap type, i.e. when the extent of "intrinsic" catalysis is negligible. This circumstance would be reflected in similar k_{et} values for a given reaction following an outer-sphere route as for the inner-sphere pathway (see Sect. 4.6).

Strong-overlap electrochemical reactions are clearly expected to be rapid, at least at potentials where the free-energy driving force for the electron-transfer step is not highly unfavorable. As H_{AB} increases, ΔG^{*}_{int} will correspondingly decrease and ultimately can, in principle, become smaller than $k_B T$ (0.6 kcal mol^{-1} at room temperature). Under these conditions, the oxidized and reduced forms of the adsorbed redox couples can lose their separate identity, even on the vibrational time scale, and the electron can be considered to be *delocalized*. This circumstance has been referred to as "Class III" behavior [51], originally in the context of optically induced electron transfer in mixed-valence compounds (i.e. homogeneous intramolecular systems) [56]. Strong-overlap systems having a significant electron-transfer barrier (i.e. the "valence-trapped" case) are labelled "Class II", whereas weak-overlap systems are termed "Class I" reactions under this categorization [56].

Such concepts are as applicable to electrochemical systems involving adsorbed (or "surface-attached") reactants as to homogeneous intramolecular processes, even though little is yet known of the extent to which strong-, rather than weak-, overlap reactions occur for the former. This question is clearly of central importance in electrocatalysis; experimental evidence on this matter is discussed in Sect. 4.6.

3.5.2 Work terms. "Double-layer" effects

Although the foregoing electron-transfer theory is preoccupied with describing the electron-transfer step itself, in order to understand the kinetics of overall reactions it is clearly also important to provide satisfactory models for the effective free energy of forming the precursor and successor states from the bulk reactant and product, w_p and w_s, respectively. As outlined in Sect. 2.2, it is convenient to describe the influence of the precursor and successor state stabilities upon the overall activation barrier using relations such as

$$\Delta G^{\ddagger} = \Delta G^{*}_{corr} + [w_p + \alpha_{et}(w_s - w_p)] \tag{7a}$$

The first term in brackets in eqn. (7a), w_p, is that associated with the precursor stability constant [eqns. (4) and (13)], which relates k_{ob} and k_{et} [eqn. (10)]. The second term, $\alpha_{et}(w_s - w_p)$, accounts for the effect of the double layer on the driving force for the electron-transfer step. Thus, when

$w_p \neq w_s$, the driving force for the electron-transfer step at the interface will differ from that in bulk solution.

Taken together, these two terms, comprising the conventional "double-layer" effect, can be thought of as the influence of the surface upon the transition-state stability, presuming that the reactant–surface interactions in the transition state are an *approximately weighted mean* of those in the adjacent precursor and successor states. (The "appropriate" weighting factor is the transfer coefficient α_{et}.) This therefore constitutes the "thermodynamic catalytic" influence of the surface, as distinct from the "intrinsic catalytic" effect as defined above. The former, but not the latter, is conventionally termed the "double-layer" effect, even though both, in fact, involve surface environmental influences upon the transition state stability.

The theoretical estimation of w_p and w_s for inner-sphere pathways is clearly difficult since the formation of chemical bonds is necessarily involved. However, at least for strongly adsorbed reactants, K_p is commonly sufficiently large so that it (and hence w_p) can be obtained experimentally from measurements of Γ_p [eqn. (10a)], even though some extrapolation of the K_p values from potentials where the reactant is stable to those of kinetic interest is normally required [20, 57]. Experimental estimates of K_s are usually more difficult to obtain, although it can sometimes be assumed that $K_p \sim K_s$ [12a, 57].

Few outer-sphere electrode reactions have precursor-state concentrations that are measurable [21] so that it is usual to estimate w_p and w_s from double-layer models. The simplest, and by far the most commonly used, treatment is the Frumkin model embodied in eqns. (8) and (8a) whereby, as noted in Sect. 2.2, the sole contributor to w_p and w_s is presumed to be electrostatic work associated with transporting the reactant from the bulk solution to the o.H.p. at an average potential ϕ_d. This potential is usually calculated from the Gouy–Chapman (GC) theory [58].

Several more sophisticated treatments have nevertheless been formulated. Firstly, statistical mechanical and Monte Carlo models of the diffuse layer have been developed which account for factors such as ion sizes, limitations of the Poisson–Boltzmann theory, and the presence of discrete solvent molecules [59–61]. By and large, the diffuse-layer potentials calculated from the GC model are typically within ca. 20–30% of, although systematically larger than, the values obtained from these latter treatments for commonly encountered diffuse-layer charges ($|q^m| \lesssim 30\,\mu\text{C cm}^{-2}$). Moderate deviations are seen between the ionic concentration profiles determined from the GC model and those obtained from the so-called hypernetted chain (HNC) approximation [61]. [Note that, in the HNC approach, the ionic concentration and average potential profiles are not related simply via a Boltzmann relation as they are in the simple GC model, as in eqn. (8), since the ionic concentration responds to a local, rather than an average, potential.] The deviations become larger as the charges on the ions, as well as the electrode charge density, is increased. These calculations suggest that the

GC model might be significantly in error for calculating concentration–distance profiles, at least for multicharged reactants or for systems where the size or charge type of the reacting species and supporting electrolyte ions differ. The latter type of system does not, as yet, appear to have been treated using more sophisticated diffuse-layer models.

Another approach employed for estimating interfacial reactant concentrations involves corrections to the usual GC–Frumkin (GCF) treatment from so-called "discreteness-of-charge" effects [62]. These involve (i) a "self-image" potential arising from the attractive interaction between the reacting species and its image in the metal surface, and (ii) an "exclusion disk" potential arising from the effect of the size and charge of the reacting ion on the spatial distribution of surrounding ions. These effects are estimated to be most likely to be significant for reaction sites inside the o.H.p. and where an important contribution to the double-layer charge is provided by specifically adsorbed ions [62]. nevertheless, the self-image effect appears to be small for ions at the o.H.p. or in the diffuse layer [62].

Other possible contributions to the work terms for outer-sphere reactions include the likelihood that the ion–solvent interactions will differ in the interfacial region from that in bulk solution resulting from the influence of the metal surface upon the local solvent structure. As noted in Sect. 2.4, this effect may be significant even for ions in the diffuse layer since the perturbation upon the solvent structure is liable to extend several layers out from the metal surfaces [19] (see also Sect. 4.6).

The ensemble of reaction sites with different spatial geometries that contribute to the observed rate constant for outer-sphere reactions (Sect. 3.1) may have not only differing values of κ_{el} and ΔG^*_{et} but also of w_p and w_s. At the high ionic strength supporting electrolytes commonly employed in electrochemical kinetics, the diffuse layer will be molecularly thin ($\lesssim 5$ Å) so that the effective "reaction zone thickness", δr (section 3.1), may be influenced as much by the variation of the cross-sectional reactant concentration as by the dependence of κ_{el} on the reactant–electrode distance. Stated more precisely, the spectrum of reactant geometries that contribute to the overall rate constant, featuring various reactant–electrode distances and reactant orientations, are commonly anticipated to involve large variations in w_p and hence Γ_p, as well as in κ_{el}. The usual notion of an "average" reaction site, associated with single values of w_p and w_s as well as κ_{el} and ΔG^*_{et}, is therefore strictly an oversimplification.

Strictly, then, we can write for outer-sphere electrochemical reactions (cf. ref. 28)

$$k_{ob} = \int_0^\infty g_i(r)\, k'(r)\, \mathrm{d}r \qquad (33)$$

where $k'(r)$ is the "local" rate constant at a given distance r from the electode surface and $g_i(r)$ is the corresponding electrode–ion correlation function. This latter quantity represents the local reactant concentration normalized to that in the bulk solution. Provided that the reaction is non-adiabatic even at the plane of closest approach (i.e. for $r = \sigma$), from eqn. (27) we can express the dependence of k' upon r as

$$k'(r) = k'(\sigma) \exp[-\beta(r - \sigma)] \tag{34}$$

In terms of $k'(r)$, the effective value of δr according to eqn. (34) equals β^{-1}; thus if $\beta = 1.5$ Å, then $\delta r \approx 0.7$ Å. If the reaction is adiabatic at the plane of closest approach, larger values of δr will arise since $k'(r)$ will then only decrease exponentially with increasing r for $r > \sigma$, i.e. for sufficiently large electrode–reactant distances so that non-adiabaticity is encountered. However, in terms of eqn. (33), the effective δr values can be affected substantially by the dependence of $g_i(r)$ upon r. For attractive electrode–reactant interactions, δr will be decreased since the reactant concentration will decrease sharply as r increases, whereas for repulsive interactions, δr will be increased.

Equation (34) neglects the dependence of ΔG^*_{et} upon r, arising from the distance dependence of ΔG^*_{os} as predicted by eqn. (19). However, given that $R_e = 2r$ [eqn. (19)] and probably that $r \gtrsim 6$ Å (Sect. 4.4), we expect that $R_e \gtrsim 4a$, which yields only a mild dependence of ΔG^*_{os} upon r. The physical origin of this predicted ΔG^*_{os}–r dependence is the additional stabilization afforded to the transition state by reactant–electrode self-image interactions compared with such interactions in the precursor and successor states. The former interactions are generally expected to be larger since the non-equilibrium nature of the ionic atmosphere surrounding the reacting species both just before and after electron transfer will diminish the extent of screening of the electrode–image interactions by these surrounding charges [63, 64].

However, since insufficient information is available regarding the dependence of g_i, and especially k', upon r, eqn. (33) is of only formal rather than practical value at the present time. Since the reaction zone thickness is liable to be of molecular dimensions, the conventional procedure of estimating the effective work terms for a single reaction site at the expected plane of closest approach is probably acceptable, although it is nonetheless important to recognize its limitations.

3.6 INTRINSIC VERSUS THERMODYNAMIC CONTRIBUTIONS TO ELECTROCHEMICAL REACTIVITY

At several points in the foregoing description of electron-transfer models, it was found convenient to make a distinction between so-called "intrinsic" and "thermodynamic" contributions to the reaction barrier. Since such a distinction is central to both the theoretical treatment and data interpretation of electron-transfer processes [6, 65], it is worthwhile to clarify further the underlying notions that are involved.

As summarized in eqn. (6a) (Sect. 3.2), the intrinsic barrier ΔG^*_{int} equals the free-energy barrier, ΔG^*, for the *elementary* electron-transfer step under conditions where the precursor and successor states are of equal free energy. The remaining "thermodynamic" portion of ΔG^*, $\alpha_{et}\Delta G^0_{et}$ [eqn. (6a)], describes the modification to ΔG^* brought about by unequal free energies of these states, i.e. by the thermodynamic driving force, ΔG^0_{et}. Two other "thermodynamic" quantities are utilized above, both in connection with work terms (Sect. 3.5.2). The first describes the component of ΔG^* brought about by the location of the reacting species in the interfacial region, rather than the bulk solution, during the elementary electron-transfer step; that is, the component associated with the difference between ΔG_{et} and the driving force for the overall reaction, ΔG^0 [eqn. (6b), Sect. 3.2]. The second describes the difference between ΔG^* and the barrier, $\Delta G^{\neq}$, for the overall reaction [eqn. (4a), Sect. 2.2]. The "intrinsic" and "thermodynamic" components of catalysis for inner-sphere mechanisms (Sect. 3.5.1) follow the same distinctions since these relationships are applicable to reactions following strong- as well as weak-overlap pathways.

The correction for the first two thermodynamic effects involve a knowledge of the transfer coefficient α_{et} as well as the thermodynamic free energies. (Somewhat confusingly, α_{et} has also been termed the "intrinsic" transfer coefficient [66] to distinguish it from the "apparent" quantity, α, which is defined in terms of the composite barrier for the overall electrode reaction [eqn. (5)]. When the reactant and product free-energy wells are equal-shaped parabolas, α_{et} describes the extent to which the nuclear configuration in the transition state reflects that of the product rather than that of the reactant. In a sense, α_{et} reflects the "effective fractional ionic charge" in the transition state relative to the reactant and product, even though, at least for weak-overlap pathways, the actual electron "movement" occurs at a single well-defined nuclear configuration.

For electrochemical reactions, the intrinsic barrier refers to the condition where the "chemical" free-energy difference, ΔG^0_{rc} (chem.), between the oxidized and reduced species located at the reaction site is balanced by the "electrical" free-energy driving force, ΔG^0_{rc} (elect.) ($= F\phi^0_m$), associated with the transferring electron, so that the net "electrochemical" driving force for heterogeneous electron transfer, ΔG^0_{et}, equals zero. For homogeneous self-exchange processes, the condition $\Delta G^0_{et} = 0$ is achieved through equal and opposite values of ΔG^0_{rc} (chem.) for the two reacting couples. It is important to recognize that the individual values of ΔG^0_{rc} (chem.) and ΔG^0_{rc} (elect.) are not required in order to understand the electron-transfer kinetics either at electrodes or in bulk solution. This is not to say that these quantities are unimportant; indeed their estimation is of great significance to the understanding of ionic solvation [67]. However, since kinetics is concerned only with energy *differences* between thermodynamically stable and transition states, the influence of such solvation and other purely thermodynamic effects is largely eliminated by considering intrinsic barriers (Sect. 3.2), thereby simplifying considerably the task of describing the activation process and facilitating the estimation of the barrier height.

The intrinsic barrier therefore denotes the portion of the additional free energy possessed by the transition state with respect to the free energies of the adjacent ground (precursor and successor) states that arises only as a consequence of the non-equilibrium properties of the former. The elucidation of intrinsic barriers, at least relative values for a series of structurally related reactions or for different surface environments, is clearly of central fundamental importance in electrochemical kinetics. Although not often perceived in such terms, a major objective is therefore the utilization of strategies that correct, or otherwise allow for, the influence of thermodynamic contributions upon the experimental kinetic parameters.

For solution reactants, the intrinsic barrier can be most directly related to the work-corrected standard rate constant for the single electron-transfer step

$$k^{s}_{corr} = A_p \exp(-\Delta G^{*}_{int}/RT) \quad (35)$$

where A_p is the pre-exponential factor defined by eqn. (28). In principle, ΔG^{*}_{int} can be extracted from the temperature dependence of k^{s}_{corr} since (Sect. 3.4)

$$k^{s}_{corr} = A_p \exp(\Delta S^{*}_{int}/R) \exp(-\Delta H^{*}_{int}/RT) \quad (30a)$$

Thus, provided A_p, ΔS^{*}_{int}, and ΔH^{*}_{int} are themselves temperature-independent, ΔH^{*}_{int} can be obtained from the ln $k^{s}_{corr}/(1/T)$ slope [eqn. (32)] and, since we anticipate that [40a] $\Delta S^{*}_{int} \approx 0$ (Sect. 3.4), this yields ΔG^{*}_{int} since then $\Delta H^{*}_{int} \approx \Delta G^{*}_{int}$.

A major difficulty with this analysis, however, is that the assumption $\Delta S^{*}_{int} \approx 0$ requires that the solvation environment of the transition state is unaffected by its proximity to the electrode surface (Sect. 3.4). Stated equivalently, it is often expected that the temperature-dependent work terms required to extract k^{s}_{corr} from k^{s}_{ob} contain large components from short-range solvation and other factors in addition to the usual "electrostatic" double-layer effects (Sect. 2.4 and 4.6). As noted in Sect. 2.3, the situation is somewhat more straightforward for surface-attached reactants since then the effects of work terms at least partly disappear. This question underscores the inevitable difficulties involved in extracting quantitative information on electron-transfer barriers from rate measurements.

In the following sections, selected comparisons between experimental kinetics for single-step electrochemical reactions and the foregoing electron-transfer models will be presented in order to characterize the physical features of the experimental systems as well as to scrutinize the applicability of the theoretical models themselves.

4. Comparisons between predictions of theoretical models and experimental kinetics

4.1 DISTINCTION BETWEEN "ABSOLUTE" AND "RELATIVE" THEORY–EXPERIMENT COMPARISONS

The extant theoretical models enable one, at least in principle, to predict reactivities for individual electrochemical reactions, given a knowledge of the relevant thermodynamic and structural parameters, for comparison with the experimental kinetics. There are several factors, however, which severely limit the applicability of such "absolute" theory–experiment comparisons, at least for the time being. Firstly, there are still relatively few redox couples for which sufficient bond length and vibrational information is available so that inner-shell barriers can reliably be calculated, although the situation has improved recently for inorganic systems (vide infra). Secondly, many electrochemical reactions proceed via inner-sphere pathways where the work terms are often unknown unless the precursor state is sufficiently stable so that the surface reactant concentrations can be evaluated directly (Sect. 3.5.2). Thirdly, as noted above (Sect. 2.2), many electrochemical reactions proceed via multistep pathways where one or more chemical transformations are coupled to the electron-transfer stage. Even when the latter is rate-determining, therefore, the lack of information on the thermodynamics of this step usually precludes the estimation of the barrier height and hence rate constants using the conventional theoretical models.

Nevertheless, numerous comparisons between theory and experimental kinetics can still be made, even for such systems, by examining the dependence of the rate parameters on systematic changes in system state, such as electrode potential, temperature, electrode material, reactant and solvent structure, and so on. Provided that the unknown quantities that would otherwise be required for the theoretical calculations can either be held constant or be arranged to cancel under the chosen experimental conditions, they will clearly not affect the predicted rate variations. Such confrontations between the theoretical predictions and experimental kinetics can be termed "relative" theory–experiment comparisons. These are of particular value when conditions are chosen such that only a few parameters, and in ideal cases only a single quantity, present in the theoretical models are influenced significantly by the imposed change of system state. Such circumstances would then enable the desired individual element of the theoretical model to be tested independently. For example, it is desirable to select a series of systems for which the spatial separation of the reactant and the electrode surface is varied under conditions where the barrier height and nuclear frequency factor are fixed, thus enabling the distance dependence of the elctron-tunneling probability (Sect. 3.3.2) to be assessed (vide infra).

The design of such strategies is common to much fundamental research in kinetics as in other branches of chemistry, having the general aim of elucidating individually the importance of the manifold molecular factors that influence the observed reactivities. In the following sections, several types of "relative" theory – experimental comparisons that are anticipated to yield such insight will be discussed, followed by an examination of the extent ot which the contemporary theoretical models can predict absolute electrochemical reactivities.

4.2 ELECTROSTATIC DOUBLE-LAYER EFFECTS

Studies of the effects of varying the double-layer structure upon the kinetics of electrode reactions have long been an active research topic, especially at the mercury–aqueous interface for which there exists a large and reliable body of data concerning the double-layer composition and structure. Work in this area prior to 1965 is reviewed in Delahay's well-known monograph [58]; unfortunately, more recent reviews are conspicuous by their absence (but see, for example, ref. 16a).

As noted in Sect. 3.5.2, it is conventional to treat double-layer effects upon electrode kinetics in terms of eqn. (8a) (Sect. 2.2); this relation is often expressed as

$$\ln k_{\mathrm{ob}} = \ln k_{\mathrm{corr}} + \frac{(Z - \alpha_{\mathrm{corr}})F\phi_{\mathrm{GC}}}{RT} \tag{36}$$

where the double-layer potential, ϕ_{d}, is assumed to be given by the average o.H.p. potential estimated from the Gouy–Chapman model, ϕ_{GC}, and the transfer coefficient for the electron-transfer step, α_{et}, is approximated by the "work-corrected" value, α_{corr}, where $\alpha_{\mathrm{corr}} = -(RT/F)(\mathrm{d}\ln k_{\mathrm{corr}}/\mathrm{d}E)$. Since it is usual to examine rate responses to variations in double-layer structure at a given electrode potential, $(\Delta \log k_{\mathrm{ob}})_E$, it is also useful to express eqn. (36) as

$$(\Delta \log k_{\mathrm{ob}})_E = (Z - \alpha_{\mathrm{corr}})F\Delta\phi_{\mathrm{GC}}/2.303RT \tag{36a}$$

where $\Delta\phi_{\mathrm{GC}}$ denotes the corresponding change in the diffuse-layer potential. Equation (36a) implicitly assumes that the entire double-layer effect is contained in $\Delta\phi_{\mathrm{GC}}$, so that k_{corr} remains constant.

A hallmark of many experimental studies at mercury electrodes is the relative success of this coupled Gouy–Chapman–Frumkin (GCF) approach in accounting for the rate responses to variations in the ionic strength and the addition of specifically adsorbed anions, even at ionic strengths up to 1 M [15a, 68]. The observed deviations between experiment and theory can often be accounted for largely by differences between the o.H.p. and the reaction site [15–17] (see Sect. 2.4). This success of the GCF treatment can be attributed, in part, to the relatively small charge densities on the metal surface, q^{m}, and in the inner layer, q', commonly encountered at mercury ($|q^{\mathrm{m}}, q'| \lesssim 20$–$30\,\mu\mathrm{C\,cm^{-2}}$), under which conditions the ϕ_{GC} values calculated simply from the sum of q^{m} and q' do not differ greatly from those obtained from more sophisticated diffuse-layer treatments (Sect. 3.5.2) [59, 60].

In this respect, our understanding of electrostatic work terms for electrochemical electron-transfer processes is on a sounder footing than for redox reactions in homogeneous solution. For the latter, large deviations in the observed rate responses to variations in the ionic strength [70a] and to additions of anions [70b] and cations are often obtained in comparison with the predictions of the usual Debye–Huckel–Brønsted (DHB) model [71]. The effect of varying temperature upon the rate–ionic strength responses for cation–cation reactions also shows marked discrepancies with the DHB predictions, the observed ionic-strength dependence being contained almost entirely in the enthalpic component rather than the entropic portion as predicted by the DHB model [70a]. Although experimental data on temperature-dependent double-layer effects are sparse, there is some evidence that the GCF treatment provides an approximately correct description [2c]. The DHB model has close similarities with the GCF approach, since both involve estimating the electrostatic work required to assemble the precursor and successor states on the basis of a Poisson–Boltzmann potential distribution. The relative success of the latter model may be associated with the greater applicability of Gouy–Chapman theory at high ionic strengths than is found for the Debye–Huckel treatment.

Nevertheless, substantial deviations of the observed double-layer effects from the predictions of the GCF model are not uncommon. One example is the rate decreases for anion-bridged reactions involving cationic complexes observed upon the adsorption of non-reacting anions [15a, 23] (Sect. 2.4), contrasting the rate increases predicted from the GCF model. More generally, our knowledge of double-layer effects at solid–metal surfaces and in non-aqueous media is still quite rudimentary. Large effects of varying the supporting electrolyte cation [17] and from systematic alterations in the reactant charge [17d] have been observed for ostensibly simple outer-sphere reactions at metal–non-aqueous interfaces, which are in marked disagreement with the GCF predictions.

These discrepancies bring to the fore the inevitable limitations of the simple GCF treatment. For such systems, the usual practice of extracting "double-layer corrected" rate constants, k_{corr}, from the observed values using eqn. (36) or similar relations is clearly unsatisfactory since markedly different k_{corr} values will be obtained depending on the particular electrolyte in which k_{ob} is evaluated. Unfortunately, electrochemical rate parameters are commonly reported in only a single supporting electrolyte, involving large (and, consequently, very uncertain) double-layer corrections in order to extract the desired estimates of k_{corr}. It is also important to recognize that, even if the GCF model describes correctly the observed rate responses to variations in double-layer structure, it may still not provide the correct *absolute* double-layer correction. This is because additional work-term components, such as those associated with differences in the bulk phase and interfacial solvation environments (Sect. 2.4 and 3.5.2), may be insensitive to alterations in the ionic double-layer composition and therefore remain buried in the "double-layer corrected" rate constants. Such effects may nonetheless be exposed by more drastic changes in the interfacial structure, brought about by alterations in the solvent or the electrode material (vide infra).

4.3 ALTERATIONS IN ELECTRODE POTENTIAL AND TEMPERATURE

Given that electrochemical rate constants are usually extremely sensitive to the electrode potential, there has been longstanding interest in examining the nature of the rate–potential dependence. Broadly speaking, these examinations are of two types. Firstly, for multistep (especially multielectron) processes, the slope of the log k_{ob}–E plots (so-called "Tafel slopes") can yield information on the reaction mechanism. Such treatments, although beyond the scope of the present discussion, are detailed elsewhere [13, 72]. Secondly, for single-electron processes, the functional form of log k–E plots has come under detailed scrutiny in connection with the prediction of electron-transfer models that the activation free energy should depend non-linearly upon the overpotential (Sect. 3.2).

Expressing eqn. (15) (Sect. 3.2) in terms of work-corrected rate constants yields the well-known relation [15]

$$\ln k_{\mathrm{corr}} = \ln A_{\mathrm{p}} + \frac{[-4\Delta G^{*}_{\mathrm{int}} \pm F(E - E^{0})]^{2}}{16\Delta G^{*}_{\mathrm{int}} RT} \qquad (37\mathrm{a})$$

or, equivalently,

$$\ln k_{\mathrm{corr}} = \ln k^{\mathrm{s}}_{\mathrm{corr}} \pm 0.5\, F(E - E^{0})/RT + \frac{F(E - E^{0})^{2}}{16\Delta G^{*}_{\mathrm{int}} RT} \qquad (37\mathrm{b})$$

where A_{p} is the combined pre-exponential factor [eqn. (28)], and the plus-minus signs refer to anodic and cathodic processes, respectively. The question of the applicability of eqn. (37) has provided a lively topic in the literature for some time. This has centered on the prediction that the log k_{corr} – potential plots should exhibit slopes (i.e. α_{corr} values) that decrease linearly with increasing overpotential.

The search for such curved Tafel plots has yielded some well-documented examples where essentially straight Tafel lines are observed, even when slight curvature is predicted from eqn. (37). In particular, this is the case for proton reduction [73] and the outer-sphere reduction of some Cr(III) aquo complexes [34] at mercury electrodes over wide overpotential ranges ($\gtrsim 600$ mV). However, the former reaction is not an outer-sphere process with symmetrical reactant and product parabolae to which eqn. (37) should apply, but rather involves the formation of an adsorbed hydrogen atom intermediate. The influence of such a mechanistic feature upon the rate–potential behavior is unclear even now [74]. The Cr(III)/Cr(II) aquo couple at mercury has also been examined over wide ranges of anodic as well as cathodic over-potentials [75]. In contrast to the cathodic behavior, marked decreases in the anodic work-corrected transfer coefficients are seen with increasing overpotential. Compatible behavior is also observed for the driving-force dependence of homogeneous redox processes involving oxidations of aquo cations [76]. A recent reappraisal has shown this asymmetric rate-driving force dependence for Cr(III)/Cr(II) to be consistent with the markedly different values of λ_f and λ_r [eqn. (18), Sect. 3.2] extracted from newly available Raman spectroscopic data [55]. Equation (37) was derived for simplicity by assuming that the oxidized and reduced forms have identically shaped parabolae. Significant deviations from this will usually occur for reactions such as Cr(III)/Cr(II) that involve large inner-shell reorganization and hence significant structural differences between the reactant and product, yielding $\lambda_f \neq \lambda_r$ (i.e. differently shaped reactant and product free-energy wells).

Interestingly, potential-dependent transfer coefficients in reasonable agreement with the predictions of eqn. (37) have been observed for organic molecule–radical anion nitro compounds in acetonitrile and dimethylformamide at mercury [17c, 77]. This distinctly different behavior from that for the metal aquo redox couples is not unexpected since the organic molecule–radical anion couples should involve only small differences in molecular geometry, yielding free-energy parabolae that are associated chiefly with outer-shell reorganization and therefore more nearly symmetrical.

However, truly sensitive tests of eqn. (37) or related theoretical expressions require rate–overpotential data at driving forces corresponding to small activation free energies, $\lesssim$ 4–5 kcal mol^{-1}, where the intersection region of the free-energy curves should become markedly non-symmetrical. Unfortunately, this circumstance usually corresponds to electrochemical rate constants, $\gtrsim$ 10 cm s^{-1}, which are beyond the accessible range of existing methods. Consequently, our experimental knowledge of the shape of the free-energy wells for electrochemical systems remains very incomplete in comparison with that for some homogeneous redox processes [78].

Therefore, for most experimental conditions, the transfer coefficient for the electron-transfer step, α_{et}, is predicted to approximate 0.5 with deviations from this value at moderate driving forces expected most often for processes featuring large inner-shell structural changes [55]. By and large, these expectations are borne out by experiment; work-corrected transfer coefficients in the range ca. 0.35–0.65 are commonly observed for simple one-electron redox couples, although the extraction of α_{et} values is often impeded by uncertainties in the double-layer corrections.

In comparison with the perturbations to rate constants induced by the electrode potential, relatively little attention has been directed to experimental examinations of temperature effects in electrochemical kinetics. This is probably due, in part, to uncertainties in how to control the electrical variable while the temperature is altered. However, as noted in Sect. 3.4, in actuality there are no more ambiguities in interpreting electrochemical activation parameters than for the commonly encountered Arrhenius parameters for chemical processes, especially if the temperature of the reference electrode is fixed by using a non-isothermal cell arrangement [2c].

At least in principle, such measurements should shed light either on the magnitude of the pre-exponential factor or on the nature of solvation within the transition state. Two types of data treatment are employed, depending on which of these aspects attention is being focussed. As noted in Sect. 3.4, estimates of A_p can be obtained from the observed activation enthalpies by estimating ΔS^*_{corr} using eqn. (31); alternatively, "experimental" values of ΔS^*_{corr} can be obtained by assuming A_p. Comparison with the "theoretical" ΔS^*_{corr} estimates from eqn. (31) can shed light on differences between interfacial and bulk solvation since eqn. (31) does not account for such effects (Sect. 3.4).

The latter approach has been utilized for the reduction of various Co(III) and Cr(III) complexes at mercury [2c] and other metal surfaces [79]. The resulting large positive ΔS^*_{corr} values for most of these reactions at mercury are comparable with, if somewhat smaller than, those estimated from eqn. (31), suggesting that the nearby metal surface has relatively little influence upon the transition-state solvation, as might be expected for "weak overlap" outer-sphere processes [2c]. However, the same processes at lead and thallium electrodes yield much smaller or even negative ΔS^*_{corr} values, suggesting that these surfaces substantially perturb the solvation environment of the transition state [79]. Unexpectedly negative "ideal" activation entropies have recently been obtained for aqueous proton reduction at mercury and for bromide oxidation in acetonitrile at carbon electrodes [80], although the appropriate entropic driving forces from which to extract ΔS^*_{corr} from eqn. (31) are difficult to estimate due to the multistep nature of these processes. This analysis, of course, requires an assumption to be made concerning the reaction adiabaticity via the presumed value of A_p; highly non-adiabatic pathways ($\kappa^0_{el} \ll 1$) can, in principle, account for unexpectedly small ΔS^*_{corr} values, since it is usual to take κ^0_{el} equal to unity when estimating A_p [eqn. (28)] (see Sect. 4.4).

Further information on this question should be obtainable from evaluating activation parameters for surface-attached (or adsorbed) reactants since unimolecular frequency factors A_{et} can be obtained directly [eqn. (28a), Sect. 3.4] and the effects of reactant–surface interactions on the work terms should largely cancel (Sect. 2.3). For Co(III) or Cr(III) reactants bound to mercury via thiocyanate or small conjugated organic bridges, A_{et} values around $10^{13}\,s^{-1}$ are obtained, suggesting that $\kappa_{el} \sim 1$ since $\Gamma_n \approx 1$ and $\nu_n \approx 1 \times 10^{13}\,s^{-1}$ [eqn. (28a)] for these processes [81]. However, markedly (10^2–10^3 fold) smaller A_{et} values were obtained for otherwise similar reactions involving non-conjugated organic bridges that roughly mirror the decreases in k_{et}, suggesting the onset of substantially non-adiabatic pathways [82]. This matter will be considered further in Sect. 4.4.

An old question [82, 83] that has recently received renewed attention [80, 84] is the finding for a number of electrochemical reactions that the average electrochemical transfer coefficient obtained from the Tafel slope often increases significantly with increasing temperature. This behavior is formally equivalent to the presence of an entropic, as well as enthalpic, component of the potential-dependent activation energy. This is unexpected on theoretical grounds, since the predominant effect of the electrode potential should be to alter the Fermi energy level of electrons in the metal, which is almost entirely enthalpic in nature. It has consequently been suggested that the observed behavior may signal a major flaw in the theoretical treatments [84].

In the context of the present discussion, it is worth noting that virtually all the experimental systems that exhibit such "anomalous" temperature-dependent transfer coefficients are multistep inner-sphere processes, such as proton and oxygen reduction in aqueous media [84]. It is therefore extremely difficult to extract the theoretically relevant "true" transfer coefficient for the electron-transfer step, α_{et} [eqn. (6)], from the observed value [eqn. (2)]; besides a knowledge of the reaction mechanism, this requires information on the potential-dependent work terms for the precursor and successor state [eqn. (7b)]. Therefore the observed behavior may be accountable partly in terms of work terms that have large potential-dependent entropic components. Examinations of temperature-dependent transfer coefficients for one-electron outer-sphere reactions are unfortunately quite limited. However, most systems examined (transition-metal redox couples [2c], some post-transition metal reductions [85], and nitrobenzene reduction in non-aqueous media [86]) yield essentially temperature-independent transfer coefficients, and hence potential-independent ΔS^{*}_{corr} values, within the uncertainty of the double-layer corrections.

It is interesting to note that some alteration of the activation entropy ΔS^{*}_{corr} with electrode potential will generally be expected from the theoretical formalisms on the basis of eqn. (31) since, as noted above, α_{et} (and hence $\Delta S^{*}_{corr} = \alpha_{et}\Delta S^{0}_{rc}$) should decrease with increasing overpotential, even though the value of α_{et} *at a given overpotential* is predicted to be independent of temperature. [This apparent contradiction is resolved by noting that, when $\Delta S^{0}_{rc} \neq 0$, the fixed non-isothermal cell potential, E_{ni}, to which ΔS^{*}_{corr} refers will correspond to different overpotentials, $(E_{ni} - E^{0}_{ni})$, as the temperature is varied since $\Delta S^{0}_{rc} = -F(\mathrm{d}E^{0}_{ni}/\mathrm{d}T)$ (Sect. 3.4).] The activation parameters for $Cr(OH_2)_6^{3+/2+}$ at mercury, which have been studied over a very wide range of anodic and cathodic overpotentials (ca. 1.4 V), exhibit a somewhat *smaller* dependence of ΔS^{*}_{corr} upon electrode potential than predicted from theory [55], although this may result, in part, from the inevitable uncertainties in the temperature-dependent double-layer corrections.

4.4 APPROACHES TO DETERMINING REACTION ADIABATICITY

A key element in the aforementioned theoretical treatments is the recognition that electron transfer may take place with significant probability over a range of reactant–electrode separations, r, (Sects. 3.1 and 3.3.2). At least for outer-sphere reactions, this has the important consequence that an integral of reaction sites corresponding to different precursor state geometries may contribute significantly to the measured reaction rate. Experimental examinations on the dependence of κ_{el} upon r are sparse; however, useful information has been extracted using the following strategies.

In principle, a simple means of obtaining such information is to examine the decrease of rate constants for suitable outer-sphere reactions caused by coating the metal surface with progressively thicker insulating "spacer" layers. Several studies have utilized anodic oxide films. The majority of these studies have employed thicker oxide films ($r > 20$ Å), such as passive iron layers, where the rate–potential characteristics are drastically altered [87c, d], suggesting the occurrence of "inelastic" (or "resonance") tunneling via sites in the oxide lattice [53]. Marked, ca. tenfold, rate decreases for Ce(IV)/Ce(III) also occur upon coating platinum with thin (< 10 Å) oxide films [87c], which are likely to be due to decreases in κ_{el} via elastic tunneling [53b]. Detailed analysis is, however, precluded by uncertainties in the double-layer effects for this system, as for other reactions examined in this manner [53b].

Another type of "spacer" group employed to investigate electron-tunneling effects is organic layers formed from the addition of strongly adsorbed non-aqueous components to aqueous media. Most studies of such systems have been concerned with metal deposition reactions, where it is likely that the transition state is formed within the organic layer, or with systems where the layer does not entirely block the metal surface [88]. However, there are a few studies for outer-sphere redox systems where the presence of the organic layer yields large rate decreases that are almost certainly due to shifts in the effective reaction site further from the metal surface [88e, 89]. Of particular interest is the influence of various quinoline adsorbates upon the kinetics of some simple electrode reactions at mercury, including $Co(NH_3)_6^{3+}$ and O_2 reduction. Very large (10^4 to 10^8 fold) rate decreases were observed in the presence of the quinolines, which forms tightly packed "condensed films" on the metal surface [90]. On the basis of molecular models, the outer-sphere reaction sites are constrained to lie up to 7–10 Å further from the metal surface in the presence of the quinoline layers. These results provide a clear indication that κ_{el} drops sharply with increasing reactant–electrode separation, although it has been argued that substantial variations in ΔG^* are also involved [91].

An alternative approach to the use of insulating films is to examine the kinetics of surface-attached species where the redox center is bound to the metal surface via organic linkages of varying length and structure. It is clearly desirable to select systems where the redox center is rigidly held and thereby prevented from approaching more closely to the metal surface so that the reactant–surface distance is well defined. Unimolecular rate constants, k_{et}, for the reduction of carboxylatopentaamminecobalt(III) bound to mercury electrodes via a number of surface-bound thiophene ligands are markedly (ca. 20–50 fold) smaller for linkages having interrupted conjugation [92a]. Similar effects are also observed for Co(III) reduction via other thioorganic anchoring groups [92c]. The likelihood that this reflects decreases in κ_{el} is supported by corresponding observed decreases in the pre-exponential factor from the rate–temperature dependence [81, 92a, c] (Sect. 4.3), although the flexibility of the thiophene linkage appears to prevent the observation of more strongly non-adiabatic pathways [92a]. However, thioalkylcarboxylate linkages at gold electrodes apparently provide a much more rigid series of bridging groups; k_{et} for reduction of pentaamminecobalt(III) adsorbed via these bridges decrease very substantially (by up to 10^6 fold) as the number of alkyl carbons, n, is progressively increased [92b]. The observed exponential decrease of k_{et} with n (and presumably the reactant–surface distance, r) is in accordance with eqn. (27) (Sect. 3.3.2) with $\beta \approx 1.45$ Å^{-1}. The results also suggest that k_{et} values corresponding to adiabatic pathways ($\kappa_{el} \approx 1$) are attained only when $r \lesssim 6$–8 Å [92b, c].

Although the nature of the redox center, the metal surface, and the intervening medium are also expected to influence the extent of donor–acceptor orbital overlap and hence κ_{el}, these various results strongly suggest that adiabatic pathways are restricted to reactant–surface separations approaching molecular dimensions, these distances being similar to those estimated for electron transfer between metal ions in homogeneous solution [49]. This has the important consequence that, for electrochemical reactions following true outer-sphere mechanisms, adiabatic pathways should usually be restricted to reaction sites within 1–2 Å of the plane of closest approach. Thus, taking a typical reactant radius of 3.5 Å and an inner-layer solvent thickness of 3–3.5 Å, yields a plane of closest approach around 6–7 Å from the metal surface. The κ_{el} values are therefore anticipated to drop sharply even within the diffuse-layer region.

As noted in Sect. 3.1, the pre-exponential factor for outer-sphere reactions is sensitive to the "effective electron-tunneling distance", $\delta r \kappa_{el}^0$. If the reaction is non-adiabatic at the plane of closest approach (i.e. if $\kappa_{el}^0 < 1$), given the functional dependence of κ_{el} upon r one deduces that $\delta r \approx 0.5$ Å if $\beta \sim 1.5$ Å^{-1} (vide supra), and hence $\delta r \kappa_{el}^0 \lesssim 0.5$ Å. If, however, adiabaticity is maintained for sites beyond the plane of closest approach (i.e. for $r > \sigma$), markedly larger $\delta r \kappa_{el}^0$ values can result.

Experimental estimates of $\delta r \kappa_{el}^0$ are relatively difficult to obtain. While they can, in principle, be extracted from temperature-dependence studies, this approach is complicated by uncertainties in the entropic term (Sect. 4.3). An alternative method has recently been described for some Cr(III) reductions which involves comparing the work-corrected rate constants, k_{corr}, with unimolecular rate constants, k_{et}, for structurally related reactants that reduce via ligand-bridged pathways [30]. Provided that the corresponding outer- and inner-sphere pathways involve the same activation barrier (Sect. 4.6) and the latter also follow adiabatic pathways, we can write [30]

$$\frac{k_{corr}}{k_{et}} \approx \delta r \kappa_{el}^0 \tag{38}$$

The resulting estimates of $\delta r \kappa_{el}^0$ for Cr(III)/Cr(II) aquo couples, ca. 0.1–0.3 Å, are close to that extracted from temperature-dependence measurements and indicate that these strongly hydrated species follow marginally non-adiabatic pathways even at the plane of closest approach [30]. Markedly larger values of $\delta r \kappa_{el}^0$, ca. 5 Å, were obtained for similar ammine complexes, as from a related comparison involving unimolecular outer-sphere reactivities [21]. This is consistent with the smaller hydrated radii of the ammines allowing a closer approach to the metal surface, whereupon $\kappa_{el}^0 \approx 1$ [15a, 30].

4.5 SOLVENT EFFECTS

Although aqueous media have been utilized for the large majority of experimental studies, the response of the kinetic parameters to systematic alterations in the solvent can clearly provide important information. In terms of the above theoretical treatments, the nature of the solvent medium can influence the reaction kinetics from at least the following four sources: (i) the driving force, ΔG_{et}^*, at a given electrode potential [eqns. (6) and (6a)]; (ii) the work terms w_p and w_s [eqn. (6)]; (iii) the outer-shell contribution to the intrinsic barrier [eqn. (19)]; and (iv) the solvent dynamical contribution to the nuclear frequency factor [eqns. (22) and (24)]. Since the examination of (iii) and (iv) are usually of greatest fundamental interest, it is desirable to eliminate (or correct for) the effects of (i) and (ii) from the experimental rate data. The influence of (i) can readily be eliminated, at least for chemically reversible single-step processes, by evaluating rate parameters at the standard electrode potential. The effect of (ii) for outer-sphere processes is more difficult to correct for (Sect. 4.2), especially in non-polar non-aqueous media where large diffuse-layer potentials are commonly anticipated along with complications from reactant ion pairing.

Nevertheless, with a suitably judicious choice of systems, the observed variations in rate constants resulting from the alteration of the solvent medium should reflect predominantly contributions from factors (iii) and possibly (iv). (Of course, it is necessary to select reactants, such as substitutionally inert complexes or aromatic species, for which the inner-shell barrier is small or remains unaltered as the solvent is varied.) As noted in Sect. 3.2, a question that has engendered considerable controversy concerns the likely applicability of the dielectric continuum model, embodied in eqn. (19) and similar relations, for estimating the outer-shell intrinsic barrier. Unfortunately, there are few experimental studies of solvent effects for which the conditions selected enable truly diagnostic tests of eqn. (19) to be made, i.e. for which only factor (iii) will contribute to the observed rate variations. A recent examination of standard rate constants for aromatic molecule–radical anion redox couples in several non-aqueous solvents at mercury electrodes yielded significant, if small, deviations from the predictions of eqn. (19) [93]. These were attributed to the influence of short-range solute–solvent interactions since the rate constants decreased slightly with increasing reactant–solute interactions as estimated from the solvent electron-accepting ability ("solvent acceptor number"). Such deviations, albeit small, are consistent with a modified form of eqn. (19) where the ε_s^{-1} "Born charging" term is replaced by a term based on redox thermodynamic information [24b] (Sect. 3.2).

The double-layer corrected standard rate constants for $Co(en)_3^{3+/2+}$ (en = ethylenediamine) at mercury decrease by over 10^3 fold when substituting aprotic solvents such as dimethylformamide and dimethylsulfoxide for aqueous media [94]. While this was originally speculated to be due to the presence of solvent reorganization associated with short-range solute–solvent interactions in the former solvents [94a], more recent evidence (including activation parameter data [94b]) shows that the observed behavior is almost certainly due to solvent-specific work terms [37a, 94b]. A number of studies of the solvent dependence of self-exchange kinetics for various organometallic and organic redox couples in homogeneous solution have been reported [95]. Some of these studies indicate reasonable agreement with the dielectric continuum predictions [95c–e], whereas others show substantial deviations [95a–c].

In view of the importance of water as a solvent in electrode kinetics, the effects of substituting D_2O for H_2O are of interest; while these media have almost identical dielectric properties, the former engages in significantly stronger hydrogen bonding [96]. Several transition-metal aquo couples yield large (up to ca. threefold) decreases in k^s_{corr} upon substituting D_2O for H_2O solvent [97]. Although such isotope effects, which are also seen for homogeneous reactions involving aquo complexes, may be due partly to nuclear tunneling involving ligand O–H vibrations [98], they do provide evidence of a contribution to the intrinsic outer-shell barrier from ligand–solvent hydrogen bonding [40b, 97, 99].

Studies of photo-induced electron transfer can provide more clear-cut tests of the dielectric continuum model since the incident light energy can be related directly to the barrier height [51]. Delahay's recent studies of electron photoemission from inorganic species in aqueous media [100] are of interest in this regard. However, the extraction of intrinsic barrier heights from such data involves correcting for the driving force since the reactant and product states are energetically non-symmetrical. The most direct tests of the dielectric continuum treatment are provided by photoinduced electron transfer within symmetrical binuclear complexes [51]. The optical absorption energy, E_{op}, is simply related to the corresponding intrinsic barrier by $E_{op} = 4\Delta G^*_{int}$ provided that the energy wells are parabolic [51a]. By choosing systems, such as Ru(III)/Ru(II) ammine couples, for which the inner-shell barrier is small and can be estimated, absolute values of ΔG^*_{int} can be obtained in this manner in a variety of solvents. By and large, these experimental estimates of ΔG^*_{int} are in approximate agreement with the dielectric continuum predictions, usually within ca. 0.5–1 kcal mol^{-1} [51a]. Comparison of some experimental data with a modified treatment accounting for specific solute–solvent interactions indicates that such interactions only provide a small (< 1 kcal mol^{-1}) contribution to ΔG^*_{int}, even for conditions where the redox thermodynamics are influenced markedly by short-range solute–solvent interactions [40b] (Sect. 3.2).

Experiments aimed at probing solvent dynamical effects in electrochemical kinetics, as in homogeneous electron transfer, are only of very recent origin, fueled in part by a renaissance of theoretical activity in condensed-phase reaction dynamics [47] (Sect. 3.3.1). It has been noted that solvent-dependent rate constants can sometimes be correlated with the medium viscosity, η [101]. While such behavior may also signal the onset of diffusion- rather than electron-transfer control, if the latter circumstances prevail this finding suggests that the frequency factor is controlled by solvent dynamics since τ_D and hence τ_L [eqn. (23), Sect. 3.3.1] is often roughly proportional to η [102].

Standard rate constants for several metallocene and metal arene redox couples at mercury electrodes have been evaluated in a series of non-aqueous solvents having systematic variations in their solvent dynamical, as well as dielectric, properties [45]. While the results are qualitatively inconsistent with the dielectric continuum predictions using a solvent-independent frequency factor, they are in quantitative agreement when the effect of solvent relaxation upon ν_n [eqn. (24), Sect. 3.3.1] is included [45]. This agreement between experiment and theory also extends to the activation parameters [45b]. These results are consistent with the solvent dynamical models summarized in Sect. 3.3.1; the organometallic couples have only small ($\lesssim 0.3\,\text{kcal mol}^{-1}$) inner-shell intrinsic barriers, so that the solvent is predicted to control the barrier-crossing frequency as well as the barrier height in the relatively high-friction media employed [45].

Such considerations also provide a rationale of the inability of the dielectric continuum model, as conventionally expressed with a fixed frequency factor, to describe the solvent-dependent kinetics of some other outer-sphere reactions [45b, 95b]. As noted in Sect. 3.3.1, the influence of solvent dynamics upon ν_n should disappear for reactions having moderate or large inner-shell barriers, the frequency factor being determined by ν_{is} instead [eqns. (22) and (25)]; this can account for the success of the conventional (fixed-frequency) dielectric-continuum treatment in describing solvent-dependent kinetics for some reactant systems [45].

There is little experimental information on possible solvent dynamical effects for electron transfer in aqueous solution. However, water is a dynamically "fast" solvent, ν_{os} being determined by "solvent inertial" effects so that the usual transition-state formula [eqn. (22)] should be applicable for determining ν_n (Sect. 3.2.1). Consequently, solvent dynamical effects in this and other "low friction" media (e.g. acetonitrile) should be controlled by the rotational frequency of individual solvent molecules and limited to reactions involving only very small inner-shell barriers (Sect. 3.3.1).

There are a number of kinetic studies of mechanistically simple electrode processes in mixed solvent systems. Of widespread interest are the effects of adding adsorbing organic solvents to aqueous solution in order to alter the interfacial solvent composition preferentially [88]. As noted in Sect. 4.4, if the adsorbate forms an impermeable adsorbed layer, the major effect may be to diminish the rate constant via decreases in the electron-tunneling probability, κ_{el}. Although substantial rate decreases are commonly found upon addition of non-aqueous solvents to water [88], an important factor determining this behavior appears to be the differences between the bulk and interfacial solvation environments since the latter is preferentially rich in the non-aqueous component [88b, c, e]. This is consistent with the rate minimum usually found for increasing mole fraction of the non-aqueous component, the rate increasing when the mole fraction is sufficiently high so that the reactant is solvated by the organic species in the bulk as well as interfacial environments [88]. This infers that reaction sites at least partly imbedded within the non-aqueous inner-layer region are preferred, albeit at the cost of a smaller precursor-state stability from less favorable reactant–solvent interactions compared with those in the bulk solution [88c, e]. Comparable rate–solvent composition curves are often obtained for inner-sphere redox couples involving only solvated species as well as metal deposition process where ion transfer through the film must occur [88, 89]. Apparently, then, the alternative outer-sphere reaction sites in the aqueous-rich region further from the metal surface may involve sufficiently smaller κ_{el}, and possibly larger ΔG^*_{os}, values so to provide less favorable electron-transfer pathways. This has been supported by some theoretical considerations [103].

4.6 EFFECTS OF ELECTRODE MATERIAL: ELECTROCATALYSES AFFORDED BY INNER-SPHERE PATHWAYS

Given that inner-sphere pathways are commonly encountered at metal–solution interfaces, as between reactants in homogeneous solution, a key question concerns the manner and extent to which the reactant–electrode interactions associated with such pathways lead to reactivity enhancements compared with weak-overlap pathways (Sect. 3.5.2). A useful tactic involves the comparison between the kinetics of structurally related reactions that occur via inner- and outer-sphere pathways. This presumes that the outer-sphere route yields kinetics which approximate that for the weak-overlap limit. For this purpose, it is desirable to estimate the work-corrected unimolecular rate constant for the outer-sphere pathway at a particular electrode potential, k_{et}^{op}, from the corresponding work-corrected measured value, k_{corr}, using [cf. eqns. (10) and (13)].

$$k_{et}^{op} = \frac{k_{corr}}{K_0} \tag{39a}$$

$$= \frac{k_{corr}}{\delta r} \tag{39b}$$

provided that reasonable estimates of the reaction zone thickness, δr, can be made (Sect. 4.4). The resulting k_{et}^{op} values coupled with the corresponding rate constant for the overall inner-sphere pathway, k_{ob}^{ip}, enable the origins of any enhancement of k_{ob}^{ip} relative to k_{corr}^{op} to be probed since

$$\frac{k_{ob}^{ip}}{k_{et}^{op}} = K_p^{ip} \frac{\kappa_{el}^{ip}}{\kappa_{el}^{op}} \exp\left[-(\Delta G_{ip}^* - \Delta G_{op}^*)/RT\right] \tag{40a}$$

where K_p^{ip} is the precursor stability constant for the inner-sphere pathway, and κ_{el}^{ip} and κ_{el}^{op}, and ΔG_{ip}^* and ΔG_{op}^* are the corresponding electron-tunneling probabilities and activation free energies for the electron-transfer steps for the inner- and outer-sphere pathways, respectively. Alternatively, one can evaluate the unimolecular inner-sphere rate constant, k_{et}^{is}, from k_{ob}^{ip} and K_p^{ip} [eqn. (10)], and compare this with k_{et}^{os} according to

$$\frac{k_{et}^{ip}}{k_{et}^{op}} = \left(\frac{\kappa_{el}^{ip}}{\kappa_{el}^{op}}\right) \exp\left[-(\Delta G_{ip}^* - \Delta G_{op}^*)/RT\right] \tag{40b}$$

Values of k_{ob}^{ip}/k_{et}^{op} that roughly equal K_p^{ip} or, equivalently, values of k_{et}^{ip}/k_{et}^{os} that are around unity, therefore indicate that the inner-sphere "electrocatalysis" that arises chiefly from an increase in the cross-sectional reactant concentration in the precursor state ("thermodynamic catalysis", Sect. 3.5.1). This circumstance appears to be a common one for ligand-bridged transition-metal processes [12, 92a, 104], indicating that the free-energy barrier for the electron-transfer step is largely unaltered by attaching the redox center to the metal surface via a polyatomic linkage. Values of k_{et}^{ip}/k_{et}^{os} substantially greater than unity are observed, however, for some reactions involving chloride and bromide bridges [12]. This may well signal the presence of "intrinsic inner-sphere catalysis" (Sect. 3.5.1), whereby the electron-transfer barrier is diminished significantly by the formation of the redox center–halide–surface linkage [i.e. $\Delta G_{ip}^* < \Delta G_{op}^*$ in eqn. (40)], although other possibilities, such as increases in electron-tunneling probability ($\kappa_{el}^{ip} > \kappa_{el}^{op}$), cannot be ruled out [12a]. Such intrinsic catalysis seems to be common for reactions between pairs of transition-metal complexes bridged by small inorganic anions [105]. Such catalysis presumably occurs by mutual decreases of the bond distortions required at each redox center to induce electron transfer [104a, 105].

Although the presence of merely thermodynamic, rather than additional intrinsic, inner-sphere catalysis can yield substantial rate enhancements, these will be limited inevitably by the availability of surface coordination sites. Thus, for example, a bulk reactant concentration of 1 mM and a maximum (i.e. monolayer) surface concentration of 5×10^{-10} mol cm^{-2} yields a maximum rate enhancement of 5×10^4 fold resulting from "thermodynamic catalysis" if $\delta r = 1$ Å for the alternative outer-sphere pathway. Greater inner-sphere rate accelerations require pathways yielding increases in κ_{el} or decreases in ΔG^*.

Given that very substantial dependences of rate constants upon the surface composition are often observed for reactions involving adsorbed intermediates [10, 12, 13, 25], it might be inferred that intrinsic as well as thermodynamic factors contribute importantly to the extent of electrocatalysis. While this may well be true, the detailed reaction mechanisms are usually known only incompletely at best [106]. In particular, reactant adsorption for many small molecule systems may bring about drastic changes in the thermodynamics of the electron-transfer step and may therefore yield large alterations in ΔG^* arising from a "thermodynamic" effect in addition to an "intrinsic" effect upon the electron-transfer step. Such effects may well be important for electrocatalytic processes involving multi-electron transfer. An interesting discussion along these lines is given in ref. 107.

Little experimental information exists regarding the possible occurrence of Class II or III behavior for electron transfer involving adsorbed redox couples (Sect. 3.5.1), i.e. the occurrence of strong intrinsic catalysis for electrochemical electron exchange. It has been suggested that electron exchange for a number of adsorbed redox couples including $Fe(CN)_6^{3-/4-}$, $Fe(oxalate)_3^{3-/4-}$, and $MnO_4^{-/2-}$ occur via mechanisms of the general type

$$Ox + \lambda e^- \rightleftharpoons Ads + (1 - \lambda)e^- \rightleftharpoons Red \tag{41}$$

whereby a single adsorbed intermediate, Ads, is formed via rate-controlling adsorption–desorption steps involving partial charge transfer to or from the metal surface [108]. This situation corresponds to Class III behavior in that the adsorbed intermediate can be conceived as a hybrid of adsorbed Ox and Red, the electron transfer being rapid even on the vibrational time scale. Evidence supporting such "valence delocalization" for $Fe(CN)_6^{3-/4-}$ adsorbed at gold under certain conditions has been obtained recently from surface-enhanced Raman spectroscopy, whereby a single C–N stretching mode is detected rather than a potential-dependent pair of bands corresponding to distinct adsorbed Fe(III) and Fe(II) states [109]. Although the phenomenon of partial change transfer is well documented for simple chemisorbed anions at metal surfaces [110], its occurrence remains, nevertheless, less well-established for the propagation of electron delocalization across several bonds, as would be required for most adsorbed redox couples.

Variation in the metal surface composition is, then, generally expected to yield large variations in the observed rate constant for inner-sphere pathways since the reaction energetics will be sensitive to the chemical nature of the metal surface. For outer-sphere reactions, on the other hand, the rate constants are anticipated to be independent of the electrode material after correction for electrostatic work terms provided that adiabatic (or equally non-adiabatic) pathways are followed. Although a number of studies of the dependence of the rate constants for supposed outer-sphere reactions on the nature of the electrode material have been reported, relatively few refer to sufficiently well-defined conditions where double-layer corrections are small or can be applied with confidence [111–115]. Several of these studies indeed

References pp. 56–60

show that the work-corrected reactivities for a number of organic and inorganic [112, 114] redox couples are indeed almost independent of the metal substrate. Nevertheless, the kinetics for several transition-metal reactants containing aquo ligands exhibit a strong sensitivity to the metal substrate, the rate constants and especially the pre-exponential factors typically become progressively smaller with increasing metal "hydrophilicity" [42, 115]. These effects seem likely to be due to differences in the reactant solvation between the bulk and interfacial environments associated with aquo ligand-solvent hydrogen bonding [115] (Sects. 2.4 and 3.5.2).

The rate constants for such outer-sphere reactions can therefore differ markedly from those corresponding to true weak-overlap pathways, even after correction for electrostatic double-layer effects. This can cause some difficulties with the operational definition of inner-sphere electrocatalysis considered above, whereby outer-sphere reactions are regarded as "noncatalytic" processes. In addition, there is evidence that inner- rather than outer-sphere pathways can provide the normally preferred pathways at metal–aqueous interfaces for reactants containing hydrophobic functional groups [116].

4.7 COMPARISONS BETWEEN REACTIVITIES OF CORRESPONDING ELECTROCHEMICAL AND HOMOGENEOUS REDOX PROCESSES

The early theoretical treatments of Marcus and Hush clearly established the close relationship between the kinetics of redox processes occurring in electrochemical and homogeneous environments [117]. The most straightforward relationship is between the kinetics of electrochemical exchange and the corresponding homogeneous self-exchange reaction for a given redox couple. In the weak-overlap limit, i.e. when the reactant–electrode and reactant–reactant interactions for the electrochemical and homogeneous reactions, respectively, can be neglected, it is expected that [117a]

$$2\Delta G_e^* = \Delta G_h^* \tag{42}$$

where ΔG_e^* and ΔG_h^* are the corresponding free-energy barriers. These equal the intrinsic barriers since the driving force is zero in both cases. This relation arises because only one redox center is activated for the electrochemical case compared with two centers in the homogeneous process. Besides work terms, deviations are expected from eqn. (42) according to the dielectric continuum treatment as a result of influences upon the outer-shell barriers from reactant–electrode imaging and reactant–reactant electrostatic interactions. On the basis of eqn. (19) (Sect. 3.2) and its equivalent for homogeneous outer-sphere reactions, the relation between ΔG_e^* and ΔG_h^* becomes [7]

$$2\Delta G_e^* = \Delta G_h^* + \frac{e^2}{4}\left(\frac{1}{R} - \frac{1}{R_e}\right)\left(\frac{1}{\varepsilon_{op}} - \frac{1}{\varepsilon_s}\right) = \Delta G_h^* + C \tag{43}$$

where R_h is the internuclear distance for the homogeneous self-exchange reaction. Expressing eqn. (43) in terms of the rate constants, k_{ex}^e and k_{ex}^h, for the electrochemical and homogeneous exchange reactions leads to [7]

$$2 \log (k_{ex}^e / A_p^e) = \log (k_{ex}^h / A_p^h) - \frac{C}{2.303RT} \tag{44}$$

where A_p^e and A_p^h are the corresponding pre-exponential factors defined in terms of eqn. (28). In the weak-overlap limit, or more generally when $R_h = R_e$ so that $C = 0$, eqn. (44) reduces to the well-known relation

$$\frac{k_{ex}^e}{A_p^e} = \left(\frac{k_{ex}^h}{A_p^h}\right)^{1/2} \tag{44a}$$

Another simple relationship can be recovered for reactions where the outer-shell reorganization provides the only component of ΔG^*. If reactant–electrode imaging can be neglected (so that $R_e^{-1} \rightarrow 0$) and the internuclear distance, R_h, is equal to twice the reactant radius, then the quantity C in eqn. (43) equals ΔG_h^*, so that simply

$$\Delta G_e^* = \Delta G_h^* \tag{45}$$

and therefore eqn. (44) becomes

$$\frac{k_{ex}^e}{A_p^e} = \frac{k_{ex}^h}{A_p^h} \tag{44b}$$

A similar relationship to eqn. (44) can also be derived for homogeneous "heteronuclear", or cross, reactions (i.e. where the reaction partners are chemically distinct [76, 119]

$$2 \log (k^e / A_p^e) \approx \log (k^h / A_p^h) - \frac{C}{2.203RT} \tag{46}$$

where k^h is the rate constant for the homogeneous cross reaction and k^e is the electrochemical rate constant at the intersection of the Tafel plots for the two constituent anodic and cathodic electrode reactions [119]. (This relation is only approximate for highly asymmetric and/or exoergic cross reactions, where a treatment involving free-energy minimization yields a more precise result [76].) Equation (44) is actually a special case of eqn. (46); the latter has the virtue that it can be applied to electrochemical reactions for which exchange kinetics (i.e. the standard rate constant) cannot be determined on account of chemical irreversibility (i.e. for which the formal potential is unknown).

A number of tests of eqns. (44) and (46) have been made [64, 76, 97, 117a, 118, 119]. Although the work terms can yield sizeable uncertainties, broadly speaking the available data show reasonable agreement with eqn. (44), both for organic molecule–radical couples in aprotic solvents [118a, b] and for inorganic couples in aqueous media [117a, 118c–e]. Deviations are seen from eqn. (44a) in that commonly $k^{e}_{ex}/A^{e}_{p} < (k^{h}_{ex}/A^{h}_{p})^{1/2}$, as expected from eqn. (44) given that C should be positive [eqn. (43)]. Equation (44b) has been found to be in accordance with experimental data for most organic molecule–radical systems [118a, b].

Although the choice of R_e and R_h values required to estimate C is somewhat tenuous, provided that "normal" outer-sphere pathways are followed, it is expected that $R_e = 2a = l$, where a is the reactant radius and l is the inner-layer thickness, and $R_h = 2d$. Usually, A^{e}_{p} and A^{h}_{p} are taken as 10^4 cm s^{-1} and 10^{11} M^{-1} s^{-1}, respectively, as derived from the collisional formalism (Sect. 3.3.4). However, somewhat different values of A^{e}_{p} and A^{h}_{p} can be obtained using the pre-equilibrium treatment described above, depending on the nuclear and electronic properties of the reacting system (Sect. 3.3.4) [7, 118c]. Thus, quite different relative, as well as absolute, A^{e}_{p} and A^{h}_{p} values will result, for example, for systems controlled by solvent dynamics rather than by inner-shell vibrations. Comparisons between the available electrochemical and homogeneous data taking into account these and other details have, however, not been undertaken and would be informative.

Similar equations may also be devised relating the activation parameters for corresponding electrochemical and homogeneous reactions [55, 64, 76]. Although such comparisons have only been explored so far for a limited number of inorganic redox processes, generally the experimental values differ markedly from theoretical expectations, the homogeneous activation entropies being substantially smaller than anticipated from the electrochemical values [64, 76]. This has been attributed to solvent polarization within the multicharged encounter complexes for the homogeneous processes [76].

In some respects, it is useful to regard electrochemical reactions as a special type of heteronuclear reaction where the electrode can be considered as a co-reactant. Unlike solution co-reactants, however, metal surfaces possess a continuously variable redox potential and a zero intrinsic barrier, even though electrodes may still exert considerable influences on the reaction energetics via reactant–surface interactions. For outer-sphere reactions, then, a close relationship is expected between the relative electrochemical rate constants, k^{e}, for the one-electron reduction (or oxidation) of a series of reactants and the corresponding rate constants, k^{h}, for the same reactions involving a given homogeneous reducing (or oxidizing) agent, the rates again being corrected for work terms. Such a relation given by Marcus [117b] can be expressed as

$$(\Delta \log k^{e})_{E} = (\Delta \log k^{h})_{R} \tag{47}$$

where the subscripts E and R refer to a constant electrode potential and a fixed reducing (or oxidizing) agent, respectively. (Since the derivation of eqn. (47) requires the condition of small or moderate driving forces where the transfer coefficient is close to 0.5, it will be only approximate for highly exoergic reactions [117b].

As for eqn. (46), this relation can be applied to chemically irreversible, as well as to reversible, electrochemical reactions since the influence of the reaction thermodynamics is cancelled out. Of the relatively few tests of eqn. (47) that have been made [64, 97, 116, 120], reasonable accordance with the experimental data has been obtained.

A major application of eqn. (47) is to diagnose the presence of catalytic, presumably inner-sphere, electrochemical pathways. This utilizes the availability of a number of homogeneous redox couples, such as $Ru(NH_3)_6^{3+/2+}$ and Cr(bipyridine)$_3^{3+/2+}$ that must react via inner-sphere pathways since they lack the ability to coordinate to other species [5]. Provided that at least one of the electrochemical reactions also occurs via a well-defined outer-sphere pathway, the observation of markedly larger electrochemical rate constants for a reaction other than that expected from eqn. (47) indicates that the latter utilizes a more expeditious pathway. This procedure can be used not only to diagnose the presence of inner-sphere pathways, but also to evaluate the extent of inner-sphere electrocatalysis (Sect. 4.6) it enables reliable estimates to be made of the corresponding outer-sphere rate parameters [12a, 116, 120c].

4.8 PREDICTION OF ABSOLUTE RATE PARAMETERS

Most of our present understanding of the various structural and dynamic factors that influence electrochemical kinetics has been derived from the various "relative" theory–experiment comparisons described in the foregoing six sections. An ultimate objective, however, is to understand and predict rate parameters for individual electrochemical reactions on the basis of the theoretical models. While such "absolute" rate calculations can involve substantial uncertainties, their comparison with the experimental values is of obvious importance in that it provides a critical overall test of the underlying models. As noted in Sect. 4.1, the scope of such "absolute" theory–experiment comparisons is limited at present in view of the quantitative structural and thermodynamic information often required in order to calculate the barrier height. Such calculations are most straightforward for redox couples, such as some organic molecule–radical systems, that involve little or no inner-shell barrier. The standard rate constants for such systems are usually large, $\gtrsim 1\,\mathrm{cm\,s^{-1}}$, as expected. Indeed, the rates often approach or even surpass the range of applicability of electrochemical relaxation methods, so that much literature data are of questionable reliability. Although systematic comparisons are lacking, the experimental rate constants for such systems are in tolerable agreement with theoretical expectations [45b, 52b, 121]. Consideration of specific supporting electrolyte effects (Sect. 4.2) and solvent dynamical effects (Sect. 4.5) do, however, complicate the issue somewhat, especially for non-aqueous systems.

Outer-sphere reactions having substantial inner-shell barriers provide interesting systems with which to make absolute theory–experiment comparisons since widely varying reactivities are encountered, depending on the structural changes involved. A recent such comparison for cationic transition-metal redox couples at metal–aqueous interfaces [122] yielded good agreement between the experimental and calculated rate constants for mercury surfaces, whereas typically $k_{ob} < k_{calc}$ at more hydrophilic surfaces such as lead and gallium. These latter discrepancies were attributed to the presence of unfavorable work terms arising from interfacial solvation effects [122] (Sect. 4.6). Roughly similar theory–experiment disparities were seen for a number of related homogeneous redox reactions involving these redox couples, which also may be due to the presence of unfavorable work terms associated with ion–solvent interactions [122]. The observed pre-exponential factors, A_{ob}, for the electrochemical reactions and especially the homogeneous processes are generally smaller than the calculated values, A_{calc} [122]. (The latter values were estimated by assuming reaction adiabaticity at the plane of closest approach, with increasing non-adiabaticity for greater internuclear distances.) The findings that $A_{ob} < A_{calc}$ as well as $k_{ob} < k_{calc}$ are therefore suggestive of the presence of significantly non-adiabatic pathways, even at the plane of closest approach.

Although such theory–experiment discrepancies can be substantial (often 10–10^3 fold disparities in rate constants) [122], even the simplified theoretical treatments yield free-energy barriers that are commonly within 10–20% of those inferred from the experimental rate parameters by estimating the frequency factor. In a global sense, therefore, the electron-transfer models are reasonably successful in predicting rate constants for both electrochemical and homogeneous outer-sphere reactions. The remaining deviations between the calculated and experimental rate parameters, although sometimes substantial, seem most likely associated with errors in estimating the work terms and pre-exponential factors rather than in the electron-transfer barrier itself [122].

5. Future prospects

The development of our understanding of redox kinetics at electrode surfaces, as in homogeneous solution, has recently undergone an interesting evolution. We now have an underlying theoretical framework which, in a formal sense, is relatively well developed, at least for weak-overlap processes. The availability of electrochemical perturbation techniques that utilize commercial instrumentation, together with the gradual development of methods for preparing and handling clean solid surfaces has enabled reliable electrochemical kinetic data to be obtained for an increasingly diverse range of systems. A number of features of the observed kinetics for one-electron transfer reactions at metal surfaces as well as in homogeneous solution are roughly consistent with the expectations of the weak-overlap theoretical models. While this forms a reasonably satisfying picture, our understanding of even these simple processes is nevertheless far from complete.

Several issues appear to warrant particular consideration in the near future. Firstly, it would be desirable to obtain more quantitative information on the manner in which the electronic transmission coefficient, κ_{el}, depends on the reactant geometry, reactant–surface distance, and the nature of the intervening medium. While some information along these lines could be obtained experimentally, there is a strong need for related theoretical work. Although ab initio estimates of κ_{el} have been reported for a few outer-sphere processes in solution [1d, 49a, b], such calculations have yet to be undertaken for heterogeneous electron transfer [54]. Secondly, there is a clear need to clarify further the role of solvent dynamical properties upon the pre-exponential factor. Apart from more detailed experimental tests of the dielectric continuum-based treatment (Sect. 3.3.1), a critical issue is the relative extent to which inner- rather than outer-shell dynamics dominate the frequency factor as the magnitude of the inner-shell barrier increases. Further experimental and theoretical work on these topics will therefore be required in order to formulate a satisfactory quantitative picture of the molecular factors influencing the pre-exponential factor.

Thirdly, it will be important to gain more direct information on the stability of outer-sphere precursor states, especially with regard to the limitations of simple electrostatic models (Sect. 4.2). One possible approach is to evaluate K_p for stable reactants by means of differential capacitance and/or surface tension measurements. Little double-layer compositional data have been obtained so far for species, such as multicharged transition-metal complexes, organometallics, and simple aromatic molecules that act as outer-sphere reactants. The development of theoretical double-layer models that account for solvation differences in the bulk and interfacial environments would also be of importance in this regard.

By comparison with single-electron outer-sphere processes, our understanding of inner-sphere, multielectron, and coupled electron–atom transfer electrochemical reactions remains relatively rudimentary. At least in suitable cases, kinetic measurements for surface-attached or adsorbed reactants can provide uniquely direct information on the electron-transfer step and, as for the analogous intramolecular processes in homogeneous solution, are particularly valuable for exploring electron-coupling and geometric effects. The treatment of most multistep electrode processes has so far been largely unaffected by theoretical developments in electron transfer, chiefly due to the lack of information on the thermodynamics of the individual steps. Nevertheless, the redox thermodynamics of an increasing number of previously intractable non-metal redox systems are being unraveled by pulse radiolysis and rapid kinetic approaches [123]. This enables the electron-transfer kinetics for such multistep systems in homogeneous solution to be compared with theoretical predictions based on molecular structural changes [9, 123]. These developments will also enable the kinetics of the corresponding electrode processes to also be interpreted on such a molecular level.

Overall, then, the treatment of electrochemical reaction kinetics is evolving from its traditional phenomenological basis to acquire an increasingly molecular-level character. These developments have been, and will continue to be, brought about not only by a strong interplay between experiment and theory, but also by a cognizance of the close relationship between redox processes at electrodes and in homogeneous media. The increasing interest in altering the chemical properties of electrodes by surface modification further underscores the virtues of such interfaces as versatile and powerful redox reagents. The further quantitative development of our understanding of heterogeneous and homogeneous redox reactions on a more unified basis than hitherto therefore seems not only desirable but also inevitable.

References

1 For recent reviews on electron-transfer reactions, see (a) N. Sutin, Prog. Inorg. Chem., 30 (1983), 441. (b) A. Haim, Prog. Inorg. Chem., 30 (1983) 273. (c) R.D. Cannon, Electron-Transfer Reactions, Butterworths, London, 1980. (d) M.D. Newton, Int. J. Quantum Chem. Symp., 14 (1980) 363. (e) R.A. Marcus and N. Sutin, Biochim. Biophys. Acta, 811 (1985) 265. (f) M.D. Newton and N. Sutin, Annu. Rev. Phys. Chem., 35 (1984) 437.

2 (a) M. Temkin, Zh. Fiz. Khim., 22 (1948) 1081. (b) M.J. Weaver, J. Phys. Chem., 80 (1976) 2645. (c) M.J. Weaver, J. Phys. Chem., 83 (1979) 1748.

3 J.T. Hupp and M.J. Weaver, J. Electroanal. Chem., 145 (1983) 43.

4 E.L. Yee, R.J. Cave, K.L. Guyer, P.D. Tyma and M.J. Weaver, J. Am. Chem. Soc., 101 (1979) 1131.

5 H. Taube, Electron Transfer Reactions of Complex Ions in Solution, Academic Press, New York, 1970.

6 R.A. Marcus, J. Phys. Chem., 72 (1968) 891.

7 J.T. Hupp and M.J. Weaver, J. Electroanal. Chem., 152 (1983) 1.

8 For example, see P. Delahay, Double Layer and Electrode Kinetics, Wiley–Interscience, New York, 1965, Chap. 7.

9 For example W.K. Wilmarth, D.M. Stanbury, J.E. Byrd, H.N. Po and C.-P. Chua, Coord. Chem. Rev., 51 (1983) 155.

10 For example, A.J. Appleby, in B.E. Conway and J. O'M. Bockris (Eds.), Modern Aspects of Electrochemistry, Vol. 9, Plenum Press, New york, 1974, Chap. 5. S. Srinivasan, H. Wroblowa and J. O'M. Bockris, Adv. Catal., 17 (1967) 351.

11 For example R.W. Murray, Acc. Chem. Res., 13 (1980) 135; in A.J. Bard (Ed.), Electroanalytical Chemistry. A Series of Advances, Vol. 13, Dekker, New York, 1984, p. 191.

12 (a) S.W. Barr and M.J. Weaver, Inorg. Chem., 23 (1984) 1657. (b) K.L. Guyer and M.J. Weaver, Inorg. Chem., 23 (1984) 1664.

13 B.E. Conway, in J. O'M. Bockris (Ed.), MTP International Review of Science, Physical Chemistry Series I, Vol. 6, Butterworths, London, 1973, p. 41.

14 (a) R. Parsons and F.G.R. Zobel, J. Electroanal. Chem., 9 (1965) 333. (b) J.A. Harrison, J.E.B. Randles and D.J. Schriffrin, J. Electroanal. Chem., 25 (1970) 197.

15 (a) M.J. Weaver and T.L. Satterberg, J. Phys. Chem., 81 (1977) 1772. (b) M.J. Weaver, H.Y. Liu and Y. Kim, Can. J. Chem., 59 (1981) 1944.

16 (a) L. Gierst, E. Nicolas and L. Tygat-Vanderberghen, Croat. Chim. Acta, 42 (1970) 117. (b) W.R. Fawcett, J. Electroanal. Chem., 22 (1969) 19.

17 (a) W.R. Fawcett and A. Lasia, J. Phys. Chem., 82 (1978) 1114. (b) A. Baranski and W.R. Fawcett, J. Electroanal. Chem., 100 (1977) 185. (c) D.A. Corrigan and D.H. Evans, J. Electroanal. Chem., 106 (1980) 287. (d) T. Gennett and M.J. Weaver, J. Electroanal. Chem., 186 (1985) 179.

18 (a) B.B. Damaskin and A.N. Frumkin, Electrochim. Acta, 19 (1974) 173. (b) S. Trassati, in B.E. Conway and J. O'M. Bockris (Eds.), Modern Aspects of Electrochemistry, Vol. 13, Plenum Press, New York, 1979, p. 81. (c) S. Trassati, Electrochim. Acta, 28 (1983) 1083.

19 For example (a) W. Drost-Hansen, Ind. Eng. Chem., 61 (11) (1969) 10. (b) B.W. Ninham, Pure Appl. Chem., 53 (1981) 2135.

20 M.J. Weaver and F.C. Anson, J. Electroanal. Chem., 58 (1975) 95.

21 M.A. Tadayyoni and M.J. Weaver, J. Electroanal. Chem., 187 (1985) 283.

22 For example (a) M.A. Tadayyoni, S. Farquharson and M.J. Weaver, J. Chem. Phys., 80 (1984) 1363. (b) M.A. Tadayyoni, S. Farquharson, T.T.-T. Li and M.J. Weaver, J. Phys. Chem., 88 (1984) 4701.

23 M.J. Weaver and F.C. Anson, Inorg. Chem., 15 (1976) 1871; J. Am. Chem. Soc., 97 (1975) 4403.

24 H. Taube, Electron Transfer Reactions of Complex Ions in Solution, Academic Press, New York, 1970, Chap. 2.

25 For example (a) R. de Levie, J. Electrochem. Soc., 118 (1971) 185C. (b) R. Parsons, J. Electroanal. Chem., 21 (1969) 35. (c) W.R. Fawcett and S. Levine, J. Electroanal. Chem., 43 (1973) 175; 65 (1975) 505. (d) M.J. Weaver and F.C. Anson, J. Electroanal. Chem., 65 (1975) 759. (e) R. Andreu, M. Sluyters-Rehbach and J.H. Sluyters, J. Electroanal. Chem., 171 (1984) 139.

26 (a) R.G. Linck, in M.L. Tobe (Ed.), MTP International Review of Science, Inorganic Chemistry, Series II, Vol. 9, Butterworths, London, 1974, p. 173. (b) R.G. Linck, Surv. Prog. Chem., 7 (1976) 89.

27 B.S. Brunschwig, J. Logan, M.D. Newton and N. Sutin, J. Am. Chem. Soc., 102 (1980) 5978.

28 B.L. Tembe, H.L. Friedman and M.D. Newton, J. Chem. Phys., 76 (1982) 1490.

29 R.A. Marcus, Int. J. Chem. Kinet., 13 (1981) 865.

30 J.T. Hupp and M.J. Weaver, J. Phys. Chem., 88 (1984) 1463.

31 For recent reviews of electron-transfer theories, see refs. 1a, 1c, 1d, 1f, and (a) P.P. Schmidt, Specialist Periodical Report in Electrochemistry, Vol. 5, The Chemical Society, London, 1975, Chap. 2; Vol. 6, 1977, Chap. 4; (b) R.A. Marcus, in P.A. Rock (Ed.), Special Topics in Electrochemistry, Elsevier, New York, 1977. (c) J. Ulstrup, Charge Transfer Processes in Condensed Media, Springer-Verlag, Berlin, 1979. (d) E.D. German and A.M. Kuznetsov, Electrochim. Acta, 26 (1981) 1595. (e) R.R. Dogonadze, A.M. Kuznetsov and T.A. Marigishvili, Electrochim. Acta, 25 (1980) 1.

32 R.A. Marcus, J. Chem. Phys., 43 (1965) 679.

33 Two distinct definitions of α_{et} should be distinguished. That in eqn. (15), consistent with eqn. (6a), refers to the *average* value of α_{et} between a given value of ΔG^0_{et} and $\Delta G^0_{et} = 0$. The alternative definition refers to the instantaneous *slope* of the ΔG^*_{et}–ΔG^0_{et} curve at a given driving force, i.e. to a tangent, rather than a chord, of this plot [34].

34 M.J. Weaver and F.C. Anson, J. Phys. Chem., 80 (1976) 1861.

35 EXAFS = extended X-ray absorption fine structure.

36 B.S. Brunschwig, C. Creutz, D.H. MaCartney, T.-K. Sham and N. Sutin, Discuss. Faraday Soc., 74 (1982) 113.

37 (a) J.T. Hupp, H.Y. Liu, J.K. Farmer, T. Gennett and M.J. Weaver, J. Electroanal. Chem., 168 (1984) 313. (b) J.T. Hupp and M.J. Weaver, J. Phys. Chem., 89 (1985) 2795.

38 For an extreme view, see J. O'M. Bockris and S.U.M. Khan, Quantum Electrochemistry, Plenum Press, New York, 1979.

39 D.L. Calef and P.G. Wolynes, J. Chem. Phys., 78 (1983) 470.

40 (a) J.T. Hupp and M.J. Weaver, J. Phys. Chem., 88 (1984) 1860. (b) J.T. Hupp and M.J. Weaver, J. Phys. Chem., 89 (1985) 1601.

41 J.O'M. Bockris and A.K.N. Reddy, Modern Electrochemistry, Vol. 1, Plenum Press, New York, 1971, Chap. 2.
42 N.B. Slater, Theory of Unimolecular Reactions, Cornell University Press, Ithaca, NY, 1959, pp. 57–59.
43 D.F. Calef and P.G. Wolynes, J. Phys. Chem., 87 (1983) 3387.
44 L.D. Zusman, Chem. Phys., 49 (1980) 295. I.V. Alexandrov, Chem. Phys., 51 (1980) 449.
45 (a) M.J. Weaver and T. Gennett, Chem. Phys. Lett., 113 (1985) 213. (b) T. Gennett, D. Milner and M.J. Weaver, J. Phys. Chem., 89 (1985) 2787.
46 H.L. Friedman and M.D. Newton, Faraday Discuss. Chem. Soc., 74 (1982) 73.
47 H. Fraunfelder and P.G. Wolynes, Science, 229 (1985) 337.
48 M. Ya Ovchinnikova, Russ. Theor. Exp. Chem., 17 (1981) 507.
49 (a) M.D. Newton, ACS Symp. Ser., 198 (1982) 255. (b) J. Logan and M.D. Newton, J. Chem. Phys., 78 (2) (1983) 4086. (c) D.N. Beretan and J.J. Hopfield, J. Am. Chem. Soc., 106 (1984) 1585.
50 (a) S. Larsson, J. Am. Chem. Soc., 103 (1981) 4034. (b) S. Larsson, J. Chem. Soc. Faraday Trans. 2, 79 (1983) 1375.
51 For example (a) C. Creutz, Prog. Inorg. Chem., 30 (1983) 1. (b) T.J. Meyer, Prog. Inorg. Chem., 30 (1983) 389.
52 (a) J.M. Hale, J. Electroanal. Chem., 19 (1968) 315. (b) J.M. Hale, in N.S. Hush (Ed.), Reactions of Molecules at Electrodes, Interscience, New York, 1971, Chap. 4. (c) R.R. Dogonadze, A.M. Kuznetsov and M.A. Vorotyntsev, J. Electroanal. Chem., 25 (1970) App. 17.
53 (a) W. Schmickler, J. Electroanal. Chem., 82 (1977) 65; 83 (1977) 387. (b) W. Schmickler and J. Ulstrup, Chem. Phys., 19 (1977) 217. (c) J. Ulstrup, Surf. Sci., 101 (1980) 564.
54 J.O'M. Bockris and S.U.M. Khan, Quantum Electrochemistry, Plenum Press, New York, 1979, Chap. 13.
55 J.T. Hupp and M.J. Weaver, J. Phys. Chem., 88 (1984) 6128.
56 M.B. Robin and P. Day, Adv. Inorg. Chem. Radiochem., 10 (1967) 247.
57 M.J. Weaver, Inorg. Chem., 18 (1979) 402.
58 P. Delahay, Double Layer and Electrode Kinetics, Wiley–Interscience, New York, 1965, Chap. 9.
59 G.M. Torrie and J.P. Valleau, J. Phys. Chem., 86 (1982) 3251. G.M. Torrie, J.P. Valleau and G.N. Patey, J. Chem. Phys., 76 (1982) 4615. J.P. Valleau and G.M. Torrie, J. Chem. Phys., 76 (1982) 4623.
60 D. Henderson and L. Blum, Can. J. Chem., 59 (1981) 1906; J. Electroanal. Chem., 132 (1982) 1. L.B. Bhurijan, L. Blum and D. Henderson, J. Chem. Phys., 78 (1983) 442.
61 M. Lozada-Cassou, R. Saavedra-Barrera and D. Henderson, J. Chem. Phys., 77 (1982) 5150. M. Lozada-Cassou and D. Henderson, J. Phys. Chem., 87 (1983) 2821.
62 W.R. Fawcett, in S. Bruckenstein, B. Miller, J.D.E. McIntyre and E. Yeager (Eds.), Proc. 3rd Symp. on Electrode Processes, Electrochemical Society, Pennington, NY, 1980, p. 213.
63 R.A. Marcus, Can. J. Chem., 37 (1959) 155. See also M.J. Weaver, J. Phys. Chem., 84 (1980) 568 (footnote 42).
64 M.J. Weaver, J. Phys. Chem., 84 (1980) 568.
65 N. Sutin, Acc. Chem. Res., 1 (1968) 225.
66 R. Parsons, Croat. Chim. Acta, 42 (1970) 281.
67 For example, see B.E. Conway, Ionic Hydration in Chemistry and Biophysics, Elsevier, Amsterdam, 1981.
68 For example (a) R. de Levie and M. Nemes, J. Electroanal. Chem., 58 (1975) 123. (b) M.J. Weaver and F.C. Anson, J. Electroanal. Chem., 65 (1975) 711, 737. (c) M.J. Weaver, J. Electroanal. Chem., 93 (1978) 231.
69 For example (a) M.L. Foresti, D. Cozzi and R. Guidelli, J. Electroanal. Chem., 53 (1974) 235. (b) S. Levine and W.R. Fawcett, J. Electroanal. Chem., 43 (1973) 175; 99 (1979) 265.

70 For example (a) G.M. Brown and N. Sutin, J. Am. Chem. Soc., 101 (1979) 883. (b) T.J. Przystas and N. Sutin, J. Am. Chem. Soc., 95 (1973) 5545.
71 For a broad-based review, see B. Perlmutter-Hayman, Prog. React. Kinet. 6 (1971) 239.
72 (a) R. Parsons, Trans. Faraday Soc., 47 (1951) 1332. (b) J.O'M. Bockris, in J.O'M. Bockris (Ed.), Modern Aspects of Electrochemistry, Vol. 1, Butterworths, London, 1954, p. 180.
73 For example A.J. Appleby, J.O'M. Bockris, R.K. Sen and B.E. Conway, in J.O'M. Bockris (Ed.), MTP International Review of Science, Physical Chemistry Series I, Vol. 6, Butterworths, London, 1973, p. 1.
74 A.M. Kuznetsov, J. Electroanal. Chem., 180 (1984) 121.
75 P.D. Tyma and M.J. Weaver, J. Electroanal. Chem., 111 (1980) 195.
76 M.J. Weaver and J.T. Hupp, ACS Symp. Ser., 198 (1982) 181.
77 J.-M. Savéant and D. Tessier, Faraday Discuss. Chem. Soc., 74 (1982) 57.
78 For example D. Rehm and A. Weller, Isr. J. Chem., 8 (1970) 259. C. Creutz and N. Sutin, J. Am. Chem. Soc., 99 (1977) 241.
79 H.Y. Liu, J.T. Hupp and M.J. Weaver, J. Electroanal. Chem., 179 (1984) 219.
80 B.E. Conway, D.E. Tessier and D.P. Wilkinson, J. Electroanal. Chem., 199 (1986) 249.
81 T.T.-T. Li, K.L. Guyer, S.W. Barr and M.J. Weaver, J. Electroanal. Chem., 164 (1984) 27.
82 J.N. Agar, Discuss. Faraday Soc., 1 (1947) 81.
83 B.E. Conway, D.J. MacKinnon and B.V. Tilak, Trans. Faraday Soc., 66 (1970) 1203.
84 For a recent review, see B.E. Conway, in B.E. Conway, J.O'M. Bockris and R. White (Eds.), Modern Aspects of Electrochemistry, Vol. 16, Plenum Press, New York, 1985.
85 S.M. Husaini, MS Thesis, Michigan State University, 1982.
86 R.A. Peterson, Ph.D. Thesis, University of Wisconsin at Madison, 1983.
87 (a) J.W. Schultze and K.J. Vetter, Electrochim. Acta, 18 (1973) 889. (b) K.J. Vetter and J.W. Schultze, Ber. Bunsenges. Phys. Chem., 77 (1973) 945. (c) P. Kohl and J.W. Schultze, Ber. Bunsenges. Phys. Chem., 77 (1973) 953. (d) J.W. Schultze and U. Stimming, Z. Phys. Chem. NF, 98 (1975) 285. See also (e) U. Stimming and J.W. Schultze, Electrochim. Acta, 24 (1979) 859.
88 For example (a) T. Biegler, E.R. Gonzales and R. Parsons, Collect. Czech. Chem. Commun., 36 (1971) 414. (b) J. Lipkowski and Z. Galus, J. Electroanal. Chem., 58 (1975) 51. (c) B. Behr, J. Taraszewska and J. Stroka, J. Electroanal. Chem., 58 (1975) 71. (d) R. Guidelli and M.L. Foresti, J. Electroanal. Chem., 77 (1977) 73. (e) J. Lipkowski, A. Czerwinski, E. Cieszynska, Z. Galus and J. Sobkowski, J. Electroanal. Chem., 119 (1981) 261.
89 J. Lipkowski, C. Buess-Herman, J.-P. Lambert and L. Gierst, J. Electroanal. Chem., 202 (1986) 169.
90 M.W. Humphreys and R. Parsons, J. Electroanal. Chem., 82 (1977) 369.
91 (a) Yu.I. Kharkats, Electrokhimiya, 15 (1979) 246. (b) Yu.I. Kharkats, H. Nielsen and J. Ulstrup, J. Electroanal. Chem., 169 (1984) 47.
92 (a) T.T.-T. Li, H.Y. Liu and M.J. Weaver, J. Am. Chem. Soc., 106 (1984) 1233. (b) T.T.-T. Li and M.J. Weaver, J. Am. Chem. Soc., 106 (1984) 6107. (c) M.J. Weaver and T.T.-T. Li, J. Phys. Chem., 90 (1986) 3823.
93 W.R. Fawcett and J.S. Jaworski, J. Phys. Chem., 87 (1983) 2972.
94 (a) S. Sahami and M.J. Weaver, J. Electroanal. Chem., 124 (1981) 35. (b) J.K. Farmer, T. Gennett and M.J. Weaver, J. Electroanal. Chem., 191 (1985) 357.
95 (a) B.A. Kowert, L. Marcoux and A.J. Bard, J. Am. Chem. Soc., 94 (1972) 5538. (b) E.S. Yang, M.-S. Chan and A.C. Wahl, J. Phys. Chem., 84 (1980) 3094. (c) M.-S. Chan and A.C. Wahl, J. Phys. Chem., 86 (1982) 126. (d) T.T.-T. Li, M.J. Weaver and C.H. Brubaker, J. Am. Chem. Soc., 104 (1982) 2381. (e) G. Grampp and W. Jaenicke, Ber. Bunsenges. Phys. Chem., 88 (1984) 325, 335.
96 E.M. Arnett and D.R. McKelvey, in J.F. Coetzee and C.D. Ritchie (Eds.), Solute–Solvent Interactions, Dekker, New York, 1967, Chap. 6.
97 M.J. Weaver, P.D. Tyma and S.M. Nettles, J. Electroanal. Chem., 114 (1980) 53.
98 M.D. Newton, Faraday Discuss. Chem. Soc., 74 (1982) 108.

99 M.J. Weaver and T.T.-T. Li, J. Phys. Chem., 87 (1983) 1153.

100 P. Delahay, Acc. Chem. Res., 15 (1982) 40. P. Delahay and A. Dziedzic, J. Chem. Phys., 80 (1984) 5793.

101 A. Kapturkiewicz and B. Behr, J. Electroanal. Chem., 179 (1984) 187.

102 N.E. Hill, W.E. Vaughan, A.H. Price and M. Davies, Dielectric Properties and Molecular Behavior, Van Nostrand Reinhold, London, 1969, Chaps. 1, 4 and 5.

103 Yu.I. Kharkats, H. Nielsen and J. Ulstrup, J. Electroanal. Chem., 169 (1984) 47.

104 (a) M.J. Weaver, Inorg. Chem., 18 (1979) 402. (b) S.W. Barr, K.L. Guyer, T.T.-T. Li, H.Y. Liu and M.J. Weaver, J. Electrochem. Soc., 131 (1984) 1626.

105 For example R.G. Linck, MTP Int. Rev. Sci: Inorg. Chem. Ser. 1, 9 (1972) 303.

106 For example E. Yeager, J. Electrochem. Soc., 128 (1981 160c.

107 H. Gerischer, Natl. Bur. Stand. (U.S.) Spec. Publ., 455 (1976) 1.

108 (a) L. Muller and S. Dietzsch, J. Electroanal. Chem., 121 (1981) 255. (b) L. Muller, S. Dietzsch, M. Tameczek and R. Sohr, Elektrokhimiya, 16 (1980) 62. (c) W.J. Plieth and K.J. Vetter, Collect. Czech. Chem. Commun., 36 (1971) 816. (d) W. Lorenz, J. Electroanal. Chem., 191 (1985) 31.

109 M.J. Weaver, P. Gao, D. Gosztola, M.L. Patterson and M.A. Tadayyoni, ACS Symp. Ser., 307 (1986) 135.

110 For example W. Lorenz and G. Salie, J. Electroanal. Chem., 80 (1977) 1.

111 S. Trassati, in H. Gerischer and C.W. Tobias (Eds.), Advances in Electrochemistry and Electrochemical Engineering, Vol. 10, Wiley, New York, 1977, p. 279.

112 S.W. Barr, K.L. Guyer and M.J. Weaver, J. Electroanal. Chem., 111 (1980) 41.

113 A. Capon and R. Parsons, J. Electroanal. Chem., 46 (1973) 215.

114 (a) N.V. Fedorovich, A.N. Frumkin and H.E. Case, Collect. Czech. Chem. Commun., 36 (1971) 722. (b) A.N. Frumkin, N.V. Fedorovich, N.P. Berezina and K.E. Kreis, J. Electroanal. Chem., 58 (1975) 189. (c) N.V. Fedorovich, M.D. Levi and S.I. Kulakovskaya, Sov. Electrochem., 13 (1976) 766. (d) A.N. Frumkin, N.V. Fedorovich and S.I. Kulakoskaya, Sov. Electrochem., 10 (1973) 313.

115 H.Y. Liu, J.T. Hupp and M.J. Weaver, J. Electroanal. Chem., 179 (1984) 219.

116 (a) V. Srinivasan, S.W. Barr and M.J. Weaver, Inorg. Chem., 21 (1982) 3154. (b) T.T.-T. Li and M.J. Weaver, Inorg. Chem., 24 (1985) 1882.

117 (a) R.A. Marcus, J. Phys. Chem., 67 (1963) 853. (b) R.A. Marcus, J. Chem. Phys., 43 (1965) 679. (c) N.S. Hush, Trans. Faraday Soc., 57 (1961) 557; N.S. Hush, Electrochim. Acta, 13 (1968) 1004.

118 (a) M.E. Peover, in N.S. Hush (Ed.), Reaction of Molecules at Electrodes, Wiley, New York, 1971, p. 259. (b) H. Kojima and A.J. Bard, J. Am. Chem. Soc., 97 (1975) 6317. (c) J.F. Endicott, R.R. Schroeder, D.H. Chidester and D.R. Ferrier, J. Phys. Chem., 77 (1973) 2579. (d) T. Saji, T. Yamada and S. Aoyagui, J. Electroanal. Chem., 61 (1975) 147; T. Sajo, Y. Maruyama and S. Aoyagui, J. Electroanal. Chem., 86 (1978) 219. (e) J.T. Hupp and M.J. Weaver, Inorg. Chem., 22 (1983) 2557.

119 M.J. Weaver, Inorg. Chem., 15 (1976) 1733.

120 (a) J.F. Endicott and H. Taube, J. Am. Chem. Soc., 86 (1964) 1686. (b) R.J. Klinger and J.K. Kochi, J. Am. Chem. Soc., 103 (1981) 5839. (c) T.L. Satterberg and M.J. Weaver, J. Phys. Chem., 82 (1978) 1784.

121 M.E. Peover, in N.S. Hush (Ed.), Reactions of Molecules at Electrodes, Wiley, New York, 1971, p. 259.

122 J.T. Hupp and M.J. Weaver, J. Phys. Chem., 89 (1985) 2795.

123 (a) D.M. Stanbury and L.A. Lednicky, J. Am. Chem. Soc., 106 (1984) 2847. M.S. Ram and D.M. Stanbury, J. Phys. Chem., 90 (1986) 3691.

Chapter 2

Semiconductor Electrochemistry

ANDREW HAMNETT

1. Introduction

Although the basic properties of semiconductors have been understood for fifty years and the first transistor was made more than thirty-five years ago, it is only comparatively recently that the electrochemical properties of a wide range of semiconductors have been extensively investigated. The first papers, on the electrochemistry of germanium, appeared in the mid-fifties; germanium was, at that time, one of the most extensively investigated semiconductors, but the first attempts to understand its electrochemistry were only partially successful. Only a decade later, after the careful exploratory work of Gerischer and his co-workers, was the importance of the chemistry of the outer surface layer in this material appreciated. By contrast, n-ZnO was clearly very much better behaved from the classical viewpoint and DeWald's papers became paradigmatic in a decade when an increasing number of compound semiconductors were studied. The basis of the electrochemical response of semiconductors was also elucidated in this period. By combining the electron-transfer theories being developed by the Russian school with the electronic structure of semiconductors, Gerischer established a universal theory of electron transfer at the semiconductor–electrolyte interface that has formed the framework for much of the more recent research.

The position in the early seventies was that of a relatively small number of research groups active in the field of semiconductor electrochemistry seeking primarily to understand the fundamental electrochemical processes involved in a rather narrow range of semiconducting materials. This picture changed radically in the next decade owing to two factors. The first was the observation by Fujishima and Honda that n-TiO_2, under a positive bias, evolved oxygen when illuminated with near-UV light. The existence of both anodic (on n-type) and cathodic (on p-type) photocurrents had been recognised for some years and the basic principles behind the phenomenon were well understood. However, for the elemental and compound n-type semiconductors that had been studied up to that time, the photoanodic currents had always been associated with corrosion and dissolution and were not thought to have any technological significance. The demonstration that stable photoelectrolysis of water was feasible immediately suggested the possibility of solar energy storage and this was given an enormous boost by the second factor, the 1973 oil crisis and its aftermath. The search for alternative energy sources, which had been proceeding in a somewhat desultory fashion until then, became imbued with a sense of urgency and the photoelectrolysis of water, to generate hydrogen and oxygen, was identified as a major goal by electrochemists the world over.

The last decade has seen an explosion of activity in the field as electrochemists have wrestled with unfamiliar, and often intractable, problems generated by the very wide range of materials investigated, difficulties often compounded by the use of polycrystalline samples whose bulk and surface properties have proved resistant to control. In addition to the elemental semiconductors and the III/V materials, a huge range of n- and p-type oxides, sulphides, selenides, and tellurides have been described and surface and bulk modifications carried out in the hope of enhancing photoelectrochemical efficiency. New theoretical and experimental tools have developed apace and our fundamental understanding of the semiconductor–electrolyte interface has deepened substantially.

However, the last few years have also seen a growing awareness of the problems inherent in using the semiconductor–electrolyte interface as a means of solar-energy conversion. Very long-term stability may not be possible in aqueous electrolytes and no oxide material has been identified that has properties suitable for use as a photoanode in a photoelectrolysis cell. Highly efficient photovoltaic cells are known, both in aqueous and non-aqueous solutions, but it is far from clear that the additional engineering complexity, over and above that required for the dry p–n junction photovoltaic device, will ever allow the "wet" photovoltaic cells to be competitive. These, and other problems, have led to something of a pause in the flood of papers on semiconductor electrochemistry in the last two years and the current review is therefore timely. I have tried to indicate what is, and is not, known at present and where future lines of development may lie. Individual semiconductors are not treated in detail, but it is hoped that most of the theoretical strands apparent in the last few years are discussed.

2. Electronic structure of semiconductors

The difference between metals and semiconductors is, fundamentally, electronic in origin. In a metal, overlap of atomic or ionic orbitals on neighbouring sites leads to a continuum of levels that are only partially occupied by electrons. There are, therefore, energy levels immediately above the topmost occupied level that are empty and easily accessible thermally and the ability of electrons to move freely into these levels gives rise to the characteristic properties associated with a metal, such as conductivity, reflectivity etc.

In the case of a semiconductor, overlap of the orbitals on each centre gives

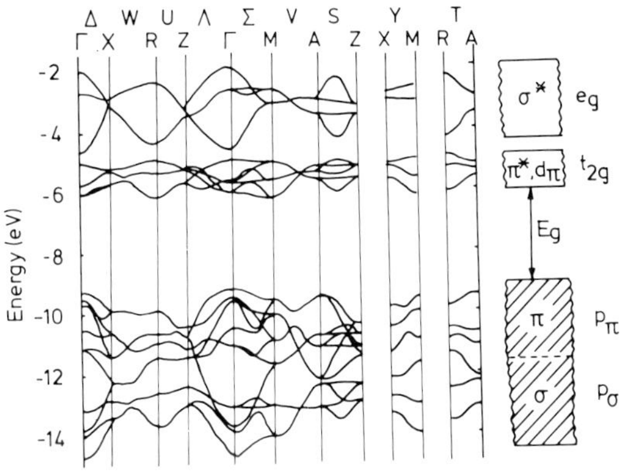

Fig. 1. The calculated band-structure of TiO_2. The left-hand side shows details of the band energies as a function of k for various directions in the Brillouin zone and the right-hand side shows the total bandwidths and orbital parentage. It can be seen that, even though TiO_2 is quite ionic, the O 2p levels are substantially broadened through O–O interactions and the Ti 3d levels by Ti–O–Ti interactions.

rise to two sets of energy levels. The lower in energy is termed the *valence band* and, in the stoichiometric semiconductor at 0 K, this band is completely occupied by electrons. The upper set of levels is termed the *conduction band* and, under the same conditions, it is completely empty. As an example of this, we may consider TiO_2 in which the valence band is formed primarily from the oxygen 2p orbitals, which overlap to give a band about 5 eV wide and whose conduction band is formed primarily from the overlap of the Ti 3d_π orbitals [1] as shown in Fig. 1. In a similar way, GaP may be thought of as having a valence band derived mainly from the P 3p (though with considerable admixture of Ga 4s) and the conduction band is mainly Ga 4s (though, again, there is a substantial admixture of P 3p). In between the conduction band (CB) and valence band (VB) there is an energy gap called the *bandgap* which in TiO_2 is ca. 3 eV wide and in GaP about 2.2 eV wide; within this energy range, extended electronic states cannot exist.

At 0 K, this picture is reasonably adequate but, as the temperature is raised, some thermal excitation of the electrons can occur from the valence band to the conduction band. Statistical calculations, beyond the scope of this article [2], lead to the conclusion that the number density of electrons at energy E in the CB, $n(E)$, is given by

$$n(E) = 2N_c(E)P_e(E) \tag{1}$$

where $N_c(E)$ is the number density of states in the conduction band [i.e. the number of available energy levels for the electrons between energies E and $E + dE$ is $N_c(E)dE$] and $P_e(E)$ is a thermal distribution function related to the Boltzmann law but modified to take account of the quantum-mechanical properties of the electrons; it is termed the Fermi function and has the form

$$P_e = \frac{1}{1 + \exp[(E - E_F)/kT]} \tag{2}$$

The reference energy E_F is called the Fermi energy and corresponds to the chemical potential of the electrons in the solid, and the factor 2 in eqn. (1) takes account of the fact that each level may be occupied by two electrons of opposite spins.

Excitation of the electron from the VB leaves behind a vacancy or "hole". Electrons in the VB can move to fill these holes leaving, in turn, further holes; conceptually, it is easier to consider the holes moving in a sea of immobile electrons and we may develop expressions for the number density of these holes in close analogy to the formulae above for the electrons. Thus, the number density of holes in the VB at energy E', $p(E')$, is given by

$$p(E') = 2N_v(E')P_h(E') \tag{3}$$

where

$$P_h = (1 - P_e) = \frac{1}{1 + \exp[(E_F - E')/kT]} \tag{4}$$

and $N_v(E)$ is the number density of states in the valence band at energy E.

The total number density of electrons excited, n_i, can be obtained by integrating over the energy range $E = E_c$ to the top of the CB and, in a similar way, the total density of holes, p_i, is obtained by integrating eqn. (3) from the top of the VB to E_v. Clearly, these integrals cannot be performed unless the functional forms of N_v and N_c are available; in practice, this is rarely so, but it is often found that the top of the valence band and the bottom of the conduction band have simple parabolic shapes of the form

$$N_c = 2\pi(2m_e)^{3/2}h^{-3}(E - E_c)^{1/2} \tag{5}$$

$$N_v = 2\pi(2m_h)^{3/2}h^{-3}(E_v - E)^{1/2} \tag{6}$$

Even though the bands may deviate substantially from this form at energies more removed from the band edges, the functional forms of eqns. (2) and (4) ensure that little contribution to the total numbers of electrons or holes will come from these regions. The deviation from free-electron-like behaviour is accounted for by the two *effective masses* m_e and m_h. For a free-electron band, m_e and m_h would both have the value of the free-electron mass. As is shown elsewhere [3], we can take into account the periodic potential in a solid in first order by incorporating into the theory an *effective mass* for the electron or hole. In terms of the overlap model, we can say, very approximately, that the larger the overlap between the orbitals on different atoms or ions, the wider the band of states becomes; the wider this band, the smaller the effective mass will be. Substituting eqns. (5) and (6) into eqns. (1) and (3) and integrating, we have, after some approximation

$$n_i = N_c \exp\left[\frac{(E_F - E_c)}{kT}\right] \qquad N_c = 2\left(\frac{2\pi m_e kT}{h^2}\right)^{3/2} \tag{7}$$

$$p_i = N_v \exp\left[\frac{(E_v - E_F)}{kT}\right] \qquad N_v = 2\left(\frac{2\pi m_h kT}{h^2}\right)^{3/2} \tag{8}$$

Since the system is neutral, we must have $n_i = p_i$ and

$$E_F = \tfrac{1}{2}(E_v + E_c) + \tfrac{3}{4} kT \log_e (m_h/m_e) \tag{9}$$

The condition that $n_i = p_i$ defines the system as being *intrinsic*. It may be helpful to insert some numbers into the above equations: for normal values of m_e and m_h (0.1–10 m_0, where m_0 is the free electron mass), N_v and N_c take the values $10^{19} \ldots 10^{20}\,cm^{-3}$. Thus, for Ge at 300 K, $n_i = p_i = 2.3 \times 10^{13}\,cm^{-3}$, assuming a bandgap of 0.6 eV. By contrast, for TiO_2 with a bandgap of 3.0 eV, $n_i = p_i = 10^{-6}\,cm^{-3}$.

The very small intrinsic carrier concentrations for TiO_2 imply that the carrier concentration in this material is likely to be dominated by ionisation of impurities. In this case, the semiconductor is described as *extrinsic*. Truly intrinsic semiconductors are rather rare unless the bandgap becomes very small ($\leqslant$0.4 eV) or heroic efforts are made to remove all impurities. Impurities may ionise in one of two ways: they may release electrons into the conduction band, or they may capture electrons, releasing holes into the valence band. The former type of impurity leads to an "n-type" semiconductor, the latter to a "p-type" material. Examples are legion: the basis of the modern electronics industry is silicon, but Si has a bandgap of 1.1 eV and the number of intrinsic carriers is therefore very low ($10^{10}\,cm^{-3}$) at room temperature. To enhance the conductivity, very small amounts of phosphorus or boron can be annealed into the sample. The former has one extra electron, which it can release into the CB forming n-Si, whereas the latter has one too few valence electrons and can capture an electron from the VB to give p-type conduction. The amount of P or B required to give a conductivity of 100 $(\Omega\,cm)^{-1}$ can be calculated to be ca. $5 \times 10^{13}\,cm^{-3}$; this represents about one part of P or B in 10^9 parts of Si. Since P must be the dominant impurity, this in turn implies that we must start with extremely pure Si, hence the need for clean room facilities in semiconductor technology.

With binary compounds, an additional source of impurity levels can arise from non-stoichiometry. Thus, many oxides are either cation or anion deficient; the former, such as $Ni_{(1-x)}O$ ($x \leqslant 0.01$) are p-type whereas the latter, such as $TiO_{(2-x)}$ ($x \leqslant 0.01$) are n-type. The extent of non-stoichiometry varies with the partial pressure of O_2 in a manner that can usually be calculated from point-defect theory, provided that x remains small.

Quantitative treatment of extrinsic semiconductors is obviously more difficult than the intrinsic case discussed above since n_i and p_i are no longer equal [4]. However, we still have

$$n = N_c \exp[(E_F - E_c)/kT] \tag{10}$$

$$p = N_v \exp [(E_v - E_F)/kT] \tag{11}$$

and so

$$n.p = N_v N_c \exp [-(E_c - E_v)/kT] = N_c N_v \exp [-E_g/kT] \tag{12}$$

where E_g is the bandgap. Thus, n and p are always related by the expression

$$n.p = n_i.p_i = n_i^2 \tag{13}$$

We now suppose that there are N_D donor levels of energy E_d below the conduction band; when occupied, these levels are neutral and when ionised, they are positively charged. The number of levels, n_d, occupied is given by

$$n_d = \frac{N_D}{1 + g_d \exp [(E_d - E_F)/kT]} \tag{14}$$

where g_d is a statistical number which takes the value $\frac{1}{2}$ for the common situation of a one-electron impurity level such as P in Si.

From the electroneutrality principle, since we have $(N_D - n_d)$ positively charges ionised impurity levels

$$n + n_d = N_D + p \tag{15}$$

where we have assumed that the number of acceptor traps, N_A, is negligible. Then

$$N_c \exp \left\{\frac{(E_F - E_c)}{kT}\right\} - N_v \exp \{(E_V - E_F)\} = \frac{N_D g_d \exp \{(E_d - E_F)/kT\}}{1 + g_d \exp \{(E_d - E_F)/kT\}} \tag{16}$$

In principle, this may be solved as a cubic in exp $(-E_F/kT)$. Rather than do this, however, we will examine three extreme solutions.

(a) At high temperatures, the L.H.S. of eqn. (16) is dominant and there is a return to the intrinsic situation described above.

(b) At lower temperatures, the second term on the L.H.S., representing the concentration of holes, becomes negligible and we have

$$N_c \exp \left\{\frac{(E_F - E_c)}{kT}\right\} = \frac{N_D g_d \exp \{(E_d - E_F)/kT\}}{1 + g_d \exp \{(E_d - E_F)/kT\}} \tag{17}$$

Commonly, $(E_c - E_d)$ is very small (ca. 0.005 eV) and $(E_d - E_F) \gg kT$, so the R.H.S. of eqn. (17) is approximately equal to N_D. Hence

$$E_F - E_c = kT \log_e (N_D/N_c) \tag{18}$$

(c) At very low temperatures, the assumption that E_d and E_F are well separated, i.e. all the donor centres are ionised, breaks down and we must solve the quadratic equation (17) explicitly to give

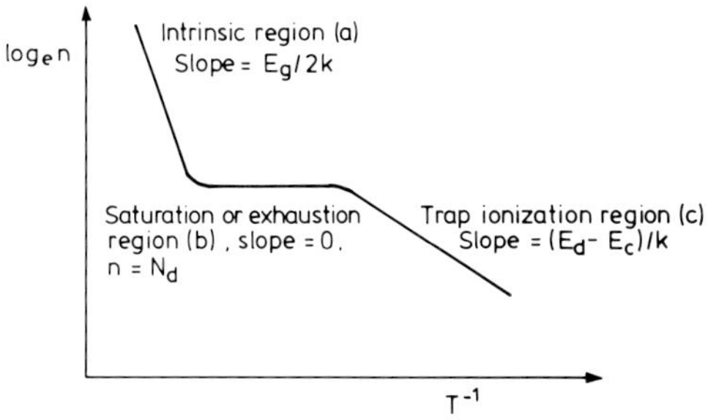

Fig. 2. Variation in the number of carriers with T^{-1} for an extrinsic semiconductor.

$$n = \left(\frac{N_D N_c}{2}\right)^{1/2} \exp\left\{\frac{-(E_d - E_c)}{kT}\right\} \tag{19}$$

If, in eqn. (15), we cannot neglect the number of acceptor centres, N_A, then N_A should be added to the L.H.S. Solving at low temperatures, we find

$$n = \left(\frac{N_D - N_A}{2N_A}\right) N_c \exp\left\{\frac{-(E_d - E_F)}{kT}\right\} \tag{20}$$

The number of carriers in an n-type extrinsic semiconductor is shown in Fig. 2 as a function of (inverse) temperature [4].

We have seen that, at finite temperature, there will always be free carriers in a semiconductor, either by virtue of intrinsic excitation or as a result of a donor or acceptor impurity. From the electrochemical viewpoint, the most important result of this carrier concentration is that the material exhibits a finite conductivity. This conductivity is given by

$$\sigma = ne_0\mu_e + pe_0\mu_h \tag{21}$$

where μ_e and μ_h are the electron and hole mobilities, respectively. The mobility, like the values of N_c and N_v, is a fundamental property of the semiconductor. It is usually a fairly weak function of the temperature and may also depend on the concentration of impurities. It varies, in band-type semiconductors, typically from 1 to 1000 $cm^2 V^{-1} s^{-1}$, though very high values are known for some materials such as GaAs, whose hole mobility has been reported as ca. 10 000 $cm^2 V^{-1} s^{-1}$ at 300 K. Values smaller than 1 $cm^2 V^{-1} s^{-1}$ may be encountered, and very small values ($\leqslant 10^{-3}$ $cm^2 V^{-1} s^{-1}$) are quite commonly found. The fact that mobilities lying between 10^{-3} and 1 $cm^2 V^{-1} s^{-1}$ are rather unusual is of considerable interest from the solid state point of view and arises because the scattering length of the electron becomes comparable with the lattice dimension of a typical crystal for $\mu = 1$ $cm^2 V^{-1} s^{-1}$, with the result that the electron hopping time from site to site becomes comparable with the optical vibration frequencies usually measured [5]. If this occurs, the lattice will have time to relax about the electron before it can hop to a neighbouring site and this relaxation process substantially increases the activation energy for the hop. There is, then, something of a discontinuity in commonly encountered mobilities below $1\ cm^2\ V^{-1}\ s^{-1}$ and materials exhibiting "small polaron" effects would be expected to possess very low mobilities. The practical consequences of this are, of course, that such "small polaron" materials usually have very low conductivities and can only be studied electrochemically in the form of thin films. This aspect of semiconductor electrochemistry may well prove to be of considerable interest though, in order to limit the coverage of this article, such studies on thin films will not be considered in detail.

Implicit in the discussion in the previous paragraph has been the assumption that the material is homogeneous. In fact, the chemist encounters comparatively few such substances, certainly on a scale suitable for electrochemical measurements, though most of the studies in the semiconductor electrochemistry literature have been on single-crystal samples. Where single-crystal samples are not available, the material must be inhomogeneous, at least to some degree. The commonest type of sample likely to be encountered is a polycrystalline pellet; such samples are composed of small crystallites compressed together such that electrical contact is made through small "necks" joining the crystallites. Provided the crystallites are larger than a few microns in diameter, the electrochemical properties of such polycrystalline samples should resemble closely those of a single crystal; however, as the size of the crystallites decreases, edge effects become more and more significant and the prediction and understanding of the electrochemistry of such samples is still at a very early stage. At the other extreme, we can conceive of samples whose long-range order has completely disappeared; such samples are termed "amorphous" and their properties have recently attracted considerable interest [6]. Surprisingly, such materials are more theoretically tractable than gross polycrystalline pellets, since some spatial averaging may be carried out. The electrochemistry of such materials is, however, still quite undeveloped, but rapid progress is likely in view of the interest shown, for technological reasons, in the properties of a-Si.

It is normally unnecessary for the electrochemist to be concerned with the mobility of carriers in most of the semiconductors whose properties have been studied, since the very low conductivity of "small polaron" samples would normally preclude their measurement. However, a proviso must be entered here in the case of binary and, more especially, ternary samples. It may well be the case that the majority carriers in a particular material are indeed itinerant (i.e. have mobilities in excess of ca. $1\ cm^2\ V^{-1}\ s^{-1}$), but there is no guarantee that this will be true of the minority carriers generated by optical absorption. Thus, the oxide $MnTiO_3$ shows a marked optical charge transfer absorption from Mn(II) to Ti(IV), the latter being the CB. The resultant holes reside on localised sites in the Mn levels, presumably as local Mn(III) centres, and are comparatively immobile. The result is that there is very little photoelectrochemistry associated with these minority carriers and appreciable photoanodic current is only observed when itinerant holes are photogenerated in the oxygen 2*p* band, a process that requires UV light.

For the vast majority of materials likely to be encountered, the mobilities of the majority carriers have been determined and tabulated. For a novel material, such measurements must be performed, normally using the Hall effect, and the details of these measurements are discussed elsewhere [4].

3. Double layer models for the semiconductor–electrolyte interface

We have seen in the previous section that the bulk semiconductor is characterised by a filled valence band and an empty conduction band with a distribution of acceptor and donor levels. Provided that there is no nett charge on the semiconductor, the position of the Fermi level in the semiconductor is then determined solely by the semiconductor statistics as shown above. However, once immersed in an electrolyte, the interfacial potential change, and therefore the charge on the semiconductor, can be externally controlled and the distribution of energy levels becomes more complex [7–15]. The general theory will be developed below, but some important

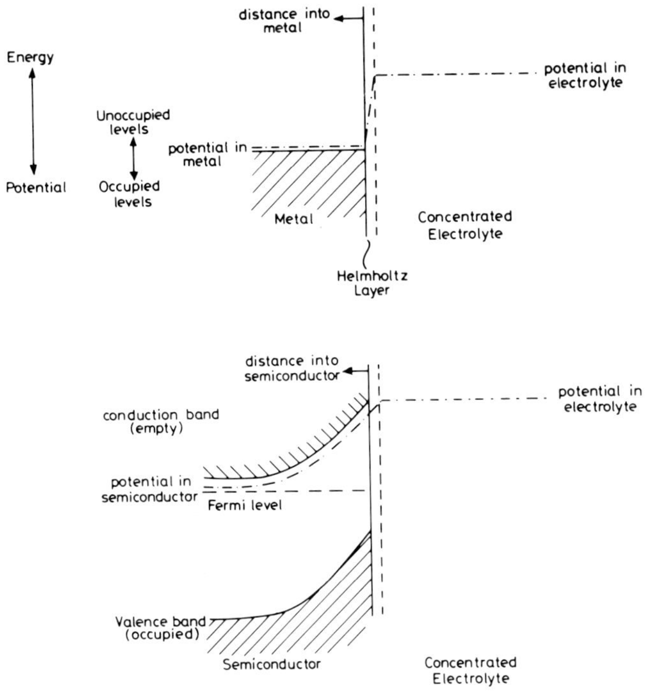

Fig. 3. Schematic potential distribution for a metal and for a semiconductor.

generalisations can be made at this stage. For a metallic conductor, the charge, and therefore the associated potential drop, is concentrated at the surface, penetrating only a distance of the order of Ångstrøms into the interior of the metal, as may be verified by consideration of the electroreflectance spectrum of gold [16]. The result is that almost all the potential drop in the interfacial region takes place in the region outside the metal; in the Helmholtz region and the electrolyte.

For a semiconductor, this pattern is not found; since charge and potential drop are related by

$$\frac{d^2\phi}{dx^2} = -\frac{\rho}{\varepsilon\varepsilon_0} \tag{22}$$

the relatively small charge densities found in semiconductors ensure that the potential drop can only be accommodated over comparatively large distances. The result is that the electric field penetrates the interior of the semiconductor, remaining finite over a depletion layer that may be microns thick. Clearly, the larger the charge density in the semiconductor, i.e. the higher the density of donor or acceptor states, the narrower this depletion layer will be and the more "metal-like" the sample will become. This situation is shown in Fig. 3.

3.1 THE SIMPLE INTERFACE

Consider the interfacial region shown in detail in Fig. 4. The total interfacial drop, ϕ_i, is composed of three contributions: ϕ_{sc}, the space–charge potential dropped inside the semiconductor, ϕ_H, the potential across the (uncharged) Helmholtz layer, and ϕ_{el}, the potential dropped in the electrolyte. To solve for the potential distribution in the interfacial region, we make use of

(a) the Poisson equation, eqn. (22),

(b) the continuity equation at the dielectric boundaries

$$\varepsilon_{sc} E_{sc} = \varepsilon_H E_H \qquad x = 0$$

$$\varepsilon_H E_H = \varepsilon_{el} E_{el} \qquad x = -d_H$$

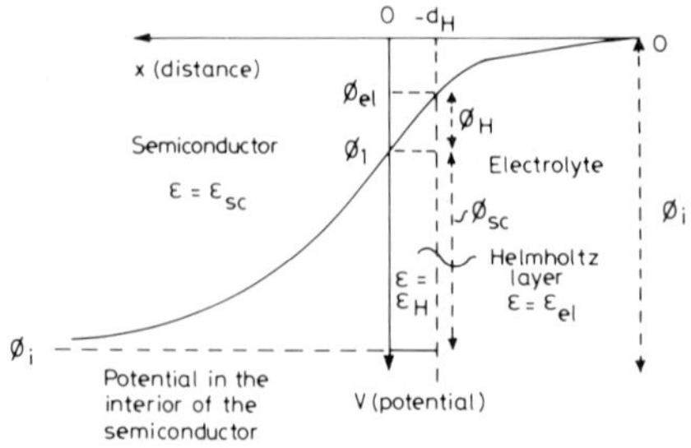

Fig. 4. Detailed potential distributiuon for a semiconductor–electrolyte interface.

(c) the charge neutrality equation

$$\Sigma q \equiv q_{sc} + q_{el} = 0 \qquad (\text{note } q_H = 0)$$

Consider first the electrolyte region. If the concentration of the ith ion, of charge $z_i e_0$, at position x is given by the Boltzmann equation

$$c_i(x) = c_{ib} \exp(-\xi_i \phi) \tag{23}$$

where

$$\xi_i = \frac{z_i e_0}{kT} = z_i \frac{F}{RT}$$

and c_{ib} is the bulk concentration of i in moles per unit volume, we find, from the Poisson equation

$$\frac{d^2\phi}{dx^2} = -\frac{F}{\varepsilon_{el}\varepsilon_0}\sum_i z_i c_{ib} e^{-\xi_i \phi} = -\frac{FC}{\varepsilon_{el}\varepsilon_0}(e^{-\xi\phi} - e^{\xi\phi}) = \frac{2FC}{\varepsilon_{el}\varepsilon_0}\sinh(\xi\phi) \tag{24}$$

for a 1:1 electrolyte. Integrating once, for the (presumed) 1:1 electrolyte

$$\left(\frac{d\phi}{dx}\right)^2 = \frac{4FC}{\varepsilon_{el}\varepsilon_0\xi}\cosh(\xi\phi) + K \tag{25}$$

We suppose that, as $x \to -\infty$, $\phi_{el} \to 0$, $(d\phi/dx) \to 0$ so $K = -4FC/\varepsilon\varepsilon_0\xi$ Taking square roots

$$\frac{d\phi}{dx} = \pm\left(\frac{4RTC}{\varepsilon_{el}\varepsilon_0}\right)^{1/2}[\cosh(\xi\phi) - 1]^{1/2} = \pm\left(\frac{8RTC}{\varepsilon_{el}\varepsilon_0}\right)^{1/2}\sinh(\xi\phi/2) \tag{26}$$

where the ambiguity of sign can be resolved by noting that, in our case, $d\phi/dx > 0$. The total charge in the electrolyte, q_{el}, is given immediately by the Gauss formula

$$q_{el} = -\varepsilon_{el}\varepsilon_0\left(\frac{d\phi}{dx}\right)_{x=-d_H} \tag{27}$$

Hence

$$q_{el} = -(8RTC\varepsilon_0\varepsilon_{el})^{1/2}\sinh(\xi\phi/2) \tag{28}$$

Integrating eqn. (26)

$$\phi = \frac{4}{\xi}\tanh^{-1}\left[\left(\tanh\left(\frac{\xi\phi_{el}}{4}\right)\right)e^{\kappa(x+d_H)}\right]; \quad \kappa = \left(\frac{2F^2C}{\varepsilon_{el}\varepsilon_0 RT}\right)^{1/2}; \quad x < -d_H \tag{29}$$

and, at $x = -d_H$, continuity gives

$$-e_0\varepsilon_{\mathrm{el}}\left(\frac{\mathrm{d}\phi}{\mathrm{d}x}\right)_{-d_{\mathrm{H}}} = -\varepsilon_{\mathrm{el}}\varepsilon_0\left(\frac{8RTC}{\varepsilon_{\mathrm{el}}\varepsilon_0}\right)^{1/2}\sinh\left(\frac{\xi\phi_{\mathrm{el}}}{2}\right) = -\varepsilon_0\varepsilon_{\mathrm{H}}\left(\frac{\mathrm{d}\phi}{\mathrm{d}x}\right)_{-d_{\mathrm{H}}} \quad (30)$$

In the Helmholtz region, the Poisson equation takes the simple form

$$\frac{\mathrm{d}^2\phi}{\mathrm{d}x^2} = 0$$

whence

$$\frac{\mathrm{d}\phi}{\mathrm{d}x} = \text{constant} \equiv K_1$$

Hence

$$\phi_1 = \phi_{\mathrm{el}} + \phi_{\mathrm{H}} \quad (31)$$

where

$$\phi_{\mathrm{H}} = K_1 d_{\mathrm{H}} = \frac{\varepsilon_{\mathrm{el}} d_{\mathrm{H}}}{\varepsilon_{\mathrm{H}}}\left(\frac{8RTC}{\varepsilon_{\mathrm{el}}\varepsilon_0}\right)^{1/2}\sinh\left(\frac{\xi\phi_{\mathrm{el}}}{2}\right)$$

Inside the semiconductor, we have

$$\frac{\mathrm{d}^2\phi}{\mathrm{d}x^2} = -\frac{e_0}{\varepsilon_0\varepsilon_{\mathrm{sc}}}[-n(x) + p(x) + N_{\mathrm{D}} - N_{\mathrm{A}}] \quad (32)$$

where n, p, N_{A}, and N_{D} were defined in Sect. 2. If we assumed that the mobile carriers n, p have concentrations given by the Boltzmann equation, then

$$n = n_{\mathrm{b}} \exp[+\xi(\phi - \phi_i)]$$

$$p = p_{\mathrm{b}} \exp[-\xi(\phi - \phi_i)]$$

where ϕ_i is the potential in the bulk of the semiconductor and n_{b} and p_{b} are the bulk electron and hole concentrations, respectively. Normally, for an n-type semiconductor, we can assume that $n_{\mathrm{b}} = N_{\mathrm{D}}$ since, for many such semiconductors, the traps are very shallow and fully ionised at room temperature. If there are deep traps present, the situation is more complex and is considered below. Integrating eqn. (32) as before leads to

$$\left(\frac{\mathrm{d}\phi}{\mathrm{d}x}\right)^2 = -\frac{2e_0}{\varepsilon_0\varepsilon_{\mathrm{sc}}}[(N_{\mathrm{D}} - N_{\mathrm{A}})(\phi - \phi_i) - p_{\mathrm{b}}/\xi\,(\mathrm{e}^{-\xi(\phi-\phi_i)} - 1)$$

$$- n_{\mathrm{b}}/\xi\,(\mathrm{e}^{\xi(\phi-\phi_i)} - 1)]$$

putting $\lambda = (p_{\mathrm{b}}/n_{\mathrm{b}})^{1/2}$ and $\xi(\phi - \phi_i) = \eta$

$$\left(\frac{\mathrm{d}\phi}{\mathrm{d}x}\right) = \pm\left(\frac{2kTn_{\mathrm{b}}}{\varepsilon_0\varepsilon_{\mathrm{sc}}}\right)^{1/2}\mathrm{F}(\eta, \lambda)$$

$$\mathrm{F}(\eta, \lambda) = [\lambda^2(\mathrm{e}^{-\eta} - 1) + (\mathrm{e}^{\eta} - 1) + (\lambda^2 - 1)\eta]^{1/2} \quad (33)$$

where the sign chosen depends on the boundary conditions as illustrated below, and the function $F(\eta, \lambda)$ is tabulated elsewhere [17, 18]. The charge inside the semiconductor is given by

$$q_{sc} = \varepsilon_0 \varepsilon_{sc} \left(\frac{d\phi}{dx}\right)_{x=0} \quad \text{[NB the sign is taken as} > 0 \text{ since } x \text{ is positive into the semiconductor]}$$

$$= (2kTn_b\varepsilon_0\varepsilon_{sc})^{1/2}F(Y, \lambda); \quad Y = \xi(\phi_1 - \phi_i) = \xi\phi_{sc}$$

Since $q_{el} = -q_{sc}$, we have

$$(2kTn_b\varepsilon_0\varepsilon_{sc})^{1/2}F(Y, \lambda) = (8RT\varepsilon_0\varepsilon_{sc}C)^{1/2} \sinh(\xi\phi_{el}/2) \tag{34}$$

and

$$\phi_{el} + \phi_H + \phi_{sc} = \phi_i$$

The rather intractable form of the function $F(\xi\phi_{sc}, \lambda)$ makes it difficult to draw very general conclusions from this result. However, there is a most important simplification that can be made for one special class of semiconductors, viz. those possessing wide bandgaps. If we consider, for the moment, an extrinsic n-type semiconductor, then $\lambda \ll \lambda^{-1}$ and the potential drop, ϕ_i, between the bulk of the semiconductor and the bulk of the electrolyte has the sign shown in Fig. 4. For this case, $\eta < 0$ and

$$F(\eta, \lambda) \simeq [\exp(\eta) - 1 - \eta]^{1/2} \simeq [-1 - \eta]^{1/2} \tag{35}$$

It follows from eqn. (34) that, provided $n_b/N_0 \ll C$ (where N_0 is Avagadro's number) and d_H is small

$$\phi_{el} \ll \phi_{sc}$$

$$\phi_H \ll \phi_{sc}$$

In other words, almost all the interfacial potential is dropped within the semiconductor, in the depletion layer. Some examples of this are shown in Fig. 5 and it is evident that, as the potential is varied, the band-edges behave as if they were energetically "pinned" at the surface, provided that $\eta < 0$. Using the approximation (35), we can further integrate eqn. (33). Remembering the definition of η for an n-type semiconductor, we can transform eqn. (33) into

$$\frac{d\eta}{dx} = \frac{\sqrt{2}}{L_D}(-1 - \eta)^{1/2}; \quad L_D = \left(\frac{\varepsilon_0\varepsilon_{sc}kT}{e_0^2 n_b}\right)^{1/2}$$

where the positive sign is chosen since η increases (towards zero) inside the semiconductor. Given that, for moderate or high exhaustion, $|\eta| \gg 1$, we have

$$\frac{d\eta}{dx} \simeq \frac{\sqrt{2}}{L_D}(-\eta)^{1/2}$$

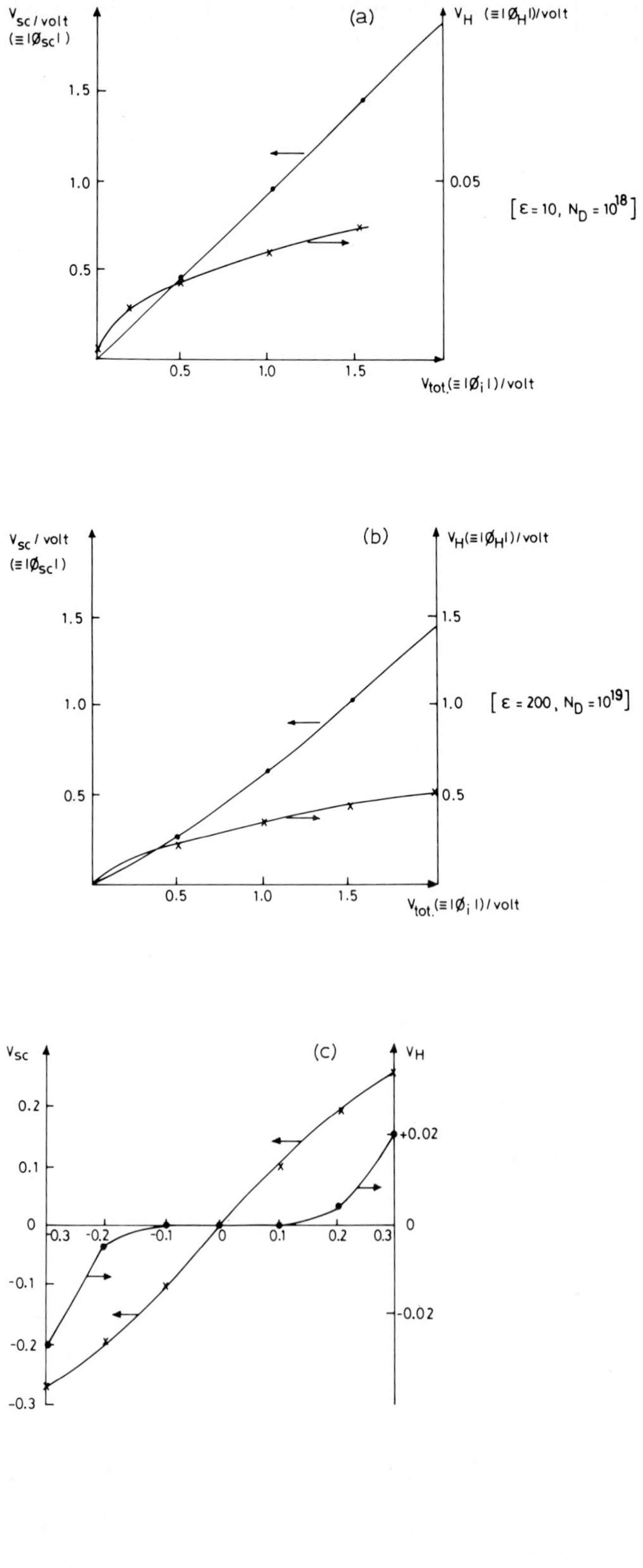
(a)
Vsc / volt
(≡ |Øsc|)
VH (≡ |ØH|) / volt
1.5
1.0
0.5
0.05
0.5
1.0
1.5
Vtot. (≡ |Øi|) / volt
[ε = 10, ND = 10^18]
(b)
Vsc / volt
(≡ |Øsc|)
VH (≡ |ØH|) / volt
1.5
1.0
0.5
1.5
1.0
0.5
0.5
1.0
1.5
Vtot. (≡ |Øi|) / volt
[ε = 200, ND = 10^19]
(c)
Vsc
VH
0.2
0.1
0
-0.1
-0.2
-0.3
+0.02
0
-0.02
-0.3
-0.2
-0.1
0
0.1
0.2
0.3

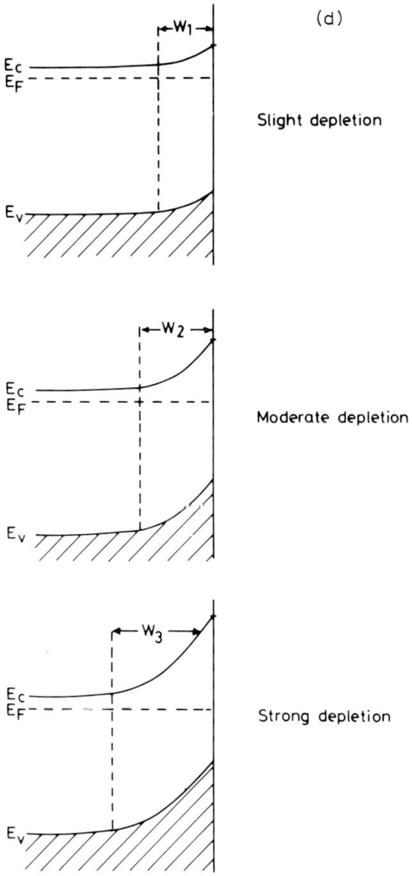

Fig. 5. Calculated interfacial potentials for (a) a moderately highly doped sample of n-GaP ($N_D = 10^{18}\ \mathrm{cm}^{-3}$, $\varepsilon_{sc} = 10$); (b) a highly doped sample of n-TiO_2 ($N_D = 10^{19}\ \mathrm{cm}^{-3}$, $\varepsilon_{sc} = 200$); (c) an intrinsic sample of Ge ($E_g \simeq 0.6\,\mathrm{eV}$) (for all samples, $d_H \equiv 3\,\mathrm{Å}$, $\varepsilon_H \equiv 6$, $\varepsilon_{el} \equiv 80$ and $C \equiv 1\,\mathrm{M}$); and (d) behaviour of the Fermi level and band edges for case (a) shown diagrammatically (W defines the width of the depletion layer).

whence

$$\frac{\sqrt{2}x}{L_D} = -2(-\eta)^{1/2} + 2(-Y)^{1/2}$$

where Y is defined above. This describes a finite layer, of width W, where

$$W = (-2Y)^{1/2} L_D \tag{36}$$

in which η is non-zero. This finite region is called the depletion layer or exhaustion region. Of course, this integration is only approximate and the full expression

$$\frac{d\eta}{dx} = \frac{\sqrt{2}}{L_D}(e^{\eta} - 1 - \eta)^{1/2}$$

should be used for $|\eta| \leqslant 4$. For $|\eta| < 1$ we can expand this to give

$$\frac{d\eta}{dx} = -\frac{\eta}{L_D}$$

whence

$$\eta = A e^{-x/L_D} \tag{37}$$

Numerical analysis shows that, for these small values of η or, equivalently, large values of x, a good approximation for A is $\kappa \exp\{(2|Y| - 1)^{1/2}\}$ where κ lies in the range 0.60–0.62. Numerical analysis also shows that the approximation (36) for the width of the depletion layer corresponds to the point at which the potential drop $|\phi - \phi_i|$ has decreased to a value of 0.01 V, accurately enough zero for most purposes.

For an n-type material, this implies that, for the approximation of eqn. (37) to be valid, the semiconductor must be biased positive of some reference potential at which $q_{sc} = 0$. Evidently, at potentials negative of this point, the semiconductor behaves in a more complex fashion.

To understand this in more detail, we can consider Fig. 5(a). As indicated in Sect. 2, the electrons in the bulk of the semiconductor are defined statistically by the Fermi level of electrochemical potential. Obviously, at equilibrium, the Fermi level must be constant throughout the semiconductor, including the depletion or exhaustion region as shown in Fig. 3. It follows that, as the potential is varied within the exhaustion region for an n-type semiconductor, the Fermi level must move up and down in the depletion region with respect to the band edges, reflecting the population changes defined in eqns. (32) and (33) and this is shown schematically in Fig. 5(d).

As the potential is made more negative, the Fermi level approaches the point at which it will intersect the conduction band at the surface, and Fermi–Dirac rather than Maxwell–Boltzmann statistics must be used; under these circumstances, the majority band at the surface will become degenerate and the potential change no longer takes place entirely within the semiconductor. At sufficiently negative potentials, most of the potential change is accommodated across the Helmholtz layer, as occurs for a metal. The situation for a p-type semiconductor is analogous: at negative potentials, a depletion layer forms within the semiconductor and most of the potential change then occurs across this layer. At positive potentials, metallic behaviour is encountered.

As shown in Fig. 5, this pattern of behaviour becomes less clear-cut as the donor density and/or the dielectric constant within the semiconductor increase; as $\varepsilon_{sc} n_b$ and $\varepsilon_{el} C N_0$ become comparable, the depletion layer shrinks and an increasing proportion of the potential drop is accommodated within the Helmholtz layer.

It is evident from Fig. 5 that, as the potential becomes yet more positive in an n-type material, the Fermi level will approach the VB. It might be expected that, under these circumstances, the minority band would become

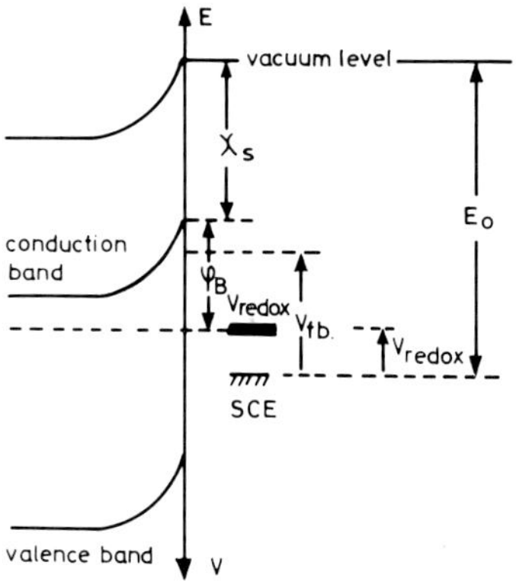

Fig. 6. Energy diagram of an n-type semiconductor–electrolyte interface defining the electron affinity χ_s and the absolute redox potential and reference electrode potential (SCE).

degenerate and there would again be a reversion to metallic behaviour. The situation in which the concentration of minority carriers at the surface exceeds that of the majority carriers is termed "inversion", but the extreme inversion characteristic of minority band degeneracy is actually rarely found in practice, for kinetic reasons, especially for bandgaps in excess of ca. 1.5 eV.

As indicated above, within the approximations used in deriving eqn. (34), there must be some potential at which $q_{sc} = q_{el} = 0$. At this point $\phi_i = \phi_{sc} = 0$. For obvious graphical reasons, this is termed the "flat-band" potential and is of considerable importance since it offers the possibility of relating the electronic energy levels in the semiconductor to the redox potentials of any couples in the electrolyte. The crux of this is that the potential in the bulk of the semiconductor can be related to the position of the Fermi level. Electrochemically, potentials are related to some reference standard such as the standard hydrogen electrode (SHE), whereas electronic energy levels are related to the "vacuum" level, i.e. the energy of an electron at a point remote from the semiconductor in vacuo. The relationship between the vacuum level and an EMF standard such as the SHE is fraught with both experimental and theoretical difficulties and a range of values for the "absolute hydrogen electrode potential" has been suggested. In this article, the value of 4.5 eV will be adopted. If this value is used in Fig. 6, then the relationship between the flat-band potential, V_{fb} and such electrophysical properties as the electron affinity and work function of the semiconductor becomes apparent. Considerable efforts have been made to understand the variation of V_{fb} with the chemical composition of the semiconductor and these efforts will be reviewed below.

Although the simple model developed in the previous paragraphs is, in many ways, quite deficient, it is of interest at this point to examine the electrochemical evidence in its favour. The most convincing measurements have undoubtedly involved the use of a.c. techniques. The reason for this becomes apparent as eqn. (34) is further developed: the total charge in the semiconductor under circumstances where eqn. (34) is expected to be valid is given by

$$q_{sc} = (2kTn_b\varepsilon_0\varepsilon_{sc})^{1/2}F(Y, \lambda) \cong (2kT\varepsilon_0\varepsilon_{sc}n_b)^{1/2}[-1 - Y]^{1/2} \quad (38)$$

If the potential on the semiconductor has imposed on it an a.c. component, then the effect is to change the accumulated charge in the depletion layer according to eqn. (38). Assuming that no faradaic current is passing, i.e. the semiconductor is deep in depletion, the capacitive response of the semiconductor layer may be approximated as

$$\xi\left|\frac{dq_{sc}}{dY}\right| = C_{sc} = \xi\left(\frac{kTn_b\varepsilon_0\varepsilon_{sc}}{2}\right)^{1/2}[-1 - Y]^{-1/2} \quad (39)$$

In series with C_{sc} are the capacitances associated with the Helmholtz layer and the electrolyte. It is easy to show that these are expected to be much larger under the circumstances that eqn. (36) is valid, and so the quadrature response of the semiconductor may be used to calculate a capacitance, C_{obs}, that is closely approximated by C_{sc}. From eqn. (39), it is seen that

$$C_{sc}^{-2} = \left(\frac{2\xi^2}{kTn_b\varepsilon_0\varepsilon_{sc}}\right)[-1 - Y] \quad (40)$$

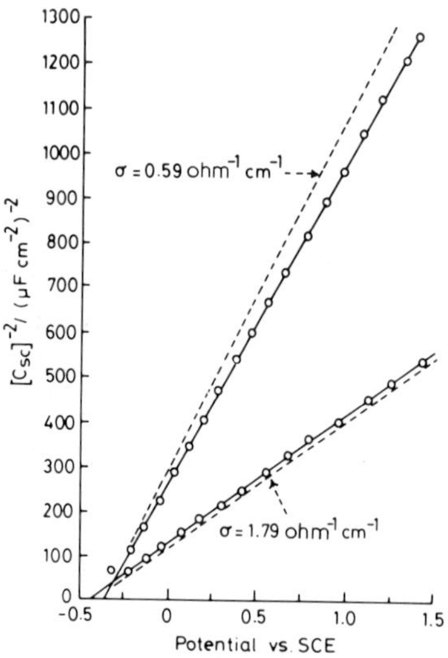

Fig. 7. Mott–Schottky plots for two different n-ZnO crystals. The experimental points form two lines intercepting the potential axis at -0.4 ± 0.05 V vs. SCE and the behaviour calculated from eqn. (40) is shown as the dotted lines.

and we should therefore find that, for an extrinsic wide-bandgap semiconductor in deep depletion

$$C_{obs}^{-2} \sim V_{applied}$$

where $V_{applied}$ is the potential applied to the semiconductor *relative to the flat-band potential.* This is the celebrated Mott–Schottky relationship, important not only as a verification of the essential correctness of the above analysis, but also because it allows us to locate, on an external reference scale, the potential at which there is no field inside the semiconductor, i.e. the potential at which the bands are "flat". Although the Mott–Schottky relationship will be exhaustively analysed below, and there are many cases in which it is not valid, there are examples in the literature, such as DeWald's classic study of n-ZnO [10], in which eqn. (40) is obeyed with considerable precision, as can be seen from Fig. 7.

If the approximation $\lambda^{-1} \gg \lambda$ is not satisfied but the surface is still sufficiently stable for the a.c. response to be measured, then the expression [7]

$$C_{sc} = \xi\left(\frac{kT\varepsilon_0\varepsilon_{sc}n_b}{2}\right)^{1/2} \frac{[-\lambda^2 e^{-Y} + e^{Y} + \lambda^2 - 1]}{[\lambda^2(e^{-Y} - 1) + (e^{Y} - 1) + (\lambda^2 - 1)Y]^{1/2}} \qquad (41)$$

must be used. This clearly predicts a minimum in C_{sc} since, at large positive potentials, $\lambda \exp[-Y] \gg \lambda^{-1} \exp[Y]$, Y and

$$C_{sc} \longrightarrow \left(\frac{kT\varepsilon_0\varepsilon_{sc}n_b}{2}\right)^{1/2} e^{|Y|/2} \lambda^{1/2} \xi \qquad (42)$$

whereas, at large negative potentials, $\lambda^{-1} e^{Y} \gg \lambda e^{-Y}$ and

$$C_{sc} \longrightarrow \left(\frac{kT\varepsilon_0\varepsilon_{sc}n_b}{2}\right)^{1/2} \lambda^{-1/2} e^{|Y|/2} \xi \qquad (43)$$

These expressions have been used to interpret the capacitive data on germanium, though it proved difficult in practice, on this material, to control the surface chemistry adequately [19].

The semiconductor–electrolyte interface normally encountered is far more complex than that reviewed here; the most significant additional complexity that must be dealt with is the presence of uncompensated charge at the surface of the semiconductor, which may have a drastic effect on the potential distribution. In principle, this surface charge may arise in the following ways.

(a) Surface electronic states may exist that arise from the interruption in the lattice periodicity caused by the surface. Logically, these states will exist immediately inside the semiconductor surface.

(b) Surface electronic states that arise from a chemical reaction between the semiconductor surface and the ambient medium.

(c) Surface charge that arises from the adsorption of ions into the Helmholtz layer. A simple, but very important example of this is the protonation of an oxide surface.

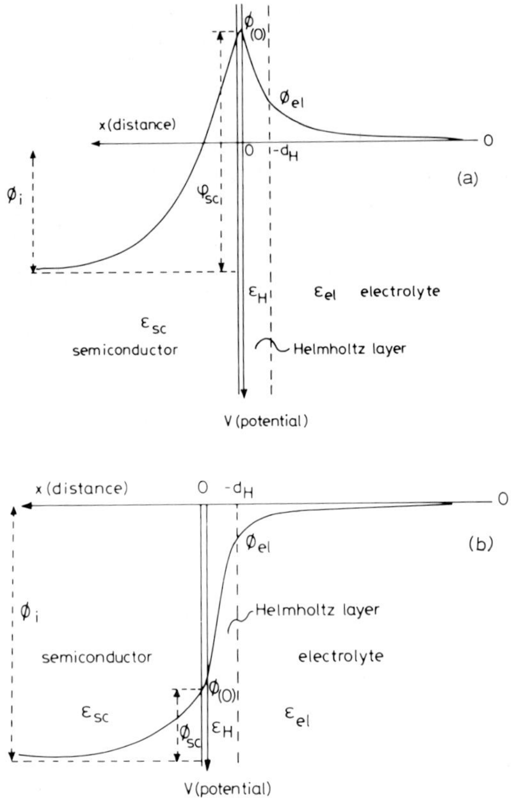

Fig. 8. Potential distribution at the semiconductor–electrolyte interface for (a) a negative charge density at the surface and (b) a positive charge density at the surface.

3.2 THE EFFECT OF UNCOMPENSATED CHARGE AT THE SURFACE

The effect of uncompensated charge may be discussed by reference to Fig. 8. Consider a charge density σ at the surface in a very narrow film of thickness δ. The potential distribution, as before, is determined by the Poisson equation, the continuity equations, and the charge balance.

We have

$$q_{el} = -(8RTC\varepsilon_{el}\varepsilon_0)^{1/2} \sinh\left(\frac{\xi\phi_{el}}{2}\right) \tag{44}$$

$$\phi(0 > x > -d_H) = (8RTC\varepsilon_{el}\varepsilon_0)^{1/2}\left(\frac{x + d_H}{\varepsilon_0\varepsilon_H}\right)\sinh\left(\frac{\xi\phi_{el}}{2}\right) + \phi_{el} \tag{45}$$

in the electrolyte and the Helmholtz layer. In the semiconductor

$$\varepsilon_0\varepsilon_{sc}\left.\frac{d\phi}{dx}\right|_0 = (2kTn_b\varepsilon_0\varepsilon_{sc})^{1/2}\,F(Y,\lambda) = +q_{sc} \tag{46}$$

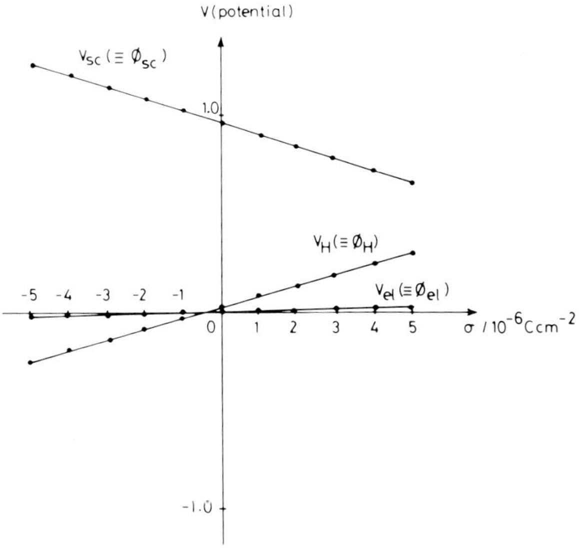

Fig. 9. Calculated potential difference for the cases shown in Fig. 8(a) and (b) for the n-GaP sample of Fig. 5(a). The value of ϕ_i is fixed at 1.0 V and the other conditions are as given in Fig. 5.

and at the surface

$$\varepsilon_0 \varepsilon_{sc} \left.\frac{d\phi}{dx}\right|_0 = \varepsilon_0 \varepsilon_H \left.\frac{d\phi}{dx}\right|_0 - \sigma \tag{47}$$

whence

$$(2kTn_b \varepsilon_0 \varepsilon_{sc})^{1/2} F(Y, \lambda) = (8RTC\varepsilon_0 \varepsilon_{el})^{1/2} \sinh\left(\frac{\xi \phi_{el}}{2}\right) - \sigma \tag{48}$$

also

$$\phi_i = \phi_{sc} + \phi(0) \tag{49}$$

$$= \phi_{sc} + (8RTC\varepsilon_{el}\varepsilon_0)^{1/2}\left(\frac{d_H}{\varepsilon_0 \varepsilon_H}\right)\sinh\left(\frac{\xi\phi_{el}}{2}\right) + \phi_{el} \tag{50}$$

The equations (48) and (50) are two independent expressions for ϕ_{sc} and ϕ_{el}, given ϕ_i. The complexity of the formulae again masks any immediate recognition of trends, but Fig. 9 shows clearly the effect of varying the surface charge density for a typical n-type semiconductor. Clearly, significant changes in the potential distribution can be induced by relatively small changes in charge density, the surface density of 10^{-6} C cm^{-2}, corresponding to 10^{13} electrons cm^{-2} or 1% of a monolayer of specifically adsorbed ions.

3.3 IONOSORPTION ON SEMICONDUCTOR SURFACES

The treatment detailed above has some important corollaries. If we consider the particular case, (c), of ions adsorbed on to the surface of the semiconductor, we may calculate, approximately, the adsorption isotherm. Evidently, as more ions adsorb, σ rises and $\phi(0)$ increases; the free energy associated with a charge layer σ is given by $\simeq 0.5\sigma\phi(0)$, since $\phi(0)$ depends linearly on σ. Now, if the coverage of charged ions is θ, and the total number of adsorption sites available per unit area is N_s, $\sigma_{max} = e_0 N_s$, and $\sigma = e_0 N_s \theta$. The differential free energy of interaction for a single ion is, at coverage θ, $\phi(0)\delta\sigma \equiv e_0\phi(0) \equiv e_0 r\theta$ where r is a proportionality constant that can be calculated from the above analysis. This free energy term must be added to the electrochemical potential of the adsorbed ions: so at equilibrium

$$\bar{\mu}_{surf} = \bar{\mu}_{soln}$$

$$\bar{\mu}_{surf} = \mu^0_{surf} + kT \ln\left(\frac{\theta}{1-\theta}\right) + e_0 r\theta$$

and

$$\bar{\mu}_{soln} = \mu^0_{soln} + kT \ln C_i \qquad (51)$$

where C_i is the concentration of ions in solution. Hence

$$\left(\frac{\theta}{1-\theta}\right) = \frac{C_i}{K} \exp\left(\frac{-e_0 r\theta}{kT}\right)$$

where

$$K \equiv \exp\{(\mu^0_{surf} - \mu^0_{soln})/kT\} \qquad (52)$$

If r is large, as suggested by the above analysis, we will have, approximately

$$e_0 r\theta \simeq kT \ln C_i \qquad (53)$$

We can calculate r quite simply in the case where $\phi_{sc} = 0$. The potential ϕ_i is now the flat-band potential, as defined above, and corresponds to the potential at which there is no depletion layer. Charge balance from eqns. (45) and (48) gives

$$\phi_i \simeq \frac{\sigma d_H}{\varepsilon_0 \varepsilon_H} \simeq V_{fb} - V^0_{fb} \qquad (54)$$

where V^0_{fb} is the flat-band potential at the point of zero surface charge.

$$V_{fb} \simeq V^0_{fb} + \frac{\sigma d_H}{\varepsilon_0 \varepsilon_H}; \qquad r \equiv \frac{e_0 N_s d_H}{\varepsilon_0 \varepsilon_H} \qquad (55)$$

If we take N_s as 2.5×10^{15} cm^{-2}, $d_H \simeq 3$ Å, and $\varepsilon_H = 6$, we find $r \simeq 10$ V, a value more than large enough to ensure that eqn. (53) holds. From this value of r, we may estimate the total charge change that occurs over a change in pH from 0 to 14 to be $\sim 25\,\mu$C cm^{-2}, a value reasonably close to the experimental data for a number of oxides [20].

For the protonation equilibria, the situation is more complex since the

TABLE 1

Point of zero zeta potential for some oxides

Oxide	pZZp	Oxide	pZZp
WO_3	0.43	SiO_2	2.0
Ta_2O_5	2.9	SnO_2	4.3
ZrO_2	6.7	Fe_2O_3	8.6
TiO_2	5.8	HgO	7.3
CuO	9.5	Fe_3O_4	6.5
ZnO	8.8	NiO	10.3
$FeTiO_3$	6.3	Cr_2O_3	8.1
CdO	11.6	Al_2O_3	9.2
Ag_2O	11.2	MgO	12.4
$SrTiO_3$	5.3	$BaTiO_3$	9.9

surface in aqueous solution is likely to be amphoteric. There are, therefore two equilibria

$$AH_2^+ = AH + H^+ \qquad K_+ = \frac{[AH][H^+]_s}{[AH_2^+]} \tag{56}$$

$$AH = A^- + H^+ \qquad K_- = \frac{[A^-][H^+]_s}{[AH]} \tag{57}$$

where [AH] etc. are surface coverages, which may be written θ_+, θ_s, and θ_- for $[AH_2^+]$, [AH], and $[A^-]$, respectively.

There will clearly be a pH at which the surface charge $e_0 N_s(\theta_+ - \theta_-)$ is zero. From eqns. (56) and (57), the $[H^+]$ concentration is

$$C_+^0 = (K_+ K_-)^{1/2} \tag{58}$$

As above, we may write the relationship between charge and potential as

$$e_0\phi(0) = e_0 r(\theta_+ - \theta_-) \tag{59}$$

and combining this with eqn. (58) gives

$$e_0 r(\theta_+ - \theta_-) = kT \ln (C_+/C_+^0) \tag{60}$$

More detailed treatments are available elsewhere [21–23], but the essential point is clear; as for the simple ionosorption case considered initially, the coverage of charged species is expected to be small and the isotherm dominated by coulombic effects owing to the large value of r. The theory also predicts that the flat-band potential will vary by 60 mV/decade concentration change of hydrogen ions in solution as the pH is moved away from that corresponding to a neutral surface. This latter pH, termed the "point of zero zeta potential" (pZZP), a term borrowed from the theory of colloids, is a fundamental property of the material, as indicated by eqn. (58) and is tabulated for a number of oxides in Table 1 [24–35].

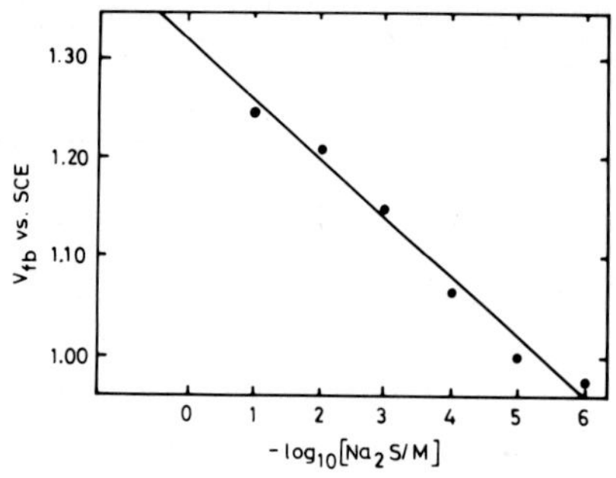

Fig. 10. Dependence of the flat-band potential (determined from a.c. measurement) of n-CdS on Na_2S concentration. The slope is 60 mV per decade Na_2S concentration which suggests that the singly charged HS^- ion is the species adsorbed.

It is evident that the data of Table 1 show considerable scatter and depend strongly on the supporting electrolyte [36]; nevertheless, there is a clear corelation between the pZZP and the "electronegativity" of the compound, χ(SC), a quantity defined as the geometric mean of the electronegativities of the constituent atoms [37, 38]. This correlation must surely reflect the acidity of the metal-ion centre.

Although this discussion of ionosorption has been concerned primarily with proton adsorption, the specific adsorption of any ion will lead to similar changes in V_{fb}. Thus, Fig. 10 shows that a similar change of 60 mV/decade concentration is found for CdS with concentration of HS^- in solution [39]. However, a caveat must be entered here; the shift of 60 mV/decade is not always found even for different faces of the same material and two examples

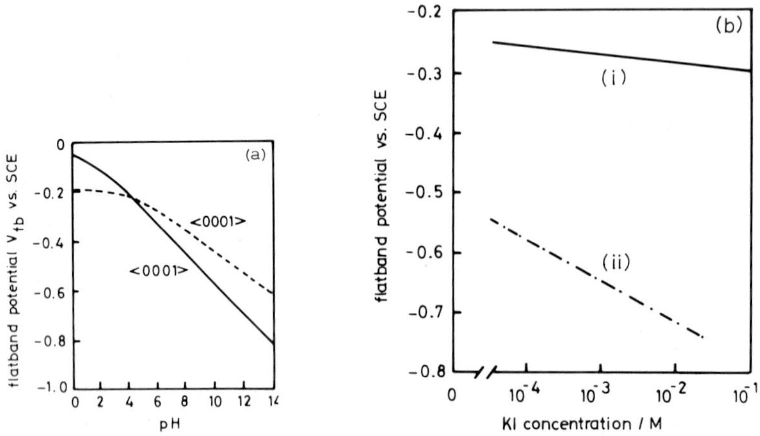

Fig. 11. (a) pH dependence of the flat-band potential of n-ZnO for two different surfaces. (b) Dependence of the flat-band potential of n-WSe_2 on KI concentration for (i) smooth (ii) stepped surfaces.

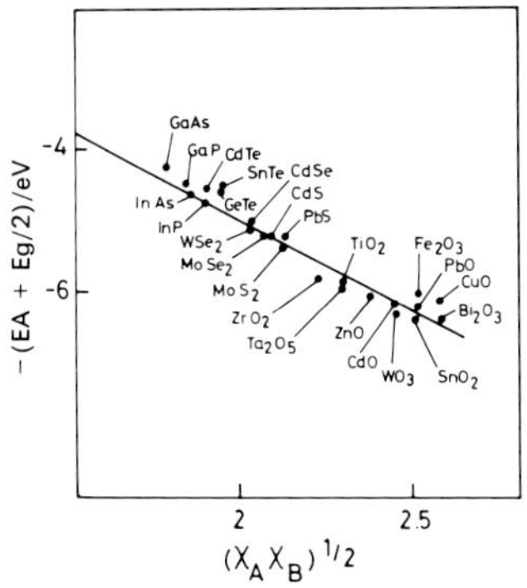

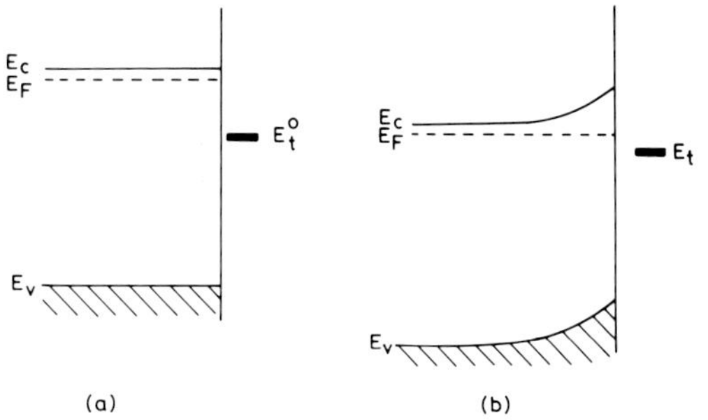

Fig. 12. Dependence of the intrinsic Fermi level, calculated as $-(\mathrm{EA} + E_g/2)$ where EA is the electron affinity (i.e. the energy of the conduction band edge, which can be obtained from the flat-band potential) and E_g the bandgap, on the mean electronegativity of A and B, $(\chi_A \chi_B)^{1/2}$ for semiconductors $A_x B_y$, where χ_A, χ_B are Pauling electronegativities.

Fig. 13. Potential distribution in the presence of a surface state at energy E_t^0 below the Fermi level (a) situation at flat-band potential (b) situation with bias present.

are shown in Fig. 11 [40]; Fig. 11(a) shows the variation of flat-band potential with pH for two different surfaces of n-ZnO and Fig. 11(b) shows similar data for two different faces of WSe_2 with specific adsorption of I^-.

The actual value of the flat-band potential depends on the energies of the valence and conduction bands in the material and can be cleary related to the vacuum level provided that the reference electrode can also be so referenced [41–44]. If we can relate $\chi(\mathrm{SC})$ to the intrinsic Fermi level E_F^i, then the electron affinity of a semiconductor can be written

$$\mathrm{EA} = C\chi(\mathrm{SC}) - \frac{E_g}{2} \qquad (61)$$

where $C \simeq 1$ if Mulliken electronegativities re used. This relationship is illustrated in Fig. 12 [45]. The data of Fig. 12 are actually uncorrected for V_{fb} variation with pH, all data being obtained at pH = 0. It is, therefore, quite remarkable that relationship (61) should be obeyed so well.

The final formula for the flat-band potential from Fig. 12 is

$$V_{fb} = \mathrm{EA} - E_0 - \Delta_{FC} - \Delta_{pH} \qquad (62)$$

where E_0 is the potential of the reference electrode vs. vacuum, $\Delta_{FC} \equiv (E_F - E_c)/e_0 \equiv RT/F \ln(N_D/N_c)$ and $\Delta_{pH} = 2.3RT/F(\mathrm{pH} - \mathrm{pH}_{pzzp})$. This formula predicts that V_{fb} will vary linearly with temperature, a result recently confirmed [46].

3.4 ELECTRONIC SURFACE STATES

The discussion of ionosorption centred on the equilibrium between the surface and the electrolyte. However, there are also electronic surface states

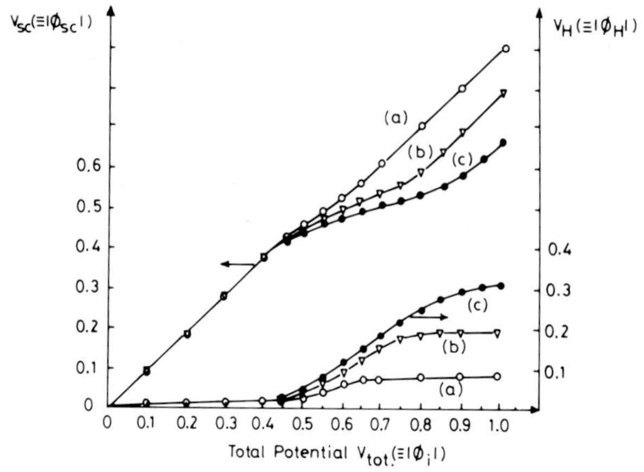

Fig. 14. Potential distribution calculated for the case of Fig. 13 with the n-GaP sample of Fig. 5 and e_0N_t taking the values (a) 10^{-6}, (b) 3×10^{-6}, and (c) 5×10^{-6} C cm^{-2}.

that, in the absence of redox couples in solution, are in equilibrium with the bulk of the semiconductor. If we suppose for simplicity, that there are N_t electronic states per unit area at energy E_t^0 below the Fermi level E_F^0 of an n-type semiconductor at flat band as shown in Fig. 13, then we have

$$\frac{n_t}{N_t - n_t} = \exp\{-(E_t^0 - e_0\phi_{sc} - E_F^0)\} \tag{63}$$

where n_t is the electron occupancy of the surface state and we recall that $\phi_{sc} < 0$. If the surface state is presumed neutral at the flat-band potential, i.e. where $n_t \simeq N$, the (positive) charge for an occupancy n_t is

$$\begin{aligned}\sigma &= e_0(N_t - n_t)\\ &= \frac{e_0N_t}{1 + \exp\{(E_F - E_t^0 + e_0\phi_{sc})/kT\}}\end{aligned} \tag{64}$$

The analysis may be carried through as before for the potential distribution; some representative results are shown in graphical form in Fig. 14. The significant variables are $\partial\phi_{sc}/\partial\phi_i$ and $\partial\phi_H/\partial\phi_i$; if there are no surface states, then, as we have seen, $|\partial\phi_{sc}/\partial\phi_i| \simeq 1$ and $|\partial\phi_H/\partial\phi_i| \simeq 0$. As the surface state concentration is increased, we find that, where $|E_F - E_t^0 + e_0\phi_{sc}| \geqslant 3kT$, the *change* in potential drop in the semiconductor with ϕ_i is much larger than $|\partial\phi_H\partial\phi_i|$; i.e. even though there may be a considerable nett charge in the surface states, it does not change as the interfacial potential alters and the system will appear to behave quasi-classically. If $|E_F - E_t^0 + e_0\phi_{sc}| \leqslant 3kT$, then the derivatives may become comparable. It is easy to show that, within the framework of the above model, the critical condition on surface state density, N_t, is that

$$N_t \geqslant \frac{\varepsilon_0 \varepsilon_H kT}{e_0^2 d_H} \tag{65}$$

For N_t values satisfying eqn. (65), there will be a range of ϕ_i values for which

$$\left|\frac{\partial \phi_{sc}}{\partial \phi_i}\right| < \left|\frac{\partial \phi_H}{\partial \phi_i}\right|$$

If $\varepsilon_H \simeq 6$ and $d_H \simeq 3$ Å, $N_t^{crit} \simeq 8 \times 10^{12}$ cm^{-2}. For a typical atomic population at the surface of 10^{15} cm^{-2}, it is clear that N_t^{crit} corresponds to a very small coverage ($\ll 1\%$) and densities of surface imperfections might be expected to exceed this value on most materials. It might therefore be imagined that no semiconductor–electrolyte interface would ever show quasi-classical behaviour; in fact, at least in aqueous electrolytes, classical behaviour is often found and there are two reasons for this.

(i) Water interacts very strongly with electronic states arising from imperfections on surfaces; LEED/EPS studies have shown that, whereas carefully prepared oxide surfaces in vacuo are often *hydrophobic*, creation of imperfections by Ar^+-ion bombardment leads to the rapid adsorption of multilayers of water molecules. This interaction is electronic in origin and leads to a marked shift of the electronic energy levels away from the band-gap region.

(ii) Water and other protic solvents permit proton adsorption/desorption equilibria of the type referred to above. The small size and high mobility of the proton allow charge compensation to occur at the interface and σ is reduced to a small fraction of the value expected on the basis of surface-state concentration. It is interesting that in non-aqueous solvents, where such compensation is not feasible owing to the large size and poor mobility of the supporting electrolyte cations and anions commonly employed, non-classical behaviour is much more frequently encountered.

It is clear from Fig. 14 that for N_t values $\geqslant 10^{13}$ cm^{-2}, there is a considerable range of ϕ_i values for which $|\partial\phi_{sc}/\partial\phi_i| \ll 1$. This is termed, rather misleadingly, "Fermi-level pinning" or, more graphically, "band edge unpinning" and its consequences have been much explored in recent years [47–50]. It must be emphasised that, whereas this phenomenon is well-known in solid-state devices such as metal–semiconductor junctions, the problems facing its exploration in electrolyte solutions are quite formidable. Indeed, even for the metal–semiconductor interface, recent evidence has thrown doubt on the original explanation offered by Bardeen [51] that *intrinsic* surface states are to blame for the pinning of barrier heights in such devices.

It was demonstrated by Cowley and Sze [52] that, under certain conditions, a rather general relationship of the form

$$\phi_B = S\phi_M + K \tag{66}$$

was obeyed for the metal–semiconductor interface, where ϕ_B is the barrier height, and is related to ϕ_{sc} above by the expression $\phi_B = \phi_{SC} + \Delta$ where

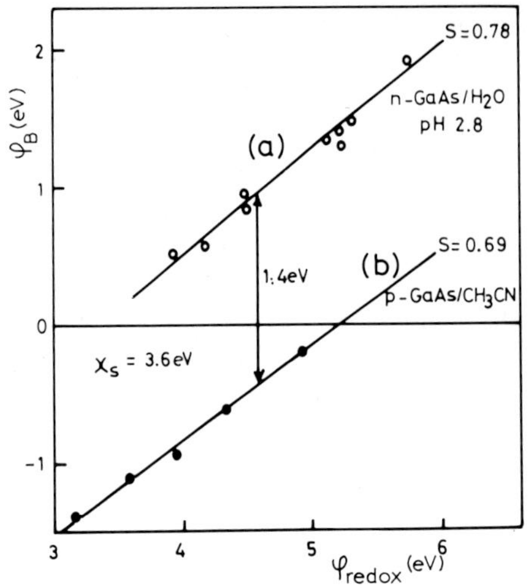

Fig. 15. Plot of ϕ_B (as defined in Fig. 6) vs. ϕ_{redox} (defined as $E_0 + e_0 V_{redox}$ as in Fig. 6) for n-GaAs/H_2O and p-GaAs/CH_3CN in the presence of a range of redox couples. The values for n-GaAs are all corrected to pH 2.8, the estimated pZZp. Good linearity between ϕ_B and ϕ_{redox} are fond in both cases and, from the equation $\phi_B = S\phi_{redox} + K$, the values of S shown are obtained. ●, Nagasubramanian et al. [197]; ○, Horowitz et al. [54].

$\Delta \equiv (E_c - E_F)_{bulk}$ or $(E_v - E_F)_{bulk}$ for an n-type or p-type semiconductor, respectively, ϕ_M is the metal work function and K and S are constants that depend on the particular interface being considered. The ideal Schottky barrier, such as that discussed above, has $S \equiv 1$, whereas perfect pinning would have $S \equiv 0$. Experimentally [53], it is found that S remains small if $\Delta\chi$ (the electronegativity difference $|\chi_A - \chi_B|$) for the binary semiconductor AB is less than 0.7 and changes rapidly towards unity for $\Delta\chi > 0.7$. Thus, for GaAs where $\Delta\chi = 0.4$, $S = 0.04$ (p-GaAs), 0.07 (n-GaAs).

The analogue for ϕ_M in solution is the redox energy of the faradaically dominant redox couple in solution, ϕ_R, defined as $E_0 + e_0 V_{redox}$, where E_0 is the energy difference between the vacuum level and the appropriate redox energy level (usually taken from the above to be 4.75 eV for SCE); at equilibrium, we assume that the Fermi level in the semiconductor will become equal to the redox energy ϕ_R or, more accurately, $\phi_R + e_0\Delta V$ where ΔV is the potential drop at the surface associated with adsorbed ionic charge in the Helmholtz layer. Obviously, by correcting to the pZZP, we may hope to eliminate this latter term provided protonation/deprotonation equilibria are the dominant sources of adsorbed charge. We may obtain, in solution, the value of ϕ_B by, for example, the a.c. techniques described in more detail below and ϕ_R is normally known, so the analogous relationship

$$\phi_B = S\phi_R + K' \tag{67}$$

can be tested [54]. This programme has been carried out for a number of semiconductors and the results for GaAs are shown in Fig. 15. It is clear that,

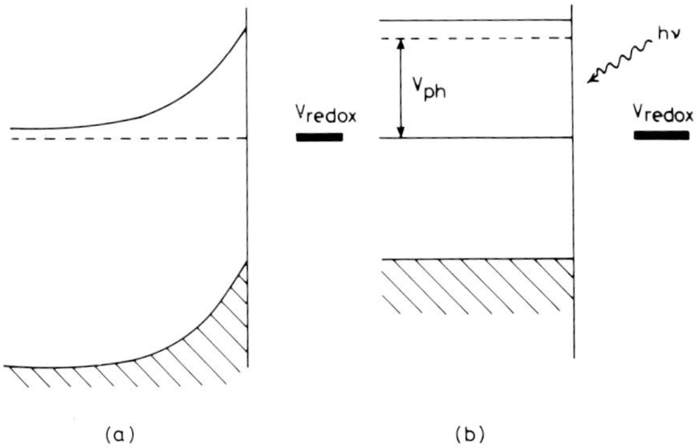

Fig. 16. Principles of the photovoltage measurement. (a) Semiconductor in the dark, with the bulk Fermi level E_F in equilibrium with V_{redox}; (b) under illumination, when the open-circuit potential adjusts so that, in principle, there is no potential drop in the depletion layer.

for n-GaAs in aqueous solution and p-GaAs in acetonitrile (ACN), eqn. (67) is satisfied and values of S much closer to unity are found.

An alternative technique for the measurement of ϕ_{sc} is to study the photopotential. The theory of this is discussed in some detail in Sect. 7, but the essential features of the measurement are shown in Fig. 16. After equilibration in the dark, when the potential of the electrode at open circuit becomes equal to the redox potential V_{redox}, the light is turned on and the electrode potential changes at open circuit in such a way that the bands become flat. There are many problems with this technique and it is considerably less reliable than a properly conducted a.c. experiment, but it may give a reasonably accurate picture if surface recombination is small (vide infra). Some results for p-GaAs in aqueous solution are shown in Fig. 17 and the S values derived are of the order 0.7, though the dispersion apparent in Fig. 17 makes a quantitative interpretation difficult.

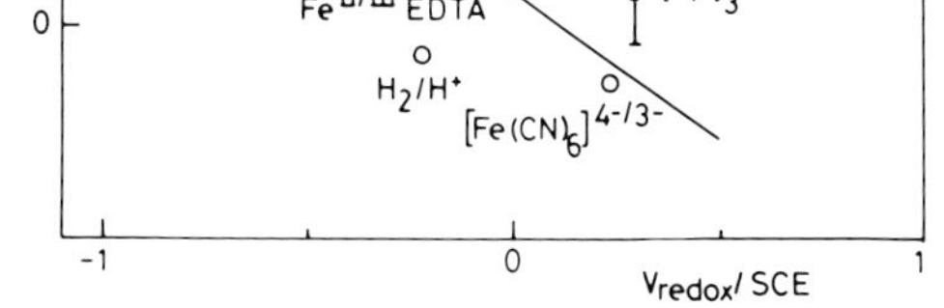

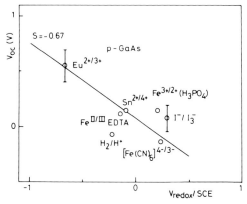

Fig. 17. Plot of $V_{oc} \equiv V_{on} - V_{redox}$ (where V_{on} is the open-circuit potential of the semiconductor with the light on) for p-GaAs/H_2O vs. V_{redox}, which corresponds to the open-circuit potential of the semiconductor in the dark. Analogously to Fig. 15, we expected $V_{oc} = SV_{redox} + K$ and S is negative here as we are dealing with a p-type semiconductor.

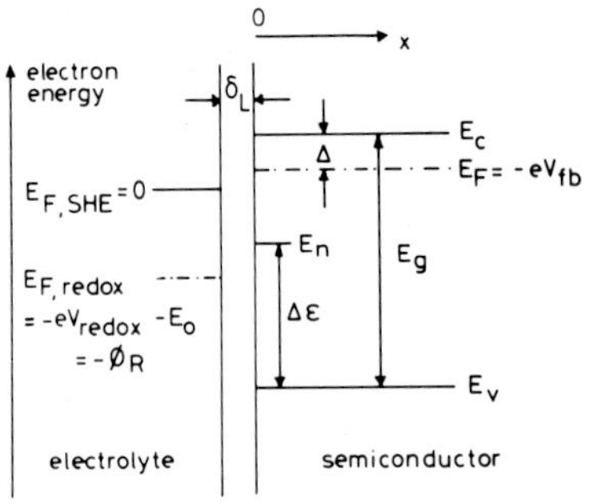

Fig. 18. Energy-level diagram where there is a continuum of surface states.

It appears, at least for GaAs, that only weak Fermi level pinning may occur. This is in marked contrast to the metal–semiconductor case and it would appear from photoemission work that the GaAs surface does *not* have a large density of intrinsic states when cleaved in vacuo, apparently because of surface rearrangements [55]. The pinning of GaAs at a metal–semiconductor interface must therefore be due to extrinsic states, possibly arising from strong metal–semiconductor interactions. It is interesting that photoemission work does show an appreciable surface-state density if a sub-monolayer oxide film forms.

The origin of the expression for ϕ_B can be understood in terms of the semiconductor–eletrolyte model developed above [56]. The charge on the surface, q_{ss}, may be considered as arising from a series of electronic energy levels of approximately constant energy density within the bandgap. There is appreciable experimental evidence that this, rather than a large density of states at a particular energy may be nearer the correct physical picture, at least in III/V materials. Let us now suppose that the dominant equilibrium is that between these surface electronic states and a fast redox couple in solution. At flat band potential, $q_{sc} = 0$, $d\phi/dx|_{x \geqslant 0} = 0$ and the potential dropped in the Helmholtz layer is $\Delta\phi = \phi(0) - \phi_{el} = \sigma d_H/\varepsilon_0\varepsilon_H$. The change in ϕ_i is therefore $-\Delta\phi$ and the flat-band potential $\Delta V_{fb} = -\Delta\phi_i = \Delta\phi$. For a redox couple of potential V_{redox}, the situation is shown in Fig. 18. The energy E_n in Fig. 15 is *defined* through the fact that when the surface states are filled to that energy, the nett surface charge $\sigma = 0$. Then

$$E_n - E_{F,redox} = \Delta\varepsilon - E_g + E_c + e_0\phi_R \tag{68}$$

$$E_c = -e_0 V_{fb} + \Delta \tag{69}$$

$$\sigma = e_0 N_s (E_n - E_{F,redox}) \tag{70}$$

where N_s is the number density of electronic surface states and has units $cm^{-2}\,eV^{-1}$. From eqns. (68)–(70), we have

$$\Delta V_{\mathrm{fb}} = \frac{d_{\mathrm{H}} e_0 N_{\mathrm{s}} (\Delta\varepsilon - E_{\mathrm{g}} + \Delta - e_0 V_{\mathrm{fb}} + e_0 \phi_{\mathrm{R}})}{\varepsilon_0 \varepsilon_{\mathrm{H}}}$$

$$\equiv V_{\mathrm{fb}} - V_{\mathrm{fb}}^0 \qquad (71)$$

where V_{fb}^0 is the flat-band potential when $\sigma = 0$. After some rearrangement, we have

$$V_{\mathrm{fb}} = \frac{1}{1 + x^{-1}} \frac{\phi_{\mathrm{R}}}{e_0} + \frac{1}{1 + x^{-1}} \left(\frac{E_{\mathrm{n}} - E_{\mathrm{F}}}{e_0} \right) + \frac{V_{\mathrm{fb}}^0}{1 + x} \qquad (72)$$

where $x = d_{\mathrm{H}} e_0^2 N_{\mathrm{s}} / \varepsilon_0 \varepsilon_{\mathrm{H}}$. Evidently, when $x = 0$, we recover the ideal Schottky barrier. The parameter S, introduced above, is given by

$$S = 1 - \frac{1}{1 + x^{-1}} = \frac{1}{1 + x}$$

It is evident that this treatment is complementary to that given above for the situation in which the dominant equilibrium was that between the bulk and surface of the semiconductor and these two physically very different situations must be carefully distinguished, especially in view of the fact that the term "Fermi-level pinning" discussed above actually bears both these meanings in the literature. In this review, we will, therefore, explicitly distinguish two types of pinning.

Type I. Equilibrium is established between the surface states and the solution redox couple but *not* between the semiconductor majority carriers and either the surface states or the solution couple. Under these circumstances, the flat-band potential will apparently change with redox couple, but any *change* in the potential is dropped across the depletion layer and the semiconductor will appear otherwise to behaving classically.

Type II. Equilibrium is established between the surface states and the majority carriers within the semiconductor. Under these circumstances, a fraction of the potential *change* will be dropped across the depletion layer and a fraction across the Helmholtz layer. If a redox couple is present in solution, and the kinetics of electron transfer between this and the surface states are also rapid, then a large dark current will be found.

It is again interesting to note the sensitivity of x to d_{H}; for a semiconductor–electrolyte junction in which d_{H} corresponds to the Helmholtz layer ($\simeq 3$ Å) and $\varepsilon_{\mathrm{H}} \simeq 6$, then $N_{\mathrm{s}} \simeq 3 \times 10^{13}$ cm^{-2} eV^{-1} for $S = 0.78$, the value found experimentally for n-GaAs. If, on the other hand, an oxide layer has grown on the surface of the crystal to a depth of, say, 20 Å and $\varepsilon_{\mathrm{H}} \simeq 3.6$, the same value of S is obtained for $N_{\mathrm{s}} \simeq 3 \times 10^{12}$ cm^{-2} eV^{-1}. The implications are evident; reproducible, clean surfaces are vital for semiconductor electrochemistry research.

3.5 DEEP BULK TRAPS

The donor sites in the semiconductors discussed above are all considered

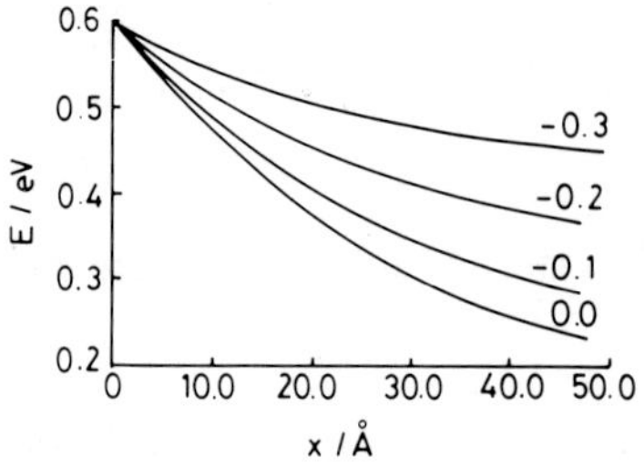

Fig. 19. The space charge barrier in an n-type semiconductor in exhaustion where the major donor site energy lies well below the conduction-band edge. The system parameters are, with respect to the Fermi level, E_c^b (energy of the CB in the bulk) = 0.2 eV, E_c^s (CB at surface) = 0.6 eV, L_D = 20.Å. E_t (vs. E_F) values are given against each curve.

to be completely ionised at 300 K and this allows us to write the charge density as in eqn. (32). However, many semiconductors, especially n-type oxides, possess trap levels that lie deep in energy below the conduction band edge and whose ionisation is far from complete at room temperature. If the energy of such traps is E_t with respect to the Fermi level in the bulk and the parameter $\eta(x)$ is defined as $\eta \equiv (E_c^b - E_c(x))/kT$, then, for an n-type semiconductor for which the deep traps are the sole donors, we have [12, 57]

$$\frac{d\phi}{dx} = \left(\frac{2kTN_D}{\varepsilon_0 \varepsilon_{sc}}\right)^{1/2} F(\eta) \tag{73}$$

where N_0 is the donor density and

$$F(\eta) = \{(1 - f)(e^\eta - \eta - 1) + \ln[1 + (e^\eta - 1)f] - \eta f\}^{1/2} \tag{74}$$

The function f is the Fermi function $\equiv [1 + \exp(E_t/kT)]^{-1}$.

This is obviously more complex than the previous formula

$$F(\eta) = \{(e^\eta - \eta - 1)\}^{1/2} \tag{37}$$

and, as shown in Fig. 19, the main effect is to extend the depletion layer.

A common situation is that there are N_D shallow donor sites and N_D^t deeper-lying levels; we then have, for an n-type semiconductor

$$\frac{d^2\phi}{dx^2} = \frac{e_0}{\varepsilon_0 \varepsilon_{sc}} [-n(x) + N_D + N_D^t(1 - f')]$$

where

$$f' = \frac{1}{1 + \exp\{\xi(E_t - e_0\phi + e_0\phi_i)\}}. \tag{75}$$

Integrating once, this gives

$$\left(\frac{d\phi}{dx}\right)^2 = \frac{2e_0}{\varepsilon_0 \varepsilon_{sc} \xi} \{n^0 e^\eta - n^0 - N_D\eta + N_D^t \ln[1 + (e^\eta - 1)f - \eta N_D^t]\} \tag{76}$$

$$n^0 = N_D^t(1 - f) + N_D$$

and the properties of this function are discussed elsewhere [12].

4. Alternating current techniques [7, 58–60]

We have seen that the behaviour of an extrinsic semiconductor electrode is quite different according to whether the potential is positive or negative of the flat-band potential. For an n-type semiconductor at potentials negative of the flat-band potential, the surface will become degenerate and a large fraction of the potential change will occur across the Helmholtz layer. At positive potentials, corresponding to the "reverse bias" configuration, the potential change is accommodated primarily across the depletion layer inside the semiconductor. The theory of the a.c. response has been worked out in considerable detail for the latter case and, as wil be seen below, analytical solutions may be obtained for a wide variety of cases. For the more general case of an intrinsic semiconductor, or for an extrinsic material under inversion, the theoretical development has been less satisfactory, though some results have been given in the literature.

From the experimental standpoint, the use of a.c. techniques offers many advantages. Sensitivity is much higher than in d.c. measurements, since phase-sensitive detection can be used and very small probe signals can be employed ($\leqslant 5\,\mathrm{mV}$). The technique is therefore a truly equilibrium one, unlike cyclic voltammetry. An alternative approach to the commonly used sinusoidal signal superimposed on the selected d.c. potential is to use a potential step and to employ Laplace transform methods. Instrumentally, this is rather more demanding and the advantages are not clear [51]. Fourier transform methods have also been considered and their use will have advantages in terms of the time-scale for an experiment, especially at very low frequencies.

4.1 THE EXTRINSIC CASE. BASIC DEVELOPMENT

In a one-dimensional n-type material, the equation of motion for the majority carriers can be written

$$\frac{\partial n}{\partial t} = D\frac{\partial}{\partial x}\left[\frac{\partial n}{\partial x} + \left(\frac{e_0}{kT}\right)n\mathscr{E}\right] \tag{77}$$

where n is the number concentration of these carriers at distance x from the surface, D their diffusion coefficient (which is related to the mobility, μ, of the carriers by the Einstein relationship $\mu = e_0 D/kT$), and $\mathscr{E}$ is the local electric field. If the potential at x is given by $\phi(x)$ and a small a.c. field is applied, then ϕ can be considered as made up of a static part ϕ_0 and a modulated part $\phi_1 \exp(i\omega t)$, where $\phi_1 \ll \phi_0$. Associated with the variation in ϕ there will be sinusoidal variations in n and we may write

$$n = n_0 + n_1 \mathrm{e}^{i\mu t} \tag{78}$$

$$\mathscr{E} = \mathscr{E}_0 + \mathscr{E}_1 \mathrm{e}^{i\omega t} \tag{79}$$

Inserting these into the transport equation, eqn. (77), we obtain, for the frequency-dependent terms

$$\frac{d^2 n_1}{dx^2} + \frac{e_0}{kT}\frac{d}{dx}(n_1 \mathscr{E}_0 + n_0 \mathscr{E}_1) = \frac{i\omega n_1}{D} \tag{80}$$

In the bulk of the semiconductor, the potential will be defined as zero. This differs slightly from the convention adopted in Sect. 3, but simplifies the resultant analysis. For a reverse-biased n-type semiconductor, it follows that the potential ϕ in the depletion layer will be negative. In this region, the Boltzmann distribution gives

$$n_0 = n_b \exp(e_0 \phi_0 / kT) \equiv N_D \exp(e_0 \phi_0 / kT) \tag{81}$$

where n_b is the bulk concentration of electrons, N_D the density of donor sites and we have assumed, following the discussion after eqn. (32), that all donor sites are fully ionised at room temperature such that $N_D = n_b$.

At the other surface of the semiconductor, i.e. away from the electrolyte, we may suppose that we have an ohmic contact. If the total width of the semiconductor is L_s, this leads to the boundary conditions

$$n_1 = 0, \quad \phi_1 = 0 \text{ at } x = L_s \tag{82}$$

As before, Poisson's equation holds.

$$\frac{d\mathscr{E}_1}{dx} = -\frac{n_1 e_0}{\varepsilon_0 \varepsilon_{sc}} \tag{83}$$

Integration on eqn. (80) gives

$$\frac{dn_1}{dx} + (n_1 \mathscr{E}_0 + n_0 \mathscr{E}_1)\frac{e_0}{kT} = \int dx \frac{i\omega n_1}{D} = -\frac{i\omega \varepsilon_{sc}\varepsilon_0}{e_0 D}\int dx \left(\frac{d\mathscr{E}_1}{dx}\right)$$

$$= -\frac{i\omega \varepsilon_{sc}\varepsilon_0 \mathscr{E}_1}{e_0 D} + C \tag{84}$$

where C is a constant whose value corresponds to the total a.c. current j_1 divided by $e_0 D$ since, in a one-dimensional system, j_1 is the same everywhere and is equal to the sum of the charge flux due to particle migration, $e_0[(dn_1/dx) + \mu(n_1 \mathscr{E}_0 + n_0 \mathscr{E}_1)]$, and the displacement current $\varepsilon_0 \varepsilon_{sc} \partial \mathscr{E}_1 / \partial t$. Hence

$$\frac{dn_1}{dx} + (n_1 \mathscr{E}_0 + n_0 \mathscr{E}_1)\frac{e_0}{kT} = -\frac{i\omega \varepsilon_0 \varepsilon_{sc} \mathscr{E}_1}{e_0 D} + \frac{j_1}{e_0 D} \tag{85}$$

Further integration is possible using the integrating factor $\exp(-e_0 \phi_0 / kT)$. Recalling that $\mathscr{E}_0 = -d\phi_0/dx$ and

$$\int dx \frac{d\phi_1}{dx}\frac{e_0 n_0}{kT} e^{-e_0 \phi_0 / kT} = \frac{e_0 N_D \phi_1}{kT} \tag{86}$$

we have, from eqn. (85)

$$n_1 - \frac{e_0 n_0 \phi_1}{kT} = -\frac{i\omega\varepsilon_0\varepsilon_{sc}}{e_0 D}\exp\left(\frac{e_0\phi_0}{kT}\right)\int_{L_s}^{x} dx\mathscr{E}_1 \exp\left(-\frac{e_0\phi_0}{kT}\right)$$

$$+\frac{j_1}{e_0 D}\exp\left(\frac{e_0\phi_0}{kT}\right)\int_{L_s}^{x} dx \exp\left(-\frac{e_0\phi_0}{kT}\right) \tag{87}$$

Returning to eqn. (85) and recalling that

$$\frac{d\mathscr{E}_0}{dx} = -\frac{e_0(N_D - n_0)}{\varepsilon_0\varepsilon_{sc}} \tag{88}$$

we have

$$\frac{dn_1}{dx} + (n_1\mathscr{E}_0 + n_0\mathscr{E}_1)\frac{e_0}{kT} \equiv \frac{d}{dx}\left(n_1 - \frac{\varepsilon_0\varepsilon_{sc}\mathscr{E}_0\mathscr{E}_1}{kT} - \frac{e_0\phi_1 N_D}{kT}\right)$$

$$\equiv \frac{i\omega\varepsilon_0\varepsilon_{sc}}{e_0 D}\frac{d\phi_1}{dx} + \frac{j_1}{e_0 D} \tag{89}$$

whence we obtain

$$n_1 - \frac{\varepsilon_{sc}\varepsilon_0\mathscr{E}_0\mathscr{E}_1}{kT} - \frac{e_0\phi_1 N_D}{kT} = \frac{i\omega\varepsilon_0\varepsilon_{sc}\phi_1}{e_0 D} + \frac{j_1}{e_0 D}(x - L_s) \tag{90}$$

provided that $\mathscr{E}_0 = 0$ at $x = L_s$, which will be valid if $L_s \gg W$, where W is the width of the depletion region. Equations (87) and (90) provide the basic a.c. response theory for a semiconductor electrode that is reverse biased. They can be developed for a variety of boundary conditions at the semiconductor–electrolyte interface and the resultant a.c. response can, in most cases, be expressed by means of an equivalent circuit.

4.2 CLASSICAL SEMICONDUCTOR WITH NO SURFACE STATES

We consider first the simplest possible case of an interfacial potential similar to that shown in Fig. 4. If a small potential modulation is applied to the semiconductor in depletion, its effect can be calculated by eliminating n_1 between eqns. (87) and (90) to obtain, for $x < W$

$$\frac{\varepsilon_0\varepsilon_{sc}\mathscr{E}_0\mathscr{E}_1}{kT} = -\frac{e_0\phi_1(N_D - n_0)}{kT}$$

$$+\frac{i\omega\varepsilon_0\varepsilon_{sc}}{e_0 D}\left\{-\phi - J_1 + \phi_1(W)\exp\left(\frac{e_0\phi_0}{kT}\right)\right\}$$

$$+\frac{j_1}{e_0 D}\left\{(L_s - x) + J_2 - (L_s - W)\exp\left(\frac{e_0\phi_0}{kT}\right)\right\} \tag{91}$$

where

$$\left.\begin{aligned} J_1 &= \exp\left(\frac{e_0\phi_0}{kT}\right)\int_W^x \mathscr{E}_1 \exp\left(-\frac{e_0\phi_0}{kT}\right)\mathrm{d}x \\ J_2 &= \exp\left(\frac{e_0\phi_0}{kT}\right)\int_W^x \exp\left(-\frac{e_0\phi_0}{kT}\right)\mathrm{d}x \end{aligned}\right| \quad (92)$$

and we have assumed that, for $x > W$, $\phi_0 \to 0$ and $\phi_0(L_s) = \phi_1(L_s) = 0$. Since kT/e_0 is small (ca. 25 mV at 300 K), the value of $\exp(e_0\phi_0/kT)$ is exceedingly small in the depletion layer (since $\phi_0 < 0$). The terms $\phi_1(W)\exp(e_0\phi_0/kT)$ and $(L_s - W)\exp(e_0\phi_0/kT)$ can therefore be neglected. As we approach the surface, at which $E_1 \to E_{1s}$, the integrals J_1 and J_2 can be calculated straightforwardly.

Consider J_2; if we use the formula for ϕ_2 derived from eqn. (36)

$$\frac{e_0\phi_0}{kT} \approx -\frac{(W - x)^2}{2L_D^2} \quad (93)$$

where, as above, $L_D = (\varepsilon_{sc}\varepsilon_0 kT/e_0^2 N_D)^{1/2}$ and $W = (-2e_0\phi_0/kT)^{1/2}L_D$. At the surface

$$\begin{aligned} J_2(0) &= \exp\left(\frac{e_0\phi_0}{kT}\right)\int_W^0 \exp\left(-\frac{e_0\phi_0}{kT}\right)\mathrm{d}x \\ &\approx \exp\left(-\frac{W^2}{2L_D^2}\right)\int_W^0 \exp\left[\frac{(W - x)^2}{2L_D^2}\right]\mathrm{d}x \\ &\approx -\exp(-W^2/2L_D^2)\int_0^{W/(\sqrt{2}L_D)} \sqrt{2}L_D \exp(\zeta^2)\,\mathrm{d}\zeta \\ &\approx -\frac{L_D^2}{W} \quad \text{if} \quad \frac{W}{\sqrt{2}L_D} \gtrsim 1 \end{aligned} \quad (94)$$

Similarly, $J_1(0) \simeq -\mathscr{E}_{1s}L_D^2/W$, since most of the contribution to the integral comes in the region where $x \simeq 0$. Inserting these values into eqn. (91), we obtain

$$\begin{aligned} \varepsilon_0\varepsilon_{sc}\mathscr{E}_{1s}\left[1 - \frac{i\omega kTL_D^2}{e_0 D\mathscr{E}_{0s}W}\right] &= -\frac{e_0\phi_{1s}(N_D - n_{0s})}{\mathscr{E}_{0s}} - \frac{i\omega\varepsilon_0\varepsilon_{sc}kT\phi_{1s}}{e_0 D\mathscr{E}_{0s}} \\ &\quad + \frac{j_1 kT}{e_0 D\mathscr{E}_{0s}}\left(L_s - \frac{L_D^2}{W}\right) \end{aligned} \quad (95)$$

where ϕ_{1s} is the total a.c. potential drop across the semiconductor and, generally, the subscript s in eqn. (95) refers to the values of the variables at the surface. Now, if there are no faradaic processes occurring at the semiconductor and there are no surface states emptying and filling, then the current at the surface must be wholly displacive; hence

$$j_1 \quad = \quad i\omega\varepsilon_0\varepsilon_{sc}\mathscr{E}_{1s} \tag{96}$$

Multiplying eqn. (95) by $i\omega$ and inserting this into the expression for j_1, we find

$$j_1(1 + i\omega A) \quad = \quad (i\omega C_{sc} - \omega^2 A')\phi_{1s} \tag{97}$$

where $A = -L_s/\mathscr{E}_{0s}\mu$, $A' = -\varepsilon_{sc}\varepsilon_0/\mathscr{E}_{0s}\mu$ and $C_{sc} = -e_0(n_B - n_{0s})/\mathscr{E}_{0s} \simeq -e_0 n_B/\mathscr{E}_{0s}$. Expression (97) corresponds to the current expected for the equivalent circuit

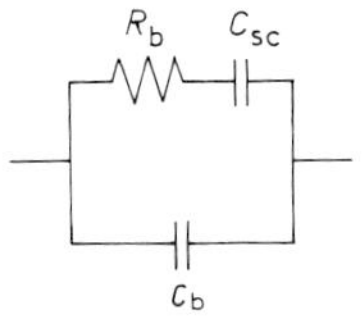

The admittance of this circuit is

$$Y \quad = \quad i\omega C_b + \frac{1}{(R_b + 1/i\omega C_{sc})} \tag{98}$$

Comparing this with eqn. (97), assuming $C_{sc} \gg C_b$, and writing $\mathscr{E}_{0s} = -kTW/e_0 L_D^2$, we have

$$C_b = \frac{\varepsilon_0\varepsilon_{sc}}{L_s}; \qquad R_b = \frac{L_s}{\mu e_0 N_D};$$

$$\text{and} \quad C_{sc} = \left(\frac{e_0^2\varepsilon_0\varepsilon_{sc}N_D}{2kT}\right)^{1/2}\left(-\frac{e_0\phi_{1s}}{kT}\right)^{-1/2} = \frac{\varepsilon_0\varepsilon_{sc}}{W} \tag{99}$$

Provided that L_s is large compared with W, C_b may be neglected and we arrive at the very simple equivalent circuit

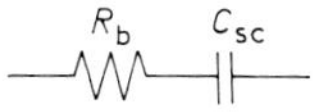

where R_b is the bulk resistance of the semiconductor per unit area, and $C_{sc} = \varepsilon_{sc}\varepsilon_0/W$ is the capacitance of a dielectric layer of width W.

4.2.1 Experimental tests of the classical model

The simplicity of the two-component equivalent circuit can be expected to be reflected in practice provided

(a) the cell is designed so that all other capacitances in series with C_{sc} are very large,

(b) the amplitude of the modulation is small compared with kT/e_0,

(c) there are no ongoing faradaic processes or surface state effects in the frequency range studied; i.e. eqn. (96) is satisfied, and

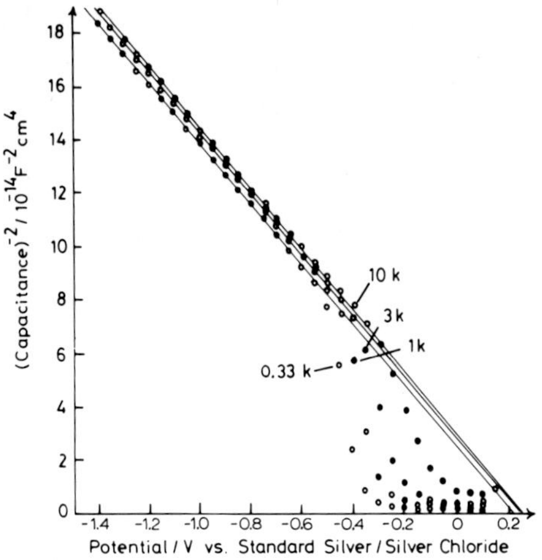

Fig. 20. Series capacitance measurements for (100) p-GaAs in 0.5 M H_2SO_4 at the frequencies shown. The electrode was pre-etched in CH_3OH/Br_2.

(d) the electrolyte is sufficiently concentrated that the potential dropped in the Gouy layer is negligible.

Under these circumstances, we might expect that the whole of the a.c. component of the potential should appear in the bulk of the semiconductor, across the depletion layer. In addition, we may replace $-\phi_{0s}$ by $\phi_s - \phi_{fb}$, where ϕ_{fb} is the flat-band potential and ϕ_B is the actual d.c. potential applied to the semiconductor.

Hence, the formula for C_{sc} becomes

$$C_{sc} = \left(\frac{e_0^2 \varepsilon_0 \varepsilon_{sc} N_D}{2kT}\right)^{1/2} \left(\frac{e_0[\phi_B - \phi_{fb}]}{kT}\right)^{-1/2} \tag{100}$$

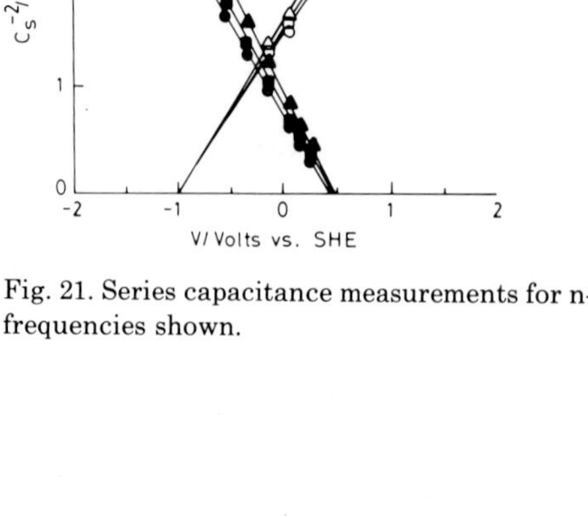

Fig. 21. Series capacitance measurements for n-GaAs and p-GaAs in 1 M KCl–0.01 M HCl at the frequencies shown.

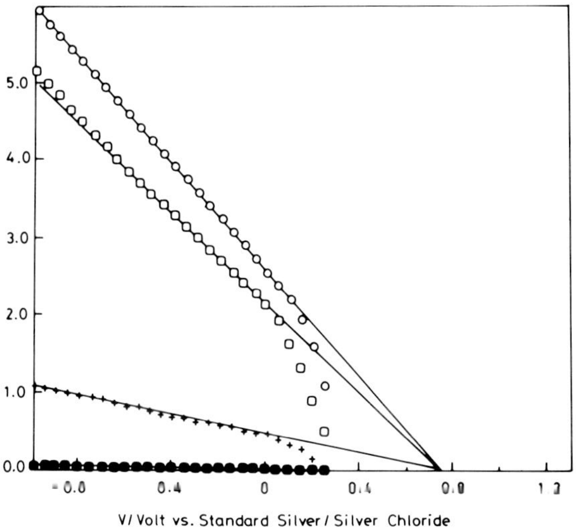

Fig. 22. Series capacitance measurements for (100) p-GaP in 0.5 M H_2SO_4 at ○, 10 kHz; □, 1 kHz; +, 330 Hz; and ■, 87 Hz.

whence

$$C_{sc}^{-2} = \left(\frac{2}{\varepsilon_0 \varepsilon_{sc} e_0 N_D}\right)[\phi_B - \phi_{fb}] \tag{101}$$

Thus, a plot of $(C_{sc})^{-2}$ vs. ϕ_B should be linear with an intercept at the flat-band potential. This linearity can be used as an immediate test of the theory and an example is shown in Fig. 20 for p-GaAs [62]. The capacitance derived from the two-component circuit is almost independent of frequency for potentials more than 0.5 V from the flat-band potential and the intercept is clearly defined. Similar results for both n-GaAs and p-GaAs are shown in Fig. 21 [63] and an important check on the theory is that the flatband potentials for the n- and p-type electrodes differ by ca. 1.4 V, corresponding to a bandgap of 1.4 eV for GaAs. This is expected as the Fermi level will be very close to the band edges in the bulk for these wide-band materials.

Regrettably, such ideal behaviour as that exhibited in Figs. 20 and 21 is very rare. In almost all materials studied, deviations from the simple theory are encountered. There are four commonly observed cases:

(i) The capacitance shows frequency dispersion but C^{-2} still varies linearly with ϕ_B and all the lines intersect at a common point on the origin. This effect is shown for p-GaP at potentials remote from flatband (Fig. 22) [59].

(ii) The C^{-2}/ϕ_B plots are linear, as in (i), and lines obtained at different frequencies are approximately parallel; an example is shown for n-CdSe in Fig. 23 [64].

(iii) The C^{-2}/ϕ_B plots are linear, but neither intersect on the axis nor are they parallel; an example is shown for p-GaP $(\bar{1}\bar{1}\bar{1})$ in Fig. 24 [62].

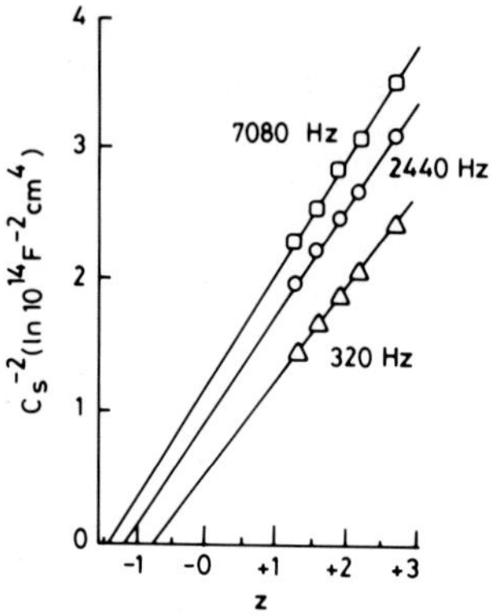

Fig. 23. Series capacitance measurements for (000$\bar{1}$) n-CdSe, $N_D \simeq 10^{15}\,cm^{-2}$ in 0.25 M K_2SO_4–0.05 M CH_3COOH–0.05 M $CH_3COO^-\,Na^+$ at the frequencies shown.

(iv) The C^{-2}/ϕ_B plots not only show dispersion but are not linear, either being curved or consisting of two or more linear portions; an example of the former type of behaviour is shown for n-Fe_2O_3 in Fig. 25 [147] and of the latter for n-TiO_2 in Fig. 26 [66].

4.3 NON-CLASSICAL BEHAVIOUR

A large number of corrective treatments has been derived for the semiconductor–electrolyte interface and we shall consider these below. The complexity of some of the models considered has, however, led to considerable doubt as to the appropriateness of some of the treatments.

(a) *Finite Helmholtz capacitance*

Associated with the double layer at the interface there will be a Helmholtz capacitance whose value may be very roughly estimated from the

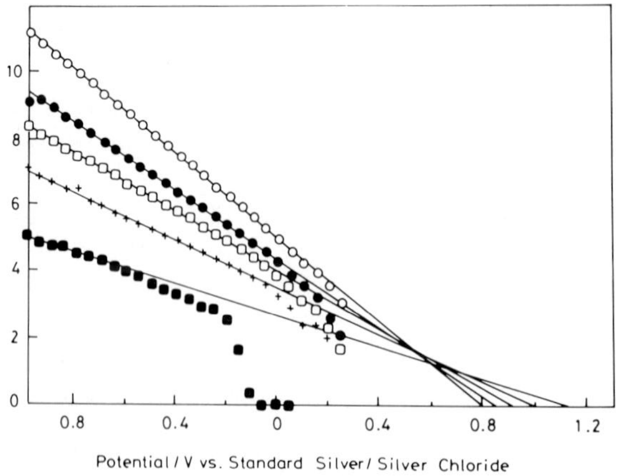

Fig. 24. Series capacitance measurements for ($\overline{111}$) p-GaP in 0.5 M H_2SO_4 at ○, 10 kHz; ●, 3 kHz; □, 1 kHz; +, 330 Hz; and ■, 87 Hz.

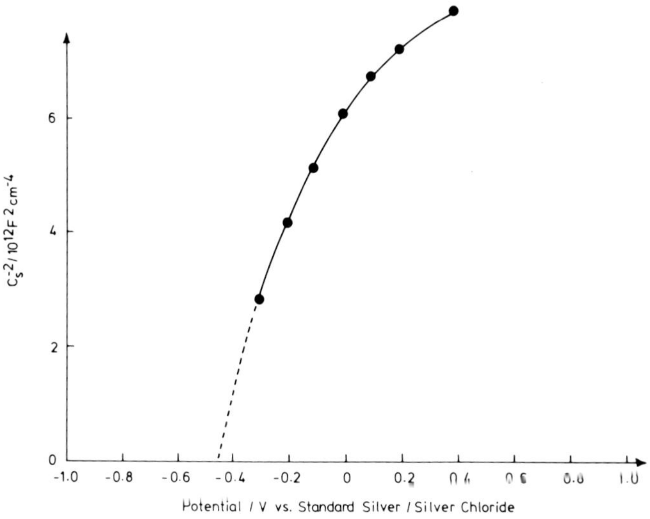

Fig. 25. Series capacitance measurements for n-Fe_2O_3 in 0.1 M NaOH at 200 Hz.

formula $\varepsilon_H \varepsilon_0 / d_H$. If we take $\varepsilon_H \simeq 6$ and $d_H \simeq 3$ Å, $C_H \simeq 20\,\mu F\,cm^{-2}$. We recall, from eqns. (31) and (36), that, in the absence of *charge* in the double layer, $\phi_H \simeq d_H q_{sc} / \varepsilon_H \varepsilon_0 = q_{sc} / C_H$.

Within the Mott–Schottky approximation, $q_{sc} = (2\varepsilon_0 \varepsilon_{sc} e_0 N_D)^{1/2} [\phi_{sc} - (kT/e_0)]^{1/2}$ and $C_{sc} = (\varepsilon_{sc} \varepsilon_0 e_0 N_D / 2)^{1/2} [\phi_{sc} - (kT/e_0)]^{-1/2}$. Clearly, for large values of $\varepsilon_{sc} N_D$, ϕ_H / ϕ_{sc} can become significant and some examples are shown in Fig. 27 [67] for a value of ε_{sc} appropriate to TiO_2 ($\simeq 173$). It can be seen that ϕ_H is an appreciable fraction of the total potential, especially near flat-band potential. From the formula above, the equivalent circuit is

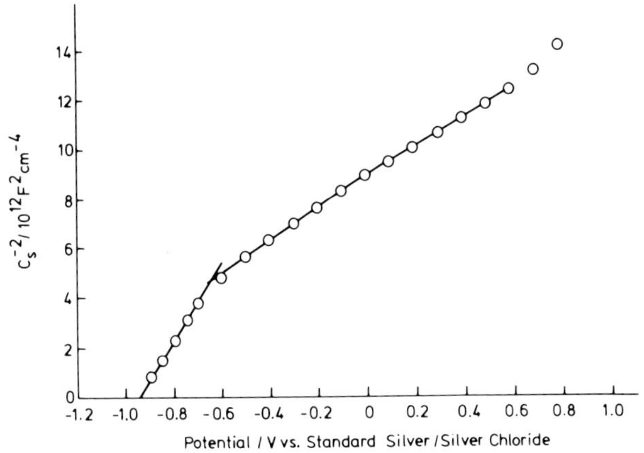

Fig. 26. Series capacitance measurements for n-TiO_2 in 1 M NaOH at 100 Hz.

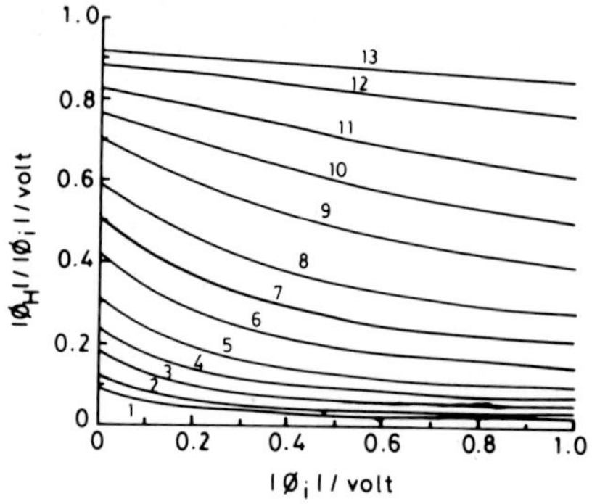

Fig. 27. The ratio of the potential drop in the Helmholtz layer $|\phi_{\mathrm{H}}|$, to the total potential drop $|\phi_{\mathrm{i}}| \equiv |\phi_{\mathrm{sc}} + \phi_{\mathrm{H}}|$ as a function of $|\phi_{\mathrm{i}}|$ for $\mathrm{TiO_2}$, $\varepsilon_{\mathrm{sc}} = 173$, $d_{\mathrm{H}} = 5\,\text{Å}$, $\varepsilon_{\mathrm{H}} = 6$, for the following donor densities (cm^{-3}): (1) 10^{16}, (2) 2×10^{16}, (3) 5×10^{16}, (4) 10^{17}, (6) 5×10^{17}, (7) 10^{18}, (8) 2×10^{18}, (9) 5×10^{18}, (10) 10^{19}, (11) 2×10^{19}, (12) 5×10^{19}, and (13) 10^{20}.

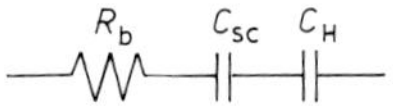

The derived series capacitance is given by

$$C_{\mathrm{s}}^{-1} = C_{\mathrm{sc}}^{-1} + C_{\mathrm{H}}^{-1} \tag{102}$$

and it is easy to show that

$$C_{\mathrm{s}}^{-2} = C_{\mathrm{H}}^{-2}\left\{1 + \frac{2C_{\mathrm{H}}^{2}}{\varepsilon_0 \varepsilon_{\mathrm{sc}} e_0 N_{\mathrm{D}}}(V - V_{\mathrm{fb}})\right\}. \tag{103}$$

It follows that a Mott–Schottky plot of C_{s}^{-2} vs. V will still be linear, with the correct slope, but the intercept is no longer V_{fb} but $V' = V_{\mathrm{fb}} - e_0 \varepsilon_{\mathrm{sc}} \varepsilon_0 N_0 / 2C_{\mathrm{H}}^2$ [68]. As an example, for $\varepsilon_{\mathrm{sc}} N_{\mathrm{D}} \simeq 10^{21}\ \mathrm{cm}^{-3}$ and $C_{\mathrm{H}} \simeq 10\,\mu\mathrm{F\,cm}^{-2}$, the intercept according to the above analysis is shifted by ca. 70 mV. As Uosaki and Kita show by direct numerical calculation [67], for extreme values of $\varepsilon_{\mathrm{sc}} N_{\mathrm{D}}$ the Mott–Schottky plots are actually curved near flat-band.

(*b*) *Presence of surface states*

The main effect of the filling and emptying of surface states is that the current at the surface is no longer purely displacive, but contains a contribution from the particle flux across the surface. If we suppose that there are surface states distributed with a distribution $g(e_{\mathrm{t}}')$, where e_{t}' is the energy of the surface state in eV relative to the Fermi level at flat-band. The total number of surface states is given by $N_{\mathrm{t}}(e_{\mathrm{t}}') = N_{\mathrm{t}} g(e_{\mathrm{t}}')$ at energy e_{t}'. The function $g(e_{\mathrm{t}}')$ is normalised such that $\int_{E_{\mathrm{c}}}^{E_{\mathrm{v}}} g(e_{\mathrm{t}}')\mathrm{d}e_{\mathrm{t}}' = 1$ and the occupancy (i.e. the number of levels actually occupied by electrons) is given by $n_{\mathrm{t}}(e_{\mathrm{t}}')$. We then have, for any energy e_{t}'

$$\frac{\partial n_t(e_t')}{\partial t} = k_f(e_t')n_s[N_t(e_t') - n_t(e_t')] - k_b(e_t')n_t(e_t') \tag{104}$$

where n_s is the electron concentration at the surface, $k_f(e_t')$ and $k_b(e_t')$ are the forward and back rate constants in units of $cm^3\ s^{-1}$ and s^{-1}, respectively, and the total particle flux is

$$\int_{E_c}^{E_V} \frac{\partial n(e_t')}{\partial t} g(e_t')de_t' \tag{105}$$

Individual balance at each energy e_t' gives

$$\frac{n_t(e_t')}{N_t(e_t') - n_t(e_t')} = \exp\left\{(-e_t' + \phi_{0s})\frac{e_0}{kT}\right\} \tag{106}$$

and

$$k_f(e_t') = \frac{k_b(e_t')}{N_D}\exp\left(-\frac{e_0 e_t'}{kT}\right) \tag{107}$$

where the e_t' is taken as negative if the state lies *below* E_F and ϕ_{0s} is the potential and is negative for the reverse-biased n-type semiconductor considered

The flux balance is then

$$\frac{\partial n(e_t')}{\partial t} = k_b(e_t')\left\{\frac{n_s[N_t(e_t') - n_t(e_t')]\exp(-e_0 e_t'/kT)}{N_D} - n_t(e_t')\right\} \tag{108}$$

and the total flux at the surface

$$j_1 = i\omega\varepsilon_{sc}\varepsilon_0\mathscr{E}_{1s} + i\omega e_0\int_{E_c}^{E_V} n_{1t}(e_t')g(e_t')\,de_t' \tag{109}$$

Using $n_t = n_{0t} + n_{1t}\exp(i\mu t)$, we have

$$n_{1t}(e_t') = \frac{n_{1s}k_b(e_t')N_t(e_t')}{f_V(e_t')N_D(f_r(e_t') + i\omega)} \tag{110}$$

where

$$\left.\begin{aligned} f_V &= \exp\left(\frac{e_0 e_t'}{kT}\right) + \exp\left(\frac{e_0\phi_{0s}}{kT}\right) \\ f_r &= k_b(e_t')f_V\exp\left(-\frac{e_0 e_t'}{kT}\right) \end{aligned}\right\} \tag{111}$$

In order to use these expressions, we must find a better approximation to the two integrals in eqns. (87) and (90) and we have, after some algebra

$$\exp\left(\frac{e_0\phi_{0s}}{kT}\right)\int\limits_{L_s}^{0}\mathscr{E}_1\exp\left(-\frac{e_0\phi_0}{kT}\right)\mathrm{d}x \approx -\frac{\mathscr{E}_{1s}L_D^2}{W} \tag{112}$$

$$\exp\left(\frac{e_0\phi_{0s}}{kT}\right)\int\limits_{L_s}^{0}\exp\left(-\frac{e_0\phi_0}{kT}\right)\mathrm{d}x \approx -\frac{n_{0s}(L_s - W)}{N_D} - \frac{L_D^2}{W} \tag{113}$$

whence, from eqns. (87) and (90)

$$\frac{n_{1s}e_0DW}{L_D^2} - K_3\phi_{1s} = i\omega\varepsilon_0\varepsilon_{sc}\mathscr{E}_{1s} - j_1 - K_2j_1 \tag{114}$$

where

$$\left.\begin{aligned} K_2 &= \frac{n_{0s}W(L_s - W)}{N_D L_D^2} \\ K_3 &= \frac{e_0^2DWn_{0s}}{kTL_D^2} \end{aligned}\right| \tag{115}$$

and from eqn. (109)

$$i\omega\varepsilon_0\varepsilon_{sc}\mathscr{E}_{1s} - j_1 = -i\omega e_0\int\limits_{E_c}^{E_V} n_{1t}(e_t')g(e_t')\,\mathrm{d}e_t' \tag{116}$$

$$= -\frac{i\omega e_0 n_{1s}N_t}{N_D}\int\limits_{E_c}^{E_V}\frac{k_b(e_t')g(e_t')\,\mathrm{d}e_t'}{f_V(e_t')[f_r(e_t') + i\omega]} \tag{117}$$

Combining this with eqn. (87) gives

$$j_1\left(1 + i\omega A + \frac{i\omega K_2}{L + i\omega M}\right) = i\omega C_{sc}\phi_{1s} + \frac{i\omega K_3\phi_{1s}}{L + i\omega M} \tag{118}$$

where

$$L = \mathrm{Re}\left\{k_s\bigg/\int\frac{k_b(e_t')g(e_t')\,\mathrm{d}e_t'}{f_V(e_t')(f_r(e_t') + i\omega)} + i\omega\right\} \tag{119}$$

and

$$\omega M = \mathrm{Im}\left\{k_s\bigg/\int\frac{k_b(e_t')g(e_t')\,\mathrm{d}e_t'}{f_V(e_t')(f_r(e_t') + i\omega)} + i\omega\right\} \tag{120}$$

It can be shown, after some manipulation, that this corresponds, approximately, to an equivalent circuit of the form [65]

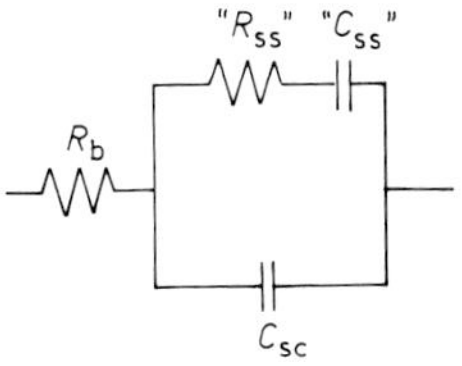

where $''C_{ss}'' \equiv K_3/L$ and $''R_{ss}'' \equiv M/K_3$. The associated time constant $\tau_{ss} \equiv C_{ss}R_{ss} = M/L$ and we define a rate constant $k_s = D_W W N_D (N_t L_D^2)$.

For a surface state distribution concentrated about an energy e_t^0, $g(e_t') \equiv \delta(e_t' - e_t^0)$ and we have

$$L = \frac{k_s f_V(e_t^0) f_r(e_t^0)}{k_b(e_t^0)} \qquad M = \frac{k_s f_V(e_t^0)}{k_b(e_t^0)} + 1 \tag{121}$$

$$\tau_{ss} = \frac{1}{f_r(e_t^0)} + \frac{k_b(e_t^0)}{k_s f_V(e_t^0) f_r(e_t^0)} \tag{122}$$

The form of this expression is typical for two rate processes in series; in fact, the first term reflects the kinetics of electron or hole transfer to a surface state whereas the second term, which reduces to

$$\frac{1}{k_s f_V(e_t^0) \exp(-e_0 e_t^0/kT)}$$

contains only rate constants characteristic of the bulk of the semiconductor and refers specifically to majority carrier transport within the semiconductor. It is unlikely that the second of these kinetic processes could be rate-limiting and the approximation is usually made of neglecting the second $i\omega$ term in eqns. (119) and (120). Under these circumstances

$$L \approx \mathrm{Re}\left\{k_s \Bigg/ \int \frac{k_b(e_t')g(e_t')\,\mathrm{d}e_t'}{f_V(e_t')[f_r(e_t') + i\omega]}\right\} \tag{123}$$

$$\omega M \approx \mathrm{Im}\left\{k_s \Bigg/ \int \frac{k_b(e_t')g(e_t')\,\mathrm{d}e_t'}{f_V(e_t')(f_r(e_t') + i\omega)}\right\} \tag{124}$$

so, finally

$$Y = \frac{i\omega K_3}{k_s} \int \frac{k_b(e_t')g(e_t')\,\mathrm{d}e_t'}{f_V(e_t')[f_r(e_t') + i\omega]} = \frac{i\omega e_0^2 n_{0s} N_t}{kTN_D} \int \frac{k_b(e_t')g(e_t')\,\mathrm{d}e_t'}{f_V(e_t')[f_r(e_t') + i\omega]} \tag{125}$$

The precise form of this expression again depends very much on the form of $g(e_t')$ and the other kinetic variables. We have, from the above expressions for f_V and f_r

$$f_{\mathrm{V}} = \exp\left(\frac{e_0 e_t'}{kT}\right) + \exp\left(\frac{e_0 \phi_{0s}}{kT}\right) \tag{126}$$

$$f_{\mathrm{r}} = k_{\mathrm{b}}(e_t') \exp\left(-\frac{e_0 e_t'}{kT}\right) \cdot f_{\mathrm{V}} \tag{127}$$

For $|e_t'| < |\phi_{0s}| \, f_{\mathrm{V}} \simeq \exp(e_0 e_t'/kT)$

$$f_{\mathrm{r}} \simeq k_{\mathrm{b}} \tag{128}$$

and for $|e_t'| < |\phi_{0s}| \, f_{\mathrm{V}} \simeq \exp(e_0 \phi_{0s}/kT)$

$$f_{\mathrm{r}} \simeq k_{\mathrm{b}} \exp\{-e_0(e_t' - \phi_{0s})/kT\} \tag{129}$$

The simplest possible model, other than the δ function discussed above, is found in the case that k_{f} is a constant and $g(e_t')$ a slowly varying function of e_t'; i.e. there is a relatively uniform distribution of surface states. We may then write

$$k_{\mathrm{b}} = K' \exp(e_t' e_0/kT) \tag{130}$$

where K' is a constant. Then

$$\begin{aligned}\int_{E_c}^{E_v} \frac{k_{\mathrm{b}}(e_t') g(e_t') \, de_t'}{f_{\mathrm{V}}(e_t')[f_{\mathrm{r}}(e_t') + i\omega]} &= \int_{E_c}^{\phi_{0s}} \frac{g(e_t') K' \exp(e_0 e_t'/kT) \, de_t'}{\exp(e_0 e_t'/kT)[k_{\mathrm{b}}(e_t') \exp(e_0 e_t'/kT) + i\omega]} \\ &\quad + \int_{\phi_{0s}}^{E_v} \frac{g(e_t') K' \exp(e_0 e_t'/kT) \, de_t'}{\exp(e_0 \phi_{0s}/kT)[\exp(e_0 \phi_{0s}/kT) + i\omega]} \\ &\approx K' \int_{E_c}^{\phi_{0s}} \frac{g(e_t') \, de_t'}{(k_{\mathrm{b}} \exp(e_0 e_t'/kT) + i\omega)} \\ &\quad + \frac{K' \exp(-e_0 \phi_{0s}/kT)}{(\exp(e_0 \phi_{0s}/kT) + i\omega)} \\ &\quad \times \int_{\phi_{0s}}^{E_c} g(e_t') \exp(e_0 e_t'/kT) \, de_t'\end{aligned} \tag{131}$$

These integrals will have contributions chiefly from regions where $e_t' \simeq \phi_{0s}$ and so the simple equivalent sub-circuit

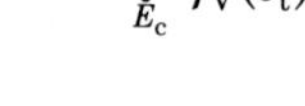

is an adequate description for the surface state network, where $C_{\mathrm{ss}} = e_0^2 N_t / kT$.

Another common situation that may arise is that of an exponential distribution of surface states: $g(e_t') \equiv C' \exp\{(1 - \beta) e_t' e_0/kT\}$. If we also assume that $k_{\mathrm{b}} \equiv K'' \exp(\gamma e_t' e_0/kT)$ then

$$\int_{E_c}^{E_v} \frac{k_b(e_t')g(e_t')\,de_t'}{f_v(e_t')(f_r(e_t') + i\omega)}$$

$$\approx \int_{E_c}^{\phi_{0s}} \frac{C'K'' \exp\{(\gamma + 1 - \beta)e_0 e_t'/kT\}\,de_t'}{\exp(e_0 e_t'/kT)\{K'' \exp(\gamma e_t' e_0/kT) + i\omega\}}$$

$$+ \int_{\phi_{0s}}^{E_c} \frac{C'K'' \exp\{(\gamma + 1 - \beta)e_0 e_t'/kT\}\,de_t'}{\exp(e_0\phi_{0s}/kT)[K'' \exp\{[(\gamma - 1)e_0 e_t'/kT] + [e_0\phi_{0s}/kT]\} + i\omega]} \tag{131}$$

If $|\phi_{0s}|$ is relatively large, we can assume that all the contribution comes from the first of the two integrals; replacing $e_0 e_t'$ by $-E$, and putting $E_c = 0$, we have

$$\int_{E_c}^{\phi_{0s}} \frac{C'K'' \exp\{(\gamma - \beta)e_0 e_t'/kT\}\,de_t'}{[K'' \exp(\gamma e_t' e_0/kT) + i\omega]} \approx \int_0^{\infty} \frac{C'' \exp(\beta E/kT)\,dE}{(1 + i\omega\tau)};$$

$$C'' = C'/e_0, \qquad \tau = (K'')^{-1} \exp\left(\frac{\gamma E}{kT}\right) \tag{133}$$

This is in essence, the formula used by McCann and Badwal [69]. The significance of β and γ must be sought at the molecular level. Some mathematical manipulation now gives, under the circumstances that $\omega K''^{-1} \to 0$

$$Y = \frac{e_0^2 n_{0s} N_t}{kTN_D}\left\{\frac{C''kT\pi(K'')^{\beta/\gamma}}{2\gamma \sin(s\pi)}\omega^n + \frac{iC''kT\pi(K'')^{\beta/\gamma}}{2\gamma \sin(r\pi)}\omega^n\right\} \tag{134}$$

where

$$s = \frac{\gamma - (\beta/2)}{\gamma}; \qquad n = 1 - \frac{\beta}{\gamma}; \qquad r = \frac{\gamma - \beta}{2\gamma} \tag{135}$$

There are, then, at least two possible types of behaviour and considerable care is necessary in analysing the increasingly complex equivalent circuits that result. In practice, provided that little faradaic current flows, it is usually found that a circuit of the form

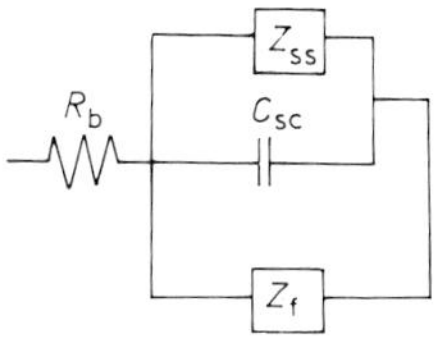

will effectively model many semiconductor–electrolyte interfaces in aqueous solution [70]; Z_f represents the faradaic impedance. For frequencies above a few Hz, it is reasonable to model Z_f by a pure resistor; its exact value

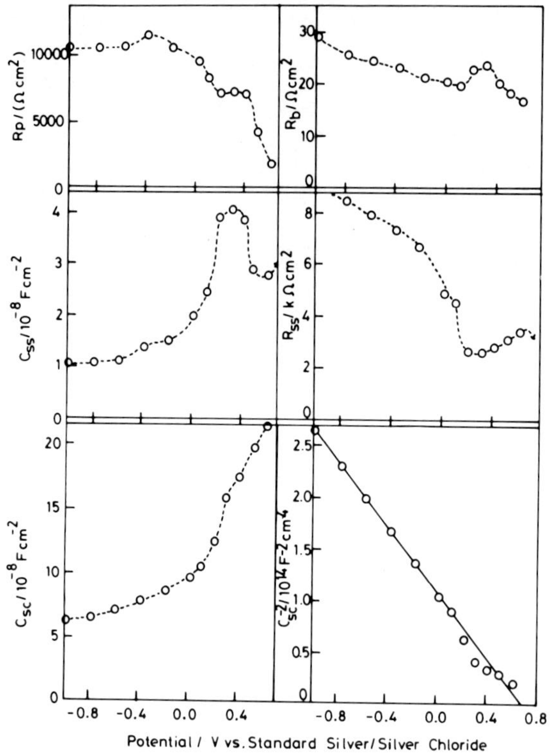

Fig. 28. Data of Fig. 24 analysed according to the five-component circuit above.

will be discussed below [71]. The value of C_H is also, as indicated above, very large and a five-component circuit of considerable general utility is

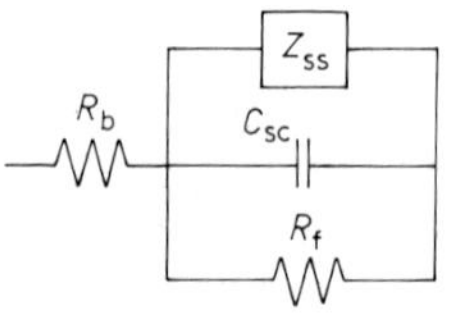

As an example, this circuit with Z_{ss} equivalent to

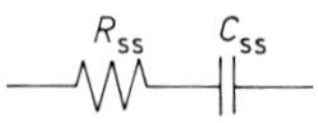

was used to analyse the a.c. data for p-GaP in the frequency range 40–10 000 Hz in aqueous solution. The results are shown in Fig. 28 and it can be seen that the capacitance C_{ms} derived from fitting the a.c. data still obeys the Mott–Schottky relationship; furthermore, the donor density, derived from the slope of the MS plot, is in reasonable agreement with that determined from the Hall measurement. The a.c. data in Fig. 28 were fitted using a least squares fit and 8–12 individual frequencies; the difficulties and dangers of this approach are well-documented in the literature and the

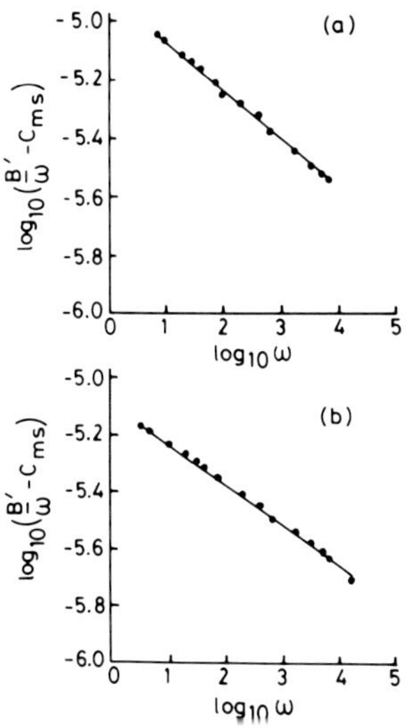

Fig. 29. Plot of $\log_{10}(B'/\omega - C_{sc})$ vs. $\log_{10}\omega$ for n-Fe_2O_3(1 m/o TiO_2) in 1 M NaOH at (a) −0.3 V and (b) 0.0 V vs. SCE where B' is the susceptance of the parallel Z_{ss}/C_{sc} circuit and Z_f is assumed to be very large. $(B'/\omega - C_{sc})$ represents the capacitance equivalent of the surface-state impedance Z_{ss}.

utmost care must be taken to ensure that the algorithms used do indeed find the true minimum [72].

An example of the type of behaviour encountered for exponential surface-state distributions is provided by n-Fe_2O_3 in 1 M NaOH [69]. The equivalent conductance and susceptance of the circuit comprising Z_{ss} in parallel with C_{sc} clearly show a power law dependence on ω as shown in Fig. 29. The C_{sc} values obtained from this model again obeyed the Mott–Schottky relationship, although the donor density of 8×10^{18} cm^{-3} and dielectric constant of 25 suggest that the true flat-band potential may lie rather positive of the value given.

Power law behaviour has also been observed by Dutoit et al. [73] and ascribed to more general relaxation processes within a narrow layer at the surface of the semiconductor. It is, of course, not possible to distinguish by a.c. techniques alone the model put forward by Dutoit et al. [73] and that described above since the mathematical development is the same and the differences may, in any case, be largely semantic. Nevertheless, Dutoit et al.'s analysis is of considerable interest. An equivalent circuit of the form

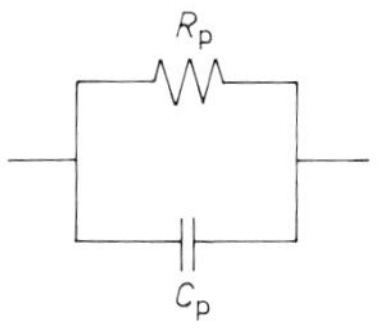

is used; comparing this with the circuit above, and neglecting R_b, we have

$$\left.\begin{aligned} C_p &= C_{sc} + K'\omega^{n-1} \\ R_p^{-1} &= K''\omega^n \end{aligned}\right| \tag{136}$$

If $\beta/\gamma \to 1$, then $R_p^{-1} \simeq \omega$ and C_p becomes nearly frequency-independent. This would correspond to the uniform distribution case discussed above. The series circuit

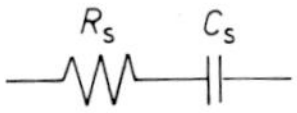

is related by $R_s \simeq R_p^{-1}C_p^{-2}\omega^{-2}$ provided, as is usually true, $(\omega R_p C_p)^2 \gg 1$. Hence, $R_s \simeq \omega^{-1}$, a result confirmed by Fig. 30.

if we examine the frequency dependence of C_p more closely, we have

$$C_s \approx C_p = (C_{sc} + K'\omega^{-\delta}) \tag{137}$$

where $\delta = -(n - 1)$.

$$\begin{aligned} C_s^{-1} &\approx (C_{sc} + K'\omega^{-\delta})^{-1} \\ &\approx (C_{sc} + K' - \delta \ln \omega)^{-1} \\ &\approx (C_{sc} + K')^{-1}\left[1 + \frac{\delta}{C_{sc} + K'} \ln \omega\right] \end{aligned} \tag{138}$$

A plot of C_s^{-1} vs. $\ln \omega$ should, therefore, also be linear if $R_s \simeq \omega^{-1}$ and this is borne out by the data for CdSe shown in Fig. 31 [73].

Closely related to the above analysis, though possibly arising from a fundamentally different physical origin, is the use by 't Lam et al. of a "constant phase angle" (cpa) element to replace Z_{ss} [74, 75]. In fact, "Z_{ss}" was represented by the circuit

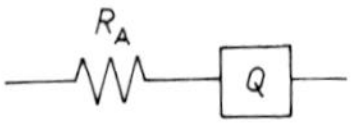

where $Q \equiv k_j(i\omega)^{-\alpha_j}$. For $\alpha_j = 1$, Q reduces to a capacitor and for $\alpha_j = 0$, to a resistor. Even more complex circuits were used by these authors to model surface roughness.

(c) *Presence of deep traps*

Hitherto, we have considered the semiconductor to have a single set of traps that are shallow and essentially completely ionised at room temperature. For such familiar semiconductors as p-GaP and n-Si, this is a very good first approximation, but for many oxide semiconductors, where an oxygen vacancy may be doubly occupied and the second ionisation relatively difficult, we must consider the possibility that ionisation is not complete.

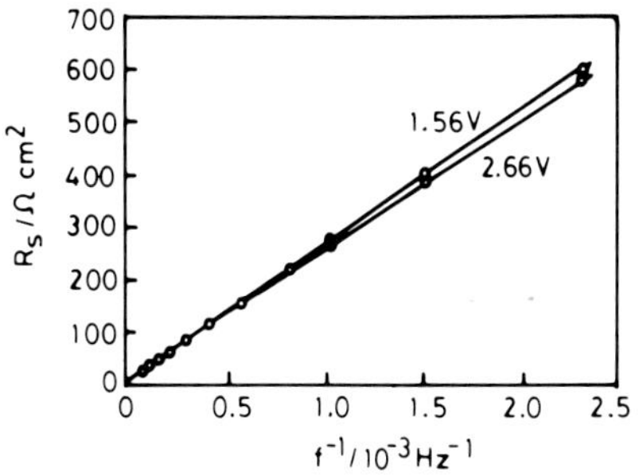

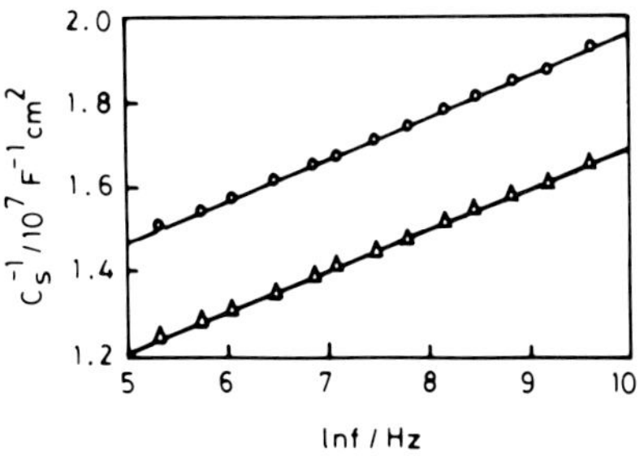

Fig. 30. Plot of series resistance R_s vs. inverse frequency f^{-1} for n-CdSe(000$\bar{1}$), $N_D \simeq 10^{15}$ cm^{-3} in 0.25 M K_2SO_4–0.05 M CH_3COOH–0.05 M NaOAc at the two frequencies given (vs. SCE).

Fig. 31. Plot of inverse series capacitance C_s^{-1} vs. ln(f) for the n-CdSe electrode of Fig. 30.

The complexity afforded by these traps arises from the fact that the frequency range normally used in a.c. studies, 1–10^5 Hz, is likely to contain the relaxation frequency appropriate to such trap sites. To explore this concept, we may consider Fig. 32 [76]. We identify a deep trap level of energy E_D relative to the conduction band, lying below the Fermi level in the bulk of the semicondutor and therefore ionised only up to a distance l into the depletion layer. For $x > l$, only the shallow traps are ionised at equilibrium. Within the spirit of the Mott–Schottky approximation, we obtain (provided $V \geqslant V_0$)

$$V_l = \frac{e_0 N^S}{2\varepsilon_0 \varepsilon_{sc}} [l^2 + 2l(W - l)] + \frac{e_0 N^D}{2\varepsilon_0 \varepsilon_{sc}} l^2 \tag{139}$$

and

$$V - V_l = \frac{e_0 N^S}{2\varepsilon_0 \varepsilon_{sc}} (W - l)^2 \tag{140}$$

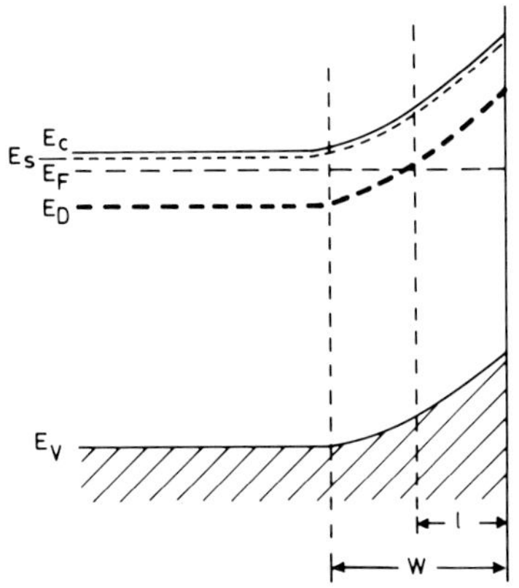

Fig. 32. Energy–potential diagram for a semiconductor with deep traps of energy, E_D, and shallow traps of energy, E_s. The distances l and W are defined in eqns. (139) and (140).

where V_l is the potential dropped in the region $0 < x < l$, W is the full depletion layer width, N^S is the density of shallow traps, N^D the density of deep traps and V the total potential dropped.

$$V = \frac{e_0 N^S}{2\varepsilon_0 \varepsilon_{sc}} W^2 + \frac{e_0 N_D}{2\varepsilon_0 \varepsilon_{sc}} l^2 \qquad (141)$$

If the charge associated with the deep traps is Q_D, then we may expect a relaxation behaviour

$$\frac{\partial Q_D}{\partial t} = -\frac{(Q_D - Q_D^0)}{\tau} \qquad (142)$$

By using small-signal theory, the equivalent circuit now becomes

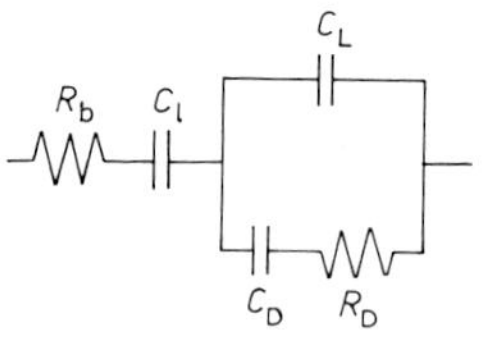

where

$$C_l = \left[\frac{e_0 \varepsilon_{sc} \varepsilon_0 (N^S + N^D)}{2V_l}\right]^{1/2}; \quad C_D = \left[\frac{e_0 \varepsilon_0 \varepsilon_{sc} N^S}{2(V - V_l)}\right]^{1/2}; \quad C_D = \alpha C_L \qquad (143)$$

where $\alpha = N^D/N^S$.

From the geometry of Fig. 32, we have

$$e_0(V - V_l) = E_D - \delta \qquad (144)$$

where δ is the energy difference between E_c and E_F in the bulk of the semiconductor. Evidently, unless $V > V_0 \equiv (E_D - \delta)/e_0$, we may treat the system without invoking deep trap ionisation. Under these circumstances, the equivalent circuit is just

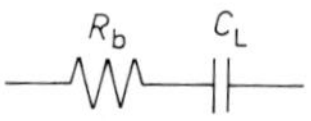

The presence of deep-lying traps gives rise to a characteristic kink in the Mott–Schottky plots, especially at higher frequencies. The results of Fig. 33 show this effect and need some further discussion. Firstly, it will be seen that, save at very low frequency the slopes of the lines do not reflect the donor density. Second, the dispersion depends strongly on α; the larger α, the larger the dispersion. Thirdly, the inflection point is not related to the trap depth. These results are highly intriguing and suggest that great care must be taken with non-linear Mott–Schottky plots. In the presence of surface

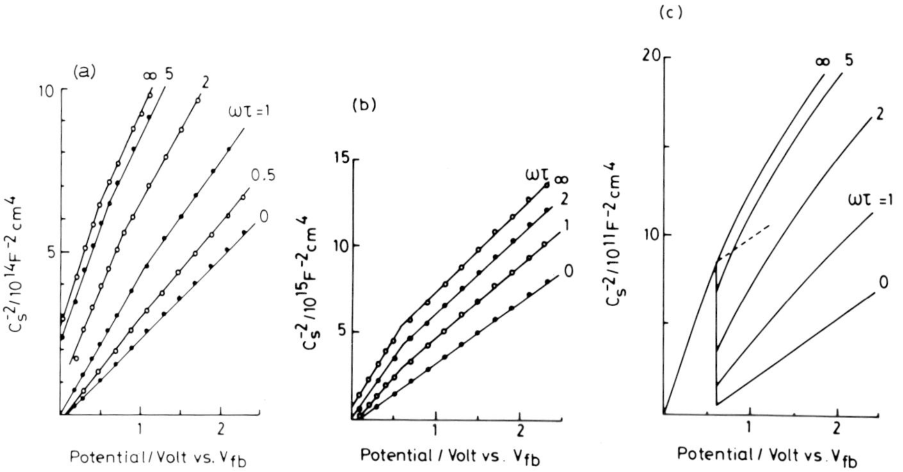

Fig. 33. Calculated response of the inverse-square capacitance, C_s^{-2}, for the case when deep donor levels are ionised. The frequencies are given by $\omega\tau$ where τ is defined in eqn. (142). (a) $N^s = 10^{15}\,cm^{-3}$, $\alpha = 10$, $\varepsilon_{sc} = 50$, $E_D = 0.3\,eV$, $\delta = 0.2\,eV$; (b) as for (a) except that $\varepsilon_{sc} = 10$, $\alpha = 3$; (c) $N^s = 10^{18}\,cm^{-3}$, $\varepsilon_{sc} = 100$, $\alpha = 3$, $E_D = 0.8\,eV$, $\delta = 0.2\,eV$.

states, the system may still be modelled provided C_{sc} is replaced by the equivalent circuit shown above. An estimate of τ can be obtained from

$$\tau \simeq \tau_0 \exp(E_0/kT) \tag{145}$$

where $\tau_0 \simeq 1/(v_{th}\sigma_{th}N_c)$ in which v_{th} is the thermal velocity of the electron, σ_{th} the capture cross-section, and N_c the effective density of states in the conduction band. Values of τ_0 in the range 10^{-8} to 10^{-12} s may be anticipated and, for $E_D = 0.5\,eV$, τ lies in the range 0.5 ms to 5 s. For much deeper traps, the kinetics are correspondingly slower; thus, for $E_D = 0.8\,eV$, τ will lie above 80 s and the effect of studying the a.c. response in the normally used frequency range will be to give a Mott–Schottky plot with an ill-defined curvature that may prove difficult to interpret, as shown in Fig. 33(c), where the broken line shows the expected total donor density.

Experimentally, both Fe_2O_3 and TiO_2 frequently exhibit non-linear Mott–Schottky plots. In the case of Fe_2O_3, the existence of deep-lying traps seems well-established but for TiO_2, the interpretation is complicated by possible inhomogeneities in the surface dopant concentration. Very strong oxidising etches are usually employed to prepare the surface of TiO_2 single crystals and these etches may lead to a smaller concentration of donor centres in the surface region. If we may approximate the oxidised region as a uniform layer of thickness l, in which the donor concentration is N^D_{surf} and the remainder of the semiconductor has a donor concentration N^D_{bulk}, and we assume that there are no deep traps or surface states, then the capacitive response may be calculated in the same manner as above, with the proviso that $\tau \rightarrow 0$ and $\alpha \equiv (N^D_{bulk}/N^D_{surf}) - 1$. For $V < V_0$, where $V_0 \equiv e_0 N^D_{surf} l^2/2\varepsilon_{sc}\varepsilon_0$, the slope of the Mott–Schottky plot reflects N^{D}_{surf}; above V_0, it reflects N^{D}_{bulk} and there is a kink.

It can be seen that, both for the case of inhomogeneous doping and for the case of deep traps, the intercept of the Mott–Schottky plots at high reverse bias do not yield the flat-band potential or any simple function thereof.

(d) *Fermi-level pinning*

At higher densities of surface states, it may be expected that the emptying and filling of surface states will cause a significant change in the potential within the depletion layer. Provided the dominant kinetics are those between surface state and semiconductor interior, we may then analyse the situation as a case II Fermi-level pinning problem. The total potential dropped in the interfacial region

$$\phi_\tau = \phi_s + \phi_{ss} \tag{146}$$

where ϕ_{ss} is the potential dropped across the Helmholtz region by virtue of the surface state electron occupancy. If C_H is the Helmholtz capacitance, then

$$\phi_{ss} \cong -\frac{e_0(N_t - n_t)}{C_H} \tag{147}$$

where n_t is the electron concentration in the surface state and we suppose that the surface states are neutral when occupied; as before, the zero of potential is taken in the bulk of the semiconductor. Differentiating with respect to time and using small-signal a.c. theory

$$\phi_{ss} = \phi_{0ss} + \phi_{1ss}\exp(i\omega t) \tag{148}$$

$$\phi_{1ss} = \frac{e_0 n_{1t}}{C_H} \tag{149}$$

We have, from the kinetic considerations given above, assuming a single energy surface state

$$n_{1s} = n_{1t}Q \tag{150}$$

where

$$Q = \frac{(f_r + i\omega)f_v N_D}{k_b N_t} \tag{151}$$

and also

$$j_1 = i\omega\varepsilon_0\varepsilon_{sc}\mathscr{E}_{1s} + i\omega e_0 n_{1t} \tag{152}$$

and

$$e_0 n_{1t} = \frac{(-j_1 R_b/R_p) + (\phi_{1s}/R_p)}{(QDW/L_D^2) + i\omega} \tag{153}$$

where $R_b = L_s/e_0\mu N_D$ and $R_p = W/e_0 n_{0s}\mu$.

From eqns. (87) and (90) writing

$$K = i\omega + \frac{QDW}{L_D^2}$$

$$j_1\left(1 + \frac{i\omega R_b}{KR_p} + i\omega C_{sc} R_b\right) = \phi_{1s}\left(i\omega C_{sc} + \frac{i\omega}{KR_p}\right) \tag{154}$$

Now

$$\begin{aligned} \phi_{1\tau} &= \phi_{1s} + \phi_{1ss} = \phi_{1s} + \frac{(\phi_{1s}/R_p) - (j_1(R_b/R_p))}{KC_H} \\ &= \phi_{1s}\left(1 + \frac{1}{KC_H R_p}\right) - j_1 \frac{R_b}{KC_H R_p} \\ &= j_1\left(1 + \frac{1}{KC_H R_p}\right) Z_s \quad j_1 \frac{R_b}{KC_H R_p} \end{aligned} \tag{155}$$

so

$$\begin{aligned} Z_\tau &= Z_s + \frac{1}{KC_H R_p}(Z_s - R_b) \\ &= Z_s + \frac{C_{ss}}{C_H} \frac{1}{(1 + i\omega C_{ss} R_{ss})}(Z_s - R_b) \\ &= R_b + Z_s'\left(1 + \frac{C_{ss}}{C_H} \frac{1}{1 + i\omega C_{ss} R_{ss}}\right) \end{aligned} \tag{156}$$

where Z_s' is the impedance of the five-component circuit (Z_s) less the bulk resistance term.

If C_{ss}/C_H is relatively small, or $\omega C_{ss} R_{ss} \gg 1$, this will approximate to our original formula, since either the effect of the surface states on the potential distribution is insignificant or the kinetics are too slow to allow the occupancy of the surface state to be significantly altered during a potential cycle. If, however, $\omega C_{ss} R_{ss} \ll 1$, the equivalent circuit will again resemble that discussed above, but the factor $[(C_{ss}/C_H) + 1]$ will premultiply the capacitive part. The effect will be to reduce the apparent admittance by this factor, which will, in turn, reduce the apparent capacitance by $[1 + (C_{ss}/C_H)]$ and increase the apparent resistances by the same factor.

One point that should be emphasised here is that the a.c. data will only sample surface states for which $\omega R_{ss} C_{ss} \simeq 1$. For $\omega R_{ss} C_{ss} \ll 1$, the system reacts too rapidly for the a.c. probe and for $\omega R_{ss} C_{ss} \gg 1$, the system will be apparently classical save that the apparent value of C_{sc} will reflect the d.c. voltage drop across the depletion layer. If this d.c. potential is determined by slow surface states, then the value deduced for C_{sc} will be quite different from that expected. To give an example, if the surface is pinned by surface states that are either in slow equilibrium with the interior of the semiconductor or for which the dominant kinetics are with the semiconductor rather than any redox couple in solution, then only a small fraction of the change in the d.c. potential will appear across the depletion layer. If the material is pinned in this way close to flat-band potential, then C_{sc} will be much larger than expected; if, in addition, the fraction of the d.c. potential dropped across the depletion layer is a constant fraction of the total potential drop, C_{sc}^{-2} would still, apparently, obey the Mott–Schottky relationship, but with too small a slope.

It is clear, therefore, that knowing C_{sc} and having an *independent* measure of the donor density, we might hope to calculate the degree to which a semiconductor is pinned under d.c. conditions as long as some knowledge of the surface state kinetics is available. Unfortunately, the precision with which the donor density is known may be relatively poor and, more problematically, the determination of C_{sc} from a circuit analysis may prove very difficult, especially if C_{ss} is large. An examination of the equivalent circuit of section (b) shows that, if C_{ss} is large and R_{ss} small, then, for any reasonable dopant density, C_{sc} will be effectively "shorted out" and, with the experimental precision normally available, it may prove impossible to determine it with any accuracy.

(e) *Surface films*

It is all but impossible to prepare any semiconductor electrode without some surface film being present. The III/V semiconductors, for example, will normally possess oxide films whose thickness will vary from less than 10 Å to more than 40 Å after exposure to air and similar observations have been reported for silicon [77]. Although the capacitance of these films will normally be considerably larger than that of the depletion layer, the film may affect the a.c. response both by virtue of the analysis leading to eqn. (72) and, if C_{ss} becomes sufficiently large, that the impedance of the depletion layer falls to a value comparable with that of the film. If the film has a finite resistivity, which may be ionic in character, then the equivalent circuit takes the form

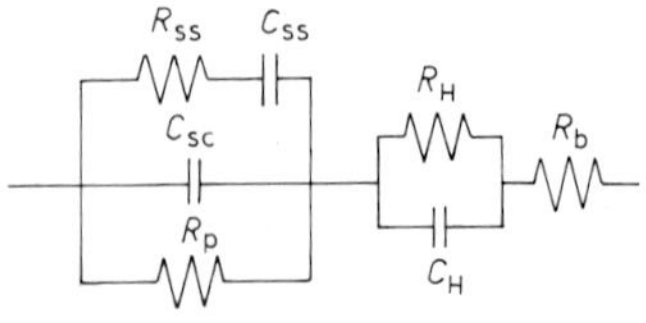

This is frequently found to be a good description of III/V semiconductor electrodes in non-aqueous solutions.

If a thicker layer of oxide or other quasi-insulating film forms on the surface [78], then a more extended circuit can be devised to take into account the change in surface charge on the film due, for example, to proton movement

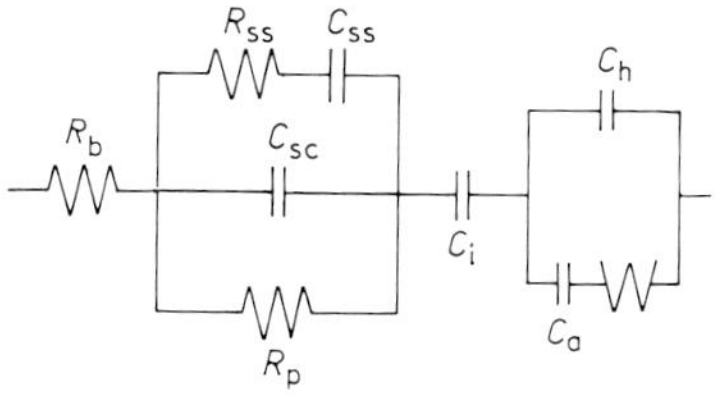

where C_h is the oxide–electrolyte double-layer capacitance and the element

can be calculated from the relationships (58) and (59) for an amphoteric oxide surface for which there are two pK values, pK_+ and pK_- and a surface charge $\sigma_0 = e_0 N_s(\theta_+ - \theta_-)$ [78]

$$2.303\,(\mathrm{pH_{pzzc}} - \mathrm{pH}) = \frac{e_0\phi_0}{kT} + \sinh^{-1}\left(\frac{\sigma_0}{\delta e_0 N_s}\right) \tag{157}$$

where $\delta = 2 \times 10^{-\Delta pK/2}$ and $\mathrm{pH_{pzzc}} = \frac{1}{2}(pK_+ + pK_-)$. Assuming proton transfer is not rate-limiting, we have

$$C_a^{-1} = -\frac{d\phi_0}{d\sigma_0} = \frac{kT}{\delta e_0^2 N_s}\left[1 + \left(\frac{\sigma_0}{\delta e_0 N_s}\right)^2\right]^{-1/2} \tag{158}$$

and Z_w is a Warburg impedance

$$Z_w = \frac{kT}{e_0^2(2\omega)^{1/2}}\left[\frac{1 - i}{(C_{H^+} D_{H^+}^{1/2} + C_{OH^-} D_{OH^-}^{1/2})}\right] \tag{159}$$

At lower frequencies and reasonably high $[H^+]$ or $[OH^-]$ concentration Z_w is very small compared with $1/\omega C_a$ and the equivalent circuit reduces to

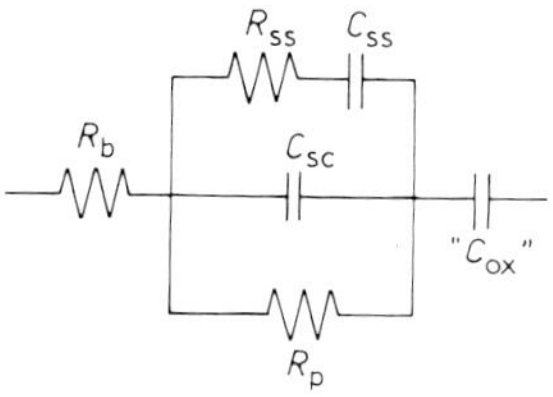

where "C_{ox}" is equivalent to

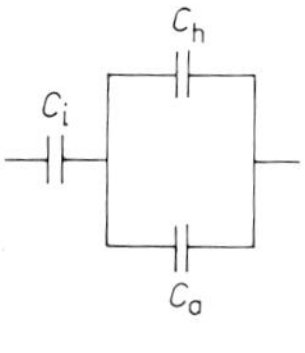

(f) Surface roughness

For rough surfaces, an approximate formula has been given as [79]

$$C_{sc}^{-2} = \frac{1}{\varepsilon_0 \varepsilon_{sc} e_0 N_D A_0^2}\left[2\left(1 - \frac{2\Delta A}{A_0}\right)(V - V_{fb}) + \frac{4\alpha}{A_0}(V - V_{fb})^2\right] \quad (160)$$

where the geometrical surface area is A_0 and the actual surface area $A \equiv A_0 + \Delta A$. The quantity $\alpha \equiv (\partial A/\partial V)_{V=V_{fb}}$ and the Mott–Schottky plot is predicted to curve upwards, rendering it extremely difficult to determine N_D.

Where the electrode is rough owing to inadequate etching after a mechanical polish, the equivalent circuit may become very complex. An example, for n-TiO_2 in 1 M NaOH, is [80]

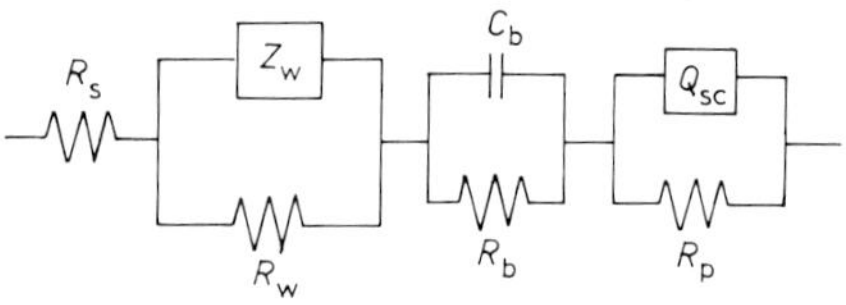

where Z_w is a Warburg impedance $\sigma(i\omega)^{-1/2}$ and Q_{sc} is a cpa element of the form $K(i\omega)^{-\alpha}$. Further improvement is obtained if Q_{sc} is replaced by

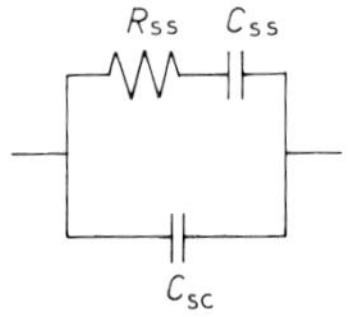

though difficulties were experienced with the latter circuit in determining R_{ss} and C_{ss} accurately. The values of C_{sc} determined for this equivalent circuit did seem to correlate well with other data on the semiconductors employed, but the interpretation of the Warburg and parallel $C_b R_b$ networks remains obscure.

4.4 INTRINSIC AND NARROW BANDGAP SEMICONDUCTORS

We have seen that, provided the Fermi level lies at least $3kT$ from either the valence or conduction bands, the depletion-layer capacitance will take the form [7]

$$C_{sc} = \frac{\varepsilon_0 \varepsilon_{sc}}{\sqrt{2} L_{Di}} \frac{|-\lambda \exp(-Y) + \lambda^{-1}\exp(Y) + (\lambda - \lambda^{-1})|}{[\lambda\{\exp(-Y) - 1\} + \lambda^{-1}\{\exp(Y) - 1)\} + (\lambda - \lambda^{-1})Y]} \quad (161)$$

where

$$L_{Di} = \left(\frac{\varepsilon_0 \varepsilon_{sc} kT}{e_0^2 n_i}\right)^{1/2}, \quad \lambda = \frac{p}{n_i}, \quad Y = \frac{e_0 \phi_{os}}{kT}$$

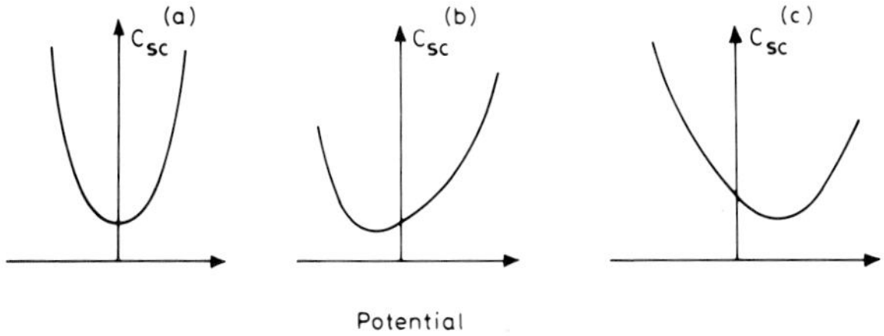

Fig. 34. Calculated capacitance for a narrow-band semiconductor with (a) *intrinsic* conductivity, (b) p-type *extrinsic* conductivity, and (c) n-type *extrinsic* conductivity.

and we recall that $Y < 0$ if the semiconductor is biased *positive* of the flat-band potential. The behaviour of this function for *intrinsic*, n-type *extrinsic* and p-type *extrinsic* is shown in Fig. 34 and some data are given for positively prepolarised Ge in Fig. 35 [81].

At flat-band potential

$$C_{sc} \longrightarrow \frac{\varepsilon_0 \varepsilon_{sc}}{L_{Di}} (\lambda + \lambda^{-1})^{-1/2}$$

and for an extrinsic n- or p-type semiconductor, $C_{sc} \rightarrow \varepsilon\varepsilon_0/L_D$.

For narrow bandgap materials, the Fermi level will approach the minority band edge during moderate to strong depletion; under these circumstances, the surface region may contain an excess of minority carrier and *inversion* is said to occur. There are, in fact, two extreme cases.

(a) The minority carrier concentration is always low owing to their rapid consumption at the surfce by fast faradaic processes. Under these circumstances, the space charge capacity is determined solely by the majority carriers and a Mott–Schottky relationship can be obtained. As an example, n-Ge shows no inversion [81]; holes are consumed by anodic dissolution and the Mott–Schottky relationship is found, even in regions where strong inversion might be expected.

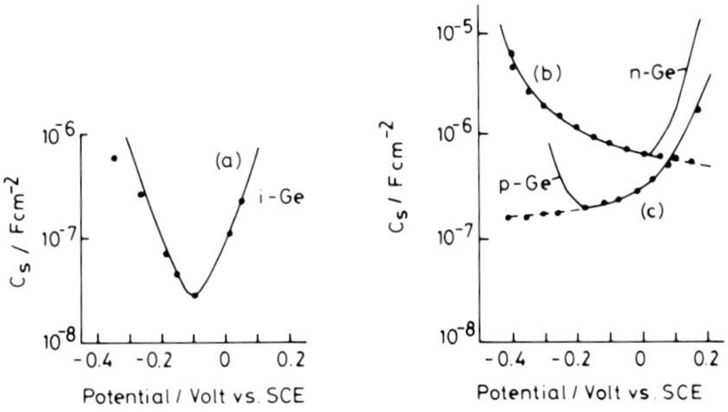

Fig. 35. Series capacitance data for the (111) faces of (a) near intrinsic (i-Ge) germanium, (b) n-Ge (0.01 Ω cm), and (c) p-Ge (0.1 Ω cm) in 0.1 M Na_2SO_4 after etching in CP4. The electrodes were all polarised anodically at ca. 100 μA cm^{-2} for a few minutes and the capacitance was measured immediately (within seconds) after polarising the electrode to the appropriate potential.

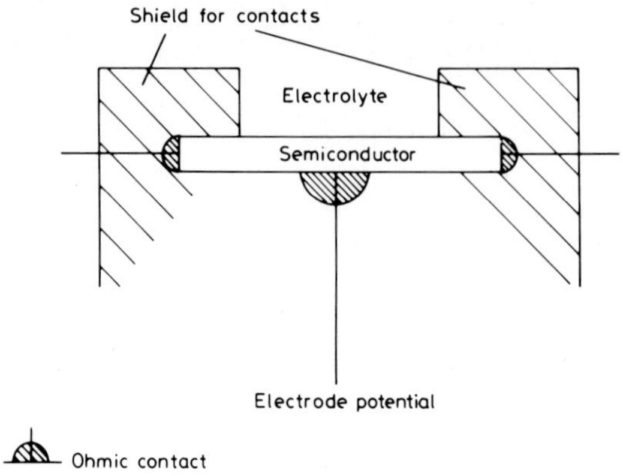

Fig. 36. Experimental arrangement for the study of semiconductor-surface conductivity changes during polarisation.

(b) The minority carriers are not extracted by faradaic processes and build up to constitute an inverted layer. The capacitance of the interface is dominated by these carriers at low frequency. At higher frequencies, however, the p/n junction that forms in the depletion layer acts as a kinetic barrier and only the minority carriers can give any conductivity [82]. The measured C_{sc} value then decreases and eventually becomes independent of potential.

4.5 SURFACE CONDUCTIVITY

In addition to the rather indirect a.c. techniques discussed above, a more direct technique for the electrical characterisation of the depletion region is available; even though this is, primarily a d.c. technique, it is included here since the information is frequently highly complementary. In this technique, ohmic contacts are made to the working surface of the semiconductor electrode and then shielded from the electrolyte as shown in Fig. 36; they are then used as probes to measure the surface conductivity. Ideally,

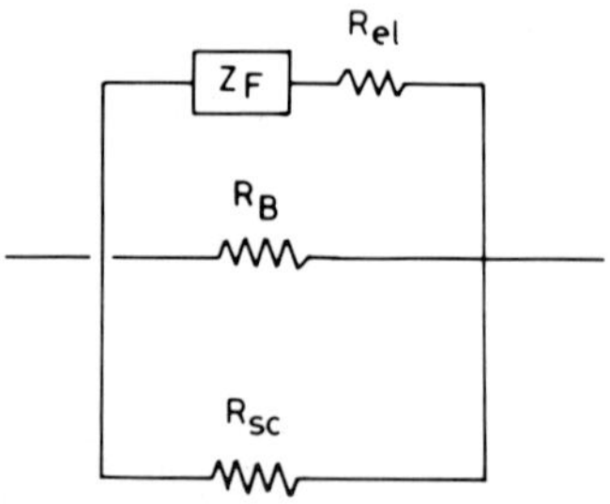

Fig. 37. Equivalent circuit for surface conductivity studies. Z_F = faradaic impedance, R_{el} = electrolyte resistance, R_B = bulk resistance of the semiconductor, and R_{sc} = surface resistance of semiconductor.

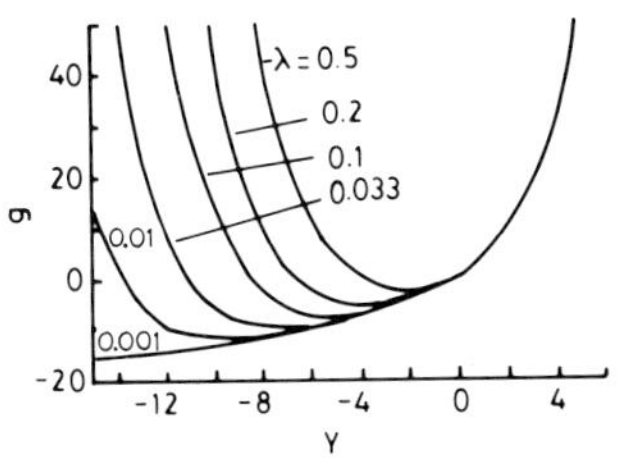

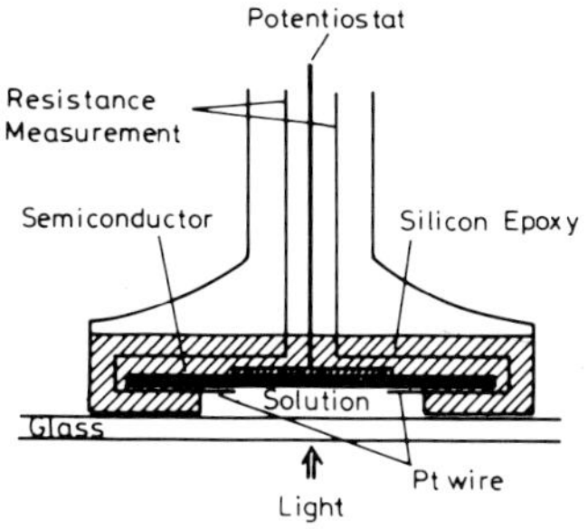

Fig. 38. Plot of eqn. (166) for an n-type semiconductor (note that $Y < 0$ corresponds to a depletion or exhaustion region).

Fig. 39. Electrode configuration for the measurement of the conductivity in a thin layer of electrolyte in contact with the electrode surface.

thin samples are used to optimise the ratio of surface-to-bulk conductivity, though samples cannot be too thin since large polarisation losses will then be incurred owing to high sample resistivity.

The equivalent circuit of Fig. 37 clearly demonstrates the main experimental difficulties encountered in determining R_{sc}; it is evident that only d.c. measurements are likely to prove practical else Z_F will be too small and the semiconductor will be shunted by R_{el} (which is likely to be very small). The bulk resistor R_B is only larger than R_{sc} for intrinsic semiconductors and it has proved difficult to extend the technique to extrinsic materials as R_{sc} becomes effectively shunted by R_B. Evidently, only R_{sc} and R_{el} vary with potential applied across the semiconductor between the back contact and the reference electrode in solution; however, the change in R_{el} is normally much smaller than R_{sc} as the mobility of the ions in solution is so much smaller than that of the carriers in the semiconductor.

The surface conductivity, σ_s, may be defined as [7]

$$\sigma_s = e_0 \mu_n \Gamma_n + e_0 \mu_p \Gamma_p \tag{162}$$

where Γ_n and Γ_P are the surface excesses of electrons and holes defined as

$$\Gamma_n = \int_0^\infty (n - n_B)\,dx \qquad \Gamma_p = \int_0^\infty (p - p_B)\,dx \tag{163}$$

By inserting the relevant expressions for n and p, together with the transformation

$$\Gamma_n = \int_0^\infty \frac{(n - n_B)}{(d\phi/dx)} \tag{164}$$

and using the expression (34) for $d\phi/dx$, we have

$$\sigma = Ag(Y) \tag{165}$$

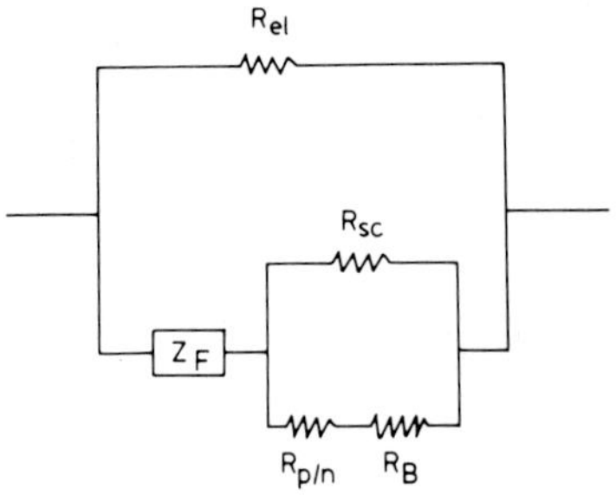

Fig. 40. Equivalent circuit appropriate to the geometry of Fig. 40. The resistance $R_{p/n}$ represents the resistance of the p/n junction formed under strong depletion or inversion.

where $Y = \xi(\phi_1 - \phi_i)$ and $\lambda = (p_B/n_B)^{1/2}$

$$\left.\begin{aligned} A &= \frac{1}{\sqrt{2}} e_0 \mu_p n_i L_D \lambda^{-1/2}, \\ g(Y) &= \lambda^{1/2} \int_Y^0 \frac{\lambda\{\exp(-Y) - 1\} + b\lambda^{-1}\{\exp(Y) - 1\}}{F(Y, \lambda)} dy, \\ b &= \mu_n/\mu_p \end{aligned}\right| \tag{166}$$

and n_i is the intrinsic concentration of carriers.

The behaviour of g is shown for an n-type semiconductor ($\lambda < 1$) in Fig. 38 [83]. Evidently there is a region immediately positive of the flat-band potential when $g < 0$ and R_{sc} will increase.

Serious problems arise experimentally in this technique for extrinsic samples, as has already been indicated. These problems arise because of the effective shunting of R_{sc} by R_B; even for the near-intrinsic samples of Fig. 38, there will be a resistive p–n junction formed at positive potentials between the n-type bulk and the inverted p-type region that is subsumed into R_{sc} in the schematic circuit of Fig. 37. For larger bandgaps, the traditional geometry cannot be used and an a.c. technique based on contacts in the *electrolyte* has been suggested [84]. The set-up is shown in Fig. 39 and the a.c. equivalent circuit is now shown in Fig. 40, assuming the faradaic impedance between the probes and the solution is negligible. Provided R_{el} is large (implying low electrolyte concentrations) and Z_F small (likely for frequencies in excess of 1 kHz), then R_{sc} may be measured.

5. Faradaic currents on semiconductors

Electron transfer at semiconductors can be contrasted with that at metal electrode surfaces. In the latter case, as shown in Fig. 41(a), electron transfer normally takes place to or from energy levels within a few kT of the Fermi

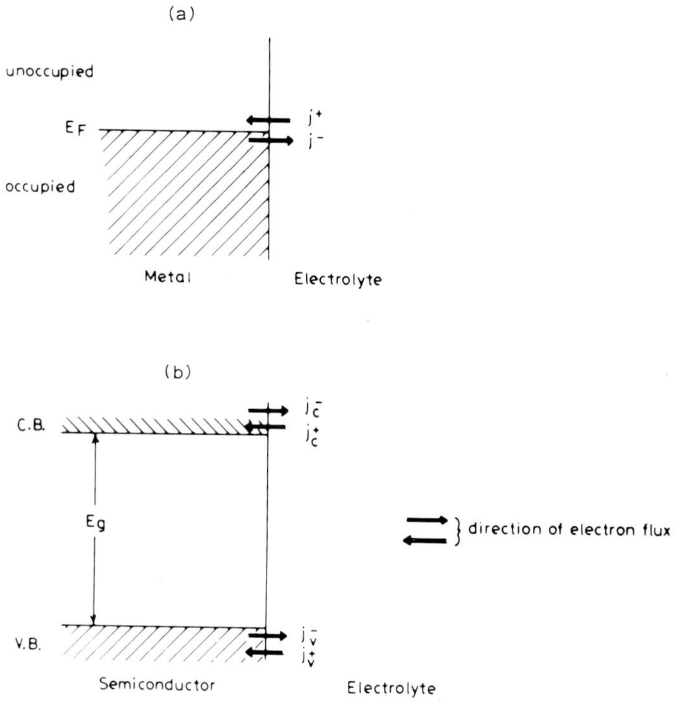

Fig. 41. (a) The major fluxes during faradaic processes on a metal surface. (b) The four components of the faradaic flux on a semiconductor.

level. However, for a non-degenerate semiconductor, the Fermi level will lie within the bandgap and the density of allowed electronic levels at the Fermi level itself will therefore be very low. It follows that electron transfer can only occur to or from levels within the conduction band or the valence band, as shown in Fig. 41(b).

A second distinction between metals and semiconductors is also found. For metals, the concentration of carriers is so high that the transport of these carriers to the surface is rarely, if ever, a limiting factor. The situation for semiconductors is quite different: even the concentration of majority carriers may be sufficiently low for transport to constitute a severe limitation on the maximum current attainable and, if the current is primarily due to minority carriers, the generation and transport of these will almost always be rate limiting.

One further distinction needs to be drawn, though this distinction also exists on a metal electrode surface: faradaic processes may involve both weak and strong interactions. The former would be typified by outer-sphere transfer from electrode to solution redox couple and theoretical progress in understanding this type of process has been considerable in the last two decades. Strong interactions between the surface and the redox species include such processes as anodic or cathodic corrosion of a metal or semiconductor and processes in which strongly adsorbed intermediates are formed. Theoretical progress has been significantly less for this type of reaction and, although corrosion is considered at a later point in this section, many of the results are phenomenological rather than firmly based in microscopic models.

In this section, we first consider a general model of the faradaic processes occurring at the semiconductor–electrolyte interface due to Gerischer [11]. From Gerischer's model, using the potential distribution at the interface, we may derive a Tafel-type description of the variation of electron transfer with potential and we will then consider the transport limitations discussed above. We then turn to the case of intermediate interactions, in which the electron transfer process is mediated by surface states on the semiconductor and, finally, we consider situations in which the simple Gerischer model breaks down.

5.1 GERISCHER'S MODEL

To proceed further in our analysis of the kinetics of faradaic reactions on semiconductors, a model is necessary for the redox couple in the solution and for the electronic structure of the semiconductor. Although single ground state energies may be defined at 0 K for both donor (reduced) and acceptor (oxidised) solution species, finite temperatures lead to the occupation of a range of vibronic levels associated both with the librational motion of the water molecule dipoles and the internal vibrations of the (complex) ion. The effect, as discussed elsewhere, is to give rise to a spread of accessible energies for both donor and acceptor species. The two types of vibronic energy level may be characterised by a single parameter in the theory, termed the reorganisation energy, λ, which is defined as the activation energy for the process of transforming the solvation shell and internal structure from the equilibrium situation in the reduced or oxidised species to the most probable energy of the other form [85].

An example will clarify this: if ${}^0E_R$ is the redox energy for the process $Ox^+ + e^-_{vac} \rightarrow Red$, ${}^0E_1$ the redox energy for $Ox^{*+} + e^-_{vac} \rightarrow Red$, where Ox^{*+} is the oxidised complex, distorted to have the same structure and solvation geometry as Red, and ${}^0E_2$ the redox energy for $Ox^+ + e^-_{vac} \rightarrow Red^*$, we have, from

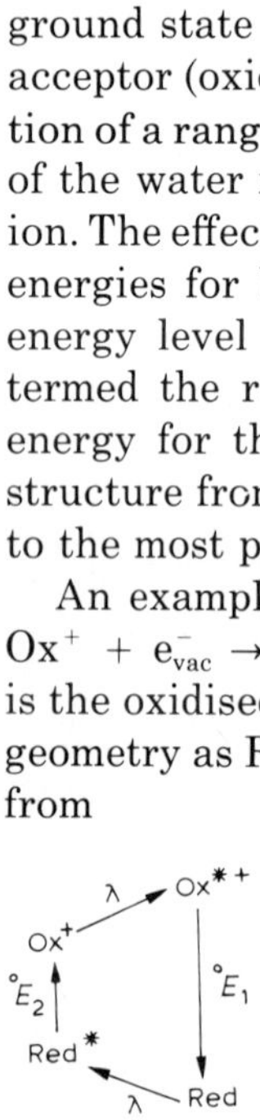

${}^0E_2 - {}^0E_1 = 2\lambda$, where we remember that ${}^0E_1$, ${}^0E_2$ and ${}^0E_R$ are all negative.

Clearly, ${}^0E_1$ corresponds to the most probable energy of the reduced form and ${}^0E_2$ to the most probable energy of the oxidised form. The midpoint

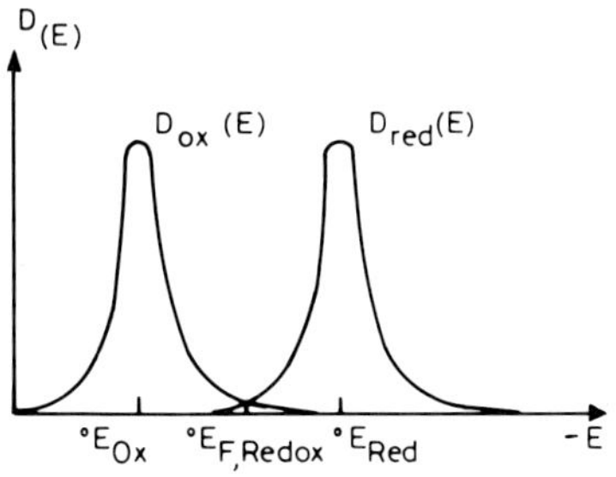

Fig. 42. Distribution functions $\mathscr{D}_{ox}(E)$ and $\mathscr{D}_{red}(E)$ for electronic energy levels of a redox couple in the electrolyte.

between ${}^0E_1$ and ${}^0E_2$ will correspond to the absolute redox potential energy $e_0 U^0_{redox}$. We also have

$$ {}^0E_1 = {}^0E_R - \lambda \tag{167} $$

$$ {}^0E_2 = {}^0E_R + \lambda \tag{168} $$

Using the harmonic approximation, the distribution of energies is given by

$$ \mathscr{D}_{0x}(E) = \frac{1}{2\sqrt{\pi\lambda kT}} \exp\left\{-\frac{(E - {}^0E_2)^2}{4\lambda kT}\right\} \tag{169} $$

$$ \mathscr{D}_{red}(E) = \frac{1}{2\sqrt{\pi\lambda kT}} \exp\left\{-\frac{(E - {}^0E_1)^2}{4\lambda kT}\right\} \tag{170} $$

and these are plotted for a typical λ value in Fig. 42 [86].

For the weak interaction case, where the redox ion in solution remains coordinated only to solvent molecules and at no time becomes adsorbed on to the surface, we have, for the cathodic and anodic currents from one band of the semiconductor

$$ j^- = + e_0 d C_{0x} \int_{-\infty}^{\infty} W_{if}(\varepsilon)\rho(\varepsilon)\mathrm{n}(\varepsilon)\, d\varepsilon \tag{171} $$

$$ j^+ = - e_0 d C_{Red} \int_{-\infty}^{\infty} W_{if}(\varepsilon)\rho(\varepsilon)[1 - \mathrm{n}(\varepsilon)]\, d\varepsilon \tag{172} $$

where ε is the energy of the carrier in the semiconductor with respect to the Fermi level, j^+ and j^- are the anodic and cathodic currents, $\rho(\varepsilon)$ is the density of the electronic levels in the semiconductor, $\mathrm{n}(\varepsilon)$ is the Fermi function, $W_{if}(\varepsilon)$ is the transition probability, and d is the distance from the semiconductor surface at which electron transfer is most probable.

The transition probability, $W_{if}(\varepsilon)$, contains an electronic matrix element, coupling the electronic states in the semiconductor to those of the complex in solution, and the thermal distribution functions $\mathscr{D}_{ox}$ and $\mathscr{D}_{red}$. The electronic matrix element is a product of the probability of tunneling through the space charge layer (if $\varepsilon < \varepsilon_s$) and a coupling term arising from perturbation theory

$$|\langle\psi_e|\mathscr{H}_i|\psi_s\rangle|^2 = \frac{2\pi}{\hbar}|V|^2$$

where ψ_e and ψ_s are the wave functions of the electron in the complex in solution and in the semiconductor, respectively, and $\mathscr{H}$, is an interaction Hamiltonian whose form is discussed elsewhere.

The integrals defining j^+ and j^- may be evaluated by substituting the expressions for $\mathscr{D}_{red}$ and $\mathscr{D}_{ox}$ from eqns. (169) and (170), $\rho(\varepsilon)$ from eqns. (5) and (6), and by assuming that $|U|^2$ is independent of ε. For fairly lightly doped semiconductors, we can assume that tunneling through the space-charge layer is very unlikely, so that transfer to and from the CB will only occur for $\varepsilon > \varepsilon_s$ and to and from the VB for $\varepsilon < \varepsilon_s$. Considering the two bands separately, which will be justified under a wide range of conditions, and remembering that $n(\varepsilon)$ is a rapidly decreasing function of ε, we find

$$j_c^+ = -\frac{2\pi}{\hbar}e_0 d|V|^2 C_{red} N_c \bar{\mathscr{D}}_{red}(E_c) \tag{173}$$

$$j_c^- = +\frac{2\pi}{\hbar}e_0 d|V|^2 C_{ox} n_s \bar{\mathscr{D}}_{ox}(E_v) \tag{174}$$

$$j_v^+ = -\frac{2\pi}{\hbar}e_0 d|V|^2 C_{red} p_s \bar{\mathscr{D}}_{red}(E_v) \tag{175}$$

$$j_v^- = +\frac{2\pi}{\hbar}e_0 d|V|^2 C_{ox} N_v \bar{\mathscr{D}}_{ox}(E_v) \tag{176}$$

where N_c and N_v are the effective density of states at the bottom of the conduction band and the top of the valence band, respectively, n_s and p_s are the electron and hole concentrations at the surface, and the effective densities of state for the complexes Ox and Red in solution are

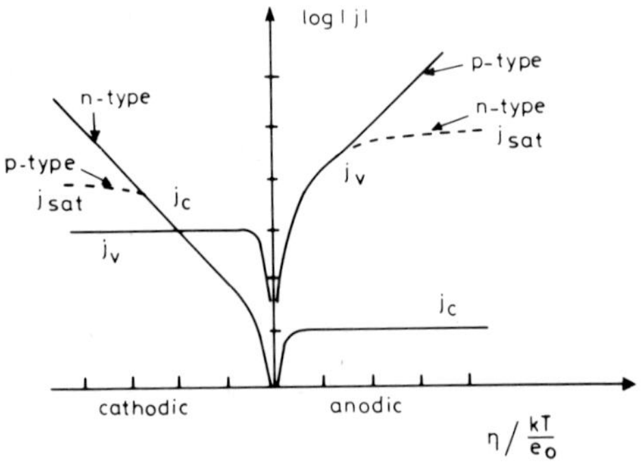

Fig. 43. Current voltage curves for the partial currents in valence and conduction bands in the Gerischer model.

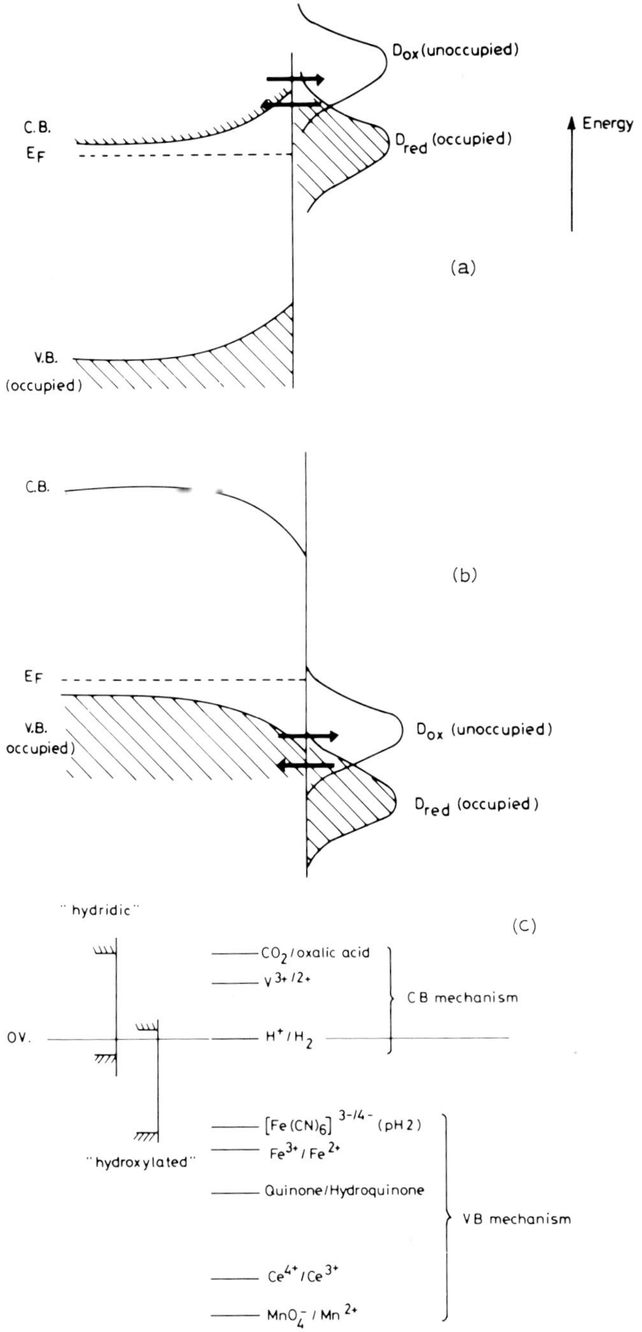

Fig. 44. (a) Faradaic processes involving a redox couple dominated by a CB mechanism on an n-type semiconductor. (b) As (a) for a VB mechanism on a p-type semiconductor. (c) Redox couples and electrochemistry on Ge.

$$\bar{\mathscr{D}}_{red} \simeq \left(\frac{kT}{\pi\lambda}\right)^{1/2} \exp\left\{-\frac{(E_c - {}^0E_{red})^2}{4\lambda kT}\right\} \qquad {}^0E_{red} = {}^0E_1 \text{ above} \tag{177}$$

$$\bar{\mathscr{D}}_{ox} \simeq \left(\frac{kT}{\pi\lambda}\right)^{1/2} \exp\left\{-\frac{(E_c - {}^0E_{ox})^2}{4\lambda kT}\right\} \qquad {}^0E_{ox} = {}^0E_2 \text{ above} \tag{178}$$

with similar expressions for $\bar{\mathscr{D}}_{red}(E_v)$ and $\bar{\mathscr{D}}_{ox}(E_v)$. The nett cathodic current through the conduction band, $J_c = j_c^- - j_c^+$ and, for the valence band, $J_v = j_v^- - j_v^+$. Plots of $j_{c(v)}^{\pm}$ are given in Fig. 43. It must be re-emphasised that E_c, ${}^0E_1$, ${}^0E_2$, and ${}^0E_R$ are all negative, whereas $\lambda > 0$. The conventional redox couple U^0_{redox} is defined as $-({}^0E_R/e_0) + C$, where C is a constant correction from vacuum to the specific reference electrode used as discussed above. If U_c is defined, similarly, as $-(E_c/e_0) + C$, then $E_c - {}^0E_2 = E_c - {}^0E_R - \lambda = -e_0(U_c - U^0_{redox}) - \lambda$ and, similarly, $E_c - {}^0E_1 = -e_0(U_c - U^0_{redox}) + \lambda$.

It is clear , from the above model and from Fig. 44(a) and (b), that, if ${}^0E_{red}$ is closer to E_c than ${}^0E_{ox}$ is to E_v, then electron transfer involving the conduction band E_{ox} is closer to E_v than ${}^0E_{red}$ to E_c, hole transfer involving the VB will be the main vehicle of the redox process. An example of this is shown for Germanium in Fig. 44(c) where a fairly clear-cut division can be made.

Gerischer's model gives an attractive insight into the basic redox processes at a semiconductor surface; it is clearly, however, that it represents a half-way house. Implicit in the assumptions are that electron or hole transfer can only occur if there is energetic overlap between the thermally broadened redox couple levels in solution and the semiconductor valence or conduction bands, but no allowance has been made in the model for the possibility of tunneling across the depletion layer. In addition, the theory, as it stands, cannot be applied to semiconductor processes such as anodic film growth or dissolution. These latter involve strong surface interactions and the treatment is significantly more complex.

5.2 VARIATION OF CURRENT WITH POTENTIAL

The expressions for the components of the total faradaic current at the semiconductor surface as given in eqns. (173)–(176) show that this current is given as the product of factors intrinsic to the electron transfer process taking place, to the concentration and thermal energy distribution of the redox couple, and to the concentration of carriers or the density of states. If we restrict attention to an n-type semiconductor and assume that only electron transfer to and from the conduction band is significant, then the nett current can be written

$$j = e_0(k_f C_{ox} n_s - k_r C_{red}) \tag{179}$$

where k_f and k_r are the rate constants, defined as $2\pi d/h|V|^2\bar{\mathscr{D}}_{ox}(E_c)$ and $2\pi d/h|V|^2\bar{\mathscr{D}}_{red}(E_c)$, respectively, C_{ox} and C_{red} the concentrations of oxidised and reduced components of the redox couple in solution (or, more precisely, at the surface of the semiconductor, and, as before, n_s is the concentration of electrons at the surface.

The potential distribution at the surface of the semiconductor is such that the bulk of the potential change is accommodated within the depletion layer. It follows, as discussed in Sect. 4, that n_s will be a strong function of the applied potential. However, the corollary of this is that the matrix element $|V|$ and the thermal distribution parameters $\bar{\mathscr{D}}_{ox}(E_c)$ and $\bar{\mathscr{D}}_{red}(E_c)$ will be much weaker functions of potential. Although, therefore, we would expect to find an exponential or Tafel-like variation of current with potential for a faradaic reaction on a semiconductor, the underlying situation is quite different from that of a metal. In the latter case, the exponential behaviour arises from the nature of the thermal distribution function $\bar{\mathscr{D}}$ and the concentration of carriers at the surface of the metal varies little with potential. To see this more clearly, we may expand eqn. (179) assuming that the reverse process of electron injection into the CB can be neglected; eqn. (179) then reduces to

$$j \;=\; e_0 k_f C_{ox} n_s \;\equiv\; e_0 k_f C_{ox} n_B \exp\,(+e_0\phi_{0s}/kT) \tag{180}$$

where ϕ_{0s} is the potential difference between the bulk and surface of the semiconductor and is *negative* for a reverse-biased n-type material. If the major part of the applied potential appears across the depletion-layer, then

$$j \;=\; e_0 k_f C_{ox} n_B \exp\,(-e_0 V_{app}/kT) \tag{181}$$

which corresponds to a Tafel slope of 60 mV/decade. An example of this type of behaviour can be seen for silicon; p-silicon in 10 M HF appears to exhibit a near perfect surface which behaves classically. The *anodic* current at such an electrode is the corrosion reaction involving the dissolution of silicon as $[SiF_6]^{2-}$; this reaction depends on the surface hole concentration $p_s = p_s^0 = \exp(e_0 V_{app}/kT)$ and the predicted 60 mV/decade Tafel slope is, in fact, observed [87]. Similarly, n-ZnO, in the presence of a variety of oxidants in solution, exhibits cathodic current Tafel slopes near 60 mV/decade and linear dependence of the current on the concentration of dopant [88, 89] and the reduction of Fe^{3+} at n-CdS has also been found to show a 60 mV slope [90].

5.3 MAJORITY FLUX LIMITATIONS

The extent to which dark currents are limited by transport of majority carriers across the interface can be gauged by the following argument [91].

The electron flux is given by

$$\frac{j}{e_0} \;=\; +\,D_n \frac{dn}{dx} - \mu_n n \frac{d\phi}{dx} \tag{182}$$

where D_n is the electron diffusion coefficient ($\equiv kT\mu_n/e_0$), ϕ is the potential, and we assume, as before, that $x = 0$ corresponds to the semiconductor–electrolyte interface and the flux j is defined as being in the $-x$ direction.

In the absence of any current, the potential is given by

$$\phi = -\frac{kT}{e_0}\frac{(W-x)^2}{L_D^2} \tag{183}$$

within the Mott–Schottky approximation. In the presence of a current flux, the surface concentration n_s is no longer equal to $n_B \exp(e_0\phi_{0s}/kT)$; instead, the flux equation may be explicitly integrated, using the integrating factor $\exp(-e_0\phi/kT)$

$$\int j \exp(-e_0\phi/kT) = +D_n[n \exp(-e_0\phi_0/kT)] \tag{184}$$

where we have assumed that currents are sufficiently small for Einstein's relation $D_n = kT\mu_n/e_0$ to hold. If we introduce the Mott – Schottky relationship (183) and integrate from $x = W$ to $x = 0$, we obtain

$$\int_W^0 \frac{j}{e_0} \exp\left\{\frac{(W-x)^2}{2L_D^2}\right\} dx = D_n\left[n_s \exp\left\{\frac{-e_0\phi_{0s}}{kT}\right\} - n_B\right]$$

whence

$$-\frac{\sqrt{2}L_D j}{e_0} \exp\left\{\frac{e_0\phi_{0s}}{kT}\right\} \int_0^{\sqrt{-e_0\phi_{0s}/kT}} e^{y^2} dy$$

$$= D_n\left[n_s - n_B \exp\left\{\frac{e_0\phi_{0s}}{kT}\right\}\right] \tag{185}$$

and so

$$j = -\frac{D_n e_0}{\sqrt{2}L_D}\frac{[n_s - n_B \exp(e_0\phi_{0s}/kT)]}{\text{Daw}(\sqrt{-e_0\phi_{0s}/kT})} \tag{186}$$

where Daw(x) is Dawson's integral $[\equiv \exp(-x^2)\int_0^x \exp(t^2)\,dt]$.

Since we also have

$$j = e_0[k_f C_{0x} n_s - k_r C_{red}] \tag{187}$$

If we make the assumption that the semiconductor behaves classically and there is an applied voltage V_0 at which $j = 0$ and $n_s^0 = n_B \exp(-e_0 V_0/kT)$

$$k_r C_{red} = k_f C_{ox} n_s^0 \tag{188}$$

$$j = e_0 k_f C_{ox}[n_s - n_s^0] \tag{189}$$

and

$$j = -\frac{e_0 D_n}{\sqrt{2}L_D}\frac{[n_s - n_s^0 \exp\{e_0(V_0 - V)/kT\}]}{\text{Daw}(\sqrt{-e_0\phi_{0s}/kT})} \tag{190}$$

Eliminating n_s, we have

$$\frac{j}{e_0} = -\frac{n_s^0[1 - \exp\{e_0(V_0 - V)/kT\}}{1 + [D_n/\{\sqrt{2}L_D k_f C_{ox}\,\text{Daw}(\sqrt{-e_0\phi_{0s}/kT})\}]} \times$$

$$\times \frac{D_n}{\sqrt{2}L_D \operatorname{Daw}(\sqrt{-e_0\phi_{0s}/kT})}$$

$$= \frac{n_B[\exp(-V_0e_0/kT) - \exp(e_0V/kT)]}{(\sqrt{2}L_D/D_n)\operatorname{Daw}(\sqrt{-e_0\phi_{0s}/kT}) + 1/k_fC_{ox})} \tag{191}$$

where the structure of eqn. (191) is typical of two rate processes in series. In this case, the rate processes correspond to transport through the semiconductor depletion layer and electron transfer at the interface. Clearly, if

$$\frac{\sqrt{2}L_D}{D_n}\operatorname{Daw}(\sqrt{-e_0\phi_{0s}/kT}) \gg \frac{1}{k_fC_{ox}} \tag{192}$$

the main limitation on the current is majority carrier transport. The conditions under which this might occur evidently depend on the value of k_fC_{ox}, but for low-mobility materials, the limitation implied by eqn. (182) is marked; thus, for TiO_2 at a positive bias of 0.5 V from flat-band potential, the saturation current may be approximated by

$$\frac{n_B\exp(-e_0V/kT)}{(\sqrt{2}L_D/D_n)\operatorname{Daw}(\sqrt{-e_0\phi_{0s}/kT})} \tag{193}$$

if $\exp(-e_0V_0kT) \ll \exp(-e_0V/kT)$. If we take values for the transport properties appropriate to TiO_2, $\mu_n \simeq 1\,\text{cm}^2\,V^{-1}\,\text{s}^{-1}$, $n_B \simeq 10^{17}\,\text{cm}^{-3}$, and $L_D \simeq 5 \times 10^{-6}$ cm, then

$$j_{max} \simeq 1\,\mu\text{A cm}^{-2} \tag{194}$$

Although this is clearly an extreme case, and mobilities are commonly much higher than this and Debye lengths smaller, nevertheless, oxide semiconductors may well be limited in practical application by carrier transport. Naturally, if the bias is such that the semiconductor is not in depletion, there will be no restriction arising from transport; indeed, the semiconductor will behave like a metal under these circumstances.

5.4 LIMITATIONS DUE TO MINORITY CARRIER TRANSPORT

In the previous section, we considered electron transfer processes involving only the majority carrier band. For semiconductors with lower bandgaps, either doped or intrinsic, both bands may be involved and we consider two current contributions arising from the two bands as shown in Fig. 41. In practice, it is found that the dominant contribution usually arises from only one of the two bands; if the majority carrier band is involved, the treatment will reduce to that given above, but if the minority carrier band is the more significant, then the current may be limited by the rate of thermal generation and transport of the majority carriers to the surface.

Let us consider the problems of an n-type semiconductor that is biased positive and which must sustain a faradaic reaction through hole transfer

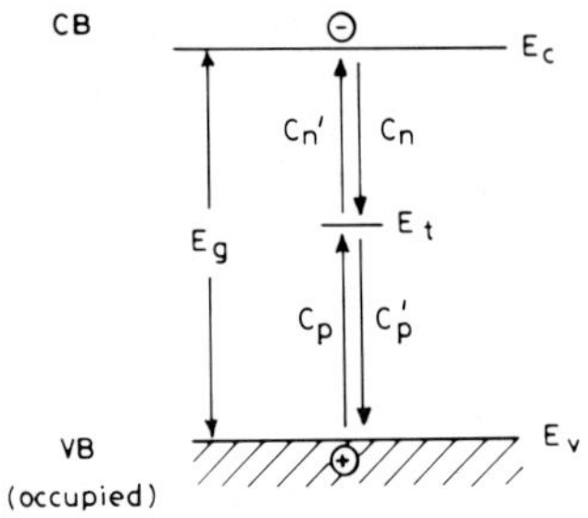

Fig. 45. Kinetic parameters in the Shockley–Read model.

via the valence band; examples would include anodic oxide formation on silicon and other related materials. Under these circumstances, the reaction will consume holes at the surface at a rate kp_s; effectively, this denudes a small region near the surface of the semiconductor of holes, which must then be supplied by thermal generation and transport. We first address ourselves to the problem of generation, and use Shockley–Read statistics to form an impression of the generation law.

5.4.1 Shockley–Read statistics [2, 7]

The concentration of both majority and minority carriers has been treated in the earlier sections of this review purely from the equilibrium standpoint. This treatment clearly must be extended if we wish to include kinetic effects in our models and extensive efforts have been made to understand the routes by which non-equilibrium concentrations of carriers, whether generated by faradaic or photofaradaic effects, can relax to the equilibrium values. Elementary estimates have indicated that the most effective means by far for the semiconductors commonly encountered by electrochemists is through impurity or "recombination" sites located within the bandgap. The reason for this is that thermal energy can only be transported in the crystal by lattice phonons; these have energies small compared with the bandgap energy and very large numbers of such phonons would have to be generated or captured if an electronic transition involving the whole bandgap were envisaged. To avoid this problem, Shockley and coworkers suggested that intermediate energy sites, although present in small concentrations, would play a dominant role in carrier equilibration. Their approach utilises simple kinetic arguments and is reproduced below. Let there exist a trap site of energy E_t between the valence and conduction bands, then there are four transfer reactions possible, as shown in Fig 45, that correspond to electron exchange with the conduction band (CB) and hole exchange with the valence band (VB). Consider first the hole equilibrium:

$$\frac{\partial p}{\partial t} = -C_p p f_t + C_p'(1 - f_t) \qquad (195)$$

where p is the concentration of holes in the VB, f_t is the Fermi function, reflecting the electron population of the traps, and C_p, C_p' are the rate constants for the hole capture and release processes and contain the number of trap sites, the capture cross-section, and the thermal velocity of the holes. At equilibrium

$$C_p' = C_p \frac{p^0 f_t^0}{1 - f_t^0} \tag{196}$$

where the superscript 0 refers to the equilibrium state. Now,

$$1 - f_t^0 = \frac{1}{1 + \exp\{(E_F - E_t)/kT\}}$$

$$= \frac{N_V \exp\{(E_V - E_F)/kT\}}{N_V \exp\{(E_V - E_F)/kT\} + N_V \exp\{(E_V - E_t)/kT\}} \tag{197}$$

$$= \frac{p^0}{p^0 + p_t} \tag{198}$$

where p_t is **defined** as $N_V \exp\{(E_V - E_t)/kT\}$ and represents the hole concentration in the trap site if the Fermi level were to coincide with E_t. Hence

$$f_t^0 = \frac{p_t}{p^0 + p_t} \tag{199}$$

and

$$C_p' = p_t C_p \tag{200}$$

so, finally

$$\frac{\partial p}{\partial t} = C_p\{-pf_t + p_t(1 - f_t)\}. \tag{201}$$

Similarly, for electrons, we have

$$\frac{\partial n}{\partial t} = -C_n\{n(1 - f_t) - n_t f_t\} \tag{202}$$

where n_t is defined as

$$n_t = N_c \exp\{(E_t - E_c)/kT\}$$

At steady state, the hole and electron fluxes into the trap state must balance and

$$f_t = \frac{nC_n + p_t C_p}{C_n(n + n_t) + C_p(p + p_t)} \tag{203}$$

$$\frac{\partial p}{\partial t} = -\frac{C_n C_p (np - n_i^2)}{C_n(n + n_t) + C_p(p + p_t)} \equiv g(x) \tag{204}$$

where n_i^2 is given by $N_C N_V \exp\{(E_v - E_c)/kT\}$.

If we regard this as a generating term, we have

$$\frac{\partial p}{\partial t} = D_{\mathrm{p}} \frac{\partial}{\partial x}\left[\frac{\partial p}{\partial x} - \left(\frac{e_0}{kT}\right)p\mathscr{E}\right] + \mathrm{g}(x) \tag{205}$$

where $\mathscr{E}$ is the electric field. For the n-type semiconductor, we may divide the integration region into two.

$0 < x < W$ depletion layer region;

$$\mathscr{E} = -\frac{(W - x)kT}{e_0 L_{\mathrm{D}}^2} \tag{206}$$

$x > W$ bulk region;

$$\mathscr{E} = 0 \tag{207}$$

Consider the bulk region; since we are dealing with an n-type semiconductor, $n \gg n_{\mathrm{t}}, p, p_{\mathrm{t}}$ and

$$\mathrm{g}(x) \simeq \frac{C_{\mathrm{p}}(n_{\mathrm{i}}^2 - n_{\mathrm{B}}p)}{n_{\mathrm{B}}} \tag{208}$$

where n_{B}, the bulk electron density, is assumed equal to the donor density N_{D}.

Since

$$n_{\mathrm{i}}^2 \approx n_{\mathrm{B}}p^0; \qquad \mathrm{g}(x) \simeq C_{\mathrm{p}}(p^0 - p) \simeq \frac{p^0 - p}{\tau_{\mathrm{p}}} \tag{209}$$

and given $\mathscr{E} = 0$, we have

$$D_{\mathrm{p}}\frac{\partial^2 p}{\partial x^2} - \frac{p}{\tau_{\mathrm{p}}} = -\frac{p^0}{\tau_{\mathrm{p}}} \tag{210}$$

and from the boundary condition $p \to p^0$, $x \to \infty$ ξ

$$\left.\begin{aligned} p &= p^0 + A\exp(-x/L_{\mathrm{p}}) \\ L_{\mathrm{p}}^2 &= D_{\mathrm{p}}\tau_{\mathrm{p}} \end{aligned}\right| \tag{211}$$

We now apply the theory derived in the preceding paragraphs to the situation in which minority carriers are generated thermally. In the depletion region, the concentration of holes will be substantially reduced from the equilibrium value by the current flow; in addition, n^0 is now substantially reduced from the bulk value. If we assume that n_{t} is now dominant and that $n \times p \ll n_{\mathrm{i}}^2$, $\mathrm{g}(x) = Cn_{\mathrm{i}}^2/n_{\mathrm{t}}$. Since this is independent of x, we may immediately integrate the flux equation to give

$$\frac{\partial p}{\partial x} + \left(\frac{W - x}{L_{\mathrm{D}}^2}\right)p = -\int \frac{C_{\mathrm{p}}n_{\mathrm{i}}^2}{D_{\mathrm{p}}n_{\mathrm{t}}}\,\mathrm{d}x \tag{212}$$

The expression on the left-hand side of eqn. (212) is the hole flux. At the surface ($x = 0$), this is the sole contribution to the current j and we may write

$$\frac{\partial p}{\partial x} + \left(\frac{W - x}{L_D^2}\right)p = -\frac{C_p n_i^2 x}{D_p n_t} + \frac{j}{D_p e_0} \tag{213}$$

Equation (213) may be integrated further by the integrating factor exp $\{-(W - x)^2/2L_D^2\}$ = exp $(e_0\phi_0/kT)$ where ϕ_0 is the potential in the depletion layer. We find

$$p \exp\{-(W - x)^2/2L_D^2\} - p_W$$
$$= \int_W^0 \exp\{-(W - x')^2/2L_D^2\}\left[-\frac{C_p n_i^2 x'}{n_t D_p} + \frac{j}{e_0 D_p}\right] dx' \tag{214}$$

After some manipulation, this yields, at $x = 0$, approximately

$$p_s \exp\{-W^2/2L_D^2\} - p_W = -\frac{jL_D}{e_0 D_p}\sqrt{\frac{\pi}{2}} + \frac{C_p n_i^2 L_D}{n_t D_p}\left(W\sqrt{\frac{\pi}{2}} - L_D\right) \tag{215}$$

At the surface, we have

$$k_f p_s = \frac{j}{e_0} \tag{216}$$

where k_f is the faradaic rate constant and p_s the surface hole concentration. In addition, we can match the *flux* and the *hole concentration* at the point $x = W$, since

$$p_W = p^0 + A \exp(-W/L_p) = p_s \exp(-W^2/2L_D^2)$$
$$+ \frac{jL_D}{e_0 n_t D_p}\sqrt{\frac{\pi}{2}} - \frac{C_p n_i^2 L_D}{n_t D_p}\left(W\sqrt{\frac{\pi}{2}} - L_D\right) \tag{217}$$

and

$$\frac{j}{e_0 D_p} - \frac{C_p n_i^2 W}{D_p n_t} = -\frac{A}{L_p}\exp(-W/L_p) \tag{218}$$

Eliminating A and p_W, we find, since $L_p^2 = D_p/C_p$

$$\frac{j}{e_0}\left(\frac{1}{D_p} + \frac{\exp(-W^2/2L_D^2)}{k_f L_p} + \frac{L_D}{L_p D_p}\sqrt{\frac{\pi}{2}}\right)$$
$$= \frac{C_p n_i^2}{D_p n_t}\left[W\left(1 + \sqrt{\frac{\pi}{2}}\right) - L_D\right] + \frac{p^0}{L_p} \tag{219}$$

$$j = \frac{e_0\{(C_p n_i^2/n_t)[W(1 + \sqrt{\pi/2}) - L_D] + C_p L_p p^0\}}{\{1 + (D_p/k_f L_p)\exp(-W^2/2L_D^2) + L_D/L_p\sqrt{\pi/2}\}} \tag{220}$$

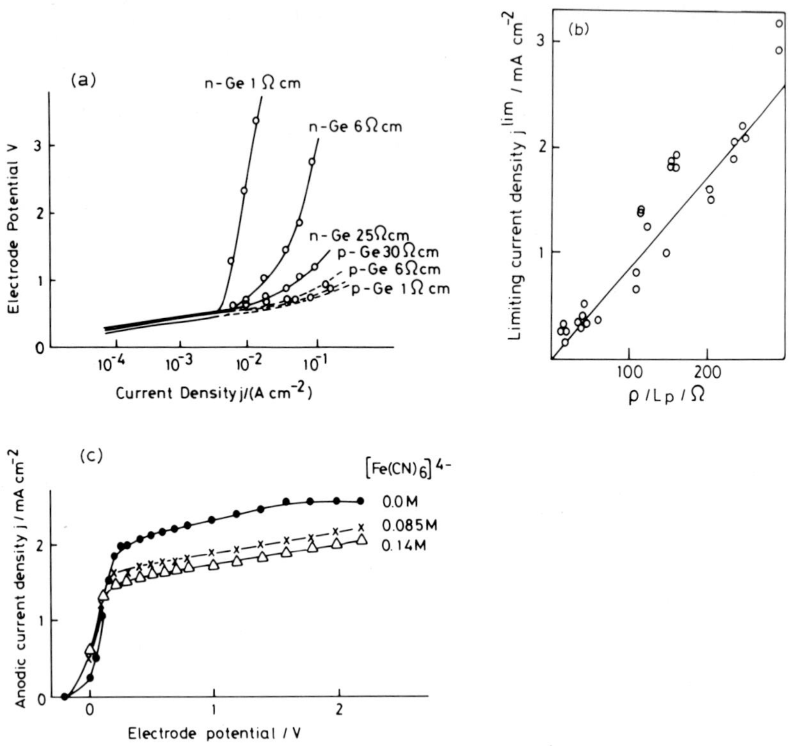

Fig. 46. (a) Anodic polarisation curves for Ge in 0.1 M HCl. (b) Relationship of j^{sat} to ρ/L_p for Ge. (c) Effect of addition of $[Fe(CN)_6]^{4-}$ on the limiting currents for Ge in 0.05 M H_2SO_4.

Physically, eqn. (220) has the following significance. The first term in the numerator is the depletion layer generating term and the second the bulk generating term. In the denominator, the first term is the interfacial faradaic term and the second a transport term across the depletion layer. There are a number of important specific cases that can be extracted from the formula. If we have a *narrow* bandgap material, then bulk generation will tend to dominate and we have

$$j \approx \frac{e_0 C_p L_p p^0}{1 + (D_p/k_f L_p)\exp(-W^2/2L_D^2) + (L_D/L_p)\sqrt{\pi/2}}$$

$$\simeq \frac{j^{\text{lim}}}{1 + (D_p/k_f L_p)\exp(e_0\phi_0/kT)} \tag{221}$$

where $j^{\text{lim}} = e_0 C_p L_p p^0 \equiv e_0^2 D_p n_i^2 \mu_n \rho / L_p$, ρ is the resistivity of the n-type material, and L_D is assumed to be smaller than L_p. If ϕ_0 is small

$$j \cong j^{\text{lim}} \frac{k_f L_p}{D_p} \exp\left(-\frac{e_0\phi_0}{kT}\right) \tag{222}$$

and, remembering that $\phi_0 \leqslant 0$, the current increases with a Tafel slope of 60 mV/decade. Above a critical potential, the faradaic term drops below unity in the denominator and $j \to j^{\lim}$. The best known example of behaviour of this general type is found for the anoidc dissolution of n-Ge, first investigated by Brattain and Garrett [92–94] and later by others [95–98]. It must be emphasised that, in this material, there is considerable evidence that, for anodic dissolution and other anodic reactions, conduction band electrons are also involved, leading to a larger, and redox-couple dependent, value of $j^{\lim}$. Some examples of limiting-current behaviour for Ge are shown in Fig. 46; Fig. 46(a) demonstrates the effect of increasing the bulk hole concentration, Fig. 46(b) shows the expected linear relationship between $j^{\lim}$ and ρ/L_p for n-Ge [eqn. (221)], and Fig. 46(c) shows that the oxidation of $[Fe(CN)_6]^{4-}$ proceeds entirely via the valence band, in constrast to anodic dissolution. Exactly analogous will be cathodic reactions on p-type semiconductors, though the most thoroughly investigated, hydrogen evolution, has proved far from easy to understand. The reason for this is that hydrogen atoms formed during the evolution process may not only undergo a second electrochemical reaction

$$e^- + H^{\cdot}_{ads} + H^+ = H_2$$

but may also diffuse into the electrode, altering its electrophysical properties very dramatically. However, *dynamic* electrode sweeps give a clearly defined limiting current that appears to be entirely associated with the minority carriers (electrons) in the conduction hand.

Another specific case that may be extracted from the above case is that, for wider bandgap materials, where p^0 becomes small and the current is dominated by space-charge generation

$$j \approx \frac{e_0 C_p n_i^2 W/n_t}{1 + (D_p/k_f L_p)\exp(e_0\phi_0/kT)} \tag{223}$$

This will be valid when $L_p/n_B \ll W/n_t$ and $j^{\lim} \propto W \propto 1/(n_B)^{1/2} \propto (\rho)^{1/2}$. At low voltages, we can again expect a Tafel region and limiting currents should be proportional to the square root of the applied potential. Silicon appears to satisfy these observations and anodic dissolution and/or passivation of silicon has been studied in a large number of papers [99–104]. The results for n-Si polarised anodically in HF are shown in Fig. 47 and the n-type samples clearly reach a limiting current in contrast to the p-type samples discussed above. Above $j \simeq 10^{-5}\,\mathrm{A\,cm^{-2}}$, a change to quadratic behaviour, $j \sim V^{1/2}$, is seen in the figure and, from the slope, values of C_p/n_t may be calculated, in reasonable agreement with the literature data. Similar data have also been obtained for n-GaAs and n-CdS [7].

As for narrower-band materials discussed above, analogous reactions on p-type materials can be studied and generation of hydrogen on p-Si has been discussed in the literature [99]. Study of the dependence of the limiting current on resistivity [105] has shown that $j^{\lim}$ apparently falls with ρ; it has

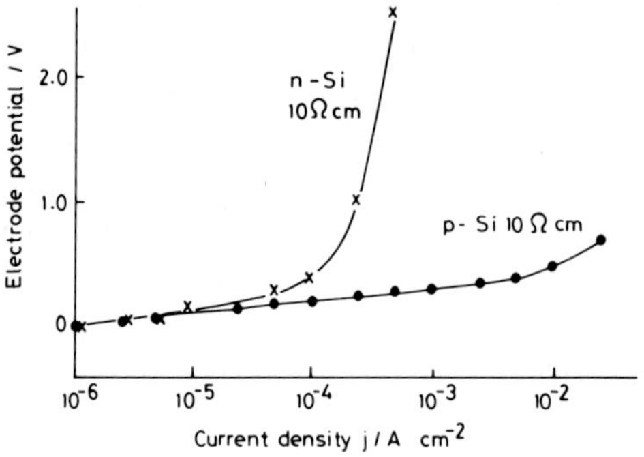

Fig. 47. Anodic polarisation curve for Si in 2.5 M HF solution.

been suggested that C_p and n_t may also change with alteration in doping, giving rise to a more complex dependence of j^{lim} on ρ. Complications with atomic hydrogen will also be present.

The expected Tafel slope of 60 mV/decade is not always found. There are a number of reasons for this, aside from kinetic effects in the bulk of the semiconductor. The kinetic effects associated with faradaically active surface states is of considerable significance, as shown below, but another common problem is that part of the potential change may appear across the Helmholtz layer rather than across the depletion layer. A well-known case in point is germanium, for which the surface is slowly converted from "hydride" to "hydroxylic" forms as the potential is ramped anodically. This conversion gives rise to a change in the surface dipole and hence $|\Delta\psi| \neq \Delta V$. In fact, the anodic dissolution of p-germanium is found to follow a law [106]

$$\frac{\mathrm{d}V}{\mathrm{d}\,(\ln i)} = \frac{RT}{\alpha F} \tag{224}$$

where $\alpha \equiv 1 - \mathrm{d}\psi_H/\mathrm{d}V$. By adding HF, α can be increased to unity, presumably by stabilisation of the hydroxylated surface by H-bonding to F^-.

5.5 INTERMEDIATE INTERACTIONS. THE ROLE OF SURFACE STATES

In addition to the models considered above, all of which have involved direction electron transfer from the surface to the redox couple in solution, it is possible for the electron transfer to be mediated by local electronic surface states on the semiconductor, which may be intrinsic or extrinsic. Transfer to these states may be classical or quantum mechanical and we will consider first the classical models.

Several groups have investigated the cathodic dark current on n-TiO_2 and n-$SrTiO_3$ in the presence of such oxidising agents as $[Fe(CN)_6]^{3-}$, Fe^{3+} and $[IrCl_6]^{2-}$ [107, 108]. They found that, at high current densities and at very high concentrations of redox couple in solution, the observed Tafel slopes

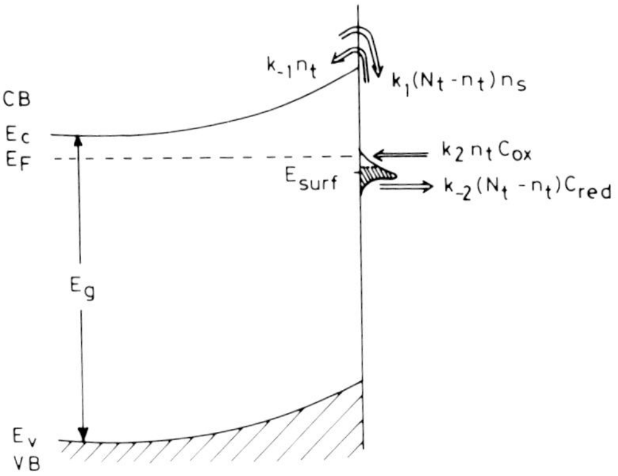

Fig. 48. Kinetic parameters for surface-mediated electron-transfer processes.

were close to 60 mV, but that, in this region, the dependence of the current j on concentration, of oxidising agent C_{oc}, had the form

$$j = \frac{j_{sat} C_{ox}}{A + C_{ox}} \tag{225}$$

At lower concentrations of oxidising agent, the Tafel slope was found to increase and this can be understood with reference to Fig. 48 in which a surface state at energy E_{surf} is filled and emptied from the CB and from the redox couple in solution. If the electron concentration is n_t in the surface states (whose total surface concentration is N_t per unit area) we have

$$\frac{\partial n_t}{\partial t} = k_1(N_t - n_t)n_s - k_{-1}n_t - k_2 n_t C_{ox} + k_{-2}(N_t - n_t)C_{red} \tag{226}$$

At steady state, $dn_t/dt = 0$

$$\frac{n_t}{N_t} = \frac{k_1 n_s + k_{-2}C_{red}}{k_{-1} + k_2 C_{ox} + k_1 n_s + k_{-2}C_{red}} \tag{227}$$

and the current density is

$$\begin{aligned} j &= e_0[k_2 n_t C_{ox} - k_{-2}(N_t - n_t)C_{red}] \\ &= e_o \left\{ N_t k_1 n_s C_{ox} - \left(\frac{k_{-1}k_{-2}}{k_2}\right) C_{red} N_t \right\} \Big/ \\ &\quad \times \left\{ k_{-1} + k_1 n_s + k_{-2}\frac{C_{red}}{k_2} + C_{ox} \right\} \end{aligned} \tag{229}$$

The structure of eqn. (229) is typical of any serial process and may be made more transparent by inverting

$$\frac{e_0}{j} = \frac{1}{k_2 N_t C_{ox}} + \frac{1}{N_t k_1 n_s} \tag{230}$$

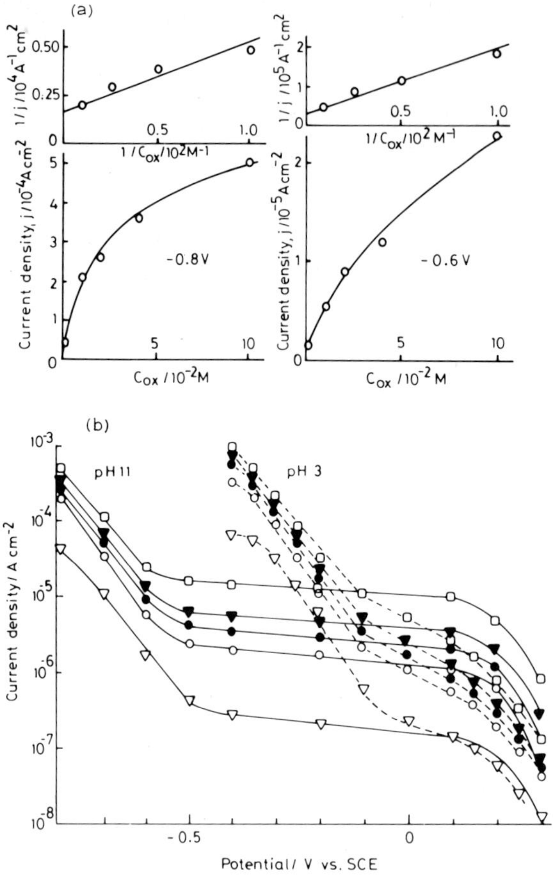

Fig. 49. (a) Cathodic current density j vs. $[Fe(CN)_6]^{3-}$ at pH 11 for two potentials: -0.8 V/SCE and -0.6 V/SCE showing linearity of $1/j$ vs. $1/C_{ox}$ (b) Tafel plots of $\log_{10} j$ vs. potential at pH 3 and 11 for concentrations, C_{ox}, of $[Fe(CN)_6]^{3-}$ of □, 10^{-1}M; ▼, 4×10^{-2}M; ●, 2×10^{-2}M; ○, 10^{-2}M; and ▽, 10^{-3}M.

where, for simplicity, we have neglected the reverse terms k_{-1} and k_{-2}. Clearly, the first term in eqn. (230) reflects the rate at which electron transfer occurs from the surface state to the redox couple and the second term the rate at which internal electron transfer within the semiconductor takes place.

The experimental data at high cathodic current densities do, indeed, have the form of eqn. (230), as shown in Fig. 49(a) with the saturation current given by $j_{sat} = e_0 N_t k_1 n_s$. In this region, j_{sat} has a Tafel slope of 60 mV, which clearly arises from the dependence of n_s on potential. In addition, the equation predicts that the saturation current will be independent of the chemical identity of the redox couple, an effect again found experimentally. As C_{ox} is lowered, the first term in eqn. (230) becomes more important and the Tafel slope rises.

Interestingly, for both TiO_2 and $SrTiO_3$, the current density at more positive potentials levels out, as shown in Fig. 49(b); this is not predicted by either eqns. (229) or (230) and both authors were driven to postulate a parallel model with two sets of surface states. For the first state, $k_2 C_{ox} \gg k_1 n_s$ and the reverse is true for the second state, so

$$\frac{j}{e_0} \sim (N_t k_1 n_s + N_t' k_2' C_{ox}) \tag{231}$$

At more negative potentials, where C_{ox} is small enough to avoid saturation, a 60 mV Tafel slope is found, but at positive potentials, $k_1 n_s$ becomes negligible and the current density is predicted to be independent of potential. More complex models have also been suggested in which a distribution of surface states with energy has been postulated, i.e. $N(E_t) \simeq N_0 \exp(aE_t/kT)$. If the forward rate constant also shows an exponential dependence on E_t and Shockley–Read statistics are assumed, then the rate law is calculated to show a non-integral dependence on C_{ox} [109].

The origin of these surface states is still uncertain, though evidence for Ti^{3+} associated with oxygen vacancies has been presented. For $[IrCl_6]^{2-}$ on n-$SrTiO_3$, the states actually appear to derive from strong interaction of the $[IrCl_6]^{2-}$ with the $SrTiO_3$ surface and similar interactions have been suggested for H_2O_2 on n-$SrTiO_3$ [110]. In the absence of any real chemical understanding of these surface states, quantum mechanical calculations can only give a qualitative insight: Schmickler has presented a rather general formalism in which the surface state plays the role of a resonant tunneling centre [111]. Model calculations for specific systems appear not to be available.

5.6 STRONG INTERACTIONS AT SEMICONDUCTOR SURFACES. ANODIC DISSOLUTION

The study of the anodic dissolution of semiconductors has played an important role in clarifying the nature of faradaic processes occurring at semiconductor surfaces [112, 113]. In principle, anodic dissolution on such semiconductors as Ge, GaAs, and GaP might proceed either by hole capture from the VB or electron injection into the CB. For Ge, for example

$$\text{Ge} + 6\,\text{OH}^- + \lambda \text{p}_{\text{vb}}^+ \rightarrow [\text{GeO}_3]^{2-} + 3\,\text{H}_2\text{O} + (4-\lambda)\text{e}_{\text{cb}}^- \qquad \text{in alkali} \tag{232}$$

$$\text{Ge} + 3\,\text{H}_2\text{O} + \lambda \text{p}_{\text{vb}}^+ \rightarrow \tfrac{1}{2}\,\text{Ge}_2\text{O}_3\cdot x\text{H}_2\text{O} + 4\,\text{H}^+ + (4-\lambda)\text{e}_{\text{cb}}^- \qquad \text{in acid} \tag{233}$$

The overall corrosion current j_{corr} is larger than the hole current j_+ by a factor $4/\lambda$. The measurement of λ is not straightforward, but we may identify and measure, with RDE techniques, the following currents at the surface of an n-type semiconductor as indicated in Fig. 50 [114].

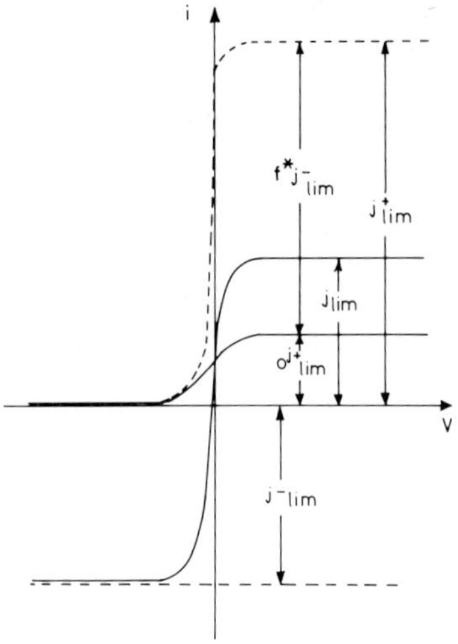

Fig. 50. Partial currents for dissolution and hole injection on semiconductors.

- ${}_0j^+_{\rm lim}$ the limiting anodic dissolution where there are no redox couples in solution.
- $j^-_{\rm lim}$ the limiting cathodic reduction current when a redox couple is present which can only be reduced by hole injection.
- $j_{\rm lim}$ the limiting anodic current observed when there is a redox couple present in the solution.

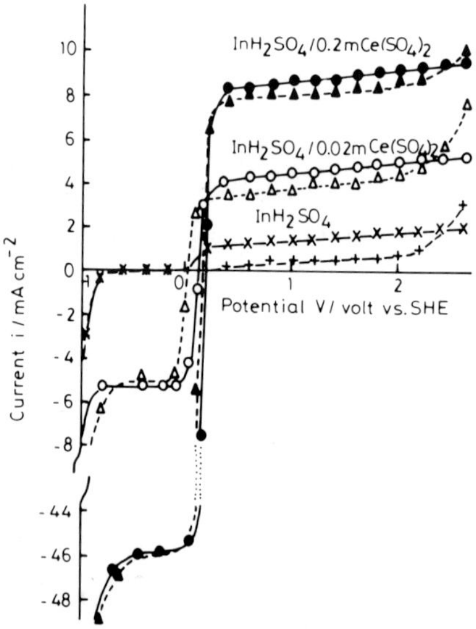

Fig. 51. Current–voltage curves for n-Ge in 0.5 M H_2SO_4–ceric sulphate solutions. Continuous curve: 14 Ω cm Ge; broken curve: 0.08 Ω cm.

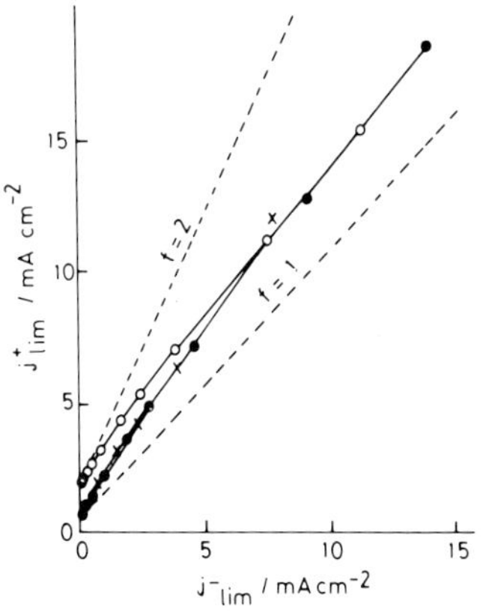

Fig. 52. Dependence of j^+_{lim} (defined in Fig. 50) vs. j^-_{lim} for 3 n-type Ge crystals for reduction of $[Fe(CN)_6]^{3-}$ in 0.1 M NaOH–1 M $NaNO_3$. The concentration of $[Fe(CN)_6]^{3-}$ varies from 0.17 to 30 mM.

The hole current due to dissolution in the last case is

$$j^+_{lim} = j_{lim} + j^-_{lim} \tag{234}$$

and the increase in dissolution hole current

$$\Delta j^+_{lim} = j_{lim} + j^-_{lim} - {}_0j^+_{lim} \tag{235}$$

From above, each hole injected through j^-_{lim} into the VB is amplified to give rise to a total current $4j^-_{lim}/\lambda \equiv f^*j^-_{lim}$; hence

$$j_{lim} = {}_0j^+_{lim} + f^*j^-_{lim} - j^-_{lim} = {}_0j^+_{lim} + j^-_{lim}(f^* - 1) \tag{236}$$

A plot of j_{lim} vs. j^-_{lim} should then be linear; from the slope, f^* and hence λ can be found. A change in j^-_{lim} may most easily be effected by varying the rotation speed or the concentration of oxidising agent and an example is shown in Fig. 51. In practice, this technique reveals that f^* is not a constant, but tends to decrease with increasing current. For Ge in the presence of $[Fe(CN)_6]^{3-}$, f^* is close to 2 at small values of j^-_{lim} but, at higher limiting current values, it deceases as shown in Fig. 52 [114]. Similar results have been reported for Ce(IV), MnO_4^- and Fe^{3+} in acid, and for O_2 and H_2O_2 in alkali; as shown in Fig. 44(c), all of these redox couples are expected to be reduced by hole injection into the VB on the hydroxylic surface and, indeed, in acid solutions, only the reduction of H^+ and V^{3+} are known to occur by electron capture from the CB. Comparatively few species are known to be reduced by both VB and CB involvement. Memming and Möllers have identified quinones as possible species [115] and have suggested that the reduction of benzoquinone proceeds as

$$BQ + H^+ + e^-_{cb} \rightarrow BQH^\cdot$$

$$BQH^\cdot + H^+ \rightarrow BQH_2 + P^+_{vb}$$

Interestingly, reduction of benzoquinone on a hydridic surface occurs only via the valence band, a result associated with the strong cathodic flat-band shift of the latter surface compared with the hydroxylated surface, as seen in Fig. 44(c).

The reduction of f^* with increasing limiting cathodic current is also found for n-GaAs in NaOCl/NaOH [116] and can be understood in terms of a simple kinetic argument: we suppose that the mechanism for Ge oxidation has the form

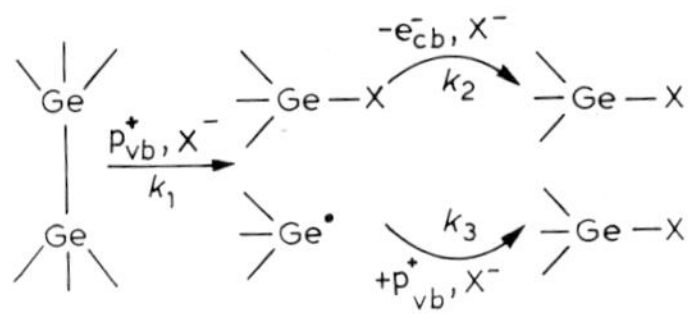

where k_2 is the rate constant for electron injection and k_3 is the rate constant for hole capture. If C_R is the radical concentration and the concentration of X^- is imagined to be incorporated into the overall rate constants, the overall current can be written

$$j = k_1 p_s + k_2 C_R + k_3 p_s C_R$$

and the electron current $j_e = k_2 C_R$. From the usual kinetic expressions

$$\frac{\partial C_R}{\partial t} = k_1 p_s - (k_2 + k_3 p_s) C_R = 0$$

so

$$C_R = \frac{k_1 p_s}{(k_2 + k_3 p_s)}$$

or

$$\frac{1}{C_R} = \frac{k_2}{k_1 p_s} + \frac{k_3}{k_1} = \frac{k_2}{j_e}$$

whence

$$\frac{1}{j_e} = \frac{2}{j} + \frac{k_3}{k_1 k_2}$$

Clearly, as j increases, $j_e \rightarrow k_3/k_1 k_2$, which is a constant. This formula has been verified for Ge and, at least qualitatively, accounts for the results on GaAs.

5.7 GENERAL DECOMPOSITION ROUTES

The most general treatment of possible decomposition routes on elemental and (A–B) semiconductors has been provided by Gerischer and Mindt

[117]. Their treatment covered both oxidative and reductive decomposition routes and may be summarised as follows.

(1) *Elemental semiconductors*

(a) Oxidative decomposition

Initiation: $(A–A)_s + X \rightarrow (AX)^+ + A_s^{\cdot} + e_{cb}^-$ electron injection (1a)

$(A–A)_s + p_{vb}^+ \rightarrow (AX)^+ + A_s^{\cdot}$ hole induced (1b)

Propagation: $A_s^{\cdot} + X \rightarrow (AX)^+ + e_{cb}^-$ (2a)

$A_s^{\cdot} + p_{vb}^+ + X \rightarrow (AX)^+$ (2b)

(b) Reductive decomposition

Initiation: $(A-A)_s + e_{cb}^- + H^+ \rightarrow AH + A_s^{\cdot}$ electron capture (1′a)

$(A–A)_s + H^+ \rightarrow AH + A_s^{\cdot} + p_{vb}^+$ hole injection (1′b)

Propagation: $A_s^{\cdot} + e_{cb}^- + H^+ \rightarrow AH$ (2′a)

$A_s^{\cdot} + H^+ \rightarrow AH + p_{vb}^+$ (2′b)

(2) *Compound semiconductors* (A–B) assuming A to be the more electropositive and the final products to be AX^+ and either BY^+ or B_n (elemental B)

(a) Oxidative decomposition

Initiation: $(A–B)_s + X \rightarrow (AX)^+ + B_s^{\cdot} + e_{cb}^-$ (3a)

$(A–B)_s + p_{vb}^+ + X \rightarrow (AX)^+ + B_s^{\cdot}$ (3b)

Propagation: $B_s^{\cdot} + Y \rightarrow BY^+ + e_{cb}^-$ (4a)

$B_s^{\cdot} + p_{vb}^+ + Y \rightarrow BY^+$ (4b)

or

$B_s^{\cdot} + B_s^{\cdot} \rightarrow (B–B)_s$ dimer (4c)

or

$B_n + B_s^{\cdot} \rightarrow B_{n+1}$ elemental form (4d)

(b) reductive decomposition

Initiation: $(A–B)_s + e_{cb}^- + H^+ \rightarrow BH + A_s$ (3′a)

$(A–B)_s + H^+ \rightarrow BH + A_s^{\cdot} + p_{vb}^+$ (3′b)

Propagation: $A_s^{\cdot} + e_{cb}^- + H^+ \rightarrow AH$ (4′a)

$A_s^{\cdot} + H^+ \rightarrow AH + p_{vb}^+$ (4′b)

or

$$A_s^{\cdot} + A_s^{\cdot} \rightarrow (A\text{–}A)_s \qquad (4'c)$$

or

$$A_s^{\cdot} + A_n \rightarrow A_{n+1} \qquad (4'd)$$

Although there is a substantial number of pathways open to the system, we can restrict attention to the most likely by the following argument. We would expect that the redox energy for steps (1a) and (3a) would lie near the VB edge since, chemically, we are breaking a bond whose energy level will be near the top cf the filled band. It follows that the activation energy difference between (1a) and (1b) or (3a) and (3b) will be roughly equal to the bandgap. For larger bandgap materials, the *initiation* of anodic decomposition by *holes* rather than electron injection is overwhelmingly favoured. Similar considerations lead to the conclusion that the initiation of reductive decomposition is most likely by electron capture rather than hole injection, i.e. (1′b) and (3′b) are favoured over (1′a) and (3′a). However, the next steps are more difficult to predict and, for narrower bandgap materials, a mixture of propagation steps (2a) and (2b) or (2′a) and (2′b) will be favoured. This situation appears to hold in the case of Ge.

5.8 TUNNELING IN SEMICONDUCTORS

In semiconductors of high-dopant density and correspondingly thin depletion layers, tunneling may occur directly between the electrons in the conduction band and the surface of the electrode provided acceptor states or redox species in solution are available. The tunneling contribution to the total current has been considered by a number of workers [118–121] and the total anodic current can be written, with some generality, as

$$j^{+} \sim \int d\varepsilon \Pr(\varepsilon, V)\, \rho(\varepsilon)\, [1 - n(\varepsilon)]\, \mathscr{D}_{\text{red}}(\varepsilon, e_0\eta_0)\, \frac{2\pi|U|^2}{\hbar} \qquad (237)$$

where the integration is over the electron energy ε with respect to the Fermi level, $\Pr(\varepsilon, V)$ is the tunneling probability at band bending V, $n(\varepsilon)$ is the Fermi function, $\rho(\varepsilon)$ the density of states in the semiconductor, and $\mathscr{D}_{\text{red}}(\varepsilon, e_0\eta_0)$ the thermalised distribution of energy levels discussed above (with η_0 as the redox potential with respect to the Fermi level). The tunneling probability, within the WKB approximation, is

$$\Pr(\varepsilon, V) = \exp\left(-2\int_{x=0}^{x_b} |k(x)|\mathrm{d}x\right) \qquad (238)$$

where x_b is defined in Fig. 53 and

$$\frac{\hbar^2 k^2(x)}{2m^*} = \varepsilon - E_c(x) \qquad (239)$$

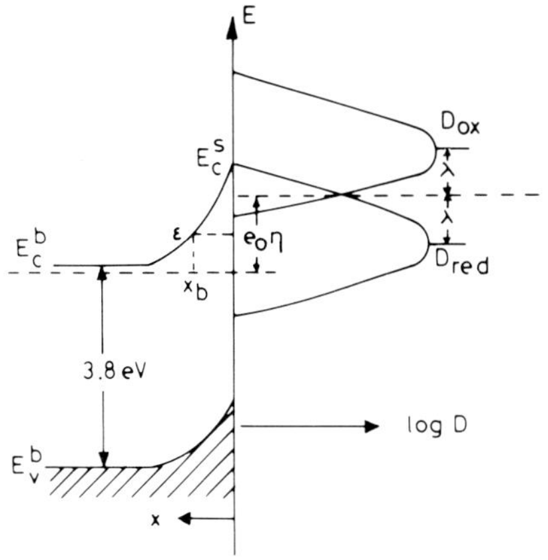

Fig. 53. Energy parameters for tunneling in n-SnO_2.

so we can write $Pr(\varepsilon, V)$ as

$$\Pr(\varepsilon, V) = \exp\left(-\frac{\sqrt{8m^*}}{\hbar}\int_0^{x_b}[E_c(x) - \varepsilon]^{1/2}\mathrm{d}x\right) \tag{240}$$

Since $E_c(x) = kT(W - x)^2/2e_0L_D^2$, the integral may be explicitly evaluated. If ε is zero, we obtain

$$\Pr(0, V) = \exp\left\{-\frac{2}{\hbar}\left(\frac{m^*\varepsilon_0\varepsilon_{sc}}{N_D}\right)^{1/2} V\right\} \tag{241}$$

The cathodic current may be calculated similarly, the only difference being that $\mathscr{D}_{red}$ must be replaced by

$$\mathscr{D}_{ox}(\varepsilon, e_o\eta_0) = \frac{1}{2\sqrt{\pi kT\lambda}}\exp\left\{-\frac{(\varepsilon - \lambda - e_o\eta_0)^2}{e\lambda kT}\right\} \tag{242}$$

and we replace $[1 - n(\varepsilon)]$ by $n(\varepsilon)$.

The final formulae for the tunneling contributions to the current are

$$j_T^a = -\frac{2\pi}{\hbar}e_0 d C_{red}|U|^2\int_{E_c^b}^{E_c^s}\rho(\varepsilon)[1 - n(\varepsilon)]\Pr(\varepsilon, V)\,\mathscr{D}_{red}(\varepsilon, e_0\eta_0)\,\mathrm{d}\varepsilon \tag{243}$$

$$j_T^c = \frac{2\pi}{\hbar}e_0 d C_{ox}|U|^2\int_{E_c^b}^{E_c^s}\rho(\varepsilon)\,n(\varepsilon)\Pr(\varepsilon, V)\,\mathscr{D}_{ox}(\varepsilon, e_0\eta_0)\,\mathrm{d}\varepsilon \tag{244}$$

where

$$\rho(\varepsilon) = \frac{1}{2\pi^2}\left(\frac{2m^*}{\hbar^2}\right)^{3/2} [\varepsilon - E_c^s + e_0(V - V_{fb})]^{1/2} \tag{245}$$

$$\Pr(\varepsilon, V) = \exp\left(-\frac{\sqrt{8m^*}}{\hbar}\int_0^{x_b} [E_c(x) - \varepsilon]^{1/2}\, dx\right) \tag{246}$$

$$E_c(x) = E_c^s + \left(\frac{N_D e_0^2}{2\varepsilon_0 \varepsilon_{sc}}\right) x(x - 2W) = E_c^s + \frac{x(x - 2W)kT}{2L_D^2} \tag{247}$$

and x_b satisfies

$$x_b(x_b - 2W) = \frac{2L_D^2}{kT}(\varepsilon - E_c^s) \qquad x_b < W$$

$$\mathscr{D}_{red}(\varepsilon, e_o\eta_0) = \frac{1}{2\sqrt{\pi kT\lambda}} \exp\left\{-\frac{(\lambda + \varepsilon - e_0\eta_0)^2}{4\lambda kT}\right\} \tag{248}$$

$$\mathscr{D}_{ox}(\varepsilon, e_0\eta_0) = \frac{1}{2\sqrt{\pi kT\lambda}} \exp\left\{-\frac{(-\lambda + \varepsilon - e_0\eta_0)^2}{4\lambda kT}\right\} \tag{249}$$

Evaluation of these integrals is then straightforward, if tedious. For the *cathodic* current, the very strongly decreasing value of $n(\varepsilon)$ leads to the result that only those electrons in the bulk of the semiconductor contribute to the tunneling current; they must tunnel right across the depletion layer, however, and the final result is

$$j_T^c = e_0 N_D d C_{ox}|U|^2 \left(\frac{\pi}{\hbar^2 kT\lambda}\right)^{1/2} \exp\left\{-\frac{2}{\hbar}\left(\frac{m^*\varepsilon_{sc}\varepsilon_0}{N_D}\right)^{1/2}(V - V_{fb})\right\}$$
$$\times \exp\left\{-\frac{(\Delta - e_0(V - V_0) - \lambda)^2}{4kT\lambda}\right\} \tag{250}$$

where $\Delta = \varepsilon_s - e_0(V - V_{fb})$ and $V - V_0 \equiv \eta_0$. If Δ is very small, the transfer coefficient $\alpha \equiv (kT/e_0)d(\ln i)/dV$ is given by

$$\alpha \simeq \frac{kT}{e_0}\frac{2}{\hbar}\left(\frac{m^*\varepsilon_0\varepsilon_{sc}}{N_D}\right)^{1/2} + \frac{e_0(V - V_0) + \lambda}{2\lambda} \equiv \alpha_{tun} + \alpha_p \tag{251}$$

and we may calculate α values for such typical values of the parameters as $N_D \simeq 10^{18}\,cm^{-3}$, $m^* = m_0$, and $\varepsilon_{sc} = 10$; this combination yields $\alpha_{tm} \simeq 4.4$; for $N_D \simeq 10^{19}\,cm^{-3}$, $\alpha_{tm} \simeq 1.3$ and for $n_s = 10^{20}\,cm^{-3}$, $\alpha_t \simeq 0.44$ at 298 K. These formulae have been extended by Kobayashi et al. to the situation where the inner coordination sphere of the redox couple makes a significant contribution to the overall electrode kinetics. If we have an angular frequency ω_c associated with the inner coordination sphere, then the electron transfer probability is given by [122]

TABLE 2

Tunneling currents for highly doped SnO_2

Carrier density/cm^{-3}	Band bending/V	j_T	
		SnO_2	TiO_2
10^{20}	1.0	0.024	1.5×10^{-87}
	0.5	1	3.9×10^{-44}
10^{19}	1.0	1.6×10^{-9}	3×10^{-275}
	0.5	1.8×10^{-4}	5.5×10^{-138}
10^{18}	1.0	2.5×10^{-32}	
	0.5	9.0×10^{-16}	

$$W_{if} = \frac{2\pi}{\hbar}|U|^2\left(\frac{1}{4\pi kT\lambda}\right)^{1/2}\sum_{m=-\infty}^{\infty}\exp\left[-\frac{(\varepsilon - \lambda_s - m\hbar\omega_c)^2}{4\lambda_s kT}\right] \times \exp\left\{-Z_c\cosh\left(\frac{\hbar\omega_c}{2kT}\right) + \frac{m\hbar\omega_c}{2kT}\right\}\mathrm{I}_m(Z_c) \tag{252}$$

where λ_s is the re-organisation energy associated with the solvent in the outer sphere, $Z_c = (\Delta_c^2/2)\ \mathrm{cosech}\ (\hbar\omega_c/2kT)$, $\Delta_c = R_{fo} - R_{io}$ (the change in the bond length in the inner coordination sphere), and $\mathrm{I}_m(x)$ is a spherical Bessel function defined by

$$\mathrm{I}_m(x) = \left(\frac{x}{2}\right)^m\sum_{k=0}^{\infty}\left[\left(\frac{x}{2}\right)^{2k}\frac{1}{k!(m+k)!}\right] \tag{253}$$

If $kT \gg \hbar\omega_c$, then by using the asymptotic expansion for $\mathrm{I}_m(x)$, we obtain the formula for W_{if} given above with $\lambda = \lambda_s + \lambda_c$ and $\lambda_c = \frac{1}{2}\hbar\omega_c\Delta_c^2$. If $kT \simeq \hbar\omega_c$, an approximate result may be obtained [123]

$$W_{if} \sim \frac{2\pi}{\hbar}|U|^2\left(\frac{1}{4\pi kT\lambda}\right)^{1/2}\exp\left\{-\frac{(\varepsilon - \lambda_s)^2}{4kT\lambda_s}\right\} \times \exp\left[-Z_c\left\{\cosh\left(\frac{\hbar\omega_c}{2kT}\right) - \cosh\left(\frac{\varepsilon\hbar\omega_c}{2kT\lambda_s}\right)\right\}\right] \tag{254}$$

and the cathodic current is given, approximately, by

$$j_v^c = e_0 N_D C_{ox} d|U|^2\left(\frac{\pi}{\hbar^2 kT\lambda_s}\right)^{1/2}\exp\left\{-\frac{2}{\hbar}\left(\frac{m^*\varepsilon_0\varepsilon_{sc}}{N_D}\right)^{1/2}(V - V_{fb})\right\} \times \exp\left\{-\frac{[e_0(V - V_0) + \lambda_s]^2}{4\lambda_s kT}\right\}\exp\left[-Z_c\left\{\cosh\left(\frac{\hbar\omega_c}{2kT}\right) - \cosh\left(-\frac{\hbar\omega_c e_0}{2\lambda_s kT}(V - V_0)\right)\right\}\right] \tag{255}$$

with a transfer coefficient

$$\alpha = \frac{kT}{e_0}\frac{2}{\hbar}\left(\frac{m^*\varepsilon_0\varepsilon_{sc}}{N_D}\right)^{1/2} + \frac{e_0(V - V_0) + \lambda_s}{2\lambda_s} + \frac{kT}{e_0}\frac{\Delta_c^2}{2}\left(\frac{\hbar\omega_c}{2kT}\right)\frac{e_0}{\lambda_s}$$
$$\times \operatorname{cosech}\left(\frac{\hbar\omega_c}{2kT}\right)\sinh\left\{-\frac{\hbar\omega_c e_0}{2\lambda_s kT}(V - V_0)\right\} \tag{256}$$

Model calculations have been carried out using a variety of parameter values and some relative values are given in Table 2 appropriate to SnO_2 ($m^* = 0.2m_0$, $\varepsilon_s = 10$) and TiO_s ($m^* = 8m_0$, $\varepsilon_s = 173$) and assuming that the last two exponential terms in eqn. (255) are unity at both potentials. It is evident from these figures that the tunneling current is very sensitive indeed to the bulk properties of the semiconductor, a result also reported experimentally [124].

For anodic currents, the approximations that lead to eqn. (250) are clearly not valid. It can be shown that an approximate expression for the tunneling probability is [125]

$$\Pr(\varepsilon, V) = \exp\left\{-\sqrt{\frac{4m^*L_D^2}{kT\hbar^2}}\left[(E_c^s - \varepsilon)^{1/2}(E_c^s - E_c^b)^{1/2} - (\varepsilon - E_b^c)\right.\right.$$
$$\left.\left.\times \ln\left(\frac{(E_c^s - E_c^b)^{1/2}(E_c^s - \varepsilon)^{1/2}}{(\varepsilon - E_c^b)^{1/2}}\right)\right]\right\} \tag{257}$$

From this, the current as a function of V may be calculated using trial values of λ for various redox couples. Some experimental results obtained by Memming and Möllers [126] are shown in Fig. 54 and it is clear that, as for the cathodic current, there is a strong dependence on dopant density. From the shape of the curves, λ values for the respective redox couples could be obtained, as shown in Fig. 54. It should be noted that these workers did, in fact, use a slightly different from of eqn. (257) derived for a triangular barrier and taking the form

$$\Pr(\varepsilon, V) = \exp\left\{-\frac{k(E_c^s - \varepsilon)^{3/2}}{(E_c^s - E_c^b)^{1/2} + (\varepsilon - E_c^b)^{1/2}}\right\} \tag{258}$$

where

$$k = \frac{8}{3\hbar}\left(\frac{m^*\varepsilon_0\varepsilon_{sc}}{e_0N_D}\right)^{1/2}$$

The values of λ derived by these authors may be compared with the values calculated by Hale using the Marcus theory. For Fe^{3+}/Fe^{2+} and Ce^{4+}/Ce^{3+}, agreement is good, but for $[Fe(CN)_6]^{3-/4-}$, agreement is much poorer, though the value obtained experimentally (0.4 ± 0.2 eV) does compare more favourably with the value of 0.65 eV estimated by Levich [127] and the range 0.65 (in base) to 1.5 eV (acid) calculated by Frese [128]. A determination of the value from n-ZnO gave $\lambda = 0.75$ eV in the pH range 4.8–12.0 [129].

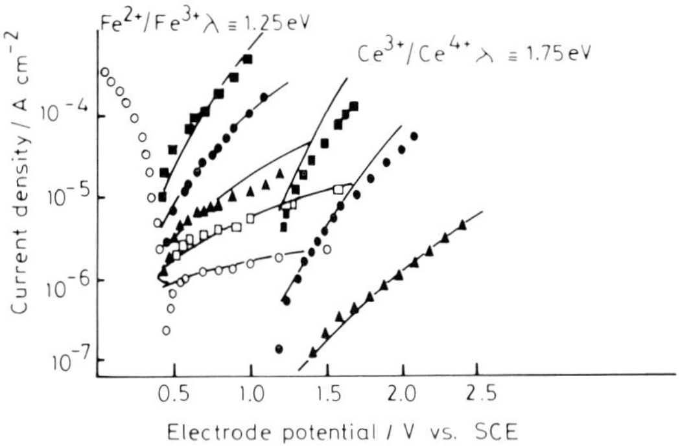

Fig. 54. Interfacial current vs. electrode potential for 0.05 M Ce^{3+}/Ce^{4+} and 0.05 M Fe^{3+}/Fe^{2+} at SnO_2 electrodes in 0.05 M H_2SO_4 at donor densities of ■, 5.2×10^{18} cm^{-3}; ●, 1.8×10^{18} cm^{-3}; ▲, 6.8×10^{18} cm^{-3}; □, 2.5×10^{18} cm^{-3}; and ○, 5.0×10^{17} cm^{-3}. The solid lines represent best fits to eqn. (258).

The tunneling through depletion layer barriers will be enormously facilitated if there are deep-lying trap states that can act as "transit stations". Tunneling involving the type of process shown in Fig. 55 is termed *resonance tunneling* and has been investigated theoretically by Schmickler [121]. If the tunneling probabilities are R_1 from electrolyte to resonance state and R_2 from this state to the CB, where

$$R_1 \sim \exp\left\{-2\int_0^l |k(x)|\,dx\right\}; \quad R_2 \sim \exp\left\{-2\int_l^{l_0} |k(x)|\,dx\right\} \tag{259}$$

and l and l_0 are the classical turning points, then the overall tunneling probability is given by

$$R = \frac{R_1 R_2}{R_1 + R_2} \tag{260}$$

The contribution of the states at E_t is then

$$\delta j_a(l) = C[1 - n(E_t)]R(E_t, e_0\eta_0)\theta(E_t - E_c^b)\mathscr{D}(E_t, e_0\eta_0) \tag{261}$$

where $\theta(x)$ is the Heaviside function.

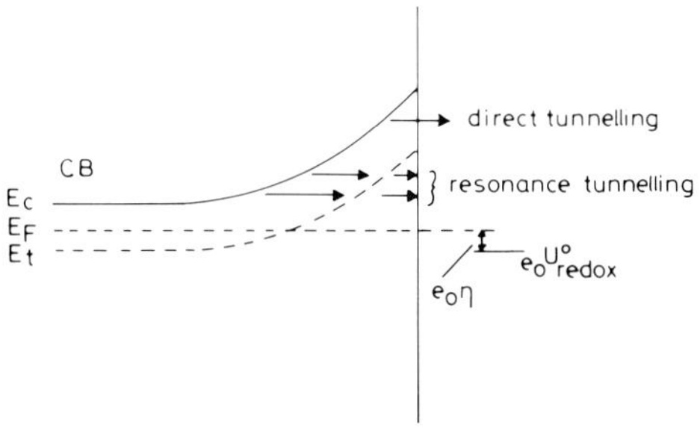

Fig. 55. Energy parameters for resonant tunnelling.

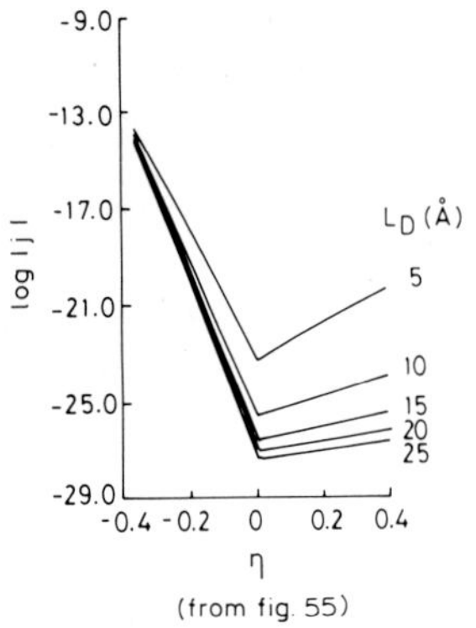

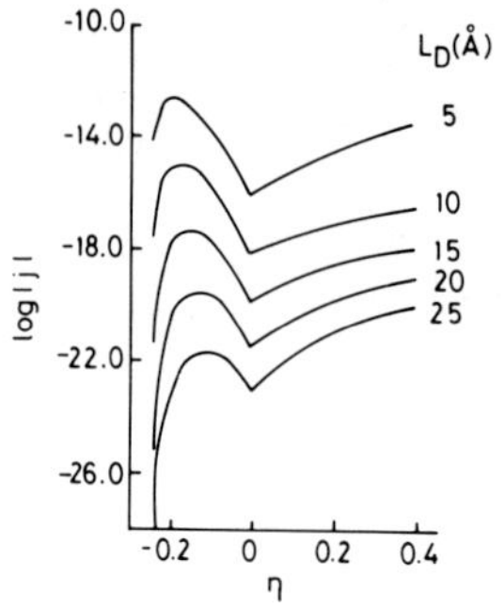

Fig. 56. Current component due to direct elastic tunnelling for a semiconductor with varying L_D values. The L_D values are given against each plot.

Fig. 57. Current component due to resonant tunnelling for the electrode of Fig. 56. The figures against the plots indicate the L_D values.

Near flat-band potential, the space-charge layer becomes considerably more transparent and the tunneling solution correspondingly more complex. By considering a monolayer of water molecules on the surface of the semiconductor that provide a square potential well, analytical solutions were obtained by Schmickler under the assumptions

(i) $E_c(x) - E_c^b = (E_c^s - E_c^b)\exp(-x/L_D)$, which is valid if $E_c^s - E_c^b \lesssim kT$ and

(ii) the water layer has thickness d and potential barrier V_B with respect to E_c^b.

The tunneling probability is now given by

$$\Pr(\varepsilon, V) = \frac{\hbar^2}{2m} \cdot 2k\mathrm{e}^{-kd} \cdot \bar{B} A_2 \tag{262}$$

$$k = \frac{1}{\hbar}\{2m^*(V_B - \varepsilon)\}^{1/2}; \quad A_2 = -\frac{2i\sqrt{b}}{k\psi_2(0) + \psi_2'(0)} \tag{263}$$

$$b = 2m\varepsilon/\hbar^2; \quad \psi_2(x) = \mathrm{e}^{ix\sqrt{b}} f_{-iL_D\sqrt{b}}(\mathrm{e}^{-x/L_D})$$

$$f_\alpha(y) = \Gamma(1 + 2\alpha)\sum_{n=0}^{\infty}\frac{a^n y^n}{n!\Gamma(4n + 2\alpha)}; \quad a = \frac{2m}{\hbar^2}(E_c^s - E_c^b) \tag{264}$$

The total tunneling current then consists of a resonance tunneling current and the direct elastic current. These two currents are shown plotted separately in Figs. 56 and 57 and it can be seen that the most important qualitative result is that the resonance tunneling mode gives a maximum in the current distribution, though this maximum may be difficult to observe in practice.

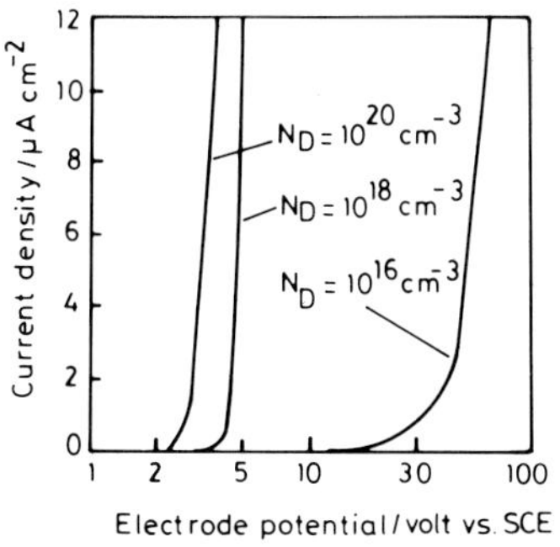

Fig. 58. Potential dependence of break-down currents at ZnO electrodes with different donor densities.

5.9 BREAKDOWN

Figure 58 shows the current density for n-ZnO at positive bias potentials [130]. It can be seen that, above a certain potential, which depends upon the donor density, there is a sharp increase in the current. For the most highly doped material, this rise occurs just above the potential at which the conduction band in the bulk has the same energy as the valence band at the surface. Higher potentials are needed with the lower N_D values and the obvious model is that holes are generated in the VB at the surface by electron tunneling from the VB to the CB. The classical tunneling distance is easily seen to be

$$\sqrt{2}\left(\frac{e_0 V}{kT}\right)^{1/2} L_D - \sqrt{2}\, L_D \left\{\frac{e_0 V - E_g}{kT}\right\} \tag{265}$$

which decreases, with increasing V, where $V = V_{app} - V_{fb}$, consistent with the data of Fig. 58.

5.10 THE A.C. THEORY OF SEMICONDUCTORS WITH FARADAIC CURRENT FLOWING

In the presence of a faradaic current, the a.c. response of a semiconductor becomes significantly more complex. Nevertheless, using the theory of Sect. 3, it is possible to derive expressions for the a.c. response of the semiconductor–electrolyte interface both in the simple case of electron transfer from CB to electrolyte and in the case where surface states play an intermediate role.

Case A. Dark current present, no surface states.

The particle flux at the surface is given by

$$j_f = k_f n_{1s} C^s_{ox} - k_b C^s_{Red} \tag{266}$$

Using small signal theory, we find that the fluxes and concentrations can be written in the form $j_f = \bar{j}_f + \tilde{j}_f e^{i\omega t}$, $n_s = \bar{n}_s + \tilde{n}_s e^{i\omega t}$ where $\bar{j}_f$ is the d.c. flux and $\tilde{j}_f$ the a.c. flux, $\bar{n}_s$ is the stationary concentration of electrons at the surface, and $\tilde{n}_s$ is the impressed a.c. variation.

At equilibrium, we will suppose that the d.c. current is very small, so that

$$\bar{j}_f \simeq 0 \qquad k_f \bar{n}_s \bar{C}^s_{ox} = k_b \bar{C}^s_{Red} \tag{267}$$

The a.c. flux is given by

$$\tilde{j}_f = k_f \bar{n}_s \tilde{C}^s_{ox} + k_f \tilde{n}_s \bar{C}^s_{ox} - k_b \tilde{C}^s_{Red} \tag{268}$$

The a.c. concentrations C_{ox} and $\tilde{C}_{Red}$ are given by

$$i\omega \tilde{C}_{ox} = D_0 \frac{d^2 \tilde{C}_{ox}}{dx^2} \tag{269}$$

$$i\omega \tilde{C}_{Red} = D_R \frac{d^2 \tilde{C}_{Red}}{dx^2} \tag{270}$$

which solves immediately to give

$$\tilde{C}_{ox}(x) = \tilde{C}^s_{ox}\, e^{-\alpha x} \tag{271}$$

and

$$\tilde{C}_{Red}(x) = \tilde{C}^s_{Red}\, e^{-\beta x} \tag{272}$$

where

$$\alpha = \left(\frac{i\omega}{D_0}\right)^{1/2} \qquad \beta = \left(\frac{i\omega}{D_R}\right)^{1/2} \tag{273}$$

Evidently there must be flux balance at the surface.

$$\text{flux of O} = -\text{ flux of R} \tag{274}$$

which leads to

$$D_0 \alpha \tilde{C}^s_{ox} = -D_R \beta \tilde{C}^s_{Red} \Rightarrow D_0^{1/2} \tilde{C}^s_{ox} = -D_R^{1/2} \tilde{C}^s_{Red} \tag{275}$$

The nett particle flux is given by

$$\tilde{j}_f = D_0 \left(\frac{d\tilde{C}_{ox}}{dx}\right)_0 = -D_R \left(\frac{d\tilde{C}_{Red}}{dx}\right)_0 = -(i\omega D_0)^{1/2}\, \tilde{C}^s_{ox}$$

$$= +(i\omega D_R)^{1/2} \tilde{C}^s_{Red} \tag{276}$$

$$= -k_f \bar{n}_s \frac{\tilde{j}_f}{(i\omega D_0)^{1/2}} + k_f \tilde{n}_s \bar{C}^s_{ox} - k_b \frac{\tilde{j}_f}{(i\omega D_R)^{1/2}} \tag{277}$$

so, finally

$$\tilde{\jmath}_{\mathrm{f}} = k_{\mathrm{f}}\tilde{n}_{\mathrm{s}}\bar{C}^{\mathrm{s}}_{\mathrm{ox}} \Bigg/ \left\{1 + \frac{k_{\mathrm{f}}\bar{n}_{\mathrm{s}}}{(i\omega D_0)^{1/2}} + \frac{k_{\mathrm{b}}}{(i\omega D_{\mathrm{R}})^{1/2}}\right\} \equiv P\tilde{n}_{\mathrm{s}} \tag{278}$$

The total current comprises the sum of the displacement and particle flux currents.

$$\tilde{\jmath} = i\omega\varepsilon_0\varepsilon_{\mathrm{sc}}\tilde{\mathscr{E}}_{\mathrm{s}} + e_0\tilde{\jmath}_{\mathrm{f}}; \quad \tilde{\mathscr{E}}_{\mathrm{s}} = \frac{1}{i\omega\varepsilon_0\varepsilon_{\mathrm{sc}}}(\tilde{\jmath} - e_0P\tilde{n}_{\mathrm{s}}) \tag{279}$$

Using the development of Sect. 4, in particular eqns. (87), (90), (112), and (103), we have

$$\tilde{n} - \frac{\varepsilon_0\varepsilon_{\mathrm{sc}}\bar{\mathscr{E}}_{\mathrm{s}}\tilde{\mathscr{E}}_{\mathrm{s}}}{kT} - \frac{e_0\tilde{\phi}_{\mathrm{s}}N_{\mathrm{D}}}{kT} = \frac{i\omega\varepsilon_0\varepsilon_{\mathrm{sc}}\tilde{\phi}_{\mathrm{s}}}{e_0D_{\mathrm{n}}} - \frac{\tilde{\jmath}L_{\mathrm{s}}}{e_0D_n} \tag{280}$$

$$\tilde{n}_{\mathrm{s}} - \frac{e_0\bar{n}_{\mathrm{s}}\tilde{\phi}_{\mathrm{s}}}{kT} = \frac{i\omega\varepsilon_0\varepsilon_{\mathrm{sc}}}{e_0D_n}\left(\frac{\tilde{\mathscr{E}}_{\mathrm{s}}L_{\mathrm{D}}^2}{W}\right) - \frac{\tilde{\jmath}}{e_0D_{\mathrm{n}}}\left(\frac{\bar{n}_{\mathrm{s}}}{N_{\mathrm{D}}}(L_{\mathrm{s}} - W) + \frac{L_{\mathrm{D}}^2}{W}\right) \tag{281}$$

Combining eqns. (279) and (281) we have, if $L_{\mathrm{s}} \gg W$

$$\tilde{n}_{\mathrm{s}} - \frac{e_0\bar{n}_{\mathrm{s}}\tilde{\phi}_{\mathrm{s}}}{kT} = \frac{L_{\mathrm{D}}^2}{e_0D_{\mathrm{n}}W}(\tilde{\jmath} - e_0P\tilde{n}_{\mathrm{s}}) - \frac{\tilde{\jmath}}{e_0D_{\mathrm{n}}}\left(\frac{\bar{n}_{\mathrm{s}}}{N_{\mathrm{D}}}L_{\mathrm{s}} + \frac{L_{\mathrm{D}}^2}{W}\right) \tag{282}$$

$$= -\frac{L_{\mathrm{D}}^2P\tilde{n}_{\mathrm{s}}}{D_{\mathrm{n}}W} - \frac{\tilde{\jmath}\bar{n}_{\mathrm{s}}L_{\mathrm{s}}}{e_0N_{\mathrm{D}}D_{\mathrm{n}}} \tag{283}$$

so that the a.c. component, $\tilde{n}_{\mathrm{s}}$, has the value

$$\tilde{n}_{\mathrm{s}} = \frac{-\tilde{\jmath}K_2 + \tilde{\phi}_{\mathrm{s}}K_3}{e_0P + (e_0D_{\mathrm{n}}W/L_{\mathrm{D}}^2)} \tag{284}$$

where the values of K_2 and K_3 are given by

$$K_2 = \frac{W\bar{n}_{\mathrm{s}}L_{\mathrm{s}}}{N_{\mathrm{D}}L_{\mathrm{D}}^2}; \quad K_3 = \frac{e_0^2\bar{n}_{\mathrm{s}}D_{\mathrm{n}}W}{L_{\mathrm{D}}^2kT} \tag{285}$$

Combining eqns. (279), (280), and (285) gives

$$\left[\frac{-\tilde{\jmath}K_2 + \tilde{\phi}_{\mathrm{s}}K_3}{e_0P + (e_0D_{\mathrm{n}}W/L_{\mathrm{D}}^2)}\right] - \frac{\bar{\mathscr{E}}_{\mathrm{s}}}{kT}\left\{\frac{-\tilde{\jmath}}{i\omega} - \frac{e_0P}{i\omega}\left[\frac{-\tilde{\jmath}K_2 + \tilde{\phi}_{\mathrm{s}}K_3}{e_0P + (e_0D_{\mathrm{n}}W/L_{\mathrm{D}}^2)}\right]\right\}$$

$$- \frac{e_0\tilde{\phi}_{\mathrm{s}}N_{\mathrm{D}}}{kT} = \frac{i\omega e_0\varepsilon_{\mathrm{sc}}\tilde{\phi}_{\mathrm{s}}}{e_0D_{\mathrm{n}}} - \frac{\tilde{\jmath}L_{\mathrm{s}}}{e_0D_{\mathrm{n}}} \tag{286}$$

whence

$$\frac{\tilde{\jmath}}{i\omega}\left\{1 + \frac{K_2[P + (i\omega kT/e_0\bar{\mathscr{E}}_{\mathrm{s}})]}{P + X_1} - \frac{i\omega kTL_{\mathrm{s}}}{e_0D_{\mathrm{n}}\bar{\mathscr{E}}_{\mathrm{s}}}\right\}$$

$$= \tilde{\phi}_{\mathrm{s}}\left\{-\frac{e_0N_{\mathrm{D}}}{\bar{\mathscr{E}}_{\mathrm{s}}} + \frac{K_3}{i\omega(P + X_1)}\left(P + \frac{i\omega kT}{e_0\bar{\mathscr{E}}_{\mathrm{s}}}\right) - \frac{i\omega\varepsilon_0\varepsilon_{\mathrm{sc}}}{e_0D_{\mathrm{n}}}\right\} \tag{287}$$

where $X_1 = D_n W/L_D^2$. From this

$$\frac{\tilde{j}}{i\omega}\left\{1 + \frac{K_2[P + (i\omega kT/e_0\bar{\mathscr{E}}_s)]}{P + X_1} - \frac{i\omega kTL_s}{e_0 D_n \bar{\mathscr{E}}_s}\right\}$$

$$= \tilde{\phi}_s\left\{-\frac{i\omega e_0 N_D}{\bar{\mathscr{E}}_s} + \frac{K_3[P + (i\omega kT/e_0\bar{\mathscr{E}}_s)]}{P + X_1} + \frac{\omega^2\varepsilon_0\varepsilon_{sc}}{e_0 D_n}\right\} \tag{288}$$

Putting

$$R_b = \frac{L_s kT}{e_0^2 D_n N_D} \equiv \frac{L_s}{e_0\mu_n N_D} \tag{289}$$

$$C_{sc} = -\frac{e_0 N_D}{\bar{\mathscr{E}}_s} \tag{290}$$

$$R_{sc} = \frac{kTL_D^2}{e_0 D_n \bar{n}_s W} \approx \frac{W}{e_0 \bar{n}_s \mu_n} \tag{291}$$

and $K_3 = 1/R_{sc}$, $K_2 = R_b K_3 = R_b/R_{sc}$. Neglecting $\omega^2\varepsilon\varepsilon_0/e_0 D_n$, this gives

$$\tilde{j}\left(1 + \frac{R_b}{R_{sc}}\left[\frac{P + (i\omega kT/e_0\bar{\mathscr{E}}_s)}{P + X_1}\right] + i\omega R_b C_{sc}\right)$$

$$= \tilde{\phi}_s\left(i\omega C_{sc} + \frac{1}{R_{sc}}\left[\frac{P + (i\omega kT/\varepsilon_0\bar{\mathscr{E}}_s)}{P + X_1}\right]\right) \tag{292}$$

If an admittance Y be defined by the expression

$$Y = \frac{1}{R_{sc}}\left[\frac{P + (i\omega kT/e_0\bar{\mathscr{E}}_s)}{P + X_1}\right] \tag{293}$$

then

$$\tilde{j}(1 + YR_b + w\omega R_b C_{sc}) = \tilde{\phi}_s(i\omega C_{sc} + Y) \tag{294}$$

and the appropriate equivalent circuit has the form

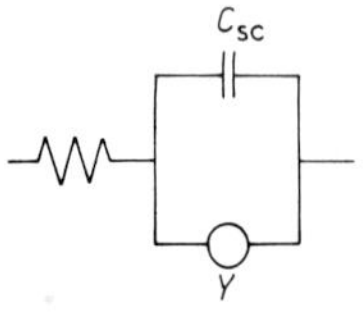

Save near flat-band potential, $i\omega kT/e_0\bar{\mathscr{E}}_s$ will be very small, having values of the order $(10^{-6}$ to $10^{-7})\omega$ cm s^{-1}. This compares with a value of typically 10^4 or greater for X_1. We can thus, without any loss of generality, assume that $X_1 \gg i\omega kT/e_0\bar{\mathscr{E}}_s$. We then have three possible cases, depending on the relative values of P and X_1, where P is deinfed in eqn. (278).

(i) $P \gg X_1$.

The current and voltage are related by

$$\tilde{j}\left(1 + \frac{R_b}{R_{sc}} + i\omega R_b C_{sc}\right) = \tilde{\phi}_s\left(i\omega C_{sc} + \frac{1}{R_{sc}}\right) \tag{295}$$

leading to the equivalent circuit

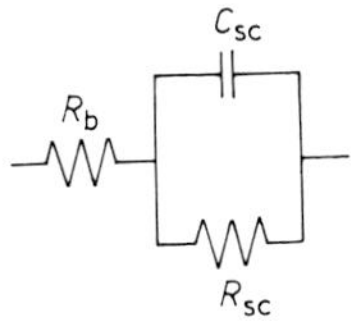

(ii) $P \ll i\omega kT/e_0$.

Under these circumstances, R_b/R_{sc} is multiplied by the small factor $i\omega kT/(e_0\bar{\mathscr{E}}_s X_1)$ and, to a good approximation, this leads to the very simple a.c. equivalent circuit

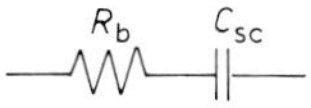

(iii) $i\omega kT/e_0\bar{\mathscr{E}}_s \ll P \ll X_1$.

Under these circumstances, the admittance Y becomes

$$Y = \frac{1}{R_{sc}}\left(\frac{P}{P + X_1}\right) \tag{296}$$

Writing the impedance $Z \equiv 1/Y$

$$\begin{aligned} Z &= R_{sc}\left(1 + \frac{X_1}{P}\right) = R_{sc} + \frac{X_1 R_{sc}}{P} \\ &= R_{sc} + \frac{X_1 R_{sc}}{k_f \bar{C}^s_{ox}}\left[1 + \frac{k_f \bar{n}_s}{(i\omega D_0)^{1/2}} + \frac{k_b}{(i\omega D_R)^{1/2}}\right] \qquad (297) \\ &= \left(R_{sc} + \frac{X_1 R_{sc}}{k_f \bar{C}^s_{ox}}\right) + \frac{R_{sc} X_1}{k_f \bar{C}^s_{ox}}\left\{\frac{k_f \bar{n}_s (1 - i)}{(2\omega D_0)^{1/2}} + \frac{k_b (1 - i)}{(2\omega D_R)^{1/2}}\right\} \qquad (298) \\ &= R_{sc} + R_{ct} + \text{—}W\text{—} \qquad (299) \end{aligned}$$

where —W— is a "Warburg" impedance and the equivalent circuit has the form

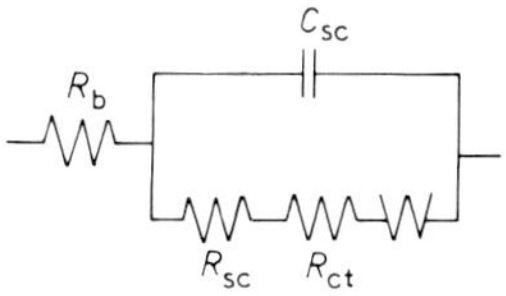

Case B. Faradic current passing through the conduction band; surface states filling and emptying independently.

The particle flux at the surface now consists of both faradic and surface state terms

$$j_f = k_{tf} n_s (N_t - n_t) - k_{tb} + k_f n_s C^s_{ox} - k_b C^s_{Red} = j^t_f + j^s_f \tag{300}$$

where k_{tf} and k_{tb} are the rate constants for electron transfer to and from the surface state, j_f^t is the nett flux to the surface state and j_f^s that to the solution. The two fluxes are essentially independent and we may write

$$j_f^t = i\omega\tilde{n}_t = k_{tb}\left\{\frac{\bar{n}_s}{N_D}\exp(e_t')(N_t - \tilde{n}_t) - \frac{\bar{n}_s \exp(e_t')\tilde{n}_t}{N_D} - \tilde{n}_t\right\} \tag{301}$$

$$= \frac{k_{tb}N_t\tilde{n}_s}{f_V N_D} - f_r\tilde{n}_t \tag{302}$$

$$j_f^s = P\tilde{n}_s \tag{303}$$

where e_t' is defined in eqn. (104) and f_r and f_V in eqn. (111). Hence we have

$$\tilde{n}_s = \frac{\tilde{n}_t(f_r + i\omega)f_V}{k_{tb}N_t} \equiv Q\tilde{n}_t \tag{304}$$

so the total current is given by

$$\tilde{j} = i\omega\varepsilon_0\varepsilon_{sc}\tilde{\mathscr{E}}_s + e_0\tilde{j}_f^t + e_0\tilde{j}_f^s = i\omega\varepsilon_0\varepsilon_{sc}\tilde{\mathscr{E}}_s + i\omega e_0\tilde{n}_t + e_0P\tilde{n}_s \tag{305}$$

$$= i\omega\varepsilon_0\varepsilon_{sc}\tilde{\mathscr{E}}_s + e_0\tilde{n}_s\left(\frac{i\omega}{Q} + P\right) \tag{306}$$

$$= e_0\tilde{n}_s(PQ + i\omega) + i\omega\varepsilon_0\varepsilon_{sc}\tilde{\mathscr{E}}_s \tag{307}$$

Inserting eqn. (305) into eqn. (281), we find

$$\tilde{n}_s - \frac{e_0\bar{n}_s\tilde{\phi}_s}{kT} = \frac{L_D^2}{e_0D_nW}(\tilde{j} - e_0P\tilde{n}_s - i\omega e_0\tilde{n}_t) - \frac{\tilde{j}}{e_0D_n}\left(\frac{\bar{n}_sL_s}{N_D} + \frac{L_D^2}{W}\right) \tag{308}$$

$$= -\frac{L_D^2}{D_nW}(P\tilde{n}_s + i\omega\tilde{n}_t) - \frac{\tilde{j}\bar{n}_sL_s}{e_0D_nN_D} \tag{309}$$

and

$$Q\tilde{n}_t - \frac{e_0\bar{n}_s\tilde{\phi}_s}{kT} = -\frac{L_D^2}{D_nW}(PQ + i\omega)\tilde{n}_t - \frac{\tilde{j}\bar{n}_sL_s}{e_0D_nN_D} \tag{310}$$

whence

$$e_0\tilde{n}_t = \frac{-\tilde{j}K_2 + K_3\tilde{\phi}_s}{(D_nWQ/L_D^2) + PQ + i\mu} \tag{311}$$

$$e_0\tilde{n}_s = \frac{Q[-\tilde{j}(R_b/R_{sc}) + (\tilde{\phi}_s/R_{sc})]}{(D_nWQ/L_D^2) + PQ + i\omega} \tag{312}$$

where K_2 and K_3 are given by

$$K_2 = \frac{\bar{n}_sL_sW}{N_DL_D^2} = R_bK_3 = \frac{R_b}{\mathrm{R}_{sc}} \tag{313}$$

$$K_3 = \frac{e_0^2 \bar{n}_s D_n W}{L_D^2 kT} = \frac{1}{R_{sc}}$$

Inserting eqns. (311) and (312) into eqn. (280) and using eqn. (307), we have

$$\frac{Q[-\tilde{j}(R_b/R_{sc}) + (\tilde{\phi}_s/R_{sc})]}{e_0[(D_n WQ/L_D^2) + PQ + i\omega]} -$$

$$\frac{\bar{\mathscr{E}}_s}{kT}\left\{\frac{\tilde{j}}{i\omega} - \frac{(PQ + i\omega)}{i\omega}\frac{[-\tilde{j}(R_b/R_{sc}) + (\tilde{\phi}_s/R_{sc})]}{(D_n WQ/L_D^2) + PQ + i\omega}\right\} - \frac{e_0\tilde{\phi}_s N_D}{kT}$$

$$= \frac{i\omega\varepsilon_0\varepsilon_{sc}\tilde{\phi}_s}{e_0 D_n} - \frac{\tilde{j}L_s}{e_0 D_n} \tag{314}$$

$$\tilde{j}\left[1 + \frac{R_b}{R_{sc}}\frac{PQ + i\omega + (Qi\omega kT/e_0\bar{\mathscr{E}}_s)}{PQ + i\omega + (D_n WQ/L_D^2)} - \frac{i\omega L_s kT}{e_0 D_n \bar{\mathscr{E}}_s}\right]$$

$$= \tilde{\phi}_s\left[-\frac{i\omega e_0 N_D}{\bar{\mathscr{E}}_s} + \frac{1}{R_{sc}}\frac{PQ + i\omega + (i\omega QkT/e_0\bar{\mathscr{E}}_s)}{PQ + i\omega + (D_n WQ/L_D^2)}\right] \tag{315}$$

Let the admittance Y' be defined by

$$Y' = \frac{1}{R_{sc}}\left[\frac{PQ + i\omega + (i\omega QkT/e_0\bar{\mathscr{E}}_s)}{PQ + i\omega + (D_n WQ/L_D^2)]}\right] \tag{316}$$

Then, the equivalent circuit can be written as

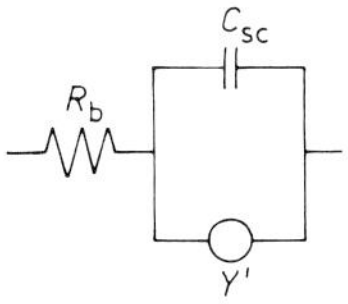

Note that if Q is large (i.e. N_t is small)

$$Y' \to \frac{1}{R_{sc}}\left[\frac{P + (i\omega kT/e_0\bar{\mathscr{E}}_s)}{P + X_1}\right] \tag{317}$$

as above [see eqn. (293)]. If $P \to 0$ and $kT/e_0\,\bar{\mathscr{E}}_s \ll 1$, then

$$Y' \to \frac{1}{R_{sc}}\left[\frac{i\omega}{i\omega + (D_n WQ/L_D^2)}\right] \tag{318}$$

and the corresponding impedance $Z' \equiv 1/Y'$ is given by

$$Z' = R_{sc}\left[1 + \frac{D_n WQ}{i\omega L_D^2}\right] = R_{sc}\left[1 + \frac{k_s f_V}{k_{tb}}\left(1 + \frac{f_r}{i\omega}\right)\right] \tag{319}$$

$$= R_{sc} + \frac{R_{sc}k_s f_V}{k_{tb}} + \frac{R_{sc}k_s f_V f_r}{i\omega k_{tb}} \tag{320}$$

References pp. 242–246

where, as above

$$k_s = \frac{D_n W}{L_D^2} \frac{N_D}{N_t}$$

so Y' can be decomposed into two series terms, a resistor R_{ss} and a capacitor C_{ss} whose values are

$$R_{ss} = R_{sc}\left(1 + \frac{k_s f_V}{k_{tb}}\right) \tag{321}$$

$$C_{ss} = \frac{k_{tb}}{R_{sc} k_s f_V f_r} \tag{322}$$

For the mixed cases, whewre both P and Q must be considered, then in the case that $kT/e_0\bar{\mathscr{E}}_s \rightarrow 0$, the value of Y' can be written

$$Y' = \frac{1}{R_{sc}}\left[\frac{PQ + i\omega}{PQ + i\omega + (D_n WQ/L_D^2)}\right] \tag{323}$$

and Y' can be decomposed to give the equivalent circuit

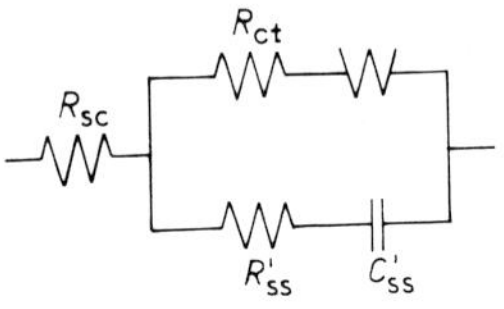

where $R_{ct} + —W— \equiv R_{sc}X_1/P$ as in eqns. (297)–(299) and $R'_{ss} + (1/i\omega C'_{ss}) \equiv QR_{sc}X_1/i\omega$ as defined in Sect. 4.

Case C. Faradic current involving surface states.

In this case, we imagine the surface state being fed from the CB and in turn feeding a redox couple in solution. The electron balance in the surface state can be written

$$\frac{\partial n_t}{\partial t} = k_{tf}(N_t - n_t)n_s - k_{tb}n_t - k_{ss}^f n_t C_{ox}^s + k_{ss}^b(N_t - n_t)C_{Red}^s \tag{324}$$

so using the usual small-signal theory

$$\begin{aligned} i\omega\tilde{n}_t &= k_{tf}(N_t - \bar{n}_t)\tilde{n}_s - (k_{tb} + k_{tf}\bar{n}_s)\tilde{n}_t - k_{ss}^f\tilde{n}_t\bar{C}_{ox}^s - k_{ss}^f\bar{n}_t\tilde{C}_{ox}^s \\ &\quad + k_{ss}^b\tilde{C}_{Red}^s(N_t - \bar{n}_t) - k_{ss}^b\bar{C}_{Red}^s\tilde{n}_t \\ &= \tilde{j}_f^s - \tilde{j}_f^r \end{aligned} \tag{325}$$

where the fluxes $\tilde{j}_f^s$ and $\tilde{j}_f^r$ are given by

$$\tilde{j}_f^s = k_{tf}\tilde{n}_s(N_t - \bar{n}_t) - (k_{tb} + k_{fb}\bar{n}_s)\tilde{n}_t \tag{326}$$

$$\tilde{j}_f^r = k_{ss}^f\bar{n}_t\tilde{C}_{ox}^s - k_{ss}^b\tilde{C}_{Red}^s(N_t - \bar{n}_t) + \tilde{n}_t(k_{ss}^f\bar{C}_{ox}^s + k_{ss}^b\bar{C}_{Red}^s)$$

Now, with the ansatz

$$\tilde{j}_{\mathrm{f}}^{\mathrm{r}} = -(i\omega D_0)^{1/2}\tilde{C}_{\mathrm{ox}}^{\mathrm{s}} = +(i\omega D_{\mathrm{R}})^{1/2}\tilde{C}_{\mathrm{Red}}^{\mathrm{s}} \tag{327}$$

we can write, for the flux

$$\tilde{j}_{\mathrm{f}}^{\mathrm{r}} = \frac{\tilde{n}_{\mathrm{t}}(k_{\mathrm{ss}}^{\mathrm{f}}\bar{C}_{\mathrm{ox}}^{\mathrm{s}} + k_{\mathrm{ss}}^{\mathrm{b}}\bar{C}_{\mathrm{Red}}^{\mathrm{s}})}{\left[1 + \dfrac{k_{\mathrm{ss}}^{\mathrm{f}}\bar{n}_{\mathrm{t}}}{(i\omega D_0)^{1/2}} + \dfrac{k_{\mathrm{ss}}^{\mathrm{b}}(N_{\mathrm{t}} - \bar{n}_{\mathrm{t}})}{(i\omega D_{\mathrm{R}})^{1/2}}\right]} \equiv P_1\tilde{n}_{\mathrm{t}} \tag{328}$$

whence

$$i\omega\tilde{n}_{\mathrm{t}} = k_{\mathrm{ft}}\tilde{n}_{\mathrm{s}}(N_{\mathrm{t}} - \bar{n}_{\mathrm{t}}) - (k_{\mathrm{tf}}\bar{n}_{\mathrm{s}} + k_{\mathrm{tb}})\tilde{n}_{\mathrm{t}} - P_1\tilde{n}_{\mathrm{t}} \tag{329}$$

$$\tilde{n}_{\mathrm{s}} = \tilde{n}_{\mathrm{t}}\frac{(f_{\mathrm{r}} + i\omega + P_1)}{q_{\mathrm{v}}} \equiv Q_1\tilde{n}_{\mathrm{t}} \tag{330}$$

and f_{r} and q_{v} are given by

$$f_{\mathrm{r}} = k_{\mathrm{tf}}\bar{n}_{\mathrm{s}} + k_{\mathrm{tb}} \tag{331}$$

$$q_{\mathrm{v}} = k_{\mathrm{tf}}(N_{\mathrm{t}} - \bar{n}_{\mathrm{t}}) \tag{332}$$

The surface electron flux is given by

$$\tilde{j}_{\mathrm{s}} = i\omega\varepsilon_0\varepsilon_{\mathrm{sc}}\tilde{\mathscr{E}}_{\mathrm{s}} + e_0\tilde{j}_{\mathrm{f}}^{\mathrm{s}} \tag{333}$$

where

$$\tilde{j}_{\mathrm{f}}^{\mathrm{s}} = q_{\mathrm{v}}\tilde{n}_{\mathrm{s}} - f_{\mathrm{r}}\tilde{n}_{\mathrm{t}} = (i\omega + P_1)\tilde{n}_{\mathrm{t}} \tag{334}$$

and finally

$$\tilde{j}_{\mathrm{f}}^{\mathrm{s}} = i\omega\varepsilon_0\varepsilon_{\mathrm{sc}}\tilde{\mathscr{E}}_{\mathrm{s}} + e_0\tilde{n}_{\mathrm{t}}(i\omega + P_1) \tag{335}$$

Putting eqn. (335) into eqn. (281), we find

$$\tilde{n}_{\mathrm{s}} - \frac{e_0\bar{n}_{\mathrm{s}}\tilde{\phi}_{\mathrm{s}}}{kT} = \frac{L_{\mathrm{D}}^2}{e_0D_{\mathrm{n}}W}[\tilde{j} - e_0\tilde{n}_{\mathrm{t}}(i\omega + P_1)] - \frac{\tilde{j}}{e_0D_{\mathrm{n}}}\left(\frac{\bar{n}_{\mathrm{s}}L_{\mathrm{s}}}{N_{\mathrm{D}}} + \frac{L_{\mathrm{D}}^2}{W}\right) \tag{336}$$

$$Q_1\tilde{n}_{\mathrm{t}} - \frac{e_0\bar{n}_{\mathrm{a}}\tilde{\phi}_{\mathrm{s}}}{kT} = -\frac{L_{\mathrm{D}}^2}{D_{\mathrm{n}}W}(i\omega + P_1) - \frac{\tilde{j}\bar{n}_{\mathrm{s}}L_{\mathrm{D}}}{e_0D_{\mathrm{n}}N_{\mathrm{D}}} \tag{337}$$

whence

$$e_0\tilde{n}_{\mathrm{t}} = \frac{-\tilde{j}K_2 + \tilde{\phi}_{\mathrm{s}}K_3}{(D_{\mathrm{n}}WQ_1/L_{\mathrm{D}}^2) + P_1 + i\omega} \tag{338}$$

and

$$e_0\tilde{n}_{\mathrm{s}} = \frac{Q_1[-\tilde{j}(R_{\mathrm{b}}/R_{\mathrm{sc}}) + (\tilde{\phi}_{\mathrm{s}}/R_{\mathrm{sc}})]}{(D_{\mathrm{n}}WQ_1/L_{\mathrm{D}}^2) + P_1 + i\omega} \tag{339}$$

Placing these values into eqn. (280), we find

$$\frac{Q_1[-\tilde{j}(R_b/R_{sc}) + (\tilde{\phi}_s/R_{sc})]}{e_0[(D_n WQ_1/L_D^2) + P_1 + i\omega]} - \frac{\bar{\mathscr{E}}_s}{kT}\left\{\frac{\tilde{j}}{i\omega} - \frac{(P_1 + i\omega)}{i\omega} \times \frac{[-\tilde{j}(R_b/R_{sc}) + (\tilde{\phi}_s/R_{sc})]}{(D_n WQ_1/L_D^2) + P_1 + i\omega}\right\} - \frac{e_0\tilde{\phi}_s N_D}{kT} = \frac{i\omega\varepsilon_0\varepsilon_{sc}\tilde{\phi}_s}{e_0 D_n} - \frac{\tilde{j}L_s}{e_0 D_n} \tag{340}$$

whence

$$\tilde{j}\left\{1 + \frac{R_b}{R_{sc}}\frac{[P_1 + i\omega + (Q_1 i\omega kT/e_0\bar{\mathscr{E}}_s)]}{[P_1 + i\omega + (Q_1 WD_n/L_D^2)]} - \frac{i\omega L_s kT}{e_0 D_n \bar{\mathscr{E}}_s}\right\}$$

$$= \tilde{\phi}_s\left\{-\frac{i\omega e_0 N_D}{\bar{\mathscr{E}}_s} + \frac{1}{R_{sc}}\frac{[P_1 + i\omega + (i\omega Q_1 kT/e_0\bar{\mathscr{E}}_s)]}{[P_1 + i\omega + (D_n WQ_1/L_D^2)]}\right\} \tag{341}$$

Let Y'' be given by

$$Y'' = \frac{1}{R_{sc}}\frac{[P_1 + i\omega + (i\omega Q_1 kT/e_0\bar{\mathscr{E}}_s)]}{[P_1 + i\omega + (D_n WQ_1/L_D^2)]} \tag{342}$$

then the equivalent circuit takes the form

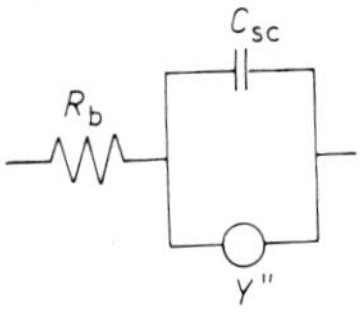

If, as usual, we allow $kT/e_0\bar{\mathscr{E}}_s \to 0$, then the impedance $Z'' \equiv 1/Y''$ takes the value

$$Z'' = R_{sc} + \frac{R_{sc}X_1 Q_1}{P_1 + i\omega} \tag{343}$$

$$X_1 = \frac{D_n W}{L_D^2}$$

After some manipulation, it can be shown that Z'' corresponds to the equivalent circuit

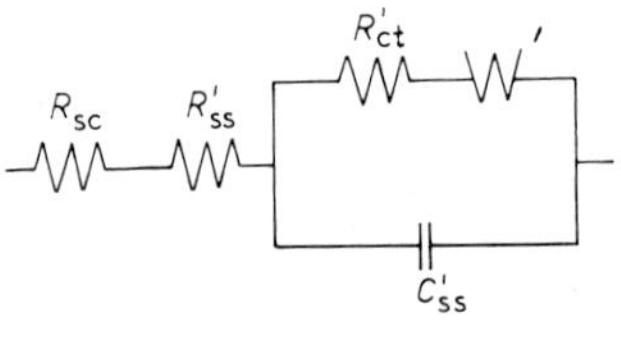

where the values of the parameters are

$$R'_{ss} = \frac{R_{sc}X_1}{q_v} = \frac{R_{sc}D_n W}{L_D^2 k_{tf}(N_t - \bar{n}_t)} \tag{344}$$

$$C'_{ss} = \frac{q_v}{R_{sc}X_1f_r} = \frac{k_{tf}(N_t - \bar{n}_t)L_D^2}{R_{sc}D_n W(k_{tf}\bar{n}_s + k_{tb})} \tag{345}$$

where $R'_{ss}C'_{ss} = 1/f_r$

$$R'_{ct} = \frac{R_{sc}X_1f_r}{q_v[k_{ss}^f\bar{C}_{ox}^s + k_{ss}^b\bar{C}_{Red}^s]} \tag{346}$$

$$-W' = \frac{R_{sc}X_1f_r}{q_v[k_{ss}^f\bar{C}_{ox}^s + k_{ss}^b\bar{C}_{Red}^s]}\left[\frac{k_{ss}^f\bar{n}_t}{(i\omega D_0)^{1/2}} + \frac{k_{ss}^b(N_t - \bar{n}_t)}{(i\omega D_R)^{1/2}}\right] \tag{347}$$

6. Photoeffects in semiconductors

There is no doubt that the major development in the electrochemistry of semiconductors in the last decade has been the further study and exploitation of often large photoeffects. The extension of these observations to the construction of efficient solar energy devices has provided an important technological thrust, complementing the theoretical framework that has developed alongside. In this section, we will concentrate on the development of the theory for wide bandgap extrinsic semiconductors, devoting most time to n-type materials. The extension to p-type semiconductors is, of course, straightforward.

The origin of the large photoeffects often seen in extrinsic semiconductors is illustrated in Fig. 59. A positive bias creates, as we have already depletion layer of thickness W. If the total bias is V volts and $v_s \equiv Ve_0/kT$, we have

$$W = (-2v_s)^{1/2}L_D \tag{348}$$

where $L_D = (\varepsilon_{sc}\varepsilon_0 kT/e_0^2 N_D)^{1/2}$ and we assume here that the potential is zero in the bulk of the semiconductor. Under illumination with radiation exceeding the bandgap, electron–hole pairs are created that can be separated by the depletion-layer field, with the minority carriers (holes in this case) being

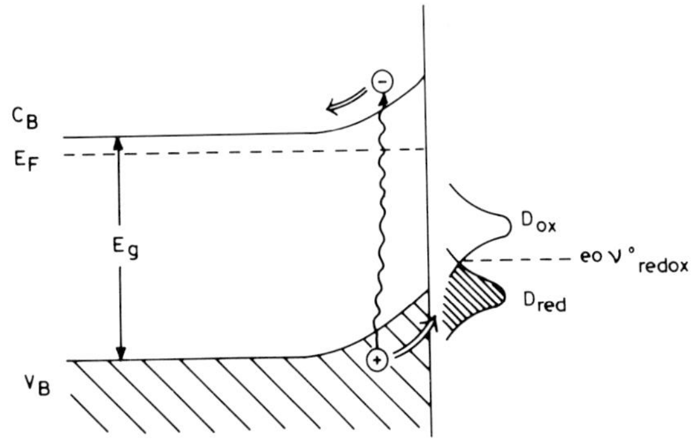

Fig. 59. Energetic parameters for a photoelectrochemical system.

driven to the surface of the semiconductor. Provided that a faradaic pathway is available, they pass into the electrolyte and the surface region becomes depleted of holes. Migration and diffusion from the interior of the semiconductor then takes place to sustain the current. The complexity of the theoretical treatment arises from:

(i) the fact that it is impossible to separate migration and diffusion in the depletion layer,

(ii) the necessity of considering both majority and minority carriers,

(iii) the difficulties of formulating the surface boundary conditions, especially if significant surface recombination occurs, and

(iv) the fact that, as the concentration of minority carriers rise, there will be an increasing effect on the potential distribution by virtue of the Poisson equation.

6.1 GENERAL CONSIDERATIONS

To describe the transport of holes, we introduce the transport equation

$$D_{\mathrm{p}} \frac{\mathrm{d}}{\mathrm{d}x}\left(\frac{\mathrm{d}p}{\mathrm{d}x} + \frac{\mathrm{d}v}{\mathrm{d}x}p\right) + g_{\mathrm{p}} = R_{\mathrm{p}} \tag{349}$$

where D_{p} is the diffusion coefficient, Einstein's relationship ($D = kT\mu/e_0$) is assumed, p is the concentration of holes, g_{p} the rate of generation of holes and R_{p} the rate of recombination of the holes. We assume that the excess hole concentration $\Delta p \simeq p$ for a wide bandgap n-type semiconductor and we also assume that Δp and Δn are small compared with N_{D}, the donor concentration, so that the potential distribution at the surface is unaltered on illumination. This latter assumption is likely to be the weakest and the conditions under which it will break down are considered below.

The generating function g_{p} has the form

$$\alpha\phi_0\theta \exp(-\alpha x) \tag{350}$$

where α is the absorption coefficient, ϕ_0 the incident light intensity at $x = 0$ and θ is the quantum efficiency for the generation of *mobile* electron–hole pairs. The presence of θ implies that the electrochemically measured absorption coefficient may differ from the optical value; in particular, at the optical threshold, the optical absorption coefficient may contain a substantial contribution that is absent or strongly reduced in the electrochemical measurement. In general, θ will be a function of x as well, but we shall not consider this additional complication.

The recombination rate R_{p} has been discussed above: there are two main routes [2]: (a) direct VB/CB transition and (b) recombination via localised states within the bandgap. The first is important in small bandgap semiconductors, but its importance decreases with increasing energy. The second is likely to be the dominant process for higher bandgap materials since it facilitates energy transfer to the lattice. The trap equilibria have already

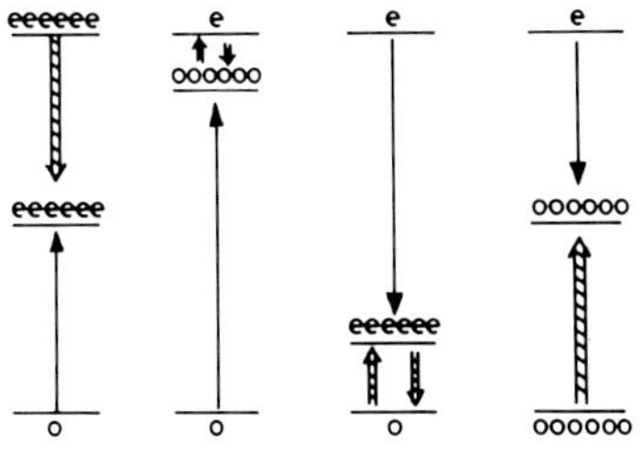

Fig. 60. Rate-limiting regime for Shockley–Read statistics; the Roman numerals denote the dominance of one particular term in eqn. (352). In terms of the denominator of eqn. (351), the regimes are defined by I, $n \gg n_t, p, p_t$; II_1, $n_t \gg n, p, p_t$; II_2, $p_t \gg n, p, n_t$; and III, $p \gg n, p, n_t$. The single arrows indicate the rate-limiting step in each case.

been explored in eqn. (195) et seq.; the rate of loss of minority carriers by recombination can be written

$$\frac{\partial p}{\partial t} = - \frac{C_n C_p (np - n_i^2)}{C_n(n + n_{tr}) + C_p(p + p_{tr})} \tag{351}$$

assuming local trap/VB and trap/CB equilibria. We can invert this formula to give, if $n_i^2 \ll np$

$$\frac{1}{R_p} = - \frac{1}{(\partial p/\delta t)} = \frac{1}{C_p p} + \frac{n_{tr}}{C_p np} + \frac{p_{tr}}{C_n np} + \frac{1}{C_n n} \tag{352}$$

where R_p is the recombination rate, $p_{tr} = N_v \exp\,[E_v - E_t)/RT]$, $n_{tr} = N_c \exp\,[(E_t - E_c)/RT]$ and the four terms are illustrated schematically in Fig. 60.

For the normal situation in the *bulk* of the semiconductor, $p \ll n$ and

$$R_p \approx C_p p \equiv \frac{p}{\tau_p} \tag{353}$$

where τ_p is the recombination time. In the depletion region, the situation is far less clear-cut and several approximations have been suggested. In the region where eqn. (353) holds, we have, for the transport equation

$$\frac{d}{dx}\left(\frac{dp}{dx} + v'p\right) + \frac{\alpha \phi_0 \theta \exp(-\alpha x)}{D_p} = \frac{p}{L_p^2} \tag{354}$$

where $L_p^2 = \tau_p D_p$ and L_p is the hole diffusion length.

The concentration of *majority* carriers in the depletion layer will depend on a quasi-thermodynamic term $N_D \exp(v)$, and a transport term. It will be shown below that the first term is likely to dominate under normal conditions [131].

6.2 PARTICULAR SOLUTIONS

Even if eqn. (353) were to hold throughout the material, solutions are still not easy to find and we first look at some approximation solutions to the problem of the variation of photocurrent with potential. The first is due to Gaertner [132] who assumed that

(1) all holes generated in the depletion layer are swept to the surface. This gives rise to a photocurrent

$$j_{\mathrm{DL}} = e_0 \int_0^W g_{\mathrm{p}}(x)\mathrm{d}x = e_0\phi_0\theta[1 - \exp(-\alpha W)] \tag{355}$$

(2) in the bulk, with $v' = 0$, eqn. (354) reads

$$\frac{\mathrm{d}}{\mathrm{d}x}\left(\frac{\mathrm{d}p}{\mathrm{d}x}\right) + \frac{\alpha\phi_0\theta \exp(-\alpha x)}{D_{\mathrm{p}}} = \frac{p}{L_{\mathrm{p}}^2} \tag{356}$$

which is straightforward to solve. The boundary conditions assumed by Gaertner were

$$\begin{aligned} p &\to 0 \quad & x &\to \infty \\ p &= 0 \quad & x &= W \end{aligned} \tag{357}$$

The justification for the second boundary condition is the assumption that, as soon as holes reach the depletion layer boundary, W, they are whisked away to the surface so that p is always very small in the region $x \leqslant W$. Solving eqn. (356) gives

$$p = \frac{\{\exp[(W - x)/L_{\mathrm{p}}]\exp(-\alpha W) - \exp(-\alpha x)\}\alpha\phi_0\theta}{[\alpha^2 - (1/L_{\mathrm{p}}^2)]D_{\mathrm{p}}} \tag{358}$$

and the diffusion current at $x = W$ is

$$j_{\mathrm{B}} = e_0 D_{\mathrm{p}}\left(\frac{\mathrm{d}p}{\mathrm{d}x}\right)_W = e_0\phi_0\theta\,\frac{\alpha L_{\mathrm{p}}}{(1 + \alpha L_{\mathrm{p}})}\exp(-\alpha W) \tag{359}$$

and the total current is

$$j_{\mathrm{B}} + j_{\mathrm{DL}} = e_0\phi_0\theta\left[1 - \frac{\exp(-\alpha W)}{1 + \alpha L_{\mathrm{p}}}\right] \tag{360}$$

It is of interest to spell out the conditions under which eqn. (360) is valid. They are

(1) recombination in the depletion layer can be neglected,

(2) the faradaic rate constant for removal of holes at the surface is much higher than any other rate process, and

(3) if the *measured* photoelectrochemical current is to be equal to $j_{\mathrm{DL}} + j_{\mathrm{B}}$, then such surface processes as recombination must be much slower.

Experimentally, eqn. (360) represents a limiting form of the observed behaviour; what is found frequently, for reasons that will be explored in more detail below, is that eqn. (360) is valid under high depletion, but that, near threshold, it almost always overestimates the photocurrent. Experimental tests in addition to the measurement of the photocurrent/potential behaviour have been carried out to verify the nature of Gaertner's derivation. It is clear, for example, that eqn. (360) and the formulae that led up to it predict that an increasing number of holes recombine rather than diffuse to the surface as x increases. This has been elegantly shown for CdSe/S, which shows a wavelength-dependent luminescence profile [133]; the wavelength can be used as a probe for x since the absorption cross-section rises rapidly from threshold, ensuring that lower wavelength light will sample regions closer to the surface.

The justification for the assumption that $p = 0$ at $x = W$ can now be seen to be a direct consequence of the fast electrochemical rate constant that keeps p very small throughout the depletion region. This is an extremely demanding condition, however; the faradaic rate constant must exceed ca. $10^4\ \mathrm{cm\,s^{-1}}$ for it to be satisfied as will be shown below. To avoid so stringent a condition, we need a flux rather than a concentration expression, and Tyagai [134] suggested the flux condition

$$D_{\mathrm{p}}\left(\frac{\mathrm{d}p}{\mathrm{d}x}\right)_{W'} = \bar{v} \tag{361}$$

where $\bar{v}$ is the hole velocity at $x = W'$ and W' is defined as the point at which the diffusion current in the bulk of the semiconductor passes into the drift current. As will be seen below, the main criticism of this approach is that such a point does not really exist. Tyagai took $\bar{v} = |\mu\mathscr{E}_0|$ where $\mathscr{E}_0$ is the electric field at W' and arbitrarily put $\mathscr{E}_0 = kT/e_0L_{\mathrm{D}}$, so $\bar{v} = D_{\mathrm{p}}/L_{\mathrm{D}}$. Defining a length of $l = \bar{v}\tau_{\mathrm{p}}$, where τ_{p} is defined in eqn. (353), the solution to eqn. (356) with boundary condition (361) is easily found. The bulk contribution to the current is then

$$j_{\mathrm{B}} = \frac{\exp(-\alpha W)e_0\phi_0\theta\alpha l}{[\alpha L_{\mathrm{p}} + 1][1 + (l/L_{\mathrm{p}})]} \tag{362}$$

and the total current

$$j_{\mathrm{T}} = e_0\phi_0\theta\left(1 - \frac{[\alpha L_{\mathrm{p}} + 1 + (l/L_{\mathrm{p}})]}{[\alpha L_{\mathrm{p}} + 1][1 + (l/L_{\mathrm{p}})]}\exp(-\alpha W)\right) \tag{363}$$

Evidently, eqn. (363) reduces to eqn. (360) if $l \gg L_{\mathrm{p}}$ or, equivalently, $L_{\mathrm{p}} \gg L_{\mathrm{D}}$.

Although Tyagasi's formula represents something of an advance, it suffers from two serious drawbacks, viz.

(a) the quantity $\bar{v}$, or equivalently l, is difficult to calculate and Tyagai's approach is quite arbitrary and unphysical and

(b) it is still assumed that every hole that reaches the edge of the depletion layer passes through the layer and out into the electrolyte.

This model was substantially improved by Wilson [135] who made two distinct changes

(1) l is treated as a parameter determined from boundary conditions imposed on the electrochemical reaction at the surface and

(2) the total flux of holes arriving at the surface is divided into recombination and charge transfer currents.

In order to understand Wilson's treatment in detail, the nature of surface recombination must be considered in more detail.

6.3 SURFACE RECOMBINATION

We saw that the bulk recombination rate was given by eqn. (351) where C_n and C_p are the trap capture rate constants (s^{-1}) and n_{tr} and p_{tr} are defined above. At the surface, the relevant parameter of interest is the surface recombination rate, S, the number of recombining electron-hole pairs per unit area per second, which can be expressed [2] by

$$S = \frac{N_t C_n^s C_p^s (n_s p_s - n_i^2)}{C_n^s (n_s + n_{tr}^s) + C_p^s (p_s + p_{tr}^s)} \tag{364}$$

where the parameter C_n^s and C_p^s are capture rate constants in $cm^3 s^{-1}$ and N_t is the surface density of recombination states. If σ_n is the capture cross-section for electrons of thermal velocity $\bar{v}_n$, we have

$$\begin{aligned} C_n^s &= \bar{v}_n \sigma_n \\ C_p^s &= \bar{v}_p \sigma_p \end{aligned} \tag{365}$$

The nett surface capture rate $\sigma \bar{v} N_t$ can be estimated by assuming σ corresponds to atomic dimensions $\simeq 10^{-16}$ cm^2; if $\bar{v} \simeq 10^7$ cm s^{-1} and $N_t \simeq 10^{13}$ cm^{-2}, we have $\sigma \bar{v} N_t \simeq 10^4$ cm s^{-1}, which is a substantial rate constant.

The expression (364) can be simplified; writing $\tau_{ps} = 1/N_t C_p^s$ and $\tau_{ns} = 1/N_t C_n^s$ and considering the situation in which we have an n-type semiconductor in which $n_i^2 \ll n_s p_s$, we find

$$S = s_t p_s = \frac{n_s p_s}{\tau_{ps}(n_s + n_{tr}^s) + \tau_{ns}(p_s + p_{tr}^s)} \tag{366}$$

Gerischer has identified three common cases [136] as indicated in Fig. 61.

(1) $p_s \ll n_s, n_{tr}^s$

$$S \sim \frac{p_s n_s}{\tau_{ps}(n_s + n_{tr}^s)} \tag{367}$$

(2) p_{tr}^s is large; then for small bias

$$S \sim \frac{p_s n_s}{\tau_{ps} n_s + \tau_{ns} p_{tr}^s} \tag{368}$$

and for larger bias

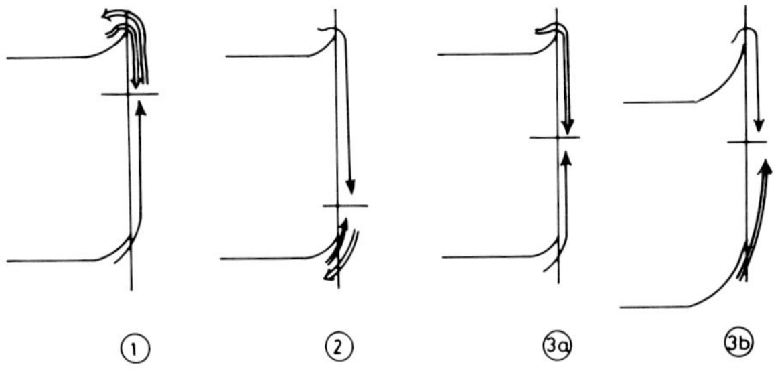

Fig. 61. Surface recombination showing the Gerischer cases (1) n^s_{tr} large, n_s variable; (2) p^s_{tr} large, n_s variable; and (3) n^s_{tr}, p^s_{tr} small; (a) n_s large, (b) n_s small.

$$S \sim \frac{p_s n_s}{\tau_{ns}(p_s + p^s_{tr})}$$

(3) in this, the least clear cut case, n^s_{tr} and p^s_{tr} are comparable but small; at low bias n_s dominates in the denominator and

$$S \sim \frac{p_s}{\tau_{ps}} \tag{369}$$

but at larger bias

$$S \sim \frac{n_s p_s}{\tau_{ps} n_s + \tau_{ns} p_s} \tag{370}$$

will be valid.

6.4 DIRECT MEASUREMENT OF SURFACE RECOMBINATION

The direct measurement of surface recombination is not a straightforward task since both theoretical and experimental assumptions must be made. In an early, but extensive discussion, Memming [137] used the following theoretical assumptions to relate surface recombination to directly measurable currents.

(a) Quasi-equilibrium conditions are established for electron and holes throughout the depletion layer; the origin of these conditions is examined critically below and they are justified provided transport limitations for the two carriers can be neglected (i.e. the density of surface traps is not so great that large fluxes of the two types of carriers is required). The condition can be written as

$$n_s p_s = n_w p_w \tag{371}$$

where the concentrations referred to are those of the perturbed system (e.g. in the presence of light) and W refers to the internal edge of the depletion layer.

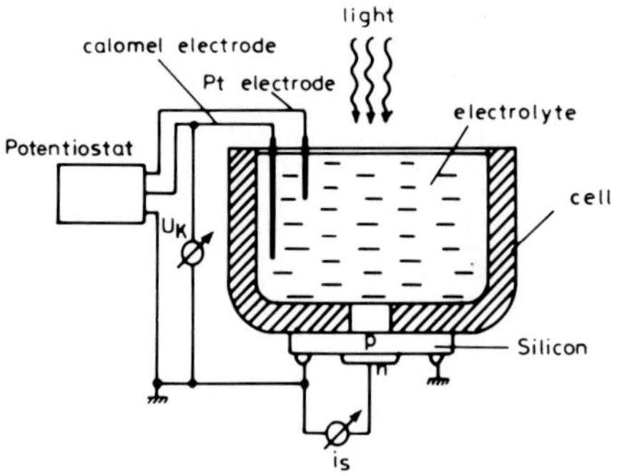

Fig. 62. Experimental arrangement for the measurement of surface recombination.

(b) For $x \geqslant W$, electrical neutrality is preserved so that

$$n_w = n_{0w} + \delta n_w \tag{372}$$

$$p_w = p_{0w} + \delta n_w \tag{373}$$

where the subscript 0 refers to the semiconductor in the absence of the perturbation (e.g. in the dark).

At the surface, the change in carrier concentrations on illumination or other perturbation, δn_s and δp_s are not equal; in fact, we assume

$$\delta n_s = \delta n_w \exp(v); \quad \delta p_s = \delta n_w \exp(-v) \tag{374}$$

and the surface recombination is given by

$$S = \frac{N_t C_n^s C_p^s (n_{0w} + p_{0w} + \delta n_w)\delta n_w}{C_n^s(n_{s0} + \delta n_s + n_{tr}^s) + C_p^s(p_{s0} + \delta p_s + p_{tr}^s)} \tag{375}$$

$$\equiv s_t\, \delta n_w$$

Inserting the values of n_{tr}^s, p_{tr}^s, δn_s and δp_s and remembering that

$$n_{s0} = N_c \exp\left(\frac{E_F - E_C}{kT}\right); \quad p_{s0} = N_v \exp\left(\frac{E_v - E_F}{kT}\right) \tag{376}$$

and defining $C_p^s/C_n^s = \exp(2v_0)$, we have, provided $\delta n_0 \ll n_0$, p_0 and $N_c/N_v = e^{2\delta}$

$$\frac{1}{s_t} = \frac{2n_i\{\cosh[v - v_0 + \{(E_F - E_i)/kT\} + \delta] + \cosh[\delta + \{(E_t - E_i)/kT\} - v_0]\}}{N_t (C_n^s)^{1/2}(C_p^s)^{1/2}(n_{0w} + p_{0w})}$$

$$+ \frac{2\cosh(v - v_0)\delta n_w}{N_t (C_n^s)^{1/2}(C_p^s)^{1/2}(n_{0w} + p_{0w})} \tag{377}$$

$$\equiv \frac{1}{s_{t,0}} + v_k \delta n_w \tag{378}$$

after some rather wearing algebra. In eqn. (377), E_i is the "intrinsic" Fermi level.

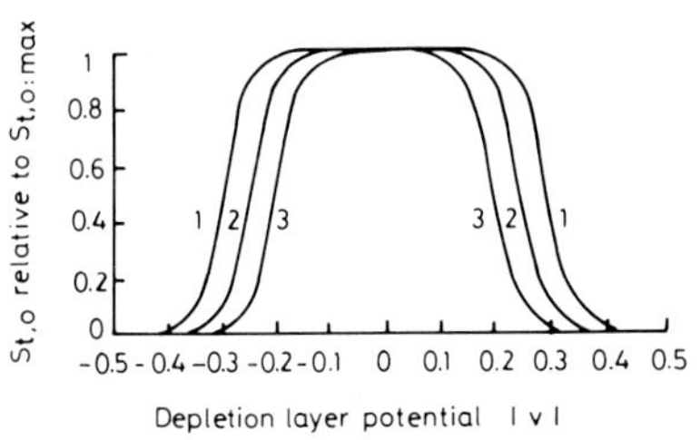

Fig. 63. Recombination rate $s_{t,0}$ from eqn. (377) vs. $|v|$ relative to $s_{t,0}$(max) for three values of $|E_t - E_i|$. (1) 0.31 eV, (2) 0.26 eV, and (3) 0.21 eV assuming an intrinsic material for which $C_p = C_n$. (N.B. the absolute magnitude of curves 1–3 will differ).

These equations predict that the effective surface recombination rate, s_t, decreases as δn_0 increases; e.g. as the light intensity increases, s_t should decrease, which corresponds to the Gerischer case (a)–(c) transition, above.

To verify this relationship, Memming constructed the ingenious cell shown schematically in Fig. 62. If an excess of minority carriers exists at $x = W$, then a short-circuit current i_s flows at the p–n junction whose magnitude is given by:

$$i_s = -e_0 D \left[\frac{\partial(\delta n)}{\partial x}\right]_{x=d} \tag{379}$$

Measuring the short-circuit current across the junction allows δn_w to be calculated. Let the flux of photons be represented by ϕ at the surface; the generation rate is

$$g(x) = g_0 \exp(-\alpha x) \tag{380}$$

where α is the absorption coefficient. We will assume that α is large, so that almost all the light is absorbed in the depletion layer. Defining $x = 0$ as the depletion layer boundary (for mathematical convenience) we have, for the minority carriers in the slab of p-type material, the following transport equation

$$D_n \frac{d^2(\delta n)}{dx^2} - \frac{\delta n}{\tau_n} \approx 0 \tag{381}$$

which solves to give

$$\delta n = A e^{-x/L_n} + B e^{x/L_n} \tag{382}$$

where $L_n^2 = D_n \tau_n$. The boundary conditions are

$$\delta n = \delta n_w \quad \text{at} \quad x = 0$$

(corresponding in Fig. 62 to the edge of the depletion layer)

$$\delta n = 0 \quad \text{at} \quad x = x_n \approx d \tag{383}$$

If $L_n \gg W$ and $\exp(2d/L_n) \gg 1$, then, using eqn. (379)

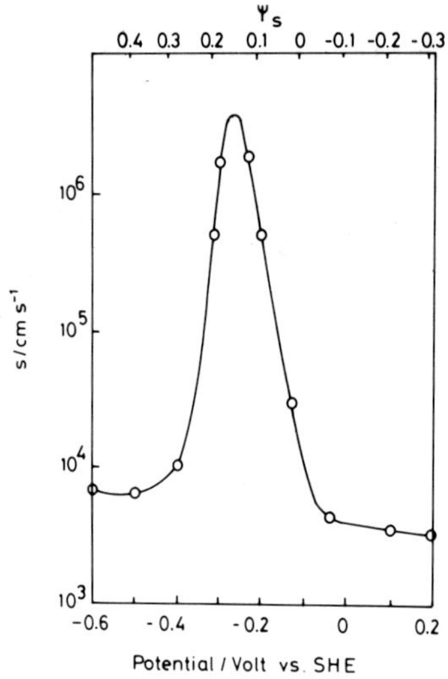

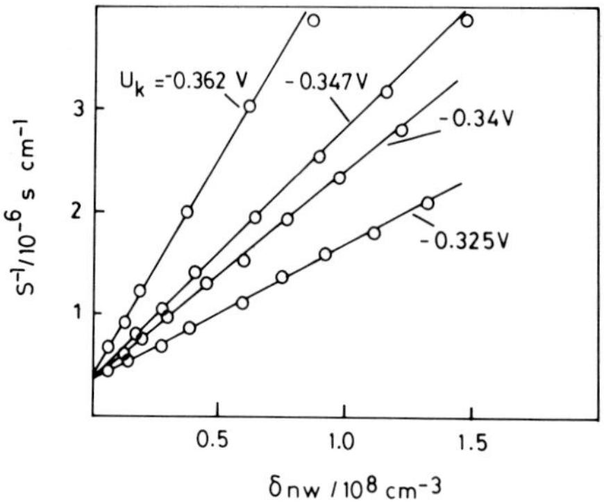

Fig. 64. Calculation of surface recombination velocity vs. electrode potential.

Fig. 65. Reciprocal surface recombination velocity vs. excess carrier density δn_w.

$$\delta n_w \simeq \frac{L_n \exp(d/L_n)}{2e_0 D_n} i_s \tag{384}$$

If the only process occurring at the surface of p-type material, at $x = -W$, is recombination and no faradaic process takes place, then the total current flowing out of the depletion layer, $-D_n[\partial(\delta n)/\partial x]_{x=0}$ together with the recombination rate $s_t \delta n_w$, must balance the total generation rate, ϕ_0, provided no recombination occurs within the depletion layer. From eqn. (381)

$$-D_n\left[\frac{\partial(\delta n)}{\partial x}\right]_0 = \frac{D_n}{L_n}\delta n_w$$

$$\frac{D_n}{L_n}\delta n_w + s_t \delta n_w = \phi_0$$

$$\delta n_w \approx \frac{\phi_0}{[s_t + (D/L_n)]} \tag{385}$$

whence, if $s_t \gg D/L_n$

$$\frac{1}{s_t} \simeq \frac{L_n \exp(d/L_n) i_s}{2e_0 D_n \phi_0}$$

and if $s_t \ll D/L_n$

$$i_s \simeq \frac{2e_0 \phi_0}{\exp(d/L_n)} \tag{386}$$

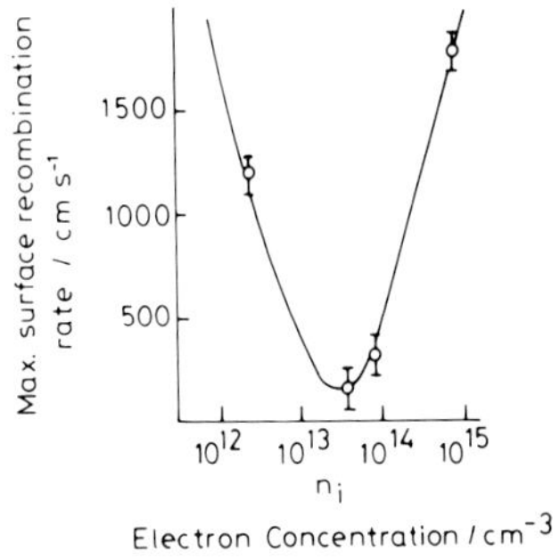

Fig. 66. Maximum surface recombination rate for Ge as a function of electron concentration n^0. The intrinsic value n_i is marked.

Physically, what this shows is that, during illumination, a certain number of electron–hole pairs are created; those that are not destroyed by surface or bulk recombination are captured by the p–n junction and constitute a short-circuit current.

The assumptions upon which Memming's analysis is based restrict its validity to small-bandgap semiconductors such as Ge and, at low light intensities, Si. The critical assumption is that $\delta n_w \ll n_{0w}, p_{0w}$ and p_{0w} decreases strongly as the bandgap increases. The shape of the surface recombination rate, s_{t0}, as a function of surface potential, ϕ_s, is shown in Fig. 63 [7]. It has a characteristic bell shape whose width depends on $|(E_t - E_i(/e_0|$ and which is centred on $\phi_s = \phi_{s0}$ [138]. In practice, only s_t can be measured directly and this is usually quite sharply peaked, as shown in Fig. 64 for p-Si [138]. From the equations above, in the depletion region, it should be possible to plot $1/s_t$ vs. δn_w to obtained s_{t0}. A typical plot is shown for p-Si in Fig. the curves in this figure show that, at least over the range of $\phi_s = 0.15$–0.27 V, s_{t0} is approximately constant. We may take $\phi_{s0} \simeq 0.2$ V, therefore, so that $c_p^s/c_n^s \simeq 10^8$ and analysis gives $E_t - E_i$ as 0.12 or 0.28 $\pm$ 0.05 eV. From the slope of the lines in Fig. 65, the change in v as a function of the applied potential (vs. a known reference, U_k) can be determined. Analysis shows that, over the 50 mV range studied, $\Delta U_k = \Delta kTv/e_0$; i.e. p-Si behaves classically in this region.

One final consequence of eqn. (377) is that, all over things being equal

$$S_{t,max} \sim (n_{0w} + p_{0w}) \tag{387}$$

and this has a minimum for an *intrinsic* material, as shown in Fig. 66 [7, 139].

Other methods for the direct measurement of S have been discussed in the literature. Closely related to Memming's technique is that of Harten, in which the photocurrent rather than the short-circuit current at the p–n junction of Fig. 62 is measured [140]. More direct methods have also been proposed: if the sample is in the form of a thin plate of width w, then the effective lifetime τ_e^{-1} is given by

$$\tau_e^{-1} = \tau_B^{-1} + \frac{2s_t}{W} \tag{388}$$

where τ_B^{-1} is the bulk lifetime measured by determining L_n separately [141]. Another technique, again using a high-surface area electrode, is to measure the decay in conductivity after the light is cut off from an illuminated electrode. The number of carriers, n, decays exponentially and the lifetime can be measured directly [141].

Harten's method [140] has now become the commonest electrochemical technique; as will be shown below, very reliable solutions are now available for the behaviour of the photocurrent over a wide variety of materials and direct methods are also available for verifying the central assumption in the theory, that the potential distribution remains classical, at least over the potential range sampled.

6.5 MORE DETAILED CALCULATIONS OF THE PHOTOCURRENT

The treatments of the surface recombination above assume that there is only one trap site. In practice, this will rarely be the case; indeed, as the analysis of the a.c. data for p-GaP above illustrated, there may be a near constant distribution of surface states throughout a substantial fraction of the bandgap. For an n-type semiconductor, the total recombination current may then be written, for Gerischer's case (a) [136] above

$$S \approx p_s/E_g \int_{\zeta}^{-E_g+\zeta} \sigma\bar{v}_p N_t(e_t) de_t \left(\frac{\exp(v_s)}{\exp(v_s) + \exp(e_t)}\right) \equiv s_t p_s \tag{389}$$

where ζ is the energy difference between Fermi level and the conduction band in the bulk of the semiconductor. The faradaic current is

$$R = k_f p_s \tag{390}$$

so the total minority (hole) carrier flux at the surface is given by

$$j_\tau = R + S = p_s(k_f + s_t) \tag{391}$$

and the observed photocurrent is

$$j_{obs} = e_0 k p_s = e_0 \left(\frac{k_f}{k_f + s_t}\right) j_\tau \tag{392}$$

In his treatment of the photoresponse in the semiconductor, Wilson assumed, as had Memming, that quasi-equilibrium conditions obtained across the depletion layer, i.e. the product np is a constant.

At the depletion layer boundary, $np \simeq N_D p_w$ and at the surface, $np \simeq p_s N_D \exp(v_s)$, where we assume, even in the depletion layer, $n \gg \Delta n$, $\Delta p \simeq p$, and p_w and p_s are the hole concentrations at $x = W$ and 0, respectively. Hence

$$p_{\mathrm{w}} \approx p_{\mathrm{s}} \exp(v_{\mathrm{s}}) \tag{393}$$

From the condition (361), replacing $l = \bar{v}\tau_{\mathrm{p}}$ and remembering that $L_{\mathrm{p}}^2 = D_{\mathrm{p}}\tau_{\mathrm{p}}$

$$\frac{\phi_0 \theta L_{\mathrm{p}}^2 \alpha \exp(-\alpha W)}{D_{\mathrm{p}}(\alpha L_{\mathrm{p}} + 1)\,[1 + (l/L_{\mathrm{p}})]} = p_{\mathrm{s}} \exp(v_{\mathrm{s}}) \tag{394}$$

Combining eqns. (363), (392), and (394), we obtain an explicit formula for l

$$l = \frac{\dfrac{L_{\mathrm{p}}^2}{D_{\mathrm{p}}}\left[(k_{\mathrm{f}} + s_{\mathrm{t}})\exp(-v_{\mathrm{s}}) - \dfrac{(1 - \exp[-\alpha W])D_{\mathrm{p}}}{\exp(-\alpha W)\{\alpha L_{\mathrm{p}}/(\alpha L_{\mathrm{p}} + 1)\}L_{\mathrm{p}}}\right]}{\left[1 + \dfrac{(1 - \exp[-\alpha W])}{\exp(-\alpha W)\{\alpha L_{\mathrm{p}}/\alpha L_{\mathrm{p}} + 1\}}\right]} \tag{395}$$

Inserting this into eqn. (363) gives an explicit expression for the total flux and hence, from eqn. (392) the observed photocurrent

$$j_{\tau} = \theta\phi_0\left(1 - \frac{\exp(-\alpha W)}{[\alpha L_{\mathrm{p}} + 1]}\right)\bigg/\left(1 + \frac{D_{\mathrm{p}}\exp(v_{\mathrm{s}})}{L_{\mathrm{p}}(k_{\mathrm{f}} + s_{\mathrm{t}})}\right) \tag{396}$$

$$j_{\mathrm{obs}} = e_0\left(\frac{k_{\mathrm{f}}}{k_{\mathrm{f}} + s_{\mathrm{t}}}\right) j_{\tau} \tag{397}$$

It is as well, at this point, to reiterate the approximations underlying Wilson's approach. They are

(1) there is no recombination in the depletion layer; all holes photogenerated in the depletion layer are swept to the surface and all holes that are generated in the bulk and reach the edge of the depletion layer are swept to the surface without further ado;

(2) the steady-state concentration of minority carriers is insufficient to disturb the potential distribution in the depletion layer in the dark; and

(3) there is a quasi-thermodynamic distribution of minority carriers within the depletion layer.

The most inaccurate assumption, especially at low band-bending, is undoubtedly the first. However, the mathematical derivation leading to eqns. (396) and (397) cannot be easily modified to take account of depletion-layer recombination owing to the way in which the depletion layer is considered. In order to develop the theory to take into account recombination in the depletion layer, it is necessary to solve explicitly the transport equation in the depletion layer as well as the bulk. If we persevere, for the moment, with the Schottky barrier model and we continue, for the moment, with the assumption that recombination does *not* occur in the depletion layer ($x < W$) then the transport equations for $x < W$, $x > W$ are

$$D_{\mathrm{p}}\frac{\mathrm{d}}{\mathrm{d}x}\left(\frac{\mathrm{d}p}{\mathrm{d}x} + \frac{\mathrm{d}v}{\mathrm{d}x}p\right) + g_{\mathrm{p}} = 0 \qquad x < W \tag{398}$$

$$D_p \frac{d}{dx}\left(\frac{dp}{dx}\right) + g_p = R_p \qquad x \geqslant W \tag{399}$$

For the case $x \geqslant W$, we can solve the transport equation explicitly as above; this time we use as the boundary conditions

$$p \rightarrow 0\ x \rightarrow \infty; \qquad p = p_w,\ x = W \tag{400}$$

whence

$$p = p_w \exp\{-(x - W)/L_p\} + \frac{\alpha\theta\phi_0}{D_p[\alpha^2 - (1/L_p^2)]} \times [\exp\{-W(\alpha - (1/L_p)\} \exp(-x/L_p) - \exp(-\alpha x)] \tag{401}$$

and flux at $x = W_+$ is

$$D_p\left(\frac{\partial p}{\partial x}\right)_{W_+} = -\frac{D_p p_w}{L_p} + \frac{\alpha\theta\phi_0 \exp(-\alpha W)}{\alpha + (1/L_p)} \tag{402}$$

For $x < W$, we may write the transport equation explicitly as

$$\frac{d}{dx}\left[\frac{dp}{dx} + \left(\frac{W - x}{L_D^2}\right)p\right] = -\frac{\alpha\theta\phi_0 \exp(-\alpha x)}{D_p} \tag{403}$$

assuming a Schottky barrier model for the depletion layer. A solution to the above equation may be obtained without difficulty under the boundary conditions

$$p = p_w, \quad x = W; \qquad p = p_s, \quad x = 0 \tag{404}$$

$$p = p_W\left[\frac{\exp[(W - x)^2/2L_D^2]}{b} \int_{(W-x)/\sqrt{2}L_D}^{W/\sqrt{2}L_D} \exp(-\mu^2)d\mu\right]$$

$$+ p_s\left[\exp\left(-\frac{W^2}{2L_D^2}\right)\exp\left(\frac{\{W - x\}^2}{2L_D^2}\right)\right]\left\{1 - \frac{1}{b}\int_{(W-x)/\sqrt{2}L_D}^{W/\sqrt{2}L_D} \exp(-\mu^2)d\mu\right\}$$

$$+ \frac{\sqrt{2}\beta}{\alpha L_D}\exp\left[\frac{(W - x)^2}{2L_D^2}\right]\left\{\int_{(W-x)/\sqrt{2}L_D}^{W/\sqrt{2}L_D} \exp(-\mu^2)\exp(\alpha\mu\sqrt{2}L_D)d\mu\right.$$

$$\left. - \frac{\int_0^{W/\sqrt{2}L_D} \exp(-\mu^2)\exp(\alpha\mu\sqrt{2}L_D)}{b}\int_{(W-x)/\sqrt{2}L_D}^{W/\sqrt{2}L_D} \exp(-\mu^2)d\mu\right\} \tag{405}$$

where

$$\beta = \frac{\alpha\theta\phi_0 L_D^2 \exp(-\alpha W)}{D_p} \quad \text{and} \quad b = \int_0^{W/\sqrt{2}L_D} \exp(-\mu^2)\mathrm{d}\mu \tag{406}$$

The flux at $x = W_-$ is

$$D_p\left(\frac{\mathrm{d}p}{\mathrm{d}x}\right)_{W_-} = \frac{D_p}{\sqrt{2}L_D}\left(\frac{p_w}{b} - \frac{\exp(-W^2/2L_D^2)p_s}{b}\right) + \frac{\beta D_p}{\alpha L_D^2}\left\{1 - \frac{\int_0^{W/\sqrt{2}L_D}\exp(-\mu^2)\exp(\alpha\mu\sqrt{2}L_D)\mathrm{d}\mu}{b}\right\} \tag{407}$$

and equating fluxes at W_+ and W_- allows us to find an expression for p_w.

$$p_W = \frac{p_s \exp(-W^2/2L_D^2)}{[1 + (\sqrt{2}bL_D/L_p)]} + \frac{\alpha\theta\phi_0 \exp(-\alpha W)}{D_p[\alpha + (1/L_p)][(1/L_p) + (1/\sqrt{2}bL_D)]} \times \left\{1 - \frac{[\alpha + (1/L_p)]}{\alpha}\left(\frac{b - \int_0^{W/\sqrt{2}L_D}\exp(-\mu^2)\exp(\alpha\mu L_D\sqrt{2})\mathrm{d}\mu}{b}\right)\right\} \tag{408}$$

It can be seen that the quasi-thermodynamic boundary condition of Wilson and Memming is satisfied if D_p is large, L_D is small and the light intensity is low. Explicitly, we have

$$\theta\phi_0 \exp(-\alpha W) \ll \frac{D_p p_s}{L_D} \tag{409}$$

From the expression for p_w and from the flux at the surface

$$\begin{aligned} j_\tau &= D_p\left[\left(\frac{\mathrm{d}p}{\mathrm{d}x}\right)_s + \frac{Wp_s}{L_D^2}\right] = \frac{D_p p_w}{\sqrt{2}bL_D} - \frac{D_p p_s \exp(-W^2/2L_D^2)}{\sqrt{2}bL_D} \\ &\quad + \frac{\beta D_p}{\alpha L_D^2}\left[\exp(-\alpha W) - \frac{\int_0^{W/\sqrt{2}L_D}\exp(-\mu^2)\exp(\alpha\mu\sqrt{2}L_D)\mathrm{d}\mu}{b}\right] \\ &\equiv -D_p\frac{p_w}{L_p} + \theta\phi_0\left[1 - \frac{\exp(-\alpha W)}{(1 + \alpha L_p)}\right] \\ &\equiv p_s(k_f + s_t) \end{aligned} \tag{410}$$

From eqns. (408) and (410), we have two expressions relating p_s and p_w. Our expression for p_s reduces to

$$p_s = \frac{\theta\phi_0\{1 - \exp(-\alpha W) + [\alpha L_p/(\alpha L_p + 1)] - \{[\alpha L_p \exp(-\alpha W)]/[1 + \alpha L_p][1 + (1/\xi)]\} \times \{1 - [(\alpha L_p + 1)/\alpha L_p][(b - \int_0^{W/\sqrt{2}L_D}\exp(-\mu^2)\exp(\alpha\mu\sqrt{2}L_D)\mathrm{d}\mu)/b\}]\}}{[k'_\Sigma + (D_p/L_p)\exp(-W^2/2L_D^2)]/(1 + \xi)} \tag{411}$$

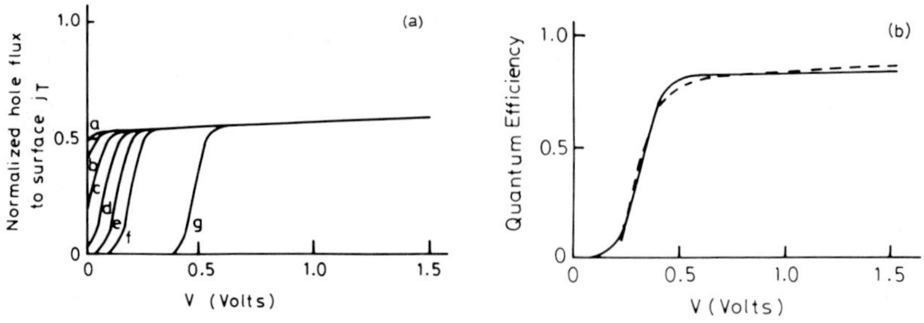

Fig. 67. (a) Calculated hole flux j_T to the surface normalised to $\theta\phi_0 = 1$ for the parameters $L_p = 10^{-4}$ cm, $\alpha = 10^4$ cm^{-1}, $N_D = 10^{18}$ cm^{-3}. The values of $k + s_t$ are a, 10^4; b, 10^3; c, 10^2; d, 10; e, 1.0; f, 0.1 and g, 10^{-6} cm s^{-1}. (b) Best fit for the quantum efficiency of TiO_2 calculated from Wilson's model using $\alpha = 10^4$ cm^{-1}, $N_D = 6 \times 10^{17}$ cm^{-3}, $L_p = 4 \times 10^{-4}$ cm, $k = 1$ cm s^{-1} and s_t is calculated from eqn. (389) using an exponential distribution of surface-recombintion states. ---, Experimental curve. $s_t = \int_0^{E_g} 10^4 e^{-20E} dE/(1 + e^{e_0(V-E)/kT})$ cm s^{-1}.

where $\xi = 2^{1/2} b L_D / L_p$ and $k'_\Sigma = k_f + s_t$. Normally, $L_D \ll L_p$ so $\xi \ll 1$. Given also that, to a good approximation

$$\int_0^{W/\sqrt{2}L_D} \exp(-\mu^2) \exp(\alpha\mu\sqrt{2}L_D)\alpha\mu \approx b + \frac{\alpha L_D}{\sqrt{2}} \tag{412}$$

the last, rather complex term in the numerator of eqn. (411) can be neglected to give

$$p_s \approx \frac{\theta\phi_0\{(1 - \exp(-\alpha W)) + [\alpha L_p/(\alpha L_p + 1)] \exp(-\alpha W)\}}{k'_\Sigma + (D_p/L_p) \exp(-W^2/2L_D^2)} \tag{413}$$

The quantum efficiency for transport to the surface, N, can be defined as

$$N \equiv \frac{j_\tau}{\phi_0} \tag{414}$$

so, if $\theta = 1$

$$N \simeq \frac{[1 - \exp(-\alpha W)] + [\alpha L_p/(\alpha L_p + 1)] \exp(-\alpha W)}{1 + (D_p/L_p k'_\Sigma) \exp(-W^2/2L_D^2)} \tag{415}$$

a formula identical to that derived by Wilson if we put $v_s = -W^2/2L_D^2$.

The structure of eqn. (415) is very revealing; the numerator consists of two *generating* terms. One, $(1 - e^{-\alpha W})$, is the generation of holes in the depletion layer and the second, $(\alpha L_p/[\alpha L_p + 1])e^{-\alpha W}$, the generation of holes in the bulk modulated by a recombination factor. The second term in the denominator represents the flux of carriers out of the depletion layer into the bulk; it therefore represents a *loss* of efficiency.

The formulae (397) and (415) have proved most valuable in understanding qualitatively the variation of observed photocurrent with bias. There are two ways in which the bias might affect $i_{obs} \equiv e_0 j_\tau k/(k + s_t)$: alteration of v_s

affects W through the formula $W = (-2v_s)^{1/2}L_D$ and secondly s_t will be a strong function of v_s as the analysis above shows. Some indications of the results obtained by Wilson are given in Fig. 67(a) [138]; obviously, the weakness of the approach is that, whatever functional form of the dependence of photocurrent on potential, it should prove possible to find a suitable energetic distribution of surface states in eqn. (389) that reproduces this dependence as indicated in Fig. 67(b). Without independent measurement of the surface state population, the model cannot adequately be tested. This is especially worrying near the photocurrent onset potential; in principle, minority carrier photocurrent should be observed as soon as there is any finite reverse, i.e. as soon as $v_s < 0$, or, equivalently for an n-type semiconductor, the potential is anodic of flat-band. In practice, no semiconductor shows a rise in photocurrent at flat-band potential as steep as that predicted by eqn. (415) with s_t assumed much smaller than k_f. This observation might be ascribed to ubiquitous surface states present just below the conduction band edge but, equally, it is in the region that neglect of depletion-layer recombination will be most serious. A second problem is that efficient solar energy devices must have N values as large as possible near threshold potential; the mathematical structure of N suggests, reasonably, that the larger the Debye length, L_D, the larger N, but there must clearly be a limit to this. Beyond some depletion layer width, recombination within the depletion layer must become significant.

The problem of recombination in the depletion layer is not at all trivial to solve. The only analytical solution that has been obtained as yet [142] is that for first-order recombination. An examination of eqn. (352) shows that this will only hold if $p \ll n$ or, equivalently, $\Delta p \ll N_D e^{v_s}$, a result only likely to be true close to flat-band potential or under low light intensity or when k'_Σ is very large. Nevertheless, since depletion layer recombination is most likely near flat-band potential, the result does have some practical importance.

The hole transport equation in the depletion layer may, assuming a Schottky barrier model, be written

$$D_p \frac{d}{dx}\left(\frac{dp}{dx} + \left(\frac{W - x}{L_D^2}\right)p\right) + g_p = \frac{p}{\tau_p} \tag{416}$$

We can make some progress by defining v as the potential

$$-v = \frac{(W - x)^2}{2L_D^2} \tag{417}$$

and rescaling p as the variable $u \equiv pe^v$; we find

$$v\frac{d^2u}{dv^2} + (\tfrac{1}{2} - v)\frac{du}{dv} + \tfrac{1}{2}\gamma u = \tfrac{1}{2}ge^v \tag{418}$$

where

$$g = \frac{\alpha\theta\phi_0 L_D^2 \exp(-\alpha W)}{D_p} \exp[\sqrt{2}\alpha L_p(-v)^{1/2}]; \quad \gamma = \frac{L_D^2}{L_p^2} \tag{419}$$

Equation (418) may be solved in two stages: first, the corresponding homogeneous equation [i.e. with the right-hand side of eqn. (418) set equal to zero] is well known as Kummer's equation and its two solutions can be written [143]

$$G_1(v) = {}_1F_1(-\gamma/2; \tfrac{1}{2}; v) \tag{420}$$

$$G_2(v) = {}_1F_1(\tfrac{1}{2} - \tfrac{1}{2}\gamma; \tfrac{3}{2}; v)(-v)^{1/2} \tag{421}$$

where ${}_1F_1$ are confluent hypergeometric functions defined in ref. 143.

Using these, we may now solve eqn. (418) by constructing a Green's function [144] using the boundary conditions $p = p_w$ at $x = W$ and $p = p_s$ at $x = 0$

$$\begin{aligned} p\exp(v) = {} & p_w G_1(v) + \frac{G_2(v)}{G_2(v_s)}\Bigg\{p_s \exp(v_2) - p_w G_1(v_s) \\ & - 2G_1(v_s)\int_0^{(-v_s)^{1/2}} g G_2(v')\mathrm{d}(-v')^{1/2}\Bigg\} \\ & + 2G_1(v)\int_0^{(-v)^{1/2}} g G_2(v')\mathrm{d}(-v')^{1/2} \\ & + 2G_2(v)\int_{(-v)^{1/2}}^{(-v_s)^{1/2}} g G_1(v')\mathrm{d}(-v')^{1/2} \end{aligned} \tag{422}$$

By following now the procedure above of matching fluxes at the depletion-layer boundary and the surface we may eliminate p_w and p_s to find, for $\theta = 1$

$$N = \frac{\exp(-\alpha W)[(\alpha L_p/\alpha L_p + 1) + \int_0^{(-v_s)^{1/2}} g' G_1(v')\mathrm{d}(-v')^{1/2} + (\sqrt{2}L_D/L_p)\int_0^{(-v_s)^{1/2}} g' G_2(v)\mathrm{d}(-v)^{1/2}]}{G_1(v_s) + (\sqrt{2}L_D/L_p)G_2(v_s) + (D_p/k'_\Sigma\sqrt{2}L_D)G'_1(v_s) + (D_p/k'_\Sigma L_D)G'_2(v_s)} \tag{423}$$

where

$$g' = \alpha\sqrt{2}L_D \exp[\alpha\sqrt{2}L_D(-v)^{1/2}];$$

$$G'_1(v) = \frac{\mathrm{d}G_1(v)}{\mathrm{d}(-v)^{1/2}} = 2\gamma(-v)^{1/2}{}_1F_1(1 - \gamma/2; \tfrac{3}{2}; v);$$

$$G'_2(v) = {}_1F_1(\tfrac{1}{2} - \tfrac{1}{2}\gamma; \tfrac{1}{2}; v) \tag{424}$$

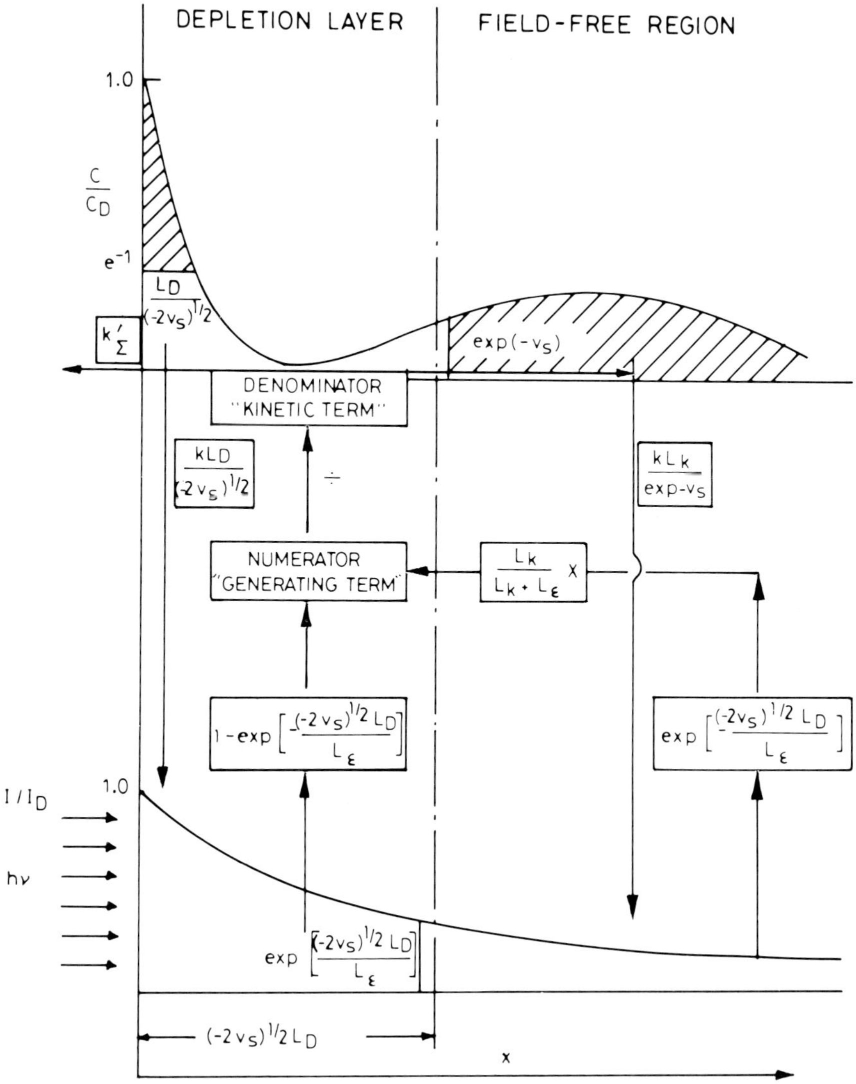

Fig. 68. The significance of the various terms in eqn. (426). The hatching indicates where the minority carriers are lost through recombination.

Although eqn. (423) is considerably more complex than eqn. (415), its basic structure is very similar, consisting as it does of generating terms in the numerator and kinetic terms in the denominator. To show this more clearly, we may use approximations for G_1 and G_2 and their derivatives, valid for $\gamma \ll 1$ and $(-v)^{1/2} \geqslant 3$.

$$G_1(v_s) \approx 1; \quad G_2(v_s) \approx (\sqrt{\pi}/2)\,\mathrm{erf}\,[(-v_s)^{1/2}]; \quad G_1'(v_s) \approx \gamma(-v_s)^{-1/2};$$

$$G_2(v_s) \simeq \exp(v_s) \tag{425}$$

which leads to the final expression for N [142]

$$N \simeq \frac{1 - \exp(-\alpha W) + [\alpha L_p/(1 + \alpha L_p)] \exp(-\alpha W)}{1 + (D_p/k_2' L_D^2)\{[L_D/(-2v_s)^{1/2}] + [L_p/\exp(-v_s)]\}} \tag{426}$$

The significance of the various terms in eqn. (426) are shown in Fig. 68. The only difference between this and the formula derived by Wilson, eqn. (415), is the presence of an additional loss term in the denominator which reflects recombination in the depletion layer. However, unlike the loss term representing diffusion *out* of the depletion layer into the bulk, the depletion recombination term decreases only slowly with increasing bias and plays a significant role even at comparatively large v_s values. However, it is clear that, at comparatively large values of γ (ca. 0.1) and reasonable values of the ratio $(D_p/k_\Sigma')L_D$, very low efficiencies are found, even at high bias potentials. This is quite unreasonable, as will be seen below.

Equation (423) represents the best that has been achieved hitherto using formal analytical procedures and the problem, as has already been emphasised, lies with the nature of the recombination formula, eqn. (352). As the band-bending increases, n must decrease to the point where a shift in the kinetic law is expected at some point in the depletion layer.

Given the difficulties facing the analytical approach, we must turn to approximate methods for the solution of the transport equation. Retaining, for the moment, the idea of a first-order recombination, we find, as above, for $x < W$

$$D_p \frac{d}{dx}\left(\frac{dp}{dx} + p\frac{dv}{dx}\right) + \alpha\theta\phi_0 \exp(-\alpha x) - kp = 0 \tag{427}$$

where $k \equiv 1/\tau_p \equiv D_p/L_p^2$. The boundary condition of immediate importance here is the surface flux condition

$$N\phi_0 = D_p\left[\left(\frac{dp}{dx}\right)_{x=0} + p_s\left(\frac{dv}{dx}\right)_{x=0}\right] = k_\Sigma' p_s \tag{428}$$

We may integrate eqn. (427) formally, with the assumption that $\theta = 1$, to give

$$D_p\left(\frac{dp}{dx} + p\frac{dv}{dx}\right) = \phi_0[N - 1 + \exp(-\alpha x)] + k\int_0^x p\,dx' \tag{429}$$

Rescaling p as before, $u \equiv pe^v$, and changing the spatial variable from x to v, where we assume the Schottky approximation, gives

$$\frac{du}{dv} = \frac{\phi_0 L_D}{D_p}\{N - 1 + \exp(-\alpha W)\exp[\alpha\sqrt{2}L_D(-v)^{1/2}]\}\frac{\exp(v)}{\sqrt{2}(-v)^{1/2}}$$

$$- \frac{\gamma \exp(v)}{\sqrt{2}(-v)^{1/2}} \int_{(-v)}^{(-v_s)} \frac{ue^{-v'}}{\sqrt{2}(-v')^{1/2}} dv' \tag{430}$$

If γ is small, the last term in eqn. (430) can be neglected. Integration then yields

$$u_{\gamma=0} = u_s - \frac{L_D \phi_0}{D_p} \sqrt{\frac{\pi}{2}} (1 - \exp(-\alpha W) - N) [\mathrm{erf}\{(-v_s)^{1/2}\} - \mathrm{erf}\{(-v^{1/2})\}] \tag{431}$$

A better approximation is obtained by using the functional form of eqn. (431), viz.

$$u_{\gamma\neq 0} = u_s - B[\mathrm{erf}\{(-v_s)^{1/2}\} - \mathrm{erf}\{(-v)^{1/2}\}] \tag{432}$$

If eqn. (432) is substituted into eqn. (430) and the latter integrated, we obtain, after some manipulation

$$B = \frac{\sqrt{\pi/2}\{(\phi_0 L_D/D_p)(1 - N - \exp[-\alpha W]) - \gamma[p_s/(-2v_s)^{1/2}]\}}{1 - \gamma \ln [1 + (-\pi v_s)^{1/2}]} \tag{433}$$

Neglecting the term $\gamma \ln [1 + (-\pi v_s)^{1/2}]$ in the numerator, we then have, for the concentration of minority carriers

$$p \approx p_s \exp(v_s - v) - \sqrt{2}\left\{\frac{\phi_0 L_D}{D_p}(1 - N - \exp[-\alpha W]) - \gamma \frac{p_s}{(-2v_s)^{1/2}}\right\} \exp(-v) \int_{(-v)^{1/2}}^{(-v_s)^{1/2}} \exp(-\mu^2)\mathrm{d}\mu \tag{434}$$

where $p_s = N\phi_0/k'_\Sigma$. By calculating p_w and $(p/x)_w$ and matching the solution at the depletion layer boundary, we once more obtain eqn. (426). However, the form of eqn. (434) is of great interest; near the surface, the dominant term is the first one, showing an exponential dependence of p on v. However, the second term is significant near the edge of the depletion layer and determines the transport into and out of the depletion layer. This is illustrated in Fig. 69, which shows the computed concentration profiles; as can be seen, a substantial concentration gradient is predicted to build up close to the depletion layer boundary, which is only offset by the exponential increase in p near the electrolyte–semiconductor boundary.

6.6 THE GENERAL SOLUTION

We now pass to a more general consideration of transport and recombination. Recalling eqn. (352)

$$\frac{1}{R_p} = -\frac{1}{(\partial p/\partial t)} = \frac{\tau_p}{p} + \frac{\tau_n}{n} + \frac{n_{tr}\tau_p}{np} + \frac{p_{tr}\tau_n}{np} \tag{435}$$

There are, in principle, three zones in which different kinetic laws may operate; following Albery and Bartlett [131] we denote these by I–III as in Fig. 60 where

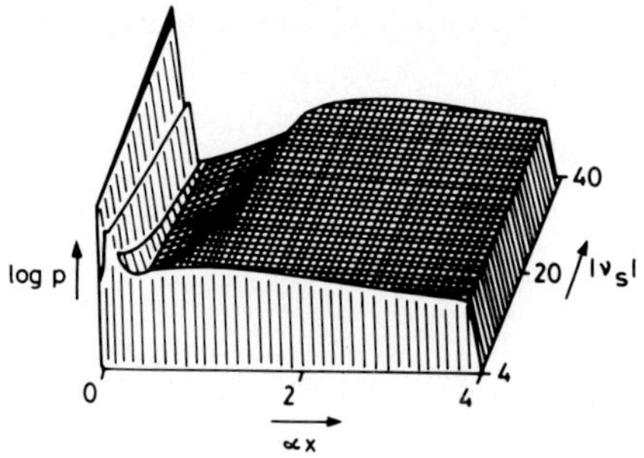

Fig. 69. Typical concentration profiles calculated from eqn. (434). The parameters used were $1/\alpha L_D = 6.90$, $L_p/L_D = 6.97$ and $k'_\Sigma L_D/D_p = 1.45 \times 10^{-2}$.

Region	Rate Law	Variation of recombination rate with x
I	$R = p/\tau_p$	Sharp maximum at the surface
II	$R = np(1/n_{tr}\tau_p + 1/p_{tr}\tau_n)$	Recombination constant assuming a Boltzmann distribution
III	$R = n/\tau_n$	Increases strongly away from surface

Examination of Fig. 69 shows that, within the depletion layer, p first falls and then, near the surface, rises rapidly. It follows from eqn. (435) that the recombination law is likely to remain first order near the depletion layer boundary and, if there is a change of mechanism, it will occur well within the depletion layer. This is an important simplification in two respects.

(1) Whatever the recombination law, the concentration of holes is likely to be given by the exponential part of eqn. (434) to a first approximation; i.e.

$$p \approx \left(\frac{NI_0}{k'_\Sigma}\right) \exp\,(v_s - v) \tag{436}$$

at least in the region where a mechanism change is likely to occur.

(2) The recombination rate will only be significant near the surface, for a value of $x < x_{RZ}$ where x_{RZ} will be calculated below. For the various possible transitions from recombination law I near the inner depletion-layer boundary to either II or III near the surface, we have

Transition	Region where recombination is significant in W
I → II	Constant $0 < x < x_{RZ}$, negligible $x > x_{RZ}$
I → III	Strong maximum on I–III boundary, $x = x_{RZ}$
I → II → III	Constant between I–II and II–III boundaries, $x'_{RZ} < x < x_{RZ}$, negligible otherwise.

The *majority* carrier equation of motion may be written

$$D_n \frac{d}{dx}\left(\frac{dn}{dx} - n\frac{dv}{dx}\right) + \alpha\phi_0 \exp(-\alpha x) = R_n(x) \tag{437}$$

At the surface, the flux of majority carriers is

$$N\phi_0\left(\frac{s_t}{s_t + k_f}\right) \equiv \rho N\phi_0 \equiv D_n\left(\frac{dn}{dx} - n\frac{dv}{dx}\right)_{x=0} \tag{438}$$

Integrating eqn. (437) with this boundary condition gives

$$D_n\left(\frac{dn}{dx} - n\frac{dv}{dx}\right) = -\phi_0(1 - \exp[-\alpha x]) + \rho N\phi_0 + \int_0^x R_n(x)dx \tag{439}$$

For $x > x_{RZ}$, the integral

$$\int_0^x R_n(x)dx \approx \int_0^{x_{RZ}} R_n(x)dx \equiv \kappa \tag{440}$$

where κ is a constant. Integration then gives

$$n = n_w \exp(v) + \frac{\sqrt{2}L_D}{D_n}\{\phi_0(1 - \exp[-\alpha W] \exp[(-2v)^{1/2}\alpha L_D]$$

$$- \rho N) - \kappa\} \exp(v) \int_0^{(-v)^{1/2}} \exp(\mu^2)d\mu \tag{441}$$

for the boundary condition $n = n_w$, $x = W$ $(v = 0)$.

The flux of *majority* carriers out of the depletion layer is, from eqn. (439)

$$-D_n\left(\frac{dn}{dx}\right)_W = \phi_0(1 - \exp[-\alpha W] - \rho N) - \kappa \tag{442}$$

We recall that the flux of *minority* carriers into the depletion layer is given by eqn. (429) as

$$D_p\left(\frac{dp}{dx}\right)_W = -\phi_0(1 - \exp[-\alpha W] - N)$$

$$+ \int k\rho\, dx \equiv -\phi_0(1 - \exp[-\alpha W] - N) + \kappa$$

using eqn. (434) where, for the first-order case, $\kappa = p_s L_D k(-2v_s)^{1/2}$. The *nett* current is then the sum of the two fluxes

$$j_{nett} = e_0\phi_0 N(1 - \rho) \tag{444}$$

In any one-dimensional system, the current is constant everywhere and it is a useful check to note that the current at the edge of the depletion layer is clearly the same as that at the semiconductor–electrolyte boundary.

The majority carrier concentration at the point x_{RZ}, where $v = v_{RZ}$, is then given by

$$n_{\mathrm{RZ}} = n_{\mathrm{w}} \exp(v_{\mathrm{RZ}}) + \frac{L_{\mathrm{D}}}{\sqrt{2}D_{\mathrm{n}}(-v_{\mathrm{RZ}})^{1/2}} \{\phi_0(1 - \exp[-\alpha x_{\mathrm{RZ}}] - \rho N) - \kappa\} \tag{445}$$

where we have used the asymptotic expansion of Dawson's integral described above in eqn. (441). The structure of eqn. (445) is again of interest; the second term represents the transport term and the first the pseudo-thermodynamic Boltzmann term. The latter must dominate near $x = W$ and the transport term may become dominant as $x \to 0$.

The most important case that we need to consider here is the I → III transition. In this case, above some critical bias potential $|v_{\mathrm{s}}|$, there exists a point in the depletion region, x_{RZ}, at which the dominant recombination mode changes from first order in p to first order in n (and zeroth order in p). The total recombination flux within the depletion layer is

$$\kappa = \int_0^W \frac{np}{(n\tau_{\mathrm{p}} + p\tau_{\mathrm{n}})} \mathrm{d}x \tag{446}$$

If we assume that the Boltzmann term dominates, so that

$$p \approx p_{\mathrm{s}} \exp(v_{\mathrm{s}} - v) \quad \text{and} \quad n \approx n_{\mathrm{w}} \exp(v) \tag{447}$$

the integrand in eqn. (446) is a maximum when $n\tau_{\mathrm{p}} = p\tau_{\mathrm{n}}$. At this point, the potential v_{RZ} is given by

$$v_{\mathrm{RZ}} = \tfrac{1}{2}v_{\mathrm{s}} + \tfrac{1}{2} \ln\left(\frac{p_{\mathrm{s}}\tau_{\mathrm{n}}}{\tau_{\mathrm{p}} n_{\mathrm{w}}}\right) \tag{448}$$

and if there is no I → III transition, $|v_{\mathrm{RZ}}| > |v_{\mathrm{s}}|$. The total recombination flux may now be written

$$\kappa = \frac{L_{\mathrm{D}}}{2}\left[\frac{p_{\mathrm{s}} n_{\mathrm{w}} \exp(v_{\mathrm{s}})}{\tau_{\mathrm{n}}\tau_{\mathrm{p}}}\right]^{1/2} \int_{v_{\mathrm{s}}-v_{\mathrm{RZ}}}^{-v_{\mathrm{RZ}}} \frac{\mathrm{d}v'}{(-2v')^{1/2}\cosh(v')} \tag{449}$$

where $v' = v - v_{\mathrm{RZ}}$. Now, $\cosh(v')$ has a strong minimum at $v' = 0$ and, to a good approximation, we may therefore integrate the expression in eqn. (449) by removing $(-2v_{\mathrm{s}})^{1/2}$ from the denominator and replacing it outside the integral by $(-2v_{\mathrm{RZ}})^{1/2}$. Assuming that $-v_{\mathrm{RZ}}$ is sufficiently positive that it may be replaced by ∞ in the upper bound of the integral, we find

$$\kappa = \frac{L_{\mathrm{D}} p_{\mathrm{s}}}{2\tau_{\mathrm{p}}} \frac{\exp(v_{\mathrm{s}} - v_{\mathrm{RZ}})}{(-2v_{\mathrm{RZ}})^{1/2}} \{\pi/2 - \tan^{-1}\sinh(v_{\mathrm{s}} - v_{\mathrm{RZ}})\} \tag{450}$$

Note that, if $|v_{\mathrm{RZ}}| > |v_{\mathrm{s}}|$, i.e. there is not transition from I to III, then we may evaluate eqn. (449) by a similar analysis, but we must replace v_{RZ} by v_{s} in the denominator. Using the expression

$$\tan^{-1}\sinh(x) \approx \pi/2 - 2\exp(-x) \quad (x > 0) \tag{451}$$

$$-\pi/2 + 2\exp(-x) \quad (x < 0)$$

valid for $|x| \geqslant 3$ and assuming $v_s - v_{RZ} \geqslant 3$, we have, for case I

$$\kappa = \frac{p_s L_D}{\tau_p} \frac{1}{(-2v_s)^{1/2}} \tag{452}$$

which is identical to that above.

We now consider the *form* of the expression for the collection efficiency in eqn. (426). It consists of a numerator $1 - e^{-\alpha W} + (\alpha L_p/[\alpha L_p + 1])e^{-\alpha W} \equiv X_{gen}$, a generating term, and a denominator that contains a depletion layer recombination term and a diffusion loss term. The last term is likely to be negligible at the higher potentials considered here and we ignore it. The recombination term in the denominator has the general form $\kappa/k'_\Sigma p_s$ and

$$N = \frac{X_{gen}}{1 + (\kappa/k'_\Sigma p_s)} \tag{453}$$

where

$$\frac{\kappa}{k'_\Sigma p_s} = \left(\frac{X_{gen}}{N} - 1\right) = \frac{L_D}{2k'_\Sigma \tau_p} \frac{\exp(v_s - v_{RZ})}{(-2v_{RZ})^{1/2}} \{\pi/2 - \tan^{-1}\sinh(v_s - v_{RZ})\} \tag{454}$$

and

$$v_{RZ} = \frac{v_s}{2} + \tfrac{1}{2}\ln\left[\frac{\tau_n \phi_0 N}{k'_\Sigma \tau_p n_W}\right] \tag{455}$$

From eqns. (454) and (455), we can calculate N and v_{RZ}. If $|v_s - v_{RZ}| \geqslant 3$, we may replace $\tan^{-1}\sinh(v_s - v_{RZ})$ by the expression in eqn. (451) [remembering that $(v_s - v_{RZ}) < 0$ if a transition is to occur]

$$\left(\frac{X_{gen}}{N} - 1\right) = \frac{\pi L_D \exp(v_s - v_{RZ})}{2k'_\Sigma \tau_p (-2v_{RZ})^{1/2}} \tag{456}$$

Substituting in eqn. (455), we obtain

$$\frac{X_{gen}^2}{N}\left(1 - \frac{N}{X_{gen}}\right)^2 = \frac{\exp(v_s)\pi^2 L_D^2 n_w}{8\phi_0 \tau_n k'_\Sigma \tau_p(-v_{RZ})} \tag{457}$$

If $N \ll X_{gen}$, we have

$$-v_{RZ} = \ln\left\{\frac{n_w L_D}{X_{gen}\phi_0 \tau_n}\right\} \tag{458}$$

Note that this value of x_{RZ} is independent of k'_Σ: it is purely a function of light intensity and bulk semiconductor properties; for values of $L_D \simeq 10^{-6}$ cm, $n_w \simeq N_D \simeq 10^{17}\,cm^{-3}$, $X_{gen} \simeq 1$, $\phi_0 \simeq 10^{16}\,cm^{-2}\,s^{-1}$, and $\tau_n \simeq 10^{-10}$ s, we have $-v_{RZ} \simeq 12$. This value, corresponding to 0.3 V, implies that, for bias values less than 0.3 V, the condition of transition from I to III is not satisfied, and

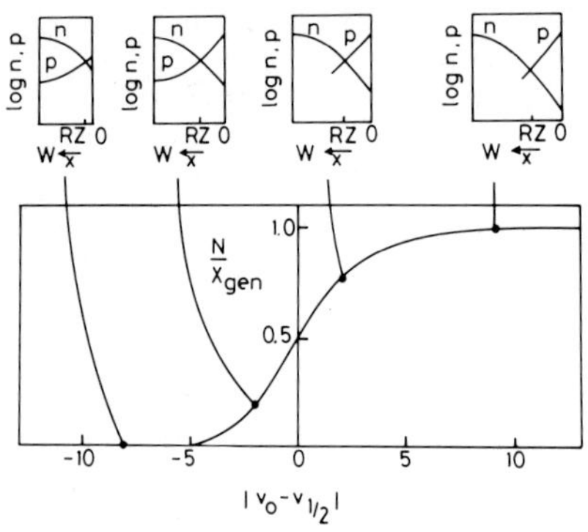

Fig. 70. Photocurrent–voltage curves calculated from eqn. (457). The insets show the calculated variation in majority (n) and minority (p) carriers with x from the surface ($x = 0$) to the bulk ($x \geqslant W$) and the position x_{RZ} at which the form of the recombination law changes.

eqn. (426) is valid. Above 0.3 V, there will be some point in the depletion layer, x_{RZ}, at which a transition to recombination case III occurs. An examination of eqns. (457) and (458) shows that at large bias voltages, $N \to X_{\mathrm{gen}}$; in other words, at large values of $|v_{\mathrm{s}}|$, the photocurrent is not limited by k'_{Σ}. The reason for this is that p_{s} can rise rapidly at large bias since there is essentially no recombination penalty to be paid in the depletion layer. Provided this rise in p_{s} does not disturb the potential distribution in the depletion layer, the overall behaviour will be as shown in Fig. 70. It can be seen that, for small values of k'_{Σ}, N is very small as long as there is no transition to regime III but, as the bias increases, there is a transition region, extending over several tenths of a volt, in which N rises to the diffusion limit corresponding to X_{gen}. The point at which $N = X_{\mathrm{gen}}/2$ could be thought of as the half-wave potential; it has the value, from eqn. (457) of

$$(-v_{1/2}) = \ln\left\{\frac{\pi^2 L_{\mathrm{D}}^2 n_{\mathrm{w}}}{4\phi_0 \tau_{\mathrm{p}} \tau_{\mathrm{n}} X_{\mathrm{gen}} k'_{\Sigma}(-v_{\mathrm{RZ}})}\right\} \tag{459}$$

If we assume $k'_{\Sigma} = 1\,\mathrm{cm\,s^{-1}}$ and conditions otherwise are as above, $(-v_{1/2}) \simeq 20$, corresponding to a bias of 0.5 V. Above this potential, it will appear that the Gaertner relationship is obeyed by the photocurrent and reliable values of L_{p} and α can be derived; examples are given in ref. 145.

Very similar conclusions to the above were reached by Reichman [146], Dare-Edwards et al. [147] and Lorenz and co-workers [148]. Dare-Edwards et al. [147] used a numerical integration procedure to show that this type of analysis was appropriate to the experimental results obtained on single-crystal n-Fe_2O_3. Reichmann [146] obtained an expression similar to eqn. (456); his formula may be derived from eqn. (456) if we assume that $v_{\mathrm{RZ}} \simeq v_{\mathrm{s}}$, so that

$$\kappa = \frac{\pi L_{\mathrm{D}}}{2\tau_{\mathrm{p}}} \frac{p_{\mathrm{s}}}{(-2v_{\mathrm{s}})^{1/2}} \tag{460}$$

which is similar, in essence, to Reichmann's result

$$\kappa = \frac{\pi k T n_i W \exp[e_0 V/2kT]}{4\tau\phi_b (p_s/p_{so})^{1/2}} \tag{461}$$

where τ is the hole lifetime, ϕ_b the band bending and V the photovoltage.

The features of all these models are (1) general recombination via intermediate energy traps is allowed inside and outside the depletion region and (2) the concentration of holes is not allowed to build up to the point at which the potential distribution is affected.

The error involved in the second of these features can be estimated by adding a term to the Poisson–Boltzmann equation. We have

$$\frac{d^2 v}{d\chi^2} = -1 - \frac{p_s}{N_D} \exp(v_s - v) \tag{462}$$

where $\chi = x/L_D$. Integrating yields

$$\frac{dv}{d\chi} = \left[2v - \frac{2p_s}{N_D} \exp(v_s - v)\right]^{1/2} \tag{463}$$

$$\sqrt{2}\chi = \int_{v_s}^{v} \left[v' - \frac{p_s}{N_D} \exp(v_s - v')\right]^{-1/2} dv' \tag{464}$$

whence

$$\chi \approx (-2v_s)^{1/2}\left[1 - \frac{p_s}{4N_D v_s^2}\right] - (-2v)^{1/2} \tag{465}$$

which compares with the Schottky approximation

$$\chi \approx (-2v_s)^{1/2} - (-2v)^{1/2} \tag{466}$$

In other words, there will be very little effect on the potential distribution provided

$$\frac{p_s}{4N_D v_s^2} \equiv \frac{N\phi_0}{4N_D k'_\Sigma v_s^2} \ll 1; \quad \text{i.e.} \ \frac{N\phi_0}{4N_D k'_\Sigma} \ll v_s^2 \tag{467}$$

For $N = 1$, $\phi_0 \simeq 10^{16}\,\text{cm}^{-2}\,\text{s}^{-1}$, $N_D \simeq 10^{17}\,\text{cm}^{-3}$, we have

$$\frac{1}{k'_\Sigma} \ll 4v_s^2; \quad k'_\Sigma \gg \frac{1}{40 v_s^2} \tag{468}$$

For the value of $v_s \simeq 10$ quoted above, $k'_\Sigma \gg 2.5 \times 10^{-3}\,\text{cm}\,\text{s}^{-1}$; values comparable with or less than this, can only be accommodated by a substantial fall in N, as discussed above.

6.7 THE ROLE OF SURFACE RECOMBINATION

We saw that the *observed* steady-state photocurrent is given by

$$j_{ph} = e_0 N\phi_0(1 - \rho) = e_0 N\phi_0 \left(\frac{k_f}{k_f + s_t}\right) \tag{469}$$

where s_t is given by the integral formulation of eqn. (389). The importance of the recombination sites at the surface clearly depends on their concentration and the relevant rate constants for faradaic vs. surface state capture. The balance between recombination and faradaic fluxes will also depend on the presence of a suitable redox couple in solution; increasing the latter will cause an increase in the photocurrent. The surface states that act as recombination centres may have two origins: (a) they may be formed as intermediates in the photo-oxidation process or (b) they may be intrinsic to the particular junction of electrolyte and semiconductor [149]. In the case of (a), we may suppose that there is a redox couple R undergoing a two-electron oxidation process of O via an intermediate R^+

$$p_{vb}^+ + R \xrightarrow{k_f} R^+$$

$$R^+ + p_{vb}^+ \xrightarrow{k_f'} O \rightsquigarrow \text{diffusion into solution}$$

$$\text{recombination}$$

$$R^+ + e_{cb}^- \xrightarrow{k_r} R$$

Then

$$\frac{\partial[R^+]}{\partial t} = p_s k_f[R] - k_f' p_s[R^+] - k_r n_s[R^+] \tag{470}$$

where [R] and $[R^+]$ are the surface concentrations of R and R^+, respectively. The total hole rate constant, k_Σ', is given by

$$k_\Sigma' p_s = p_s(k_f[R] + k_f'[R^+]) = N\phi_0 \tag{471}$$

and the rate of formation of $[R^+]$ is then

$$\frac{\partial[R^+]}{\partial t} = N\phi_0 \left(\frac{k_f[R] - k_f'[R^+]}{k_f[R] + k_f'[R^+]}\right) - k_r n_s[R^+]$$

which will be equal to zero at steady state. Putting $k_f'/k_f = \xi$ and $N\phi_0/k_r n_s = \zeta$ we have

$$\zeta\left(\frac{[R] - \xi[R^+]}{[R] + \xi[R^+]}\right) = [R^+]; \quad \xi[R^+]^2 + [R^+]([R] - \zeta\xi) - \zeta[R] = 0$$

So

$$[R^+] = \frac{1}{2\xi}\,[([R] + \zeta\xi)^2 + 4\zeta\xi[R]]^{1/2} - \frac{1}{2\xi}\,([R] + \zeta\xi) \tag{472}$$

The *nett* photocurrent is

$$j_{ph} = e_0 N\phi_0 - e_0 k_r n_s[R^+] \tag{473}$$

If $R \gg \zeta\xi$, we have

$$j_{\mathrm{ph}} \approx e_0 N\phi_0 - e_0 N\phi_0 \left(1 - \frac{\xi N\phi_0}{k_{\mathrm{r}} n_{\mathrm{s}}[\mathrm{R}]}\right) = \frac{e_0 N^2 \phi_0^2 \xi}{k_{\mathrm{r}} n_{\mathrm{s}}[\mathrm{R}]} \tag{474}$$

Thus, if n_s is large, i.e. we are close to flat-band potential and recombination is primarily via photogenerated intermediates, the photocurrent should depend on the square of the light intensity. At the other extreme, $i_{\mathrm{ph}} \simeq e_0 N\phi_0 - e_0 k_{\mathrm{r}} n_{\mathrm{s}}[\mathrm{R}]/\xi$, predicting a linear dependence on ϕ_0 but with an apparently non-zero intercept (that will be very difficult to pick up in practice).

In case (b), we may identify a surface state, T, which may be emptied by hole capture and filled by electron capture. If we suppose these two processes to be irreversible, i.e. that we have made a transition to Gerischer's recombination cases (b) or (c) and that the faradaic hole-capture route is via the reduced form of the redox couple R, then the equations describing the kinetics are

$$\mathrm{p}_{\mathrm{vb}}^{+} + \mathrm{T} \xrightarrow{k_{\mathrm{f}}} \mathrm{T}^{+}$$

$$\mathrm{n}_{\mathrm{s}} + \mathrm{T}^{+} \xrightarrow{k_{\mathrm{r}}} \mathrm{T}$$

$$\mathrm{p}_{\mathrm{vb}}^{+} + \mathrm{R} \xrightarrow{k_{\mathrm{v}}} \mathrm{O}$$

This leads to a fractional occupancy of T, f_t, as

$$f_{\mathrm{t}} = \frac{k_{\mathrm{r}} n_{\mathrm{s}}}{k_{\mathrm{f}} p_{\mathrm{s}} + k_{\mathrm{r}} n_{\mathrm{s}}} \tag{475}$$

and a recombination rate

$$S = \frac{n_{\mathrm{s}} p_{\mathrm{s}} N_{\mathrm{t}}}{(p_{\mathrm{s}}/k_{\mathrm{r}}) + (n_{\mathrm{s}}/k_{\mathrm{f}})} \tag{476}$$

where N_t is the total concentration of surface states T. We still have

$$k'_{\Sigma} p_{\mathrm{s}} = \left(k_{\mathrm{v}} + \frac{k_{\mathrm{f}} k_{\mathrm{r}} n_{\mathrm{s}}}{k_{\mathrm{f}} p_{\mathrm{s}} + k_{\mathrm{r}} n_{\mathrm{s}}}\right) p_{\mathrm{s}} \equiv N\phi_0 \tag{477}$$

and the photocurrent is still

$$j_{\mathrm{ph}} = e_0 k_{\mathrm{v}} p_{\mathrm{s}} \tag{478}$$

From eqn. (477)

$$k_{\mathrm{v}} p_{\mathrm{s}}^2 + (k_{\mathrm{v}} n_{\mathrm{s}} + k_{\mathrm{r}} k_{\mathrm{f}} n_{\mathrm{s}} - N\phi_0) p_{\mathrm{s}} - N\phi_0 n_{\mathrm{s}} = 0 \tag{479}$$

$$p_{\mathrm{s}} = \frac{[B + (B^2 + 4k_{\mathrm{v}} N\phi_0 n_{\mathrm{s}})^{1/2}]}{2k_{\mathrm{v}}} \tag{480}$$

where

$$B = N\phi_0 - k_{\mathrm{v}} n_{\mathrm{s}} - k_{\mathrm{v}} k_{\mathrm{f}} n_{\mathrm{s}} \tag{481}$$

References pp. 242–246

This unpleasant expression is only valid if the occupancy of the surface states in the dark is essentially unity, i.e. we can neglect the kinetics of the reverse reaction

$$\mathrm{T} \xrightarrow{k_d} \mathrm{T}^+ + n_s$$

If this is not so, we must incorporate the kinetics of this reverse process into our expression

$$\mathrm{p}_{\mathrm{vb}}^+ + \mathrm{T} \xrightarrow{k_f} \mathrm{T}^+$$

$$\mathrm{T}^+ + \mathrm{e}_{\mathrm{cb}}^- \underset{k_d}{\overset{k_r}{\rightleftharpoons}} \mathrm{T}$$

$$\mathrm{p}_{\mathrm{vb}}^+ + \mathrm{R} \xrightarrow{k_v} \text{products}$$

In the dark

$$e_0 k_d T_0 = e_0 k_r n_s (N_t - T_0) \tag{482}$$

where T_0 is the equilibrium dark concentration of surface states and the photocurrent is given by

$$j_{ph} = e_0 k'_\Sigma p_s + e_0 k_d T - k_r n_s T^+ e_0 \tag{483}$$

$$= e_0 k'_\Sigma p_s - k_r n_s N_t e_0 + e_0 k_r n_s N_t \left(\frac{T}{T_0}\right)$$

$$= e_0 k'_\Sigma p_s - e_0 k_r n_s N_t \left(1 - \frac{T}{T_0}\right) \tag{484}$$

From the steady-state expression for the concentration T, we have

$$\frac{\partial T}{\partial t} = 0 = -k_f T p_s + k_r T^+ n_s - k_d T \quad \textit{and} \quad T + T^+ \equiv N_t \tag{485}$$

and we have, for the flux of holes to the surface

$$k'_\Sigma p_s = (k_v + k_f T) p_s \equiv N\phi_0 \tag{486}$$

whence

$$p_s = \frac{k_r n_s (N_t - T) - k_d T}{k_f T} = \frac{N\phi_0}{k_v + k_f T} \tag{487}$$

and

$$(k_r n_s N_t - [k_r n_s + k_d] T)(k_v + k_f T) = N\phi_0 k_f T \tag{488}$$

In the dark, assuming that $p_s \ll n_s$, the concentration T_0 of occupied states is given by

$$T_0 = \frac{k_r n_s N_t}{k_r n_s + k_d} \tag{489}$$

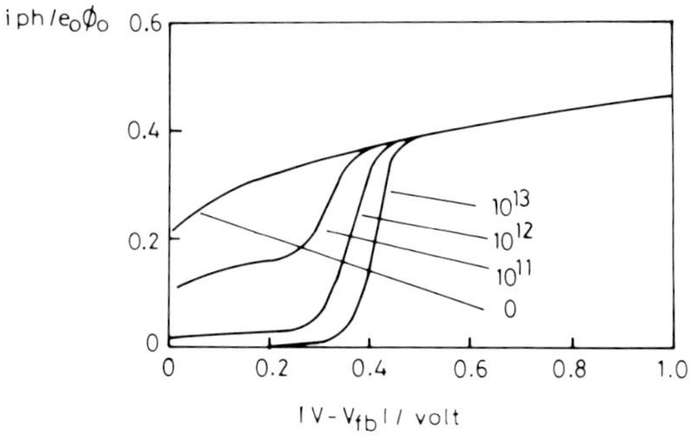

Fig. 71. Normalised photocurrent–voltage curves calculated from eqn. (492) for a p-type semiconductor with $\phi_0 = 10^{14}\,\mathrm{cm^{-2}\,s^{-1}}$, $\alpha = 3.3 \times 10^4\,\mathrm{cm^{-1}}$, $k_r = 10^7\,\mathrm{cm^3\,cm^{-1}}$, $k_f = 10^8\,\mathrm{cm^3\,s^{-1}}$, $k_v = 10^4\,\mathrm{cm\,s^{-1}}$, $N_A = 8 \times 10^{16}\,\mathrm{cm^{-3}}$, $\varepsilon_{sc} = 11$, $L_n = 7 \times 10^{-6}\,\mathrm{cm}$, N_t values ($\mathrm{cm^{-2}}$) are indicated on the figure.

whence

$$\left(1 - \frac{T}{T_0}\right)(k_v + k_f T) = \frac{N\phi_0 k_f T}{k_r N_t n_s} \tag{490}$$

$$k_f \frac{T^2}{T_0} + T\left(\frac{k_f N\phi_0}{k_r N_t n_s} + \frac{k_v}{T_0} - k_f\right) - k_v = 0 \tag{491}$$

whence T may be found in terms of T_0. The nett photocurrent is

$$j_{ph} = e_0 N\phi_0 - e_0 k_r n_s N_t\left(1 - \frac{T}{T_0}\right) \tag{492}$$

assuming that n_s is unaltered on illumination.

These formulae are best explored by the means of numerical examples and some are given in Figs. 71 and 72, which are taken from a recent paper [150] in which the above theory was worked out for a p-type semiconductor; in this paper, the energy of the surface state T was taken to be 0.3 eV above the Fermi level.

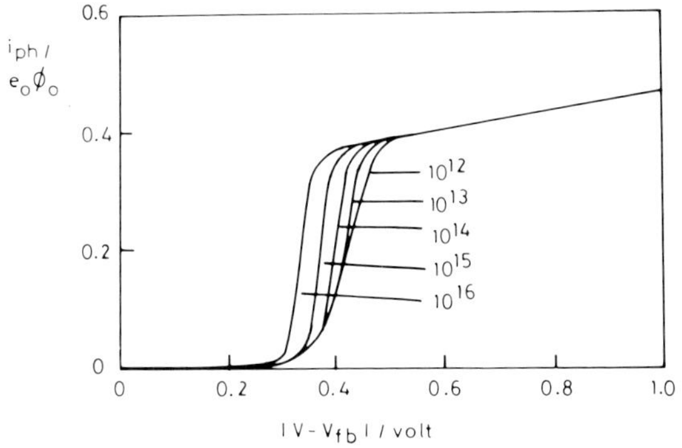

Fig. 72. Photocurrent–voltage curve calculated from eqn. (492) for various light intensities ($\mathrm{cm^{-2}\,s^{-1}}$ as shown) using the same parameters, otherwise, as Fig. 71 with N_t equal to $10^{13}\,\mathrm{cm^{-2}}$.

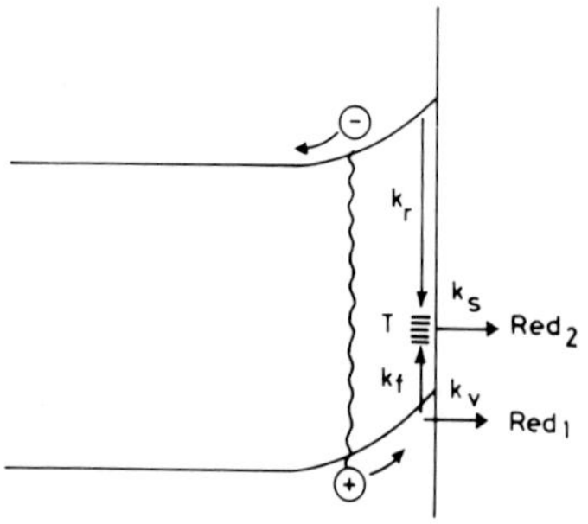

Fig. 73. Rate constants defined for eqn. (493) et seq.

An important point to note here is that, for $N_s \geqslant 10^{13}\,\mathrm{cm}^{-2}$, there will be a significant change in the potential distribution within the Helmholtz layer as the population of the surface state changes from full to empty with increased band-bending, a point discussed in some detail by Kelly and Memming [151]. To follow through their model, we consider Fig. 73 in which the major development from eqn. (477) is the possibility that holes may also be transferred from the surface state T to the solution, oxidising a redox couple R_2 with concentration C_2; as before, it is assumed that holes may be captured directly from the valence band by an appropriate redox couple in solution, here denoted by R_1, with concentration C_1. For simplicity, it can be assumed that (1) there is no reverse process to k_p or k_n, which implies that the surface state must, energetically, be well-removed from the valence or conduction band edges and (2) there is no concentration polarisation in the electrolyte.

If, as above, we denote the occupancy of the surface state by f_t, we find

$$k'_\Sigma p_s \quad = \quad N\phi_0 \quad = \quad k_v p_s C_1 \; + \; k_f p_s N_t f_t \tag{493}$$

where $f_t N_t \equiv T$ above. Also

$$N_t \frac{\mathrm{d}f_t}{\mathrm{d}t} \quad = \quad k_r n_s N_t (1 - f_t) \; - \; k_f p_s N_t p_t \; - \; k_s C_2 f_t N_t \tag{494}$$

and

$$n_s \quad = \quad n_B \exp{(v_s)} \tag{495}$$

If we assume that internal losses through depletion-layer recombination are small, we may write $N \simeq X_{gen}$; if, on the other hand, depletion layer recombination must be taken into account and $N \ll X_{gen}$, then $N \simeq k'_\Sigma F'$, where F is defined, from above, as

$$F' \quad = \quad \frac{X_{gen}(-2v_s)^{1/2}}{kL_D \exp{[(v_s \; - \; v_{RZ})\mathrm{H}(v_{RZ} \; - \; v_s)]}} \tag{496}$$

where H(x) is the Heaviside function and $k \; = \; 1/\tau_p$.

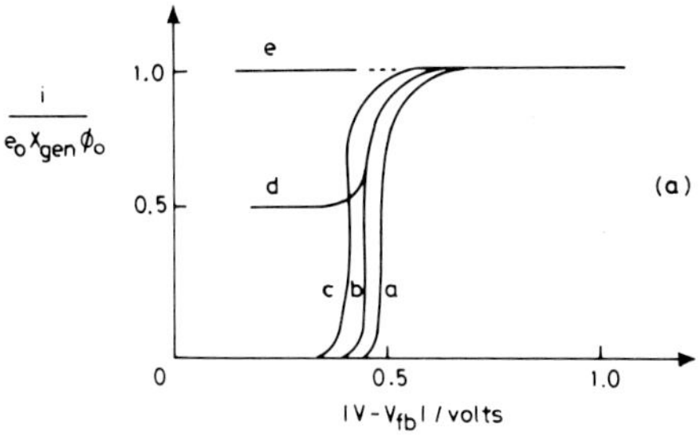

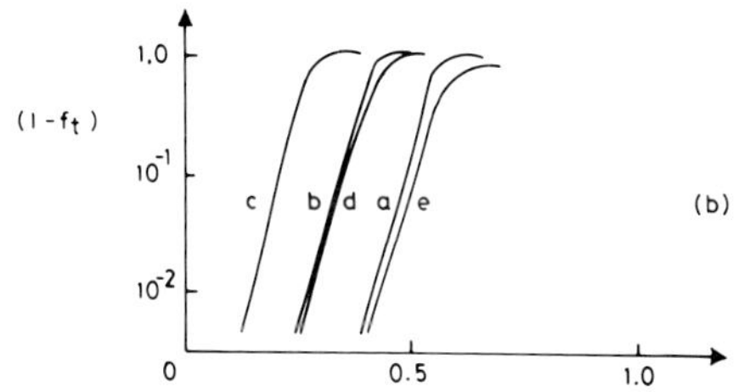

Fig. 74. (a) Photocurrent–voltage curves calculated from eqn. (503) assuming $k_s \equiv 0$ with the following parameters. Curves (a)–(c): $k_f N_t = 10^2 k_v C_1$; $X_{gen}\phi_0 = 10^{15}$, 10^{16}, and $10^{17}\,\text{cm}^{-2}\,\text{s}^{-1}$, (d): $k_f N_t = k_v C_1$, $X_{gen}\phi_0 = 10^{16}\,\text{cm}^{-2}\,\text{s}^{-1}$. Curve (e): $k_f N_t = 10^{-2} k_v C_1$, $X_{gen}\phi_0 = 10^{16}\,\text{cm}^{-2}\,\text{s}^{-1}$. (b) Hole population of surface states $(1 - f_t)$ under the same conditions as in (a).

Case (1). Depletion-layer recombination significant.

$$N \simeq k'_\Sigma F' \quad \text{and} \quad p_s \simeq F'\phi_0 \tag{497}$$

whence the fractional occupancy f_t is given by

$$f_t = \frac{k_r n_s}{k_r n_s + k_f F'\phi_0 + k_s C_2} \tag{498}$$

As might be expected, f_t decreases as ϕ_0 increases or F increases. The overall current is

$$j_{ph} = e_0 k'_\Sigma p_s - e_0 k_r n_s N_t (1 - f_t) = e_0 k_v C_1 F'\phi_0 - e_0 k_r n_s N_t + \frac{e_0 N_t (k_f F'\phi_0 + k_r n_s) k_r n_s}{k_r n_s + k_f F'\phi_0 + k_s C_2} \tag{499}$$

$$= e_0 k_v C_1 F'\phi_0 - \frac{k_s C_2 e_0 k_r n_s N_t}{k_r n_s + k_f F'\phi_0 + k_s C_2} \tag{500}$$

Case (2). Depletion-layer recombination unimportant.

$$N \simeq X_{gen}, \quad p_s \simeq \frac{X_{gen}\phi_0}{k'_\Sigma} \quad k'_\Sigma = k_v C_1 + k_f N_t f_t \tag{501}$$

The value of f_t may be calculated from

$$k_r n_s (1 - f_t)(k_v C_1 + k_f N_t f_t) = k_f X_{gen} \phi_0 f_t + k_s C_2 (k_v C_1 + k_f N_t f_t) \tag{502}$$

References pp. 242–246

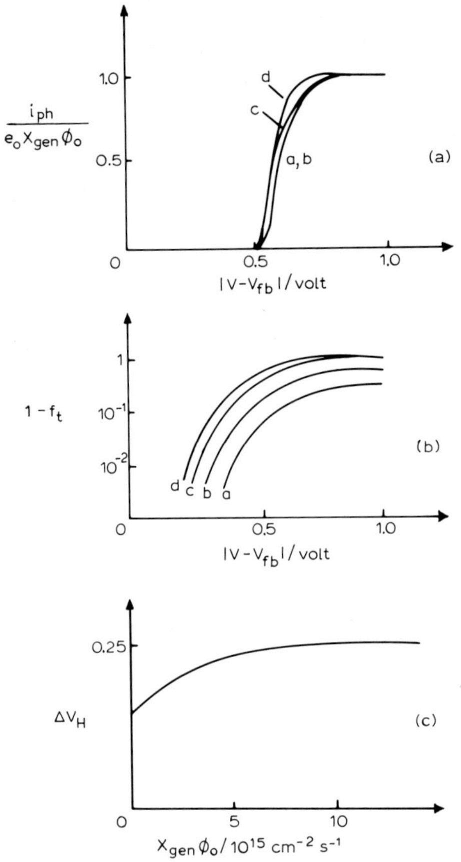

Fig. 75 (a) Photocurrent–voltage curves calculated from eqns. (502)–(505) assuming $k_v^0 C_{red} = 10^3\,\text{cm s}^{-1}$, $k_s^0 C_{red} = 10^8\,\text{cm s}^{-1}$, $N_t = 10^{13}\,\text{cm}^{-2}$, $k_r = 10^{-7}\,\text{cm}^3\,\text{s}^{-1}$, $k_f = 10^{-9}\,\text{cm}^3\,\text{s}^{-1}$, $\gamma N_t = 0.25\,\text{eV}$. Curve (a) $X_{gen}\phi_0 = 10^{14}\,\text{cm}^{-2}\,\text{s}^{-1}$; curve (b) $X_{gen}\phi_0 = 10^{15}\,\text{cm}^{-2}\,\text{s}^{-1}$; curve (c) $X_{gen}\phi_0 = 10^{16}\,\text{cm}^{-2}\,\text{s}^{-1}$; curve (d) $X_{gen}\phi_0 = 10^{17}\,\text{cm}^{-2}\,\text{s}^{-1}$. (b) Hole concentration $(1 - f_t)$ in the surface state as a function of potentials for the conditions in (a). (c) Total potential change ΔV_H in the Helmholtz layer as a function of $X_{gen}\phi_0$.

and, as before, the photocurrent is given by

$$j_{ph} = e_0 X_{gen}\phi_0 - e_0 k_r n_s N_t (1 - f_t) \qquad (503)$$

and although an analytical solution is less transparent in case (2), it can again be seen that f_t decreases strongly as n_s decreases.

Using estimates for the values of N_t ($10^{12}\,\text{cm}^{-2}$), k_r ($10^{-9}\,\text{cm}^3\,\text{s}^{-1}$), k_f ($10^{-7}\,\text{cm}^3\,\text{s}^{-1}$), and k_s ($\simeq 0$), the d.c. photocurrent–voltage curves may be calculated for case (2) [case (1) is not relevant here since $k'_\Sigma = k_f N_t$ is very large] and Fig. 74 shows the results for various values of $k_f N_t / k_v C_1$. Clearly there is a critical potential, corresponding to $k_r n_s N_t \equiv k_r n_B \exp[e_0(V - V_{fb})/kT)] \simeq X_{gen}\phi_0$ above which the states empty rapidly. Note that this potential

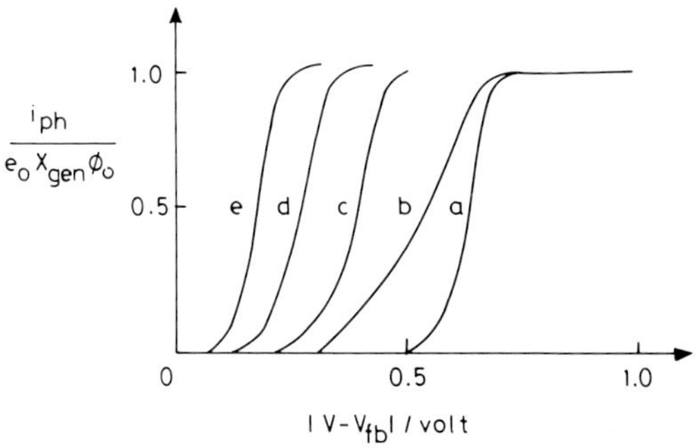

Fig. 76. Photocurrent–voltage curves calculated for $X_{gen}\phi_0 = 10^{16}\,cm^{-2}\,s^{-1}$ for addition of a second redox couple that can only be oxidised from the surface state. The parameters are all identical to Fig. 75(a) save that an additional k_s value for oxidation of C_2 is present. We assume that $k_s C_2$ is independent of ΔV_H and we have curve (a), $k_s C_2 = 0$ [as in curve (c) of Fig. 75(a)]; curve (b), $k_s C_2 = 10^3\,s^{-1}$; curve (c), $k_s C_2 = 5 \times 10^3\,s^{-1}$; curve (d), $k_s C_2 = 5 \times 10^6\,s^{-1}$.

is independent of k_v and, provided the surface state energy is such that the reverse thermal ionisation process can be neglected, it is also independent of E_t. It is, however, dependent on ϕ_0, being predicted to shift towards flat-band potential by 60 mV per decade increase in light intensity.

If N_t rises above $10^{13}\,cm^{-2}$, it may be anticipated that the complete emptying of such states will affect the potential drop in the Helmholtz layer. From above, the change in potential is $r\theta \equiv rN_t f_t/N_{ss}$, where N_{ss} is the maximum possible number of surface states. Writing $\gamma = rN_t/N_{ss}$, we may approximate γ from the estimated double-layer capacitance of 10–20 $\mu F\,cm^{-2}$; this gives $\gamma \simeq (0.08\text{–}0.16) \times 10^{-13} N_t$ V. The rate constants k_v and k_s will also depend on γf_t; specifically

$$k_v = k_v^0 \exp\left\{-\frac{(E_v - {}^0E_{red}^{(1)} + \lambda^{(1)} - \Delta V)^2}{4\lambda^{(1)}kT}\right\} \tag{504}$$

$$k_s = k_s^0 \exp\left\{-\frac{(E_{ss} - {}^0E_{red}^{(2)} + \lambda^{(2)} - \Delta V)^2}{4\lambda^{(2)}kT}\right\} \tag{505}$$

where $\Delta V = \gamma N_t f_t$ and E_{ss} is the surface-state energy.

Clearly, analytical solutions for cases (1) and (2) are no longer practicable, but numerical solutions can be obtained and some examples are shown in Fig. 75 for a species that can be oxidised both from the valence band and from the surface state [151]. One interesting feature is that the dependence of the photocurrent onset on ϕ_0 is much reduced; another effect that can be observed is that, if there is a redox couple present in solution that can only be oxidised via surface states, the rise of the photocurrent from flat-band potential becomes very sensitive to the value of $k_s C_2$ as shown in Fig. 76.

The most important insight provided by the analysis of Kelly and Memming is that the surface states may be distributed anywhere in the bandgap region, even near the minority carrier band edge, and still be incapable of

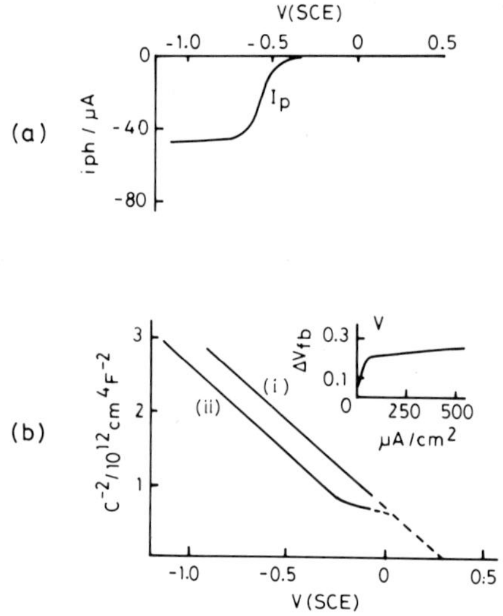

Fig. 77. Experimental data for p-GaAs showing the shift in flat-band potential, V_{fb}, under illumination. (a) Photocurrent measured in 0.5 M H_2SO_4 at 450 nm. (b) Mott–Schottky plots for p-GaAs in 0.5 M H_2SO_4 at 10 kHz (i) in the dark and (iii) illuminated to give a photocurrent of 50 μA cm^{-2}. In the inset, the shift in V_{fb} is plotted against limiting photocurrent density obtained by varying ϕ_0.

acting as effective recombination centres for comparatively modest values of the band-bending, owing to the rapid decrease in *majority* carrier density at the surface. The difficulty with the model is the paucity of evidence for such states. Alternating current impedance analysis is of little help if the electrode is not illuminated, since surface states near the minority-carrier edge will not be seen unless L_D is very small; this is clear from eqns. (104)–(120), which show that R_{ss}, and hence the rate constant, will decrease with n_s which, in turn, decreases exponentially with potential. Neither p-GaP nor p-GaAs show any large capacitive peaks in the potential region postulated by the model, at least in the frequency range normally examined (40 Hz–10 Hz), even though the model can account quantitatively for the delayed onset in the photocurrent observed for these two semiconductors. It is possible that very low frequency investigations may prove of interest here. Alternating current impedance studies of illuminated electrodes are discussed in detail below, but if it is assumed that only the majority carriers give rise to a significant quadrature (90°) response, even in the light, then the data of Fig. 77 show that the experimental photocurrent onset for p-GaAs lies close to the point at which the apparent depletion-layer capacitance shows a significant shift from that found in the dark. This shift gives rise to a linear Mott–Schottky plot at more negative potentials, but one that is shifted by up to 250 mV from the corresponding plot for the non-illuminted electrode. This is clearly strong evidence for the existence of a considerable surface state density, at least 10^{13} cm^{-2}, whose population charges affect the potential distribution in the Helmholtz layer.

A similar mechanism has been proposed on the basis of RRDE studies on p-GaP, using competitive reduction of Fe^{3+} to Fe^{2+} and water to H_2 [152]. As in the data of Memming and Kelly, the presence of Fe^{3+} strongly enhances the photocurrent near flat-band potential and the kinetic scheme proposed was supported by a variety of elegant experiments. The recombination route may go via the electronic states clearly seen in the a.c. impedance experiments reported for the unilluminated electrode; the identity of these states is still uncertain, but the long-term instability of p-GaP and p-GaAs appears to be due to strongly adsorbed hydrogen species attacking the surface or diffusing into the interior. On this basis, we may postulate a general mechanism of the sort

$h\nu \rightsquigarrow e^-_{cb} + p^+_{vb}$	(bulk or depletion layer generation)
$e^-_{cb} \rightsquigarrow (e^-_{cb})_s$	(diffusion or migration of minority carriers to the surface)
$(e^-_{cb})_s + O \rightarrow R$	(direct reduction of oxidised species (e.g. Fe^{3+}) in solution)
$(e^-_{cb})_s + S \rightarrow S^-$	(capture of surface electrons in the CB by surface states)
$S^- + H^+ \rightarrow S + H^{\cdot}_A$	(formation of weakly-bonded hydrogen atoms)
$S^- + O \rightarrow S + R$	
$(e^-_{cb})_s + H^+ \rightarrow H^{\cdot}_A$	(direct reduction of protons to weak-bonded hydrogen atoms)
$(e^-_{cb})_s + H^+ \rightarrow H^{\cdot}_D$ $S^- + H^+ \rightarrow H^{\cdot}_D + S$ $H^{\cdot}_A \rightarrow H^{\cdot}_D$	} direct or indirect formation of strongly-bonded hydrogen atoms)
$H^+ + H^{\cdot}_A + (e^-_{cb})_s \rightarrow H_2$ $H^+ + H^{\cdot}_A + S^- \rightarrow H_2 + S$ $H^{\cdot}_A + H^{\cdot}_A \rightarrow H_2$	} (formation of final product)
$p^+_{vb} + H^{\cdot}_A \rightarrow H^+$ $p^+_{vb} + H^{\cdot}_D \rightarrow H^+$ $p^+_{vb} + S^- \rightarrow S$	} (recombination routes)
$H^{\cdot}_D \rightsquigarrow (H^{\cdot}_D)_b$	(diffusion of hydrogen atoms into the interior of the crystal).

The sheer complexity of the above scheme is daunting; it is difficult to see, at the moment, how the exploration of such complex reaction sequencies can be undertaken without using a large battery of sophisticated techniques. In addition, the assumption underlying the analysis by Albery and Bartlett [152], that the potential distribution remains classical throughout the potential region explored by the techniques used, has also been questioned recently. The use of such techniques as electroreflectance has shed a disconcerting light on the detailed behaviour of p-GaAs and p-GaP near flat-band potential, as will be discussed below.

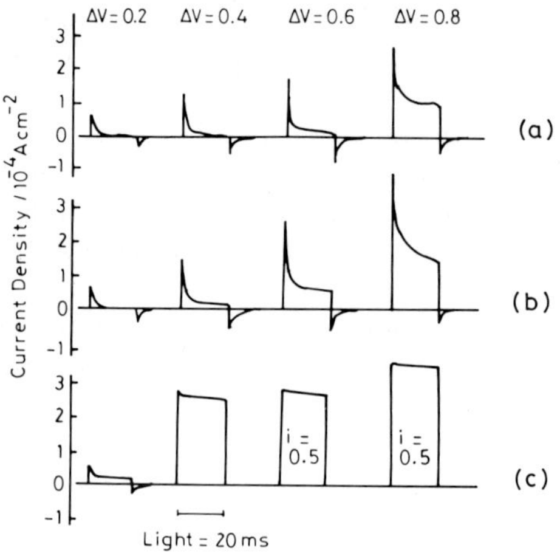

Fig. 78. Current transients during light pulses on an n-type WSe_2 electrode in contact with three different electrolytes. (a) 1 M KCl; (b) 1 M KCl and 2×10^{-2} M $K_4[Fe(CN)_6]$; (c) 1 M KCl and 10^{-2} M KI.

6.8 PHOTOCURRENT TRANSIENTS

In addition to a.c. techniques as described above, photocurrent transients have also been investigated extensively as a means of providing additional evidence in favour of models of the sort discussed above. The most typical type of behaviour, observed at relatively long time constants ($\geqslant 100$ ms),

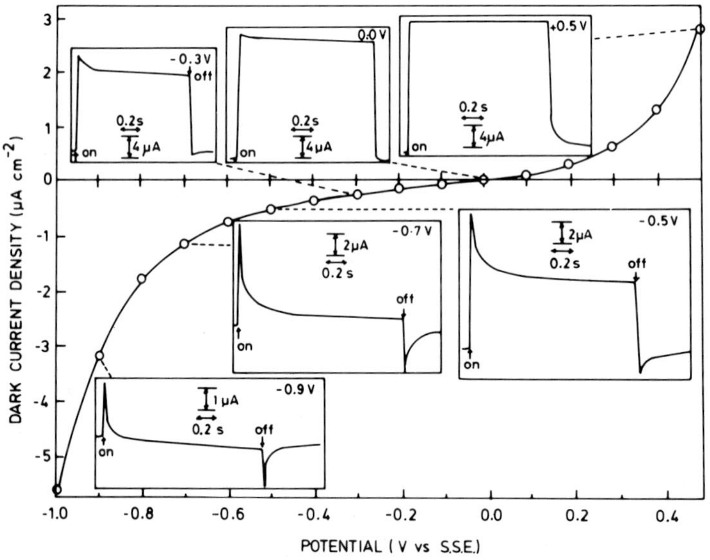

Fig. 79. Photocurrent transients with white light for n-$MnTiO_3$ in 1 M NaOH. Flat-band potential is ca. -1.0 V/SSE and the dark-current density is shown. It can be seen that the transients become essentially square for $|V - V_{fb}| \geqslant 1$ V.

consists of a rapid rise in the photocurrent followed by a relatively slow fall off to a smaller stationary current. Two examples are shown in Fig. 78 [153] and Fig. 79 [154] in which the effect of varying the potential and redox couple for two very different semiconductors is illustrated.

There are several possible fundamental causes for the observation of transient behaviour in the photocurrent.

(1) There may be an ongoing dark current whose value is altered, or "modulated" by the photocurrent.

(2) There may be a surface-adsorbed species or a significant density of surface states whose occupancy is profoundly affected by the photocurrent.

(3) There may be complex electrochemical pathways involving intermediates at the surface.

We first consider case (2), in which there is a surface-adsorbed species or a local redox couple. Let the surface concentration of the reduced species be $f_t^0 N_t$ in the dark, where N_t is the total number of surface sites per unit area of f_t^0 the Fermi occupancy. Rewriting eqn. (485), we find

$$N_t \frac{df_t}{dt} = N_t k_r n_s (1 - f_t) - k_d f_t N_t - k_f f_t p_s N_t$$

Now, N depends on f_t by virtue of the k_f term. We have, for small values of N ($\ll X_{gen}$)

$$N\phi_0 = k'_\Sigma p_s; \quad N = k'_\Sigma F'; \quad F' = \frac{X_{gen}(-2v_s)^{1/2}}{kL_D \exp[(v_s - v_{RZ})H((v_s - v_{RZ})]} \tag{506}$$

In the dark, $k_r n_s (1 - f_t^0) = k_d f_t^0$ whence

$$k_d = k_r n_s \left(\frac{1}{f_t^0} - 1\right)$$

and

$$N_t \frac{df_t}{dt} = \frac{k_r n_s N_t}{f_t^0}(f_t^0 - f_t) - k_f f_t p_s N_t \tag{507}$$

if illumination does not alter n_s. If the *only* route for consumption of holes at the surface is through *recombination* by surface-state capture, we can write $k'_\Sigma = k_f f_t N_t$, so

$$N_t \frac{df_t}{dt} = \frac{N_t n_s k_r}{f_t^0}(f_t^0 - f_t) - k_f f_t N_t F' \phi_0 \tag{508}$$

$$f_t = \frac{c_1}{c_2} + \left(f_t^0 - \frac{c_1}{c_2}\right) \exp(-c_2 t)$$

where

$$c_1 = k_r n_s; \quad c_2 = \frac{k_r n_s}{f_t^0} + k_f F' \phi_0 \tag{509}$$

The current at time t is

$$j = e_0 k_f p_s N_t f_t - \frac{e_0 k_r n_s N_t}{f_t^0}(f_t^0 - f_t) = e_0 N_t \left[k_f f_t F' \phi_0 - k_r n_s \left(\frac{f_t^0 - f_t}{f_t^0} \right) \right]$$

$$= e_0 N_t k_f F' \phi_0 f_t^0 \exp(-c_2 t) \tag{510}$$

and clearly as $t \to \infty$, $i \to 0$. The nett charge passed is

$$Q_t = \int_0^\infty j\mathrm{d}t = e_0 N_t k_f F' \phi_0 f_t^0 / \left(\frac{k_r n_s}{f_t^0} + k_f F' \phi_0 \right) \tag{511}$$

At low band-bending, F' is small and n_s large, so $Q_t \to e_0 k_f F' \phi_0 N_t (f_t^0)^2 / k_r n_s$; as the band-bending increases, n_s decreases and F' rises so $Q_t \to e_0 f_t^0 N_t$. This behaviour is illustrated in Fig. 78 in which the results of illuminating n-WSe_2 in 1 M KCl are shown. In this case, no electrolyte faradaic process is possible, save an extremely sluggish one at very high band-bending. At lower band-bending, the response is mainly transient and the overall transient charge increases with $|v_s|$ until it apparently saturates. The total charge passed at high potentials is $\simeq 10^{-7}\,\mathrm{C\,cm^{-2}}$, which suggests that $f_t^0 N_t \simeq 10^{12}\,\mathrm{cm^{-2}}$. The actual species is presumably a surface state located somewhere in the mid-bandgap region; it cannot be near the CB edge else it would be seen in the a.c. response. A surface state concentration of $10^{12}\,\mathrm{cm^{-2}}$ will not cause a large change in the Helmholtz-layer potential and we expect comparatively little re-organisation of the potential in the depletion region to occur on illumination.

If there is a separate surface recombination route via electronic surface states, this may be incorporated into the above analysis. There are three possibilities.

(1) Near threshold, case I recombination holds; $k'_\Sigma \to k_f f_t N_t + k_r$ where k_r is a recombination rate constant $\equiv 1/\tau_{ps}$ and the transient behaviour is unaltered.

(2) If case II obtains, similar conclusions can be drawn.

(3) If case III obtains at the surface, eqn. (426) must be modified. If the recombination rate is $k_r n_s$, then

$$N = \frac{X_{gen} + (k_r n_s A / k'_\Sigma \phi_0)}{1 + (A/k'_\Sigma)} \simeq \frac{k'_\Sigma X_{gen}}{A} + \frac{k_r n_s}{\phi_0} \tag{512}$$

where

$$A \equiv \left(\frac{D}{L_k} \exp(v_s) + \frac{D L_D}{L_k^2} \frac{1}{\sqrt{2 v_s}} \right) \tag{513}$$

and the total hole flux is

$$N \phi_0 = (k'_\Sigma X_{gen} \phi_0 + k_r n_s) \tag{514}$$

Here, k'_Σ is just the sum of the faradaic term and other surface recombination terms that are first order in p_s.

In the presence of an ongoing cathodic dark current, we may imagine that a faradaic route exists corresponding to the diffusion of product away from the electrode surface after reduction the surface state. If we assume that this is also first order in $f_t N_t$, we have f_t^0 in eqn. (507) replaced by

$$f_t^0 = \frac{k_r n_s}{k_r n_s + k_d + k_0} \quad (515)$$

where $k_0 f_t N_t$ is the rate of the reduction process. The current in the dark is then $-e_0 k_0 f_t^0 N_t$.

The final value of the current is

$$j = -e_0 k_0 f_t N_t + e_0 N_t k_f F' \phi_0 f_t^0 \exp(-c_2 t) \quad (516)$$

Inverse photocurrent transients are also of interest. Clearly, in cases where a surface state or surface-adsorbed species has been charged during illumination, a transient current of sign opposite to the photocurrent is expected when the light is switched off. The time constant for such as transient will not contain the light intensity as a parameter; thus, for the redox case discussed above, we have

$$N_t \frac{df_t}{dt} = k_r n_s (1 - f_t) N_t - k_d f_t N_t = N_t k_r n_s \left(1 - \frac{f_t}{f_t^0}\right) \quad (518)$$

and

$$f_t = f_t^0 + (f_t^s - f_t^0) \exp(-k_r n_s t / f_t^0) \quad (519)$$

where $f_t^s N_t$ is the stationary surface concentration of the reduced species under illumination. If this is small

$$f_t \approx f_t^0 [1 - \exp(-k_r n_s t / f_t^0)] \quad (520)$$

and

$$j = -e_0 N_t \frac{df_t}{dt} = -e_0 k_r n_s N_t \exp(-k_r n_s t / f_t^0) \quad (521)$$

Evidently, as the depletion layer potential increases, so will the time constant for the reverse transient as long as the state is sufficiently deep for $f_t^0 \simeq 1$; at large band-bending, the transient apparently vanishes since $f_t^0 \to 0$. If, on the other hand, f_t^s is not negligible compared with f_t^0, that is $n_s k_r \gg k_f F' \phi_0 f_t^0$, then $f_t^s - f_t^0 \simeq (f_t^0)^2 k_f F' \phi_0 / k_r n_s$ and the current is

$$j = -f_t^0 k_f F' \phi_0 e_0 N_t \exp(-k_r n_s t / f_t^0) \quad (522)$$

Under these circumstances, the height of the transient will depend on the light intensity and the decay constant again only on the electron concentration at the surface. As an example of this behaviour, consider the photocurrent transients for depopulation of a set of surface states on p-GaP, which are shown in Fig. 80 [152]. The statistics are poor but show an approximate linear dependence of the reverse transient on light intensity, ϕ_0, and also

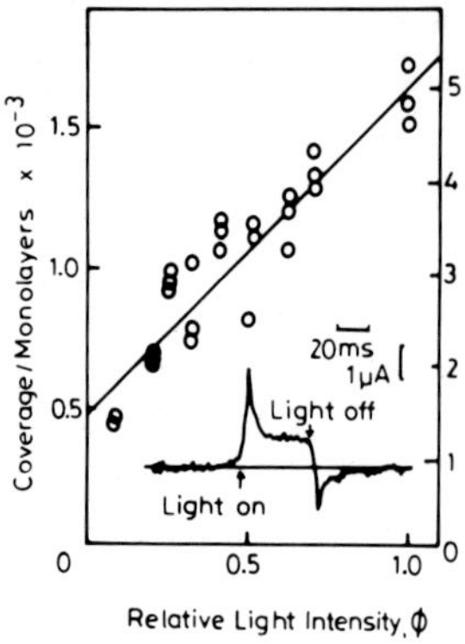

Fig. 80. Coverage of states vs. light intensity for p-GaP in 0.5 M H_2So_4.

show that the time constant is independent of light intensity. Interestingly, the decay is not exactly first order, suggesting that there may be a range of k_c and χ_0 values involved.

The redox couple will normally lie in the bandgap but it may lie sufficiently close to the valence band edge for neutralisation to be by hole injection into the valence band; such a mechanism has been suggested for the re-reduction of photo-oxidised thianthrene on n-$MoSe_2$, for which pronounced transients are seen. One further complication may arise if the density of surface states becomes large enough to effect changes in the potential distribution within the depletion layer. Under these circumstances, no analytical solutions are possible since n_s itself becomes dependent on the ratio of empty to filled surface states. However, no essentially new principles are involved and a recent discussion has been given by Peter et al. [150].

6.9 COMPETING REACTIONS AT THE SURFACE OF AN ILLUMINATED SEMICONDUCTOR

We saw above that the study of the competition between Fe^{3+} and H^+ reduction on illuminated p-GaP led to an increased understanding of the nature of surface electrochemical processes on that material. For many n-type materials, however, the most serious competing reaction with the oxidation of some redox couple in solution is the oxidative corrosion of the semiconductor itself. This has considerable practical consequencies; a photoelectrochemical device for the conversion of solar energy must be one in which the desired electrochemical route is overwhelmingly probable compared with semiconductor dissolution. So essential is this requirement, and so difficult has it proved to find satisfactory solutions for n-type semiconductors, that a substantial fraction of the recent literature on semiconductor electrochemistry has been devoted to both practical and theoretical considerations of the problem.

Detailed discussion of the various practical solutions that have been suggested for individual semiconductors is beyond the scope of this review but more general theoretical treatments, particularly from Gomes and co-workers [155–157], have shed considerable light on the basic strategies that might be followed in the design of stable systems and the kinetic consequencies of particular mechanistic regimes. We shall, therefore, follow their treatment closely.

We may define the stabilisation ratio, s, as that fraction of the total current that oxidises or reduces the solution redox couple. Specifying, for convenience, an n-type semiconductor and considering stabilisation to photogenerated holes, there are two basic classes of mechanism.

Type I. The redox couple in solution, Y/Z, is oxidized by an intermediate in the decomposition of the semiconductor. For the general surface decomposition process, we have

$$(\mathrm{SC})_{\mathrm{surf}} + \mathrm{p}_{\mathrm{vb}}^{+} \underset{k_{-1}}{\overset{k_1}{\rightleftharpoons}} \mathrm{X}_1$$

$$\mathrm{X}_1 + \mathrm{M} \underset{k_{-2}}{\overset{k_2}{\rightleftharpoons}} \mathrm{X}_2$$

where M may be another hole or another X_1 species

$$\mathrm{X}_{n-1} + \mathrm{M} \xrightarrow{k_B} \text{decomposition products}$$

also

$$\mathrm{X}_i + \mathrm{Y} \xrightarrow{k''_{-i}} \mathrm{X}_{i-1} + \mathrm{Z} \qquad (i = 1, 2, \ldots, n-1)$$

Electrons in the CB may also be captured by any intermediate

$$\mathrm{X}_i + \mathrm{e}_{\mathrm{vb}}^{-} \xrightarrow{k^{\mathrm{r}}_{-i}} \mathrm{X}_{i-1} \qquad (i = 1, 2, \ldots, n-1)$$

The rate of formation of species X_i is

$$\frac{\mathrm{d}x_i}{\mathrm{d}t} = [\text{formation rate}] - k^{\mathrm{r}}_{-i}x_i n_{\mathrm{s}} - k_{-i}x_i - k''_{-i}\,yx_i$$

$$\equiv [\text{formation rate}] - k'_{-i}x_i \qquad (523)$$

where y is the concentration of Y.

Now, the steady-state conditions achieved in most experiments imply that the *nett* formation rate of all species X_i must be the same; it is given by $(1 - s)j/n$ where j is the total current passing. Thus, for species X_1

$$\frac{(1-s)j}{n} = k_1 p_{\mathrm{s}} - k'_{-1}x_1 \qquad (524)$$

and for X_{i+1}

$$\frac{(1-s)j}{n} = k_{i+1}x_i m - k'_{-(i+1)}x_{i+1} \qquad (i = 1 \ldots n-1;\ k'_{-n} \equiv 0) \tag{525}$$

By definition, the particle flux corresponding to stabilisation can be written as

$$sj = y\sum_{i=1}^{n-1} k''_{-i}x_i \tag{526}$$

and the recombiantion rate is

$$j_{\mathrm{r}} = n_{\mathrm{s}}\sum_{i=1}^{n-1} k^{\mathrm{r}}_{-i}x_i \tag{527}$$

By incorporating these formulae into the rate equation, starting with the final intermediate X_n, we find

$$x_i = \left[\frac{(1-s)j}{n}\right]\sum_{j=1}^{n-i}\lambda_{ij}m^{-j};\quad \lambda_{ij} = \frac{1}{k_{i+j}}\prod_{l=i+1}^{i+j-1} k'_{-l}/k_l \quad (j = 2, 3, \ldots n-i) \tag{528}$$

whence

$$\left(\frac{s}{1-s}\right) = \left(\frac{y}{n}\right)\left(\sum_{i=1}^{n-1} k''_{-i}\sum_{j=1}^{n-i}\lambda_{ij}m^{-j}\right) \tag{529}$$

$$\frac{1}{1-s} = 1 + \left(\frac{y}{n}\right)\left(\sum_{i=1}^{n-1} k''_{-i}\sum_{j=1}^{n-i}\lambda_{ij}m^{-j}\right) \tag{530}$$

If M is the species X_1, it can be shown that

$$j = \left(n + y\sum_{i=1}^{n-1} k''_{-i}\sum_{j=1}^{n-i}\lambda_{ij}m^{-j}\right)\left[\sum_{j=1}^{n-1}\lambda_{1j}m^{-(j+1)}\right]^{-1} \tag{531}$$

and the hole current j_{h} is given by

$$\frac{j_{\mathrm{h}}}{j} = 1 + \left(n_{\mathrm{s}}\sum_{i=1}^{n-1} k^{\mathrm{r}}_{-i}\sum_{j=1}^{n-i}\lambda_{ij}m^{-j}\right)\left[n + y\sum_{i=1}^{n-1} k''_{-i}\sum_{j=1}^{n-i}\lambda_{ij}m^{-j}\right]^{-1} \tag{532}$$

If M is a hole, then

$$j = k_1\left(n + y\sum_{i=1}^{n-1} k''_{-i}\sum_{j=1}^{n-i}\lambda_{ij}m^{-j}\right)\left[m^{-1} + k'_{-1}\sum_{j=1}^{n-i}\lambda_{ij}m^{-(j+1)}\right] \tag{533}$$

where

$$k_1 = \left[\frac{(1-s)j}{n}\right]\left(m^{-1} + k'_{-1}\sum_{j=1}^{n-1}\lambda_{1j}m^{-(j+1)}\right)$$

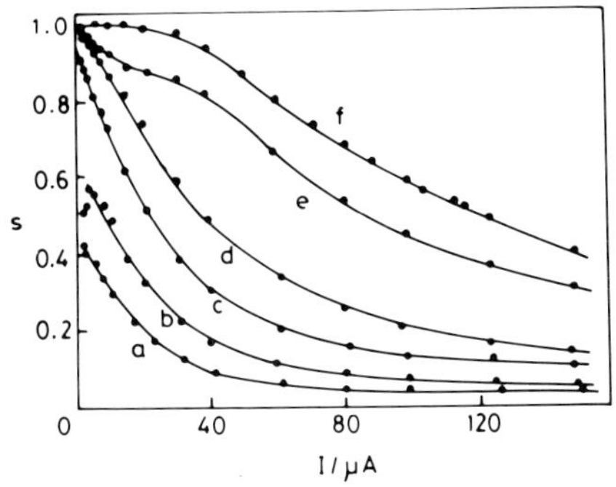

Fig. 81. Stabilisation ratio s vs. photocurrent I for $(\bar{1}\bar{1}\bar{1})$ n-InP at 0.65 V/SHE at pH 1. Concentration of stabiliser Fe^{2+} in solution is (a) 10^{-1} M, (b) 0.25 M, (c) 0.5 M, (d) 1 M, (e) 1.5 M, (f) 2 M.

with the same expression for the hole current, j_h.

Type II. Direct interaction between valence holes and Y

$$p_{vb}^{+} + Y \xrightarrow{k_0} Z \qquad \text{all } k''_{-i} \equiv 0$$

under the simplifying assumption that $m \equiv p_s$, then

$$\left(\frac{s}{1 - s}\right) = \left(\frac{k_0}{k_1}\right)\left(\frac{y}{n}\right)\left(1 + k'_{-1} \sum_{j=1}^{n-1} \lambda_{1j} m^{-j}\right) \tag{534}$$

whence

$$j = \left[nk_1 + k_0 y\left(1 + k'_{-1} \sum_{j=1}^{n-1} \lambda_{1j} m^{-j}\right)\right]\left(m^{-1} + k'_{-1} \sum_{j=1}^{n-1} \lambda_{1j} m^{-(j+1)}\right)^{-1} \tag{535}$$

and

$$\frac{j_h}{j} = 1 + \left(n_s \sum_{i=1}^{n-1} k^{r}_{-i} \sum_{j=1}^{n-i} \lambda_{ij} m^{-j}\right)\left[n + \left(\frac{k_0}{k_1}\right) y\left(1 + k'_{-1} \sum_{j=1}^{n-1} \lambda_{1j} m^{-j}\right)\right]^{-1} \tag{536}$$

General features of the mechanisms I and II can be extracted. Thus, for all cases, as $j_h \to 0$, $s \to 1$ but the value of $(\partial s/\partial j)_{y,n_s}$ may take a wide range of values from 0 to $-\infty$, depending on the case. Also, as $j \to \infty$, $(\partial s/\partial j)_{y,n_s}$ tends to a limiting value of zero but the limiting value of s is zero in mechanisms of type I but lies in the range $0 < s < 1$ for mechanisms of type II. This type of general behaviour is very typical and data for n-InP are given in Fig. 81 [157]; for this system, a detailed analysis of the results shows that they are consistent with a type I mechanism. Near flat-band potential, n_s becomes large and, again, different behaviour is found depending on details of the mechanisms. For mechanisms of type I, if all $k''_i \neq 0$ and $k^{r}_{-i} \neq 0$, then $s = 1$

for $j = 0$ but, if stabilisation and recombination through a single intermediate are assumed, then as $n_s \to \infty, j \to 0, 0 < s_{lim} < 1$.

Some special cases have been work out.

Mechanisms of type I.

(a) $k_0 = 0$, M = X, $k'_{-2} = 0$, no recombination, $k''_{-i} = 0$ save for $i \equiv 1$

$$\left(\frac{s^2}{1 - s}\right) = \frac{1}{n}\frac{k''^2_{-1}}{k_2}\left(\frac{y^2}{j}\right) \tag{537}$$

(b) $k''_{-2} \neq 0$, $k''_{-1} = 0$, $k_{-2} = k_{-3} = 0$

$$\frac{1}{n}\left(\frac{s^2}{1 - s}\right) + \frac{s^3}{(1 - s)^2} = \frac{1}{n^2}\frac{k_2 k''^2_{-2}}{k_3^2}\left(\frac{y^2}{j}\right) \tag{538}$$

(c) $k''_{-3} \neq 0$, $k_{-2} = k_{-3} = k_{-4} = 0$

$$\frac{s^2}{(1 - s)} = \frac{1}{n}\frac{k_2 k''^2_{-3}}{k_4^2}\left(\frac{y^2}{j}\right) \tag{539}$$

(d) M = p^+_{vb}, $k_{-1} = 0$, $k''_{-2} = k'_{-2} = 0$

$$\frac{s}{n} + \frac{s^2}{1 - s} = \frac{1}{n}\frac{k_1 k''_{-1}}{k_2}\left(\frac{y}{j}\right) \tag{540}$$

Mechanisms of type II.

(a) $k_{-1} \simeq 0$, M = X_1

$$s = \frac{k_0 y}{k_0 y + k_1} \tag{541}$$

(b) $k_{-1} > 0$, $k_{-2} \simeq 0 \simeq k''_{-2}$

$$\frac{s^2}{1 - s} = \frac{1}{n}\left(\frac{k^2_{-1} k_0^2}{k_1^2 k_2}\right)\left(\frac{y^2}{j}\right) \tag{542}$$

(c) M = p^+_{vb}, $k_{-1} = 0$

$$s = \frac{k_0 y}{k_1 n + k_0 y} \tag{543}$$

(d) $k_{-1} > 0$, $k_{-2} \equiv k'_{-2} = 0$

$$\frac{s^2}{1 - s} = \frac{1}{n}\left(\frac{k_{-1} k_0^2}{k_1 k_2}\right)\left(\frac{y^2}{j}\right) \tag{544}$$

It is clear that a number of mechanistic types give rise to expressions of the general form

$$\log_e\left(\frac{s^2}{1 - s}\right) \propto \log_e\left(\frac{y^2}{j}\right) \tag{545}$$

and data for n-GaP/$[Fe(CN)_6]^{4-}$ (pH = 9.2), n-GaP/Fe(II)–EDTA and n-InP/Fe(II)–EDTA (pH = 3.8), all with the $(\overline{111})$ face, show behaviour of this sort as seen in Fig. 82 [155]. Save for mechanism I(a), if eqn. (545) is satisfied, the

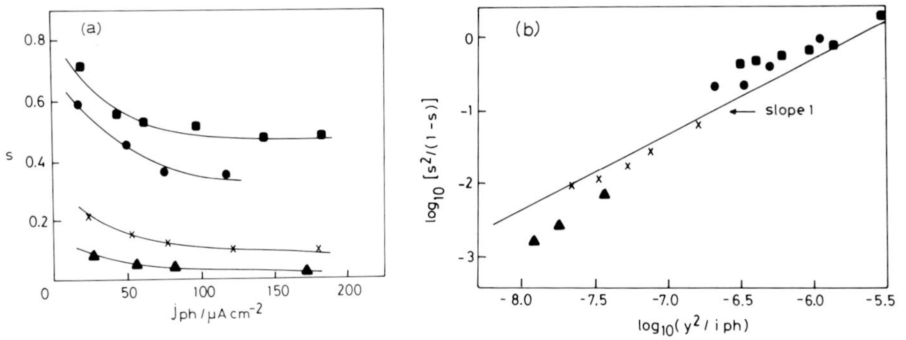

Fig. 82. (a) Stabilisation ratio s for n-GaP $(\bar{1}\bar{1}\bar{1})$ in the presence of Fe(II)–EDTA at +1.0 V/SSE: (▲) 1.0 mM; (×) 2.0 mM; (●) 5.0 mM; (■) 7.6 mM. (b) Data from (a) plotted as $\log_{10}[s^2/(1-s)]$ vs. $\log_{10}(y^2/i_{ph})$.

implication is that the first photoanodic step is reversible. Interestingly, for n-GaP/Fe(II) (pH = 0.9), the expression obeyed is

$$\log\left(\frac{s}{6} + \frac{s^2}{1-s}\right) \propto \log_e\left(\frac{y}{j}\right) \tag{546}$$

which is consistent with mechanism I(d) above (see Fig. 83 [156]).

In the region near threshold, where recombination cannot be neglected, the general form reduces to

$$\frac{s}{n} + \frac{s^2}{1-s}\left(1 + \frac{k^{\mathrm{r}}_{-1}n_{\mathrm{s}}}{k''_{-1}y}\right) = \frac{k_1k''_{-1}}{nk_2}\left(\frac{y}{j}\right) \tag{547}$$

assuming recombination takes place only through the intermediate X_1. This predicts that s *decreases* near flat-band potential, an effect actually observed.

A very similar theory has been derived by Morrison and co-workers [158, 159], save that explicit expressions were used for the rate constants, derived from Gerischer's theory as discussed above, and a type II mechanism was postulated for the system investigated. The final expression took the form

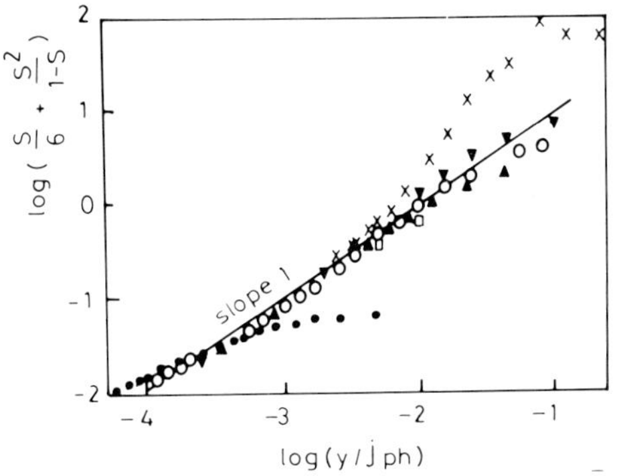

Fig. 83. Plot of log $[(s^2/6) + s^2/(1-s)]$ vs. $\log(y/i_{ph})$ for n-GaP/Fe^{2+} at pH = 0.9.

$$\frac{1}{s} = 1 + \frac{\alpha j_h}{y^2} \tag{548}$$

where

$$\alpha = \frac{k_1 k_2}{k_{-1} k_0^2} \tag{549}$$

and

$$j_h = sj \tag{550}$$

identical to II(d) above. The parameter α could be expressed in terms of Gerischer's theory; assuming that k_{-1}, k_1 and k_2 are all independent of the solution redox couple and

$$k_0 = \langle e_0 \sigma v d \rangle \left(\frac{kT}{\pi\lambda}\right)^{1/2} \exp\left[-(E_{vb} - E_F^{red} + \lambda_{red})^2/4\lambda kT\right] \tag{551}$$

Plotting $1/s$ as a function of j will yield α and hence, knowing $E_{VB} - E_F^{redox}$, which we may vary by varying the pH, λ_{red} for the appropriate redox stabiliser can be determined. This theory was checked for n-GaAs and a value of 0.7 eV for the λ value of Fe(II)–EDTA in the pH range 4–7 was found, as shown in Fig. 84.

There are other treatments of semiconductor corrosion, which have concentrated on different likely mechanisms. Tenne et al. have considered the case where the surface reactions may have chemical processes such as adsorption as rate-limiting reactions. Under these circumstances, the above treatment may be re-worked taking into account recombination [160].

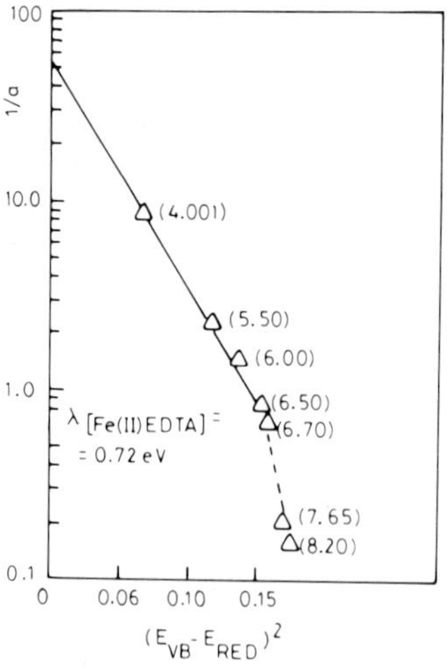

Fig. 84. Plot of $1/\alpha$ vs. $(E_{vb} - E_{red})^2$ for (100) n-GaAs/0.02 M Fe(II) in 0.1 M Na_2H_2EDTA where pH is the parametric variable as shown.

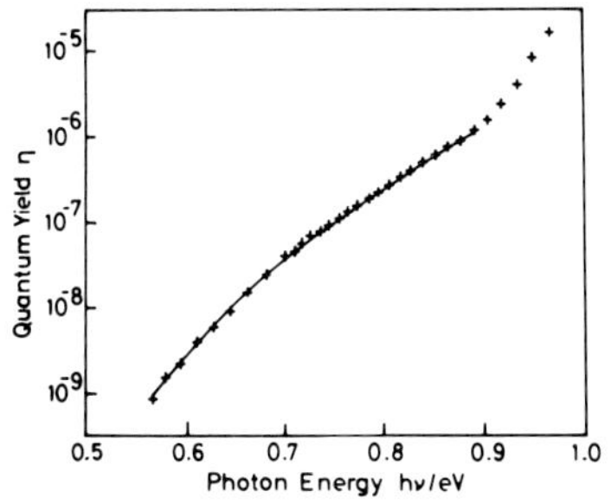

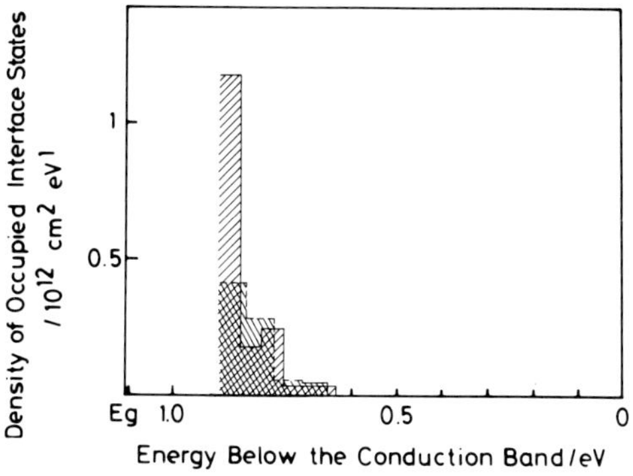

Fig. 85. Quantum yield for the sub-bandgap photocurrent of (100) n-Si in 1 M HCl at a bias potential of + 1.0 V/SCE. The sharp increase of η near 0.95 eV is related to the onset of the band-to-band transition.

Fig. 86. Density of occupied surface states as deduced from Fig. 85 using eqn. (552). The vertical state presumes an optical cross-section $\sigma \simeq \sigma_0[(\hbar - E_c + E_t)/E_g]^{3/2}$ with $\sigma_0 = 10^{-15}\,\mathrm{cm}^{-2}$. The two curves are obtained from fits with two meshes.

7. Semiconductor electrochemistry techniques involving light

In this section, we consider some techniques, other than d.c. photocurrent measurements, that use illumination as a probe. Most directly related to the previous section are techniques employing sub-bandgap radiation and we shall consider these first. We shall then treat those techniques involving the emission of radiation and then photovoltage and a.c. photocurrent measurements. Finally, we shall treat electroreflectance and thermoreflectance measurements.

7.1 SUB-BANDGAP MEASUREMENTS

Experimentally, it is found that the observed photocurrent–wavelength variation does not show a sharp onset at the bandgap threshold, but frequently a rather lengthy tail into the sub-bandgap region. An example is shown for n-Si/1 M KCl in Fig. 85 [161]. The sub-bandgap photocurrent was interpreted here in terms of an expression of the form

$$j_{\mathrm{ph}} \propto \int_{E_v}^{E_c} g(E_t)(\hbar\nu + E_t - E_c)^{\alpha}\,\mathrm{d}E_t \tag{552}$$

where the dominant contribution was ascribed to excitation from the occupied surface states to the CB rather than from the VB to empty surface states on the basis of the variation of i_{ph} with applied potential. In eqn. (552), $g(E_t)$ is, as usual, the energy distribution of surface states and α is taken to be 3/2 for reasons discussed in the original paper. The value of $g(E_t)$ convoluted into this expression is shown in Fig. 86.

Interestingly, in this study, the sub-bandgap photoresponse is reduced if the sample is simultaneously irradiated with light of energy above the bandgap; this reduces the population of the surface states and supports the assignment suggested above. Evidence that the sub-bandgap response is from surface and not from bulk states comes from the observation that it is completely quenched by a thin oxide surface layer.

The technique of sub-bandgap photocurrent measurement has also been used to probe the surface state distribution on p-GaP [162]. In this case, the sub-bandgap photocurrent *increased* after illumination with bandgap light and this was ascribed to the formation of H atoms that diffused a short way into the surface ($\simeq 10$ Å) creating "near surface" states from which excitation to the CB could be effected. These authors also found that the sub-bandgap photocurrent formed a rather ill-defined tail on the main absorption edge; however, in the presence of $[Fe(CN)_6]^{3-}$, a rather well-defined peak in the sub-bandgap photocurrent is seen at 1.6 eV, corresponding to the signal observed in the electroluminescence [163].

7.2 ELECTROCHEMICAL PHOTOCAPACITANCE SPECTROSCOPY

This technique has been elaborated as an electrochemical tool by Tench and co-workers [164–166]. Its main purpose is to explore the deep-lying bulk and surface levels and the principle of the technique is that the main role of sub-bandgap irradiation in a semiconductor will be to cause optical excitation to or from a bulk or surface state; this, in turn, will cause an alteration in the potential distribution from that existing in the dark; this alteration will manifest itself in the behaviour of the interfacial capacitance.

For bulk states, the change in electron occupancy of deep levels on illumination, Δn_d, is given by

$$\Delta n_\mathrm{d} = \frac{2V}{e_0 \varepsilon_0 \varepsilon_\mathrm{sc}} [C_0^2 - C^2] \qquad (553)$$

where V is the potential vs. V_fb, C_0 is the capacitance measured in the dark or, better, at a photon energy just below the threshold of optical absorption by the trap states, and C that just above the threshold. For an n-type material, $\Delta n_\mathrm{d} < 0$ if absorption of light causes *ionisation* of the trap (i.e. the capacitance will increase) and vice versa if electrons are excited from the VB to the empty trap sites.

The validity of eqn. (553) depends crucially on the following assumptions.

(1) Once a steady state is re-established, all the deep-lying levels within the depletion layer have been ionised if initially full or filled if initially empty.

(2) The number density of deep traps is small compared with the number density of shallow donors (which determines the potential distribution in the dark).

(3) The concentration difference, on illumination, for traps in the field-

free region of the semiconductor is negligible; this is partly a consequence of (2), but also reflects the large density of carriers in the field-free region.

Interfacial states may also be explored. Assuming that these states are *slow* and do not contribute to the measured capacitance (which can obviously be ensured, in principle, by working at a high enough frequency), they may be detected through the effects of ionisation on the potential distribution. It is easily seen that the change in surface occupancy, Δn_t (cm^{-2}), is given by

$$\Delta n_t = -\frac{\varepsilon_0 \varepsilon_{sc} N_D C_H}{2}\left(\frac{1}{C_0^2} - \frac{1}{C^2}\right) \tag{554}$$

for an n-type semiconductor, where N_D is the bulk donor concentration and C_H the Helmholtz capacitance. If the surface states are *ionised*, $\Delta n_t < 0$; thus, increasing the positive charge at the surface decreases the potential drop and *increases* C. The assumptions underlying eqn. (554) are essentially similar to eqn. (553): the validity of the assumptions will clearly increase as the depletion increases, since back-filling by tunnelling from the field-free region will be reduced. However, it is evident that, as the depletion increases, the steady-state occupancy of the surface states will decrease in the dark, thereby reducing the effectiveness of the technique, as discussed below. It is evident also that there should be no ongoing dark current of any significance, since the measured capacitance will then contain a term from the faradaic impedance which will be affected in an unpredictable manner by sub-bandgap illumination.

The technique is, therefore, to measure C as a function of $\hbar\omega$ and thereby to obtain a spectrum of bulk and surface state distributions with energy. Distinguishing between the two effects may, in principle, be achieved by examining the *potential* dependence. For bulk states, we would expect Δn_d to

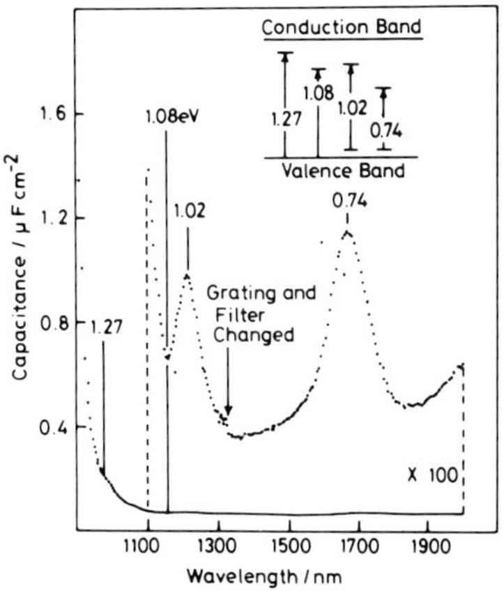

Fig. 87. Photocapacitance spectrum for p-Zn_3P_2 measured at 5 kHz and $|V - V_{fb}| = 0.3$ V in acetonitrile/0.1 M tetrabutylammonium perchlorate. The inset shows the assignment based on the spectrum.

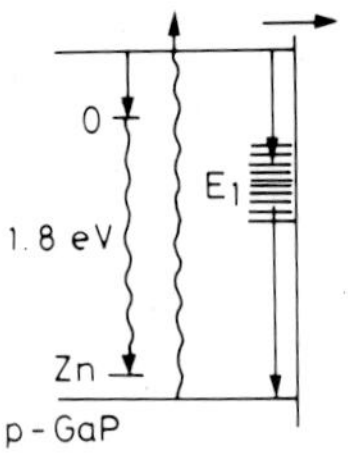

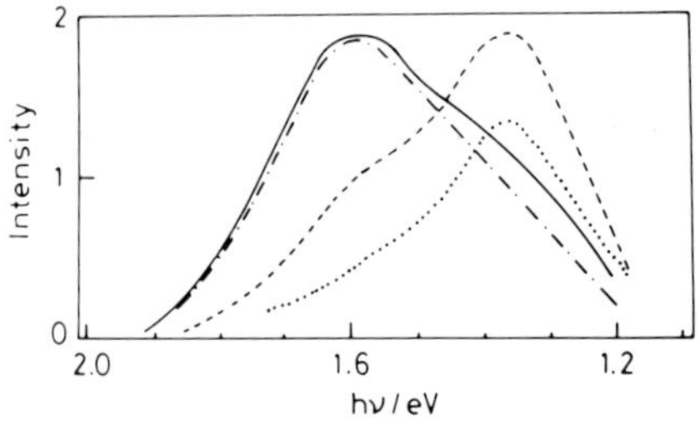

Fig. 88. Schematic energy diagram for p-GaP showing the O → Zn red luminescence. The quantum yield of photoluminescence will clearly be affected by alternative, non-luminescent recombination pathways, particularly those associated with surface states, as shown.

Fig. 89. ——, — · —, Electroluminescence yields for n-GaP (111) and n-Gap ($\bar{1}\bar{1}\bar{1}$), respectively, in 0.1 M NaOH containing 0.1 M $K_3[Fe(CN)_6]$ and 0.1 M $K_4[Fe(CN)_6]$. - - -, Electroluminescence yield for n-GaP (111) in 0.1 M NaOH/0.5 M $Na_2S_2O_8$. This latter is reduced by CB electrons to $SO_4^{-\cdot}$ which injects a "hot" hole into the VB. The main recombination processes then occur in the bulk. ···, Photoluminescence for n-GaP (111) in 0.1 M NaOH.

scale as V from eqn. (553), whereas the dependence of Δn_t on V will reflect the changing occupancy of the surface states in the dark.

Localised bulk–bulk transitions that ionise by tunnelling may also be expected. These usually give rise to fairly sharp ΔC vs. $\hbar\omega$ peaks.

Several semiconductors have been studied using this technique; the results from p-Zn_3P_2 are shown in Fig. 87 in which two localised and two VB → trap transitions (which *increase* C for a p-type semiconductor) can clearly be seen. Results for n-CdSe in acetonitrile show something of the power of the technique; acceptor states have been located at 1.40 and 1.57 eV above the VB edge: these have been ascribed, respectively, to elemental Se and reconstructed polycrystalline CdSe domains. Two donor states, at 1.04 and 1.21 eV below the CB edge were also located in this study and appeared to be associated with an oxide film. It is in non-aqueous solvent, where faradaic reactions are often extremely slow, that this technique may have its best applications.

7.3 LUMINESCENCE AND PHOTOLUMINESCENCE

Certain semiconductors, the best known being GaP, are observed to luminesce under conditions in which hole injection (n-type) or electron excitation (p-type) occurs. Luminescence may arise from bulk or surface states; in p-GaP, for example, there is a strong (photo)luminescence peak at ca. 1.8 eV due to transition from a defect level associated with O to that associated with the Zn acceptor sites, as shown in Fig. 88 [163]. This transition may also be observed through electron injection into the CB; although this is not electrochemically possible in aqueous electrolyte, owing to the fact that the CB lies above the stability range of water, it can be detected in non-aqueous solvents [163, 167].

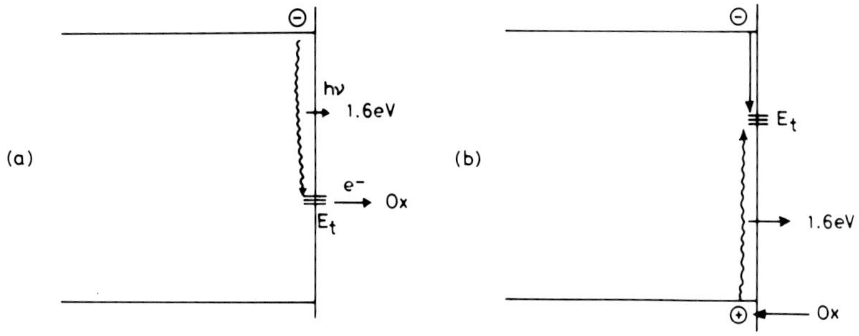

Fig. 90. Two possible models for the surface electroluminescence process on n-GaP.

Similarly, for n-GaP, photoluminescence is observed at ca. 1.3 eV. However, for n-GaP in the presence of $K_3[Fe(CN)_6]$ in the dark, weak electroluminescence was observed over a considerable spectral range, as shown in Fig. 89 [167] and a second peak at 1.6 eV appears not to be due to bulk but to surface processes.

The most serious problem with luminescence studies is the difficulty of distinguishing between luminescence processes due to transitions *to* or *from* the level. If we assume that there is just *one* level, then the luminescence peak at 1.6 eV in Fig. 89 may be due to either of the processes shown in Fig. 90.

The process shown in Fig. 90(b) is favoured by the known position of the band edges in alkali, the similarity with the electroluminescence induced by persulphate, and the energy of the $[Fe(CN)_6]^{3-/4-}$ redox couple. Either mechanism would, however, be consistent with the observation that the onset of electroluminescence coincides, approximately, with the flat-band potential, as shown in Fig. 91. Interestingly, the photoluminescence reaches a maximum at more negative potentials, a result not predicted by either model, as shown in Fig. 91 [167].

Electroluminescence at p-GaAs in the presence of $K_2S_2O_8$ has been observed [168, 198]. Unlike n-GaAs, which shows strong electroluminescence in the presence of peroxydisulphate, p-GaAs should show only photolumine-

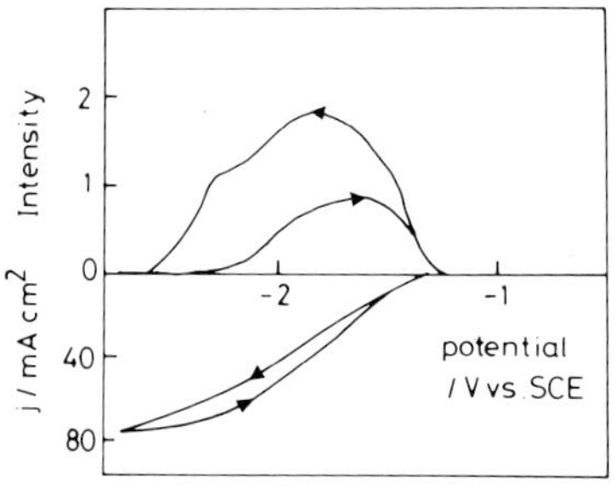

Fig. 91. Variation of the intensity of the electroluminescence signal at 1.6 eV for n-GaP (111) in 0.1 M $K_3[Fe(CN)_6]$/0.1 M $K_4[Fe(CN)_6]$ at pH 4.4 compared with the photocurrent.

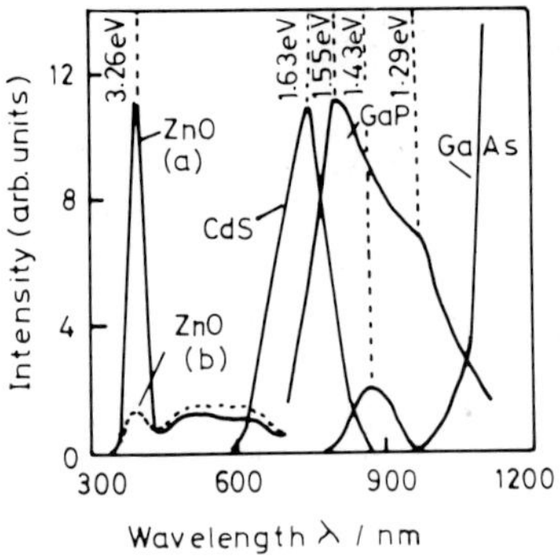

Fig. 92. Spectral distribution of the electroluminescence of four semiconductors in 5 M NaOH/ 0.05 M $Na_2S_2O_8$. The dotted curve for n-ZnO corresponds to the case of considerable surface damage.

scence, since the electron concentration in the CB is expected to be very small. However, for highly doped samples, at potentials sufficiently negative for inversion to occur, weak electroluminescence was observed [168] whose spectral distribution suggests that its origin is CB → VB. Whilst the detailed potential distribution cannot, of course, be determined by this technique, the existence of a sizeable electron concentration in the CB at sufficiently negative potentials seems well established.

It is clear, from the foregoing, that luminescence studies will be immeasurably aided by a simultaneous study of the sub-bandgap photocurrent and of the photocapacitance behaviour.

Aside from n-GaP, whose VB lies in the stability range of the aqueous solvents usually employed, materials such as n-SnO_2, n-ZnO, or n-CdS cannot be induced to show luminescence since their VBs lie below the oxidative stability limit for most solvents. However, electroluminescence can be observed if short-lived highly oxidising intermediates such as $SO_4^{-\cdot}$ and $OH^{\cdot}$ can be formed close to the electrode surface [169]; if these are reduced by hole injection, then electroluminescence is possible. Reduction of $[S_2O_8]^{2-}$ by the wide bandgap materials n-ZnO and CdS as well as GaP and GaAs leads to electroluminescence provided the potential is negative of flat-band potential. The spectral distributions for the electroluminescence of these semiconductors are shown Fig. 92; that from well-etched ZnO is dominated by a peak at 3.26 eV, close to the bandgap edge, but the size of this peak is reduced substantially, and the lower energy emission correspondingly enhanced, by leaving a damaged layer on the surface.

7.4 PHOTOVOLTAGE TECHNIQUES

The technique of photovoltage differs from the photocurrent techniques discussed above in that the electrode is held at open circuit and its equilibrium voltage is measured against a reference. The change in this equilibrium voltage on illumination is termed the *photovoltage* [170, 171]. The

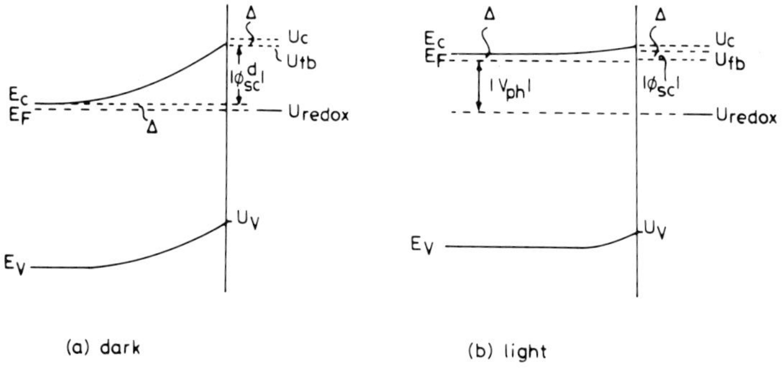

Fig. 93. The principle of photovoltage measurements. (a) Equilibrium situation in the dark, (b) illumination under open-circuit conditions.

simplest interpretation of this experiment is to assume that the potential of the semiconductor initially adjusts itself to the equilibrium position of the redox couple in solution. Provided that the dominant faradaic kinetics are indeed those involving this redox couple (in other words, a mixed potential is not set up), then this is likely to be the case. On illumination, no nett current may flow; the concentrations of majority and minority carriers, therefore, adjust to ensure that the fluxes of both to the surface are equal. This normally involves an adjustment in the potential dropped across the depletion layer and it is assumed that the depletion layer potential under illumination, ϕ^{l}_{sc}, must fall to zero. For an n-type semiconductor, we can write (see Fig. 93)

$$|V_{ph}| = U_{redox} - U_{fb} - |\phi^{l}_{sc}| \tag{555}$$

If $|\phi^{l}_{sc}| = 0$, then $|V_{ph}|$ gives a measure of the flat-band potential provided U_{redox} is known. In fact, this formula is very rarely obeyed in practice and deviations are both common and complex. Detailed theories of the potential distribution at the semiconductor–electrolyte interface have been presented, based on photovoltage measurements, but immense care needs to be taken in the interpretation of the photovoltage since kinetic effects apparently play a major role. This is especially true if surface recombination plays an important role [172].

The essential boundary condition is that the two carrier currents at the surface are equal, i.e. for the conduction and valence bands

$$j_c + j_v = 0 \tag{556}$$

For an n-type semiconductor with a redox couple O/R in solution

$$R + p_s^+ \rightarrow O$$

$$O + n_s^- \rightarrow R$$

are the faradaic reactions of importance. Writing the total hole current as j_h, we have

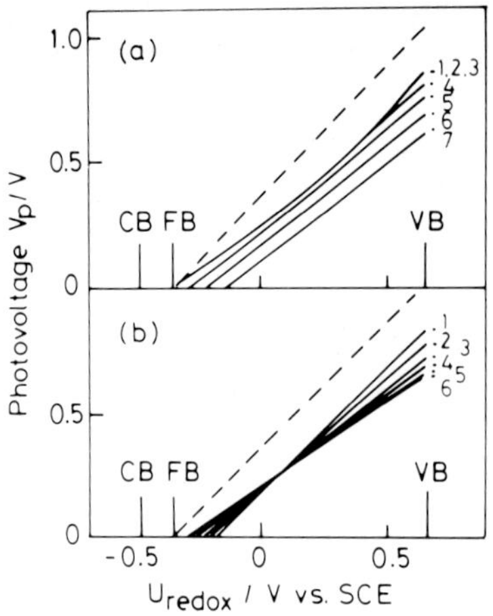

Fig. 94. Photovoltages calculated for n-WSe_2 ($N_D = 6 \times 10^{16}\ cm^{-1}$) in contact with one-equivalent redox couples. The abbreviations CB, FB, and VB refer to the positions on the energy scale with reference to SCE of the conduction band, flat-band, and valence band, respectively. For the calculation, the saturation value of j_{ph} is taken as $10\ mA\ cm^{-2}$ ($\equiv N\phi_0 e_0$), k_v is written as $k_v \equiv k_{max}\exp\{-[e_0(U^0_{redox} - U_v) + \lambda]^2/4\lambda kT\}$, where k_{max} is taken as $10^{-17}\ cm^4\ s^{-1}$ ($\simeq 10^5\ cm\ s^{-1}$) and $C_{ox} = C_R = 0.05$ M. (a) λ is taken as 1.0 eV and k_r ($/cm^4\ s^{-1}$) is taken as: (1) 10^{-20}, (2) 10^{-18}, (3) 10^{-16}, (4) 10^{-14}, (5) 10^{-12}, (6) 10^{-10}, (7) 10^{-8}. (b) k_r is taken as $10^{-11}\ cm^{-4}\ s^{-1}$ and λ(/eV) has the values: (1) 0.6, (2) 0.8, (3) 1.0, (4) 1.2, (5) 1.4, (6) 1.6. The dotted lines in (a) and (b) show the results expected with a simple model with $|\phi^1_{sc}| = 0$.

$$j_h = e_0(k_v C_R + k_r n_s)p_s \equiv j_v + e_0 k_r n_s p_s \tag{557}$$

assuming that surface recombination is of type II. Now

$$k'_\Sigma p_s = N\phi_0 = p_s(k_v C_R + k_r n_s) \tag{558}$$

whence

$$j_v = e_0 k_v C_R p_s = \frac{e_0 k_v C_R N\phi_0}{k'_\Sigma} \tag{559}$$

$$j_c = -e_0 k_c C_0 n_s \tag{560}$$

and if $j_c + j_v = 0$, then

$$n_s^2 + \frac{k_v C_R}{k_r} n_s - \frac{k_v C_R N\phi_0}{k_c C_0} = 0 \tag{561}$$

$$n_s = \frac{1}{2k_r}\left[\frac{k_v^2 C_R^2 + 4k_r k_v C_R N\phi_0}{k_c C_0}\right]^{1/2} - \frac{k_v C_R}{2k_r} \tag{562}$$

The band-bending, ϕ_{sc}, may be calculated (approximately) by

$$|\phi_{sc}| = -\frac{kT}{e_0}\ln\left(\frac{n_s}{N_D}\right) = -\frac{kT}{e_0}\ln\left\{\frac{1}{2k_r N_D}\left[\left(B^2 + \frac{Ak_v}{k_c}\right)^{1/2} - B\right]\right\} \tag{563}$$

where $A = 4k_r C_R N\phi_0/C_0$ and $B = k_v C_R$. Note that if k_r is negligible, $n_s = N\phi_0/k_c C_0$ and

$$|\phi_{sc}| = -\frac{kT}{e_0} \ln\left(\frac{N\phi_0}{k_c C_0 N_D}\right) \tag{564}$$

This predicts that the photovoltage should vary logarithmically with light intensity, a result frequently found experimentally.

As k_r increases, the photovoltage deviates from that expected on the basis of the simple theory. Some calculations are shown in Fig. 94 for n-WSe_2; whilst the linearity between V_{ph} and U_{redox} is still found, the slope is no longer unity and the intercept may be quite remote from flat-band potential.

The experimental results for the layer chalcogenides are of considerable interest, especially in non-aqueous solvents where a wide range of redox couples is easily accessible. As can be seen from Fig. 95, the photovoltage becomes essentially independent of U_{redox} if U_{redox} is greater than U_v. The significance of this is that equilibrium is now established between semiconductor and redox couple through the valence band; holes are injected by the redox couple to create a p-type or inversion layer close to the electrolyte interface. This leads to a high surface charge and a near constant ϕ_{sc}. The balance is now between the various contributions to the flux of holes in the VB, i.e. between migration and recombination, and the change in ϕ_{sc} will be a property of the semiconductor alone. The observed photovoltage will become effectively constant [173].

7.5 ALTERNATING CURRENT TECHNIQUES INVOLVING LIGHT

The factor linking the three techniques described in this section is the

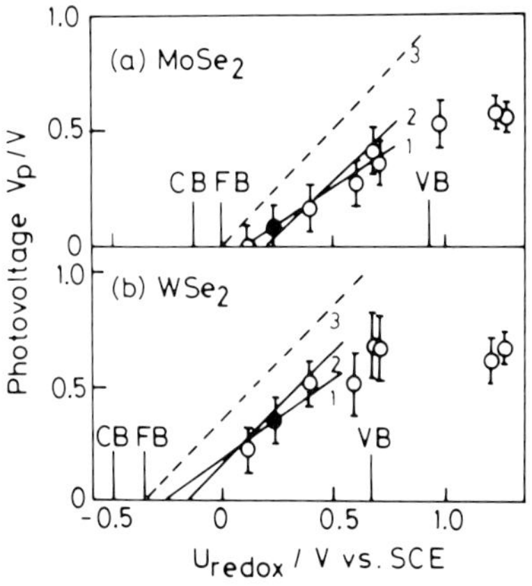

Fig. 95. Experimental data for (a) n-$MoSe_2$ and (b) n-WSe_2 in acetonitrile/0.1 M $LiClO_4$ containing a variety of one-electron redox couples. The lines 1 and 2 were calculated for $e_0 N\phi_0 = 10\,mA\,cm^{-2}$, $k_r = 10^{-10}\,cm^{-4}\,s^{-1}$, $C_{ox} = C_R = 0.05\,M$, and $\lambda = 1.4\,eV$ (1) and $0.6\,eV$ (2). Line 3 corresponds to $|\phi_{sc}^1| = 0$.

measurement of an a.c. photocurrent. All are instrumentally quite demanding and exploit the use of lock-in amplification.

7.5.1 *Oscillation of the light intensity*

This has been explored by a number of workers as a probe for surface species [152, 193–196]. If the intensity of the light has the form

$$\phi = \phi_0 + \phi_1 \exp(i\omega t) \tag{565}$$

then the photocurrent will show an a.c. response that may be examined using phase-sensitive techniques. Results have been reported for n-CdS and p-GaP. For the former case, the photocurrent was found to be in-phase with the a.c. light intensity for all potentials studied but, in the second case, rather different behaviour was seen, especially near the onset of the d.c. photocurrent response. If the photocurrent is written as $i = i_0 + i_1 e^{i\omega t}$ and

$$Z_p = \frac{i_1}{\phi_1} \tag{566}$$

we may derive a straightforward theory to describe Z_p. Consider first the case for an n-type semiconductor

$$p_{vb}^{+} + R \rightarrow O \qquad k_v$$

$$p_{vb}^{+} + T \rightarrow T^{+} \qquad k_f$$

$$T^{+} + e_{vb}^{-} \rightarrow T \qquad k_r$$

we have, assuming that migration across the depletion layer is not rate-limiting

$$k'_{\Sigma} p_{vb}^{+} = N\phi \tag{567}$$

$$k'_{\Sigma} = k_v C_R + k_f N_t f_t \text{ [see eqn. (493)]} \tag{568}$$

$$j_f = k_v C_R p_{vb}^{+} = + D_R \left(\frac{\partial C_R}{\partial x}\right)_0 \tag{569}$$

Small-signal a.c. theory gives

$$\phi = \phi_0 + \phi_1 \exp(i\omega t) \tag{570}$$

$$p_{vb}^{+} = \bar{p}_{vb}^{+} + \tilde{p}_{vb}^{+} \exp(i\omega t) \tag{571}$$

$$C_R = \bar{C}_R + \tilde{C}_R \exp(i\omega t) \tag{572}$$

$$f_t = \bar{f}_t + \tilde{f}_t \exp(i\omega t)$$

so

$$\tilde{j}_f = k_v \bar{C}_R \tilde{p}_{vb}^{+} + k_v \tilde{C}_R \bar{p}_{vb}^{+} = - C_R (i\omega D_R)^{1/2} \tag{574}$$

whence

$$\tilde{C}_R = -\frac{k_v \bar{C}_R \tilde{p}^+_{vb}}{k_v \bar{p}^+_{vb} + (i\omega D_R)^{1/2}} \tag{575}$$

and

$$(k'_\Sigma \tilde{p}^+_{vb}) = k_v \bar{C}_R \tilde{p}^+_{vb} + k_v \tilde{C}_R \bar{p}^+_{vb} + k_f \tilde{f}_t N_t \bar{p}^+_{vb} + k_f \bar{f}_t N_t \tilde{p}^+_{vb} = N\phi_1 \tag{576}$$

$$= k_f \tilde{f}_t N_t \bar{p}^+_{vb} + \frac{(i\omega D_R)^{1/2} k_v \bar{C}_R \tilde{p}^+_{vb}}{k_v \bar{p}^+_{vb} + (i\omega D_R)^{1/2}} + k_f N_t \bar{f}_t \tilde{p}^+_{vb} \tag{577}$$

The coverage by T has the time dependence

$$N_t \frac{\partial f_t}{\partial t} = -k_f N_t f_t p^+_{vb} + k_r n_s (1 - f_t) N_t \tag{578}$$

$$i\omega \tilde{f}_t = -k_f \tilde{p}^+_{vb} \bar{f}_t - k_f \bar{p}^+_{vb} \tilde{f}_t - k_r n_s \tilde{f}_t \tag{579}$$

$$\tilde{f}_t = -\frac{k_f \tilde{p}^+_{vb} \bar{f}_t}{(k_f \bar{p}^+_{vb} + k_r n_s) + i\omega} \tag{580}$$

Assuming that n_s is constant

$$N\phi_1 = \frac{k_r n_s k_f N_t \bar{f}_t \tilde{p}^+_{vb} + i\omega k_f N_t \bar{f}_t \tilde{p}^+_{vb}}{(k_f \bar{p}^+_{vb} + k_r n_s) + i\omega} + \frac{(i\omega D_R)^{1/2} k_v \bar{C}_R \tilde{p}^+_{vb}}{k_v \bar{p}^+_{vb} + (i\omega D_R)^{1/2}} \tag{581}$$

The total a.c. current is

$$e_0 (k'_\Sigma \tilde{p}^+_{vb} + k_r n_s \tilde{f}_t) = \frac{i\omega k_f e_0 N_t \bar{f}_t \tilde{p}^+_{vb}}{(k_f \bar{p}^+_{vb} + k_r n_s + i\omega)} + \frac{(i\omega D_R)^{1/2} k_v e_0 \bar{C}_R \tilde{p}^+_{vb}}{k_v \bar{p}^+_{vb} + (i\omega D_R)^{1/2}} \tag{582}$$

so

$$j_1 = \left[\frac{i\omega k_f \bar{f}_t}{k_f \bar{p}^+_{vb} + k_r n_s + i\omega} + \frac{(i\omega D_R)^{1/2} k_v \bar{C}_R}{k_v \bar{p}^+_{vb} + (i\omega D_R)^{1/2}}\right] \times \left[e_0 N\phi_1 \Big/ \left\{\frac{k_r n_s k_f N_t \bar{f}_t + i\omega k_f N_t \bar{f}_t}{k_f \bar{p}^+_{vb} + k_r n_s + i\omega} + \frac{(i\omega D_R)^{1/2} k_v \bar{C}_R}{k_v \bar{p}^+_{vb} + (i\omega D_R)^{1/2}}\right\}\right] \tag{583}$$

If $k_v = 0$, this reduces to

$$j_1 = \left(\frac{i\omega}{k_r n_s + i\omega}\right) e_0 N\phi_1 \tag{584}$$

If recombination can be neglected (i.e. $k_f = k_r = 0$)

$$j_1 \sim e_0 N\phi_1 \tag{585}$$

a situation found for n-CdS and p-GaP in the saturation region [152]. If, as proposed above, electron transfer between semiconductor and electrolyte occurs via the surface state T, then $k_v = 0$ and we have a new rate form

$$T^+ + R \rightarrow T + O \qquad k_s$$

References pp. 242–246

Then

$$\frac{\partial f_t}{\partial t} = -k_f f_t p_{vb}^+ + k_r n_s(1 - f_t) + k_s C_R(1 - f_t) \tag{586}$$

$$\tilde{j}_f = N_t k_s(1 - \bar{f}_t)\tilde{C}_R - N_t k_s \tilde{f}_t \bar{C}_R = -\tilde{C}_R(i\omega D_R)^{1/2} \tag{587}$$

$$\tilde{C}_R = \frac{N_t k_s \bar{C}_R \tilde{f}_t}{N_t k_s(1 - \bar{f}_t) + (i\omega D_R)^{1/2}} \tag{588}$$

and

$$i\omega\tilde{f}_t = -k_f\bar{f}_t\tilde{p}_{vb}^+ - k_f\tilde{f}_t\bar{p}_{vb}^+ - k_r n_s\tilde{f}_t + k_s\tilde{C}_R(1 - \bar{f}_t) - k_s\bar{C}_R\tilde{f}_t \tag{589}$$

The coverage variation is given by

$$\tilde{f}_t = \frac{-k_f\bar{f}_t\tilde{p}_{vb}^+ + k_s\tilde{C}_R(1 - \bar{f}_t)}{k_f\bar{p}_{vb}^+ + k_r n_s + k_s\bar{C}_R + i\omega} \tag{590}$$

$$= -\frac{k_f\bar{f}_t\tilde{p}_{vb}^+}{k_f\bar{p}_{vb}^+ + k_r n_s + k_s\bar{C}_R + i\omega - \{k_s^2 N_t\bar{C}_R(1 - \bar{f}_t)/[k_s N_t(1 - \bar{f}_t) + (i\omega D_R)^{1/2}]\}} \tag{591}$$

The a.c. current is now

$$N_t e_0(k_f\tilde{p}_{vb}^+\bar{f}_t + k_f\bar{p}_{vb}^+\tilde{f}_t + k_r n_s\tilde{f}_t) = e_0(N\phi_1 + k_r n_s\tilde{f}_t) \tag{592}$$

and

$$N_t(k_f\bar{f}_t\tilde{p}_{vb}^+ + k_f\tilde{f}_t\bar{p}_{vb}^+) = N\phi_1 = \tilde{p}_{vb}^+ N_t \times$$
$$\left[k_f\bar{f}_t - \frac{k_f^2\bar{f}_t\bar{p}_{vb}^+}{k_f\bar{p}_{vb}^+ + k_r n_s + k_s\bar{C}_R + i\omega - \{k_s^2 N_t\bar{C}_R(1 - \bar{f}_t)/[k_s N_t(1 - \bar{f}_t) + (i\omega D_R)^{1/2}]\}}\right] \tag{593}$$

If we can neglect transport limitations in the electrolyte, we have

$$N\phi_1 = \tilde{p}_{vb}^+\left[\frac{k_f k_r n_s\bar{f}_t + k_f\bar{f}_t k_s\bar{C}_R + i\omega k_f\bar{f}_t}{k_f\bar{p}_{vb}^+ + k_r n_s + k_s\bar{C}_R + i\omega}\right]N_t \tag{594}$$

and the current is now

$$j_1 = e_0 N\phi_1\left(\frac{i\omega + k_s\bar{C}_R}{i\omega + k_r n_s + k_s\bar{C}_R}\right) \equiv e_0 N\phi_1\left(1 - \frac{k_r n_s}{k_r n_s + k_s\bar{C}_R + i\omega}\right) \tag{595}$$

Evidently, the photocurrent will normally *lead* the light intensity, a fact that may be traced to the expression for the coverage, f_t, of T. Other, more complex expressions may, of course, be derived within the above formalism, which is mainly due to Albery and Bartlett [152]; unless low frequencies are employed, it is unlikely that transport limitations will play an important role.

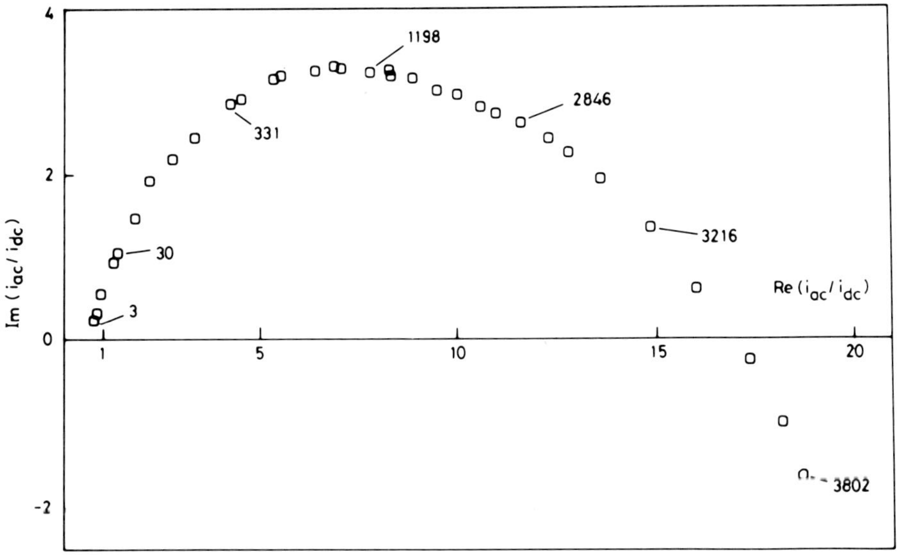

Fig. 96. Real vs. imaginary parts of the normalised a.c. photocurrent for p-GaP $(\bar{1}\bar{1}\bar{1})$ in 0.5 M H_2SO_4. The frequencies marked are in Hz.

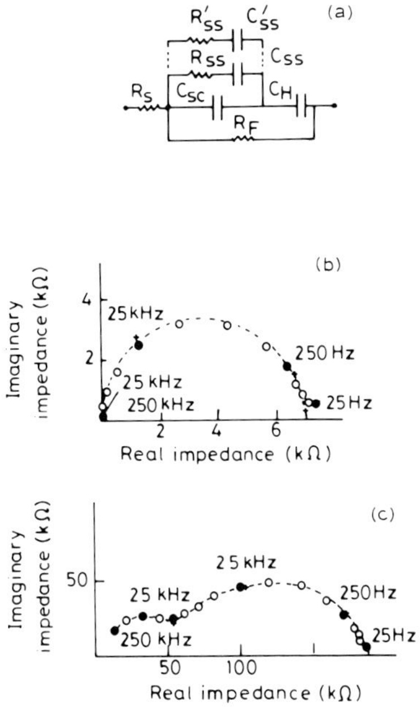

Fig. 97. (a) Electrical equivalent circuit of an illuminated semiconductor–electrolyte interface. (b), (c) Experimental impedance plots for n-GaAs/selenide under 22 mW cm^{-2} illumination at different potentials. (b) $V = -0.60$ V/SCE (in the photocurrent saturation region); (c) $V = -1.575$ V/SCE (in the onset region). The circles are experimental points and the dotted curve is the best fit to (a).

The detailed experimental data for p-GaP are shown in Fig. 96. Analysis using a close analogue of eqn. (595) for a p-type semiconductor leads, under the assumption that $k_s C_0$ is very small, to $k_h p_{vb}^+ \simeq 10^3\,\mathrm{s}^{-1}$. Very recently, the experiment has been substantially extended by Peter and co-workers [193–196] and applied to a range of semiconductor systems.

7.5.2 Oscillation of the potential

The theory of a.c. effects in the light presents formidable problems since both types of carrier may contribute, in principle, to the signal and it is very difficult indeed to disentangle the basic effects. The result is that, at the moment, the theory is still essentially phenomenological.

The most transparent treatment appears to have been carried out by Allongue and Cachet [174] who proposed an equivalent circuit of the type shown in Fig. 97. This circuit is similar to that used for the interpretation of a.c. response in the dark, but an additional $C'_{ss} R'_{ss}$ network is provided on illumination, and R_f decreases strongly.

The photocurrent is given by

$$j_{ph} = (k'_\Sigma p_s - S)e_0 \tag{596}$$

where S is given by

$$\frac{N_t p_s n_s}{\tau_p (n_s + n_t)} \qquad \text{Case I} \tag{597}$$

$$\frac{N_t p_s n_s}{(p_s/k_r) + (n_s/k_f)} \qquad \text{Case III} \tag{598}$$

If we consider the low frequency limit, then, for case I, where $k''_r \simeq N_t/\tau_p$ and $n_s \gg p_s, n_t$

$$R_f^{-1} = \frac{dj_{ph}}{dV_{applied}} = \left(k'_\Sigma \frac{dp_s}{dV_{ap}} - k''_r \frac{dp_s}{dV_{ap}}\right) C_0 \tag{599}$$

$$= e_0 k'_\Sigma \frac{dp_s}{dV_{ap}} \left(1 - \frac{k''_r}{k'_\Sigma}\right) \tag{600}$$

$$= e_0 \phi_0 \frac{dN}{dV_{ap}} \left(1 - \frac{k''_r}{k'_\Sigma}\right) \tag{601}$$

and for case III

$$R_f^{-1} = \left(k'_\Sigma \frac{dp_s}{dV_{ap}} - k'_r \frac{dn_s}{dV_{ap}}\right) C_0; \qquad k'_r \approx k_r N_t \tag{602}$$

$$\approx e_0 \phi_0 \frac{dN}{dV_{ap}} \tag{603}$$

since, in many cases, recombiantion will represent a small fraction of the total flux here. For the intermediate case, $n_t \geqslant n_s$, more complex formulae must be invoked.

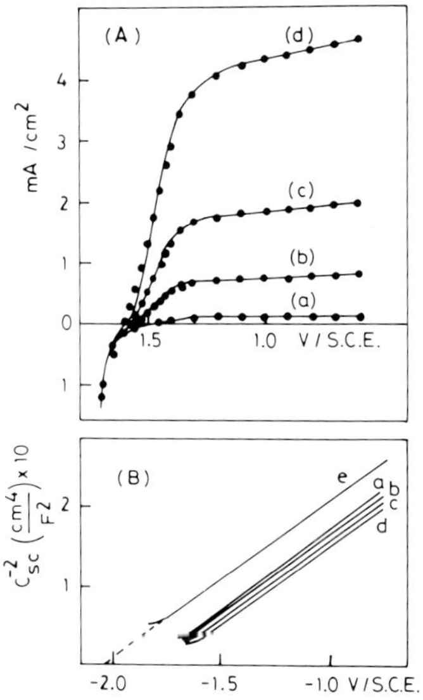

Fig. 98. (A) Current–voltage curves for n-GaAs/selenide junction under (a) 1.5 mW cm^{-2}, (b) 9 mW cm^{-2}, (c) 22 mW cm^{-2}, and (d) 50 mW cm^{-2}. (B) Mott–Schottky plots using data from the equivalent circuit of Fig. 97(a) at light intensities as in (A). The line (e) was obtained in the dark and gives $V_{fb} \simeq -2.06$ V/SCE.

Provided that neither R_f or the RC networks shunt C_{sc} too effectively, it should still be possible to extract C_{sc} by an analysis similar to that present above. In fact, it has been shown by a number of workers that Mott–Schottky behaviour can still be obtained even in the light, working at a single frequency, though a thoroughgoing analysis of the behaviour using the equivalent circuit of Fig. 97 seems only to have been carried out recently. The data on n-GaAs/selenide solution, interpreted using Fig. 97 seem to suggest, as shown in Fig. 98, that there is a marked shift, by about 250 mV, in the flat-band potential of n-GaAs on illumination in the region very close to the onset potential, as shown in Fig. 98.

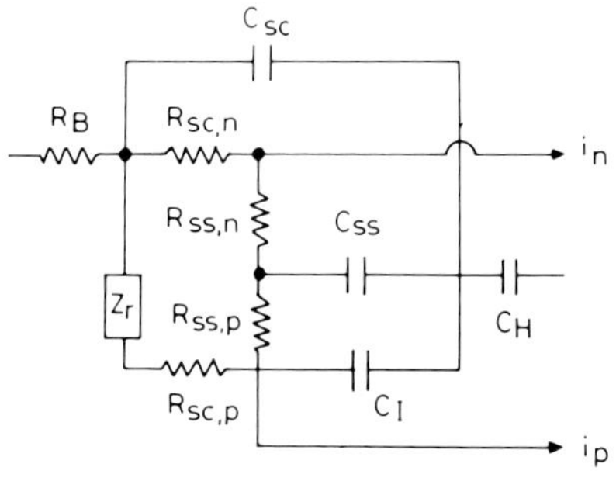

Fig. 99. Equivalent circuit for a.c. modulation under illumination when recombination in the depletion layer can be neglected.

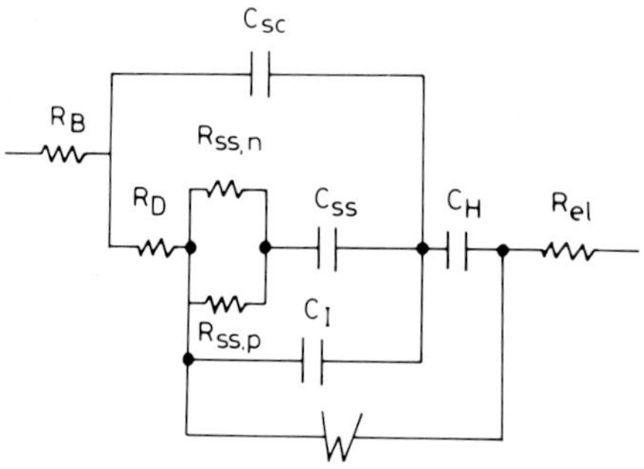

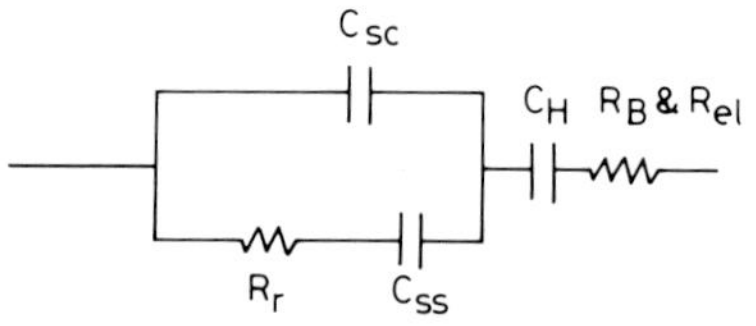

Fig. 100. Equivalent circuit for a.c. modulation under illumination when depletion-layer recombination cannot be neglected.

Fig. 101. Simplified equivalent circuit for a.c. modulation under illumination.

More complex treatments have been proposed, in which efforts have been made to separate hole and electron currents in the depletion layer [7]. The circuit of Fig. 99 illustrates the situation for a semiconductor under depletion or inversion conditions in which zero recombination occurs in the depletion layer [175]. In the figure, the suffix n refers to electrons and C_I is a capacitance associated with inversion (if this is operative). The impedance Z_r describes the generation of holes and their recombination in the bulk of the semiconductor.

If bulk recombination is important in the depletion layer, then we cannot separate hole and electron flows in the above manner and the Z_r, $R_{sc,p}$ network collapses to a frequency-independent resistor R_D, as shown in Fig. 100. In this figure W is a Warburg impedance for the hole current. This is too complex, as it stands, for analysis and a simpler case can be derived if C_{ss} is dominant and the frequency range is such that W can also be neglected. Under these circumstances, R_D, $R_{ss,n}$ and $R_{ss,p}$ further collapse to a simple resistor R_r, leading to the equivalent circuit shown in Fig. 101, which has been applied to p-GaAs under illumination and n-GaAs under hole injection.

Other workers have suggested rather different equivalent circuits, in which the photocurrent is represented by a current source in the circuit, and an inductance has been invoked at low frequency to account empirically for the results [176, 177]. The origin of this inductive behaviour is not clear and it is evident that this remains an area in which considerable theoretical advances need to be made before the technique gains general acceptance.

7.5.3 Double modulation experiments

If both the light intensity and the electrode potential are modulated at different frequencies, information may, in principle, be obtained at the sum and difference frequencies. This method has been little used; the complexities of the a.c. impedance analysis in the light have precluded all but the most heroic from attempting the experiment [178] and the only results have been presented within the framework of the Gärtner photoresponse model, i.e. in the photocurrent saturation region. We can write

$$j_{ph} = e_0 \phi_0 \left\{ 1 - \frac{\exp(-\alpha W)}{1 + \alpha L_p} \right\} \tag{604}$$

Let us modulate the light intensity and the potential

$$\phi = \phi_0 + \phi_1 \exp(i\omega_1 t) \tag{605}$$

$$V = V_0 + V_1 \exp(i\omega_2 t) \tag{606}$$

since

$$W = \left[\left(\frac{2\varepsilon_0 \varepsilon_{sc}}{e_0 N_D} \right) (V - V_{fb}) \right]^{1/2} \tag{607}$$

the depletion layer may be written

$$W = W_0 + W_1 \exp(i\omega_2 t) \tag{608}$$

where

$$W_1 = \frac{\varepsilon_0 \varepsilon_{sc} V_1}{e_0 W_0 N_D} \tag{609}$$

Inserting these expressions into eqn. (604) and using small-signal theory, we recover

$$j_{ph} = j_{ph}^0 + j_{ph}^{(1)} \exp(i\omega_1 t) + j_{ph}^{(2)} \exp(i\omega_2 t) + j_{ph}^{(12)+} \exp[i(\omega_1 + \omega_2)t] + j_{ph}^{(12)-} \exp[i(\omega_1 - \omega_2)t] \tag{610}$$

where

$$j_{ph}^{(1)} = e_0 \left[\frac{1 - \exp(-\alpha W_0)}{1 + \alpha L_p} \right] \tag{611}$$

$$j_{ph}^{(2)} = e_0 \phi_0 \left(\frac{\alpha}{1 + \alpha L_p} \right) \left(\frac{\varepsilon_0 \varepsilon_{sc} V_1}{e_0 N_D W_0} \right) \exp(-\alpha W_0) \tag{612}$$

$$j_{ph}^{(12)+} = j_{ph}^{(12)-} = \frac{e_0 \phi_1}{2} \left(\frac{\alpha}{1 + \alpha L_p} \right) \left(\frac{\varepsilon_0 \varepsilon_{sc} V_1}{e_0 N_D W_0} \right) \exp(-\alpha W_0) \tag{613}$$

It follows that, if N_D is known

$$\ln \left(\frac{j_{ph}^{(12)-}}{W_1} \right) = \ln \left[\frac{e_0 \phi_1}{2} \left(\frac{\alpha}{1 + \alpha L_p} \right) \right] - \alpha W_0 \tag{614}$$

and a plot of $\ln(j_{ph}^{(12)-}/W_1)$ yields α. Similarly, a plot of $j_{ph}^{(1)}$ vs. $e^{-\alpha W_0}$ gives a slope of $-e_0 \phi_1/(1 + \alpha L_p)$ and for $e^{-\alpha W_0} = 1$ has the value

$$[j_{ph}^{(1)}]_{W=0} = e_0 \phi_1 \left(\frac{\alpha L_p}{1 + \alpha L_p} \right) \tag{615}$$

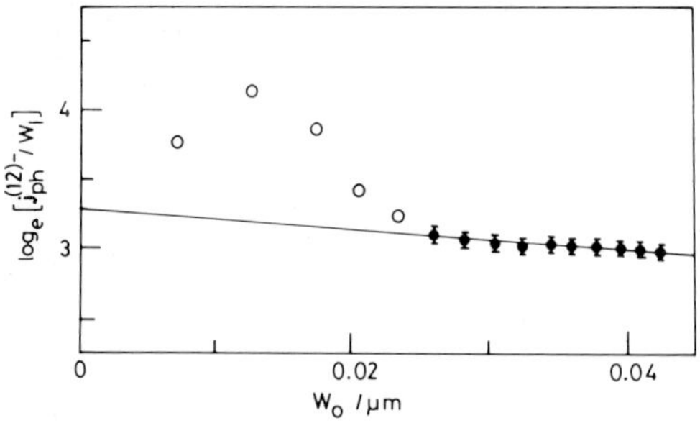

Fig. 102. Plot of $\log_e [j_{ph}^{(12)-}/W_1]$ vs. W_0 showing that, for large values of W_0, eqn. (614) is satisfied.

but

$$\left[\ln\left(\frac{j_{ph}^{(12)-}}{W_1}\right)\right]_{W\to 0} = \ln\left[\frac{e_0\phi_1}{2}\left(\frac{\alpha}{1+\alpha L_p}\right)\right] \tag{616}$$

so

$$\frac{[j_{ph}^{(1)}]_{W=0}}{[j_{ph}^{(12)-}/W_1]_{W=0}} = 2L_p \tag{617}$$

The advantage of this technique is that absolute photocurrent measurements are not needed; both α and L_p result from the same experiment. This technique has been tried for p-GaP in 0.1 M $EuCl_3$/1 M $HClO_4$ and the results, at least at potentials well removed from V_{fb} and for low light intensities, are in reasonable accord with the above analysis as shown in Figs. 102 and 103. The values of α and L_p derived from these data lie within the range of values quoted in the literature.

7.6 PHOTOPULSING MEASUREMENTS

The advent of fast pulsed lasers has enabled photo-induced transients on the picosecond timescale to be investigated. Careful analysis of the data can

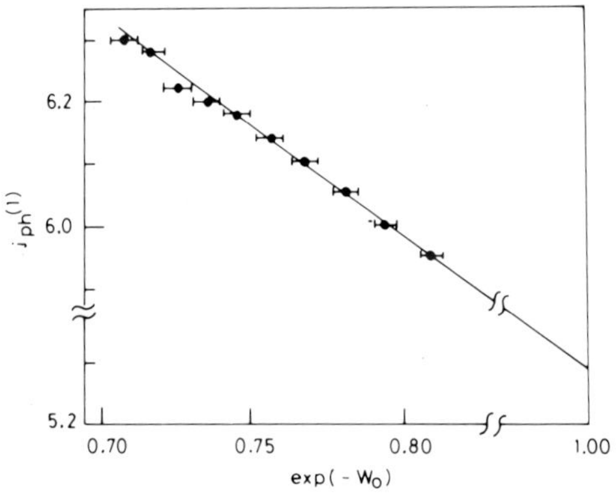

Fig. 103. Plot of $j_{ph}^{(1)}$ vs. exp $(-\alpha W_0)$ showing that eqn. (611) is obeyed.

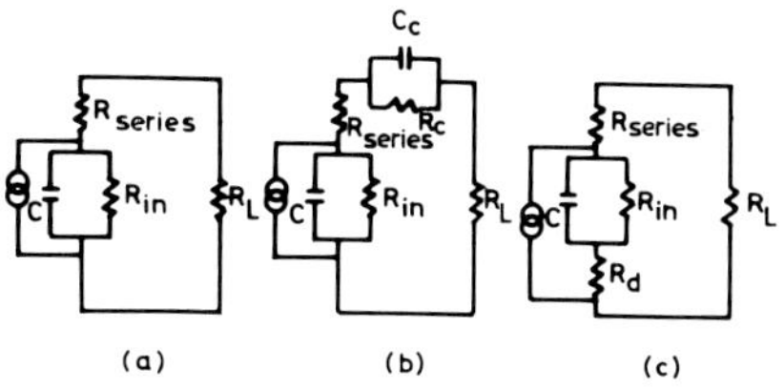

Fig. 104. Equivalent circuits for the analysis of photocurrent-decay transients. The circuit elements are: C, photocapacitor; R_{series}, total series resistance of the spectroelectrochemical cell; R_L, load resistor; R_{in}, internal leakage resistor; R_c, C_c, resistor and capacitor of counterelectrode solution interface; R_d, resistance due to damaged surface layer.

lead to a knowledge of the rate constants employed throughout the discussion of photocurrent effects. In principle, relaxation processes following a light pulse may involve [179].

(a) diffusion of minority carriers to the junction region,
(b) transit through the junction,
(c) thermalisation of the carriers,
(d) recombination lifetime of the carriers, and
(e) relaxation through the RC time constant of the cell and circuit combined

There are three types of transient measurement that may be made, viz.

(i) photocurrent transients,
(ii) photopotential transients, and
(iii) photoluminescence transients.

The first of these can be treated with a simple equivalent circuit of the form shown in Fig. 104. It is normally assumed that processes (b) and (c) are very fast compared with (d) and (e) and that, if a direct bandgap semiconductor is used, (a) can also be minimised. Under these circumstances, the initial condition is

$$Q_{sc} = Q^0_{sc}; \qquad Q_H = 0 \tag{618}$$

The rate constants k_{sc} and k_H can be defined by the expressions

$$k_{sc} = \frac{1}{C_{sc}R_{sc}} \tag{619}$$

$$k_H = \frac{1}{C_H R_H} \tag{620}$$

Physically, k_{sc} and k_H represent the rate constants for discharge of (a) the depletion layer charged by minority carriers that recombine through bulk or surface processes and (b) the faradaic transfer processes at the interface.

Starting with eqn. (618), a rather long analysis under the conditions specified leads to [180]

$$j = Q_{sc}^0 \left[\frac{(k_{sc}/C_{sc}R_T) - (k_H/C_H R_T)}{k_T - (1/C_T R_T)} \exp(-k_T t) + \frac{(k_H/C_T R_T) - (1/R_T^2 C_T C_{sc})}{k_T - (1/R_T C_T)} \exp\left(-\frac{t}{R_T C_T}\right) \right] \quad (621)$$

where

$$k_T = k_{sc} + k_H \quad (622)$$

$$R_T = R_{in} + R_{out} \quad (623)$$

$$\frac{1}{C_T} = \frac{1}{C_{sc}} + \frac{1}{C_H} \quad (624)$$

Normally, for semiconductors, $C_{sc} \ll C_H$ so $C_T \simeq C_{sc}$. R_{out} may be varied systematically and the decay of j can often be approximated by a single exponential form, i.e. $k_T \gg 1/R_T C_T$ or $k_T \ll 1/R_T C_T$. Data for n-CdS affirm that, for potentials well positive of V_{fb}, the long-time transient time-constant $\tau \simeq (R_{in} + R_{out})C_{sc}$, and a plot of τ vs. R_{load} ($\equiv R_{in} + R_{out}$) is linear, as shown in Fig. 105. Confirmation of this is obtained from the fact that $1/\tau^2$ obeys the Mott–Schottky relationship. At potentials close to V_{fb}, k_{sc} becomes much larger and the decay law more complex.

Photopotential transients have also been studied [181]. The light pulse will generate a non-stationary concentration of electrons and holes; analysis reveals that these separate rapidly in the depletion layer ($\leqslant 10^{-8}$ s), giving rise to an exponential concentration of holes near the surface (of a reverse-biased n-type semiconductor) and an exponential concentration of electrons at the inner edge of the depletion layer, as discussed above. This new charge distribution will alter the potential distribution and numerical integration for an n-type wide bandgap material shows that, if

$$\Delta Q_t^* = \frac{\Delta Q_t}{\sigma_0} \quad \text{and} \quad \Delta Q_t = \int_0^W p \, dx \quad (625)$$

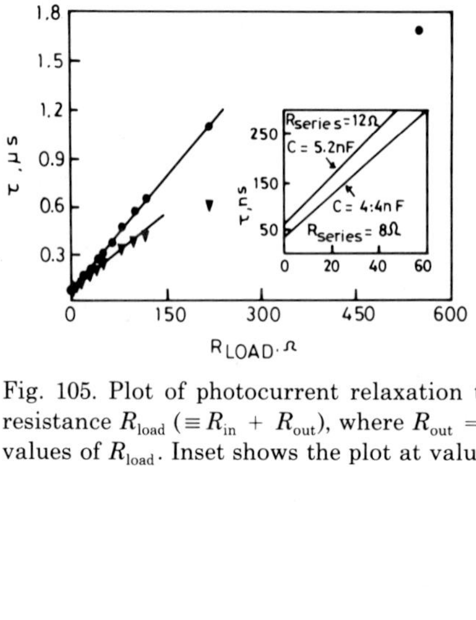

Fig. 105. Plot of photocurrent relaxation time, τ, following a ns light pulse vs. total load resistance R_{load} ($\equiv R_{in} + R_{out}$), where $R_{out} = R_{series} + R_L$, showing linearity expected at small values of R_{load}. Inset shows the plot at values of R_{load} approaching zero.

where σ_0 is the charge per unit area in the interface $\equiv WN_D$, and p is the hole concentration, assumed zero in the dark, then, with some accuracy,

$$p = p_w \exp(-v) \tag{626}$$

$$n = N_D \exp(v) \tag{627}$$

$$\Delta v_t = (\beta - 1.5)[\Delta Q_t^* - 0.5(\Delta Q_t^*)^2] \tag{628}$$

where

$$\beta = \frac{e_0^2 \sigma_0^2}{kT\varepsilon_0\varepsilon_{sc} N_D} = 2[\gamma_b \exp(-v_{sc}) + (1 + \gamma_b)\exp(v_{sc}) - v_{sc} - 1 - 2\gamma_b] \tag{629}$$

and

$$\gamma_b = \frac{p_b}{N_D} \tag{630}$$

where p_b is the bulk hole concentration.

If relaxation is via a band-to-band recombination, an unlikely event for a wide bandgap material, then

$$\frac{d(\Delta v_t)}{dt} \sim -k'\Delta v_t e^{\Delta v_t} \tag{631}$$

where

$$k' = \frac{e_0^2 N_D D_p(\alpha + \sqrt{\alpha})(2v_{sc} + \Delta v_t + 3.5)\exp(v_{sc})}{2kT\varepsilon_0\varepsilon_{sc}(v_{sc} + 1)} \tag{632}$$

and

$$\alpha = \frac{k_r N_D W^2}{D} \tag{633}$$

which, it should be noted, is not a simple exponential. A rather similar expression is obtained if the main recombination route is via Shockley–Read states in the *bulk* of the semiconductors, but the analysis has not been carried through for recombination via such states in the depletion layer.

If the main relaxation process is via surface state recombination and the rate-limiting process is assumed to be diffusion of the majority to the surface, then

$$\frac{d(\Delta v_t)}{dt} = -k''(\beta - 2\Delta v_t)[\exp(\Delta v_t) - 1] \tag{634}$$

where

$$k'' = \frac{e_0^2 N_D D_n \exp(-v_{sc})}{kT\varepsilon_0\varepsilon_{sc}} \tag{635}$$

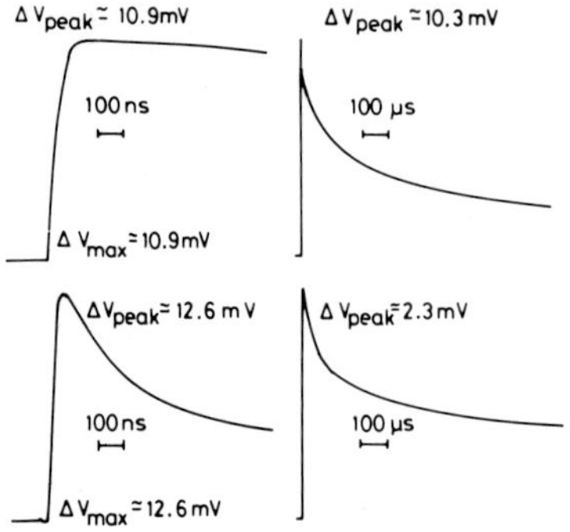

Fig. 106. Photopotential decay transients following a ns illumination pulse on n-CdSe in contact with selenide solution. The lower curves show the mechanically polished electrodes with a damaged surface layer and the upper curves the same after etching.

and once more we find that the decay is not governed by a simple exponential.

If the surface recombination process itself is slow, in particular if the rate of capture of a CB electron by an oxidised surface state is given by the rate constant k_r, then an analysis similar to that above gives

$$\frac{d(\Delta v_t)}{dt} = \left\{\frac{-e_0^2 N_D}{2(v_{sc}+1)\varepsilon_0\varepsilon_{sc}kT}\right\}^{1/2} (2v_{sc} + 2 + \Delta v_t)k_r[S^+]\exp(v_{sc})[\exp(\Delta v_t) - 1] \tag{636}$$

where $[S^+]$ is the concentration of surface recombination centres. These expressions have been used to fit the photopotential data for n-CdSe in contact with sulphide–polysulphide mixtures and typical transients are shown in Fig. 106 [182]. It is evident that for neither etched nor unetched samples can the decay be described by a single exponential. The long-time constant decay (10–100 ms) appears to be associated with diffusion-limited surface-state recombination effects and faster time constants, especially for non-etched samples, may be associated with incomplete charge separation, some photogenerated carriers being trapped at defect centres and recombining rapidly.

Very fast recombination process can be studied by photoluminescence transients [182, 199]. In this experiment, very short laser pulses are used and ultra-fast recombination studied. For a mechanically polished but unetched sample of n-CdSe, luminescence decay times of 50 ps were found due to recombination in the damaged surface layer. On etching, the quantum yield for luminescence and the decay lifetime were both substantially increased [182].

7.7 MODULATED REFLECTANCE TECHNIQUES

In contrast to the techniques that we have considered in the previous section, which have involved the measurement of an electrochemical quantity such as current or potential as a function of incident light, there exists a battery of techniques, more familiar to the physicist than the electrochemist, in which the intensity of the reflected light is measured as a function of some perturbation of the semiconductor [183, 184]. The experimental basis for these techniques is shown in Fig. 107; light from a monochromated source is incident on the semiconductor electrode in a suitable cell and the reflected light passes into a photomultiplier tube where it is amplified. A periodic perturbation, such as a sinusoidal potential as in Fig. 107 or a temperature

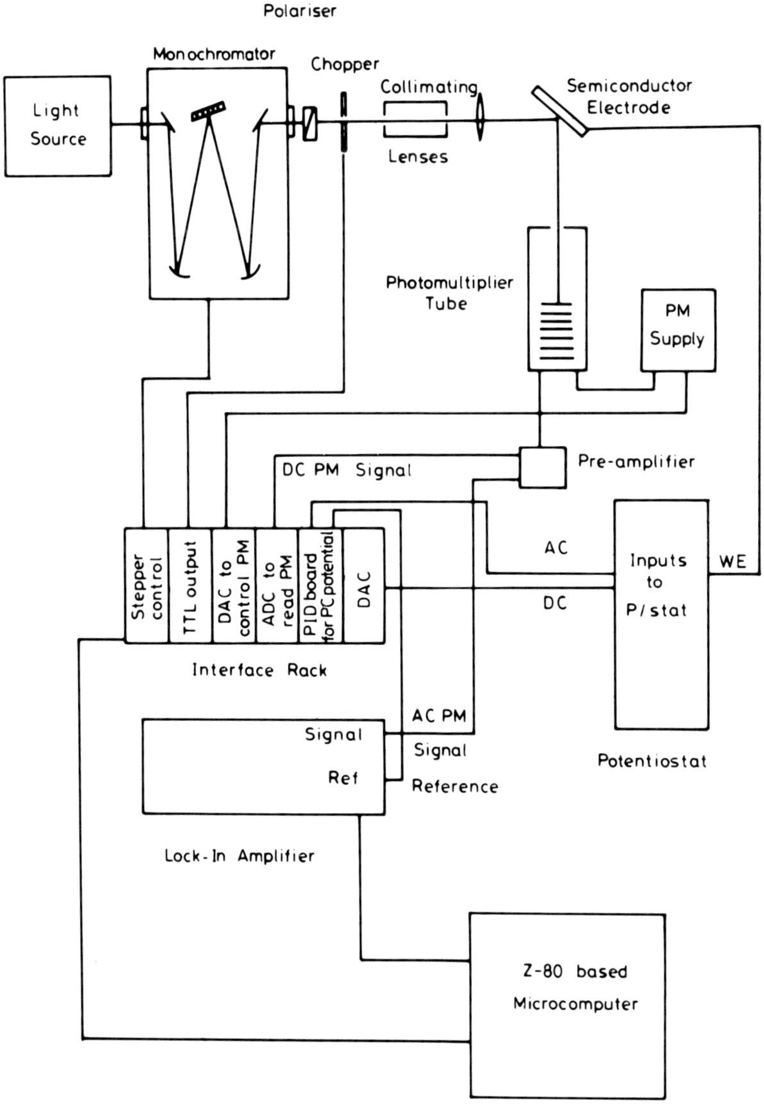

Fig. 107. Experimental arrangement for modulated electroreflectance from semiconductors.

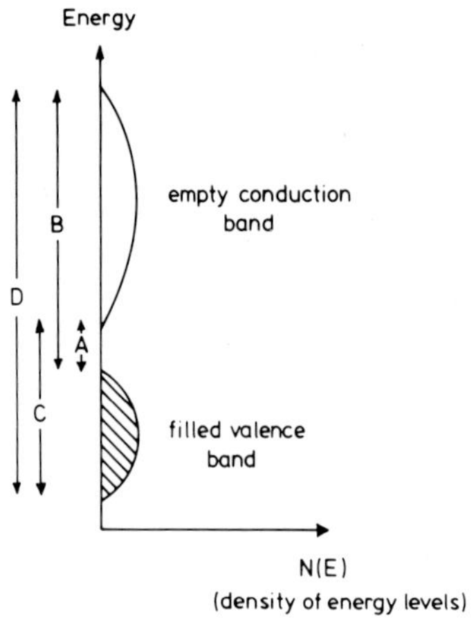

Fig. 108. Critical points A–D in the interband transition between valence and conduction bands.

step, is applied to the semiconductor and it is found that the intensity of the reflected light also shows an impressed periodic variation, which can be electronically separated from the main d.c. signal and separately processed. The effects are often very small and lock-in amplification is usually employed to extract the a.c. signal from unwanted noise; variations in reflectivity of one part in 10^6 can be measured with modern equipment.

The techniques described in the previous paragraph were initially developed as spectroscopic measurements. The reason for this can be understood from Fig. 108. As indicated in Sect. 2, the electronic structure of a simple semiconductor consists of a series of bands of energy levels. The topmost filled band is termed the valence band and the lowest unoccupied level is termed the conduction band. Transitions between these two bands will constitute the lowest-energy optical absorption processes in the material and it is evident that the optical absorption onset corresponds to A in the figure. Transitions marked B–D also correspond to energies at which new transitions are permitted or disallowed on energetic grounds and the points A–D are said to be "critical" points. Now, it is evident that, if we perturb the electronic structure in some way, then at the energies corresponding to transitions in the middle of the band, there will be comparatively little effect on the overall spectrum. However, at the critical points, where a modulation may allow or prevent a transition, the effects will be very much larger.

This may be understood more easily from Fig. 109(a) [185]. Here, we imagine a perturbation that alters the position of the bands with respect to each other vertically on the E vs. k diagram (i.e. in the Brillouin zone). Clearly, such a perturbation will alter the energy of the optical absorption onset and large differential signal will be seen at threshold. The effect is to

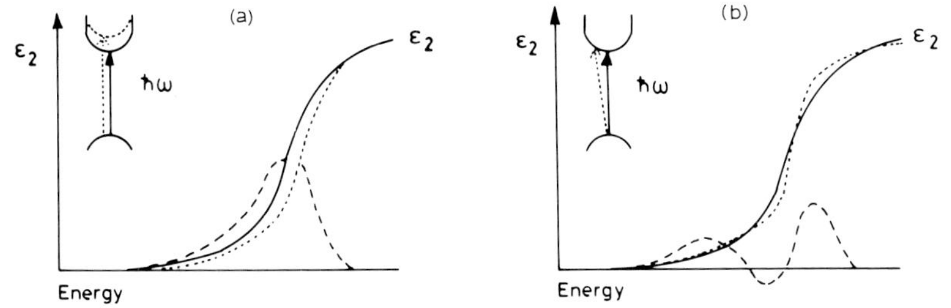

Fig. 109. Effect of (a) a thermal perturbation that causes a vertical displacement of the bands and (b) electroreflectance, causing a loss in translational symmetry with a concomitant relaxation of the requirement for vertical optical transitions in k-space. ——, Unperturbed; ---, perturbed; — — —, difference.

emphasise the bandgap energy at the expense of the optical structure above that energy. From the theoretical solid state point of view, however, it is the bandgap rather than the details of the higher energy optical structure that is usually of prime importance and techniques of this type were developed in the sixties as an experimental counterpoise to the rapidly burgeoning theories of electronic solid-state structure. An example of a perturbation that will give rise to the effects shown in Fig. 109(a) is temperature; a change in this will alter the unit cell dimensions and cause a concomitant change in the separation of bonding and antibonding states. However, thermoreflectance is not only a rather difficult technique in its own right, but it can tell us nothing about the potential distribution at the semiconductor–electrolyte interface. To explore this, we must invoke a technique that actually varies the potential and this is termed electroreflectance.

The basis of electroreflectance is more subtle than thermoreflectance; Fig. 109(b) shows that the main effect arises from the fact that the presence of an electric field destroys the translational symmetry along one of the directions of the crystal. This loss of symmetry means that k need no longer be conserved along that axis and optical transitions need no longer be vertical in the E/k diagram. As for thermoreflectance, this effect will be most marked at the critical points, again allowing the spectroscopist to extract the data of real interest from the otherwise rather shapeless absorption spectrum of the solid.

The early treatments of the electroreflectance effect concentrated on the case of uniform electric fields and zero thermal broadening and was therefore suitable only for very lightly doped samples at very low temperatures. The optical properties of a solid are contained in the dielectric function. This function is complex, with the imaginary part only non-zero if the material actually absorbs light. The imaginary part of the dielectric function, ε_2, can be written for a single band-to-band transition as [186]

$$\varepsilon_2(\omega) = \frac{\pi e_0^2}{\varepsilon_0 m^2 \omega^2}\left(\frac{2}{(2\pi)^3}\right) \int_S \mathrm{d}S_{n'n} \int_{E_{n'n}} \mathrm{d}E \frac{|\hat{\varepsilon}\cdot \mathbf{P}_{n'n}(\mathbf{k})|^2}{|\nabla_k E_{n'n}(\mathbf{k})|}\delta(E_{n'n} - \hbar\omega) \qquad (637)$$

where the integration is over the surface in k-space at which $E_{n'n} = \hbar\omega$. In eqn. (637), $P_{n'n}(k)$ is the momentum matrix element between states in the bands n and n' at wave vector $\mathbf{k}$, $\hat{\varepsilon}$ is a unit vector in the direction of the electric-field vector of the incoming electromagnetic wave, and the other symbols have their usual meanings. If the bands are parabolic, which is likely to be a good approximation near the edges, as discussed above, then explicit closed expressions for the integral may be obtained. In the presence of an electric field, $\mathscr{E} = (\mathscr{E}_x, \mathscr{E}_y, \mathscr{E}_z)$, the calculation of ε_2 must be modified. The new expression may be written

$$\varepsilon_2(\omega, \mathscr{E}) = \frac{\pi e_0^2}{\varepsilon_0 m^2 \omega^2} |\hat{\varepsilon} \cdot \mathbf{P}|^2 \left(\frac{2}{(2\pi)^3}\right) \int \mathrm{d}^3\mathbf{k} \left|\frac{1}{\hbar\Omega}\right| A_i \left(\frac{\hbar\omega_g + \Sigma_i(\hbar^2 k_i^2/2\mu_i) - \hbar\omega}{\hbar\Omega}\right) \tag{638}$$

where the index i runs over x, y, z, $A_i(x)$ is the Airy function regular at the origin and as $x \to \infty$, and the important parameter Ω is a measure of the electric field. In fact

$$\hbar\Omega = 2^{-2/3}\hbar\theta \tag{639}$$

and

$$\hbar\theta = \left\{\frac{\hbar^2 e_0^2}{2}\left[\frac{\mathscr{E}_x^2}{\mu_x} + \frac{\mathscr{E}_y^2}{\mu_y} + \frac{\mathscr{E}_z^2}{\mu_z}\right]\right\}^{1/3} \tag{640}$$

The relationship between eqns. (637) and (638) may be seen if we recall that in the limit $\Omega \to 0$

$$\frac{1}{\hbar\Omega} A_i \left[\frac{E - \hbar\omega}{\hbar\Omega}\right] \to \delta(E - \hbar\omega) \tag{641}$$

and eqn. (637) can be recovered. For parabolic bands, explicit formulae may again be obtained.

This theory suffers from two drawbacks which have made its application to the problem of determining potential distributions very difficult. The first is that all thermal broadening has been neglected and the second is that the electric field has been assumed to be constant, which is manifestly not the case in the depletion layer. The first problem is very technical, but a solution was eventually found by Aspnes [184]. Aspnes also went on to consider the case where thermal broadening is much greater than the energy $\hbar\Omega$ associated with the electric field [185]. A considerable simplification is possible under these circumstances since ε_2 reduces to

$$\varepsilon_2(\omega, \mathscr{E}, \Gamma) = \varepsilon_2(\omega, 0, \Gamma) + \mathrm{Im}\left\{\frac{(\Omega)^3}{3\omega^2}\frac{\partial^3}{\partial\omega^3}[\omega^2\tilde{\varepsilon}(\Gamma, \omega)]\right\} \tag{642}$$

Evidently, the differential signal will have a line shape corresponding to the third derivative of the optical absorption coefficient. This third derivative spectroscopy has had a considerable impact on the development of electroreflectance since the critical points are now magnified out of all proportion to the size of the rest of the spectrum. It was also realised, however, that this line shape would be independent of the d.c. potential applied to the semiconductor, which would tend to rule out electroreflectance as a means of exploring potential distributions.

In spite of the initial success of the third derivative treatment, there are a large number of experimental observations that do not appear to fit this model and recent developments have concentrated in this area. Raccah and co-workers [187, 188] have shown that the effects of the electric field may not only be felt through the Franz–Keldysh effect described by eqn. (638) but also through electrostrictive effects and through the alteration in the optical absorption around impurity centres. Assuming a homogeneous field strength inside the semiconductor, it is possible to show that the observed electroreflectance signal can be written as the sum of these three effects, all of which will give rise to different line shapes. Formally, we may write the electroreflectance magnitude as

$$\frac{\Delta R}{R} = \mathrm{Re}(Ce^{i\theta}\Delta\tilde{\varepsilon}) = -\left(\frac{2e_0 N_{\mathrm{D}} \Delta V}{\varepsilon_0 \varepsilon_{\mathrm{sc}}}\right)\mathscr{L}_{\mathrm{n}}(\hbar\omega)$$

where the subscript n refers to the particular physical effect giving rise to the line shape. The Franz–Keldysh third derivative line shape takes the form

$$\mathscr{L}_{5/2}(\hbar\omega) = \mathrm{Re}\,[C'\mathrm{e}^{i\theta}(\hbar\omega - \hbar\mu_{\mathrm{g}} + i\Gamma)^{-5/2}]$$

the second derivative line shape has the form

$$\mathscr{L}_{3/2}(\hbar\omega) = \mathrm{Re}\,[C''\mathrm{e}^{i\theta}\,(\hbar\omega - \hbar\omega_{\mathrm{g}} + i\Gamma)^{-3/2}] \qquad (643)$$

for the impurity effects and the first derivative line shape has the form

$$\mathscr{L}_{1/2}(\hbar\omega) = \mathrm{Re}\,[C'''\mathrm{e}^{i\theta}(\hbar\omega - \hbar\omega_{\mathrm{g}} + i\Gamma)^{-1/2}] \qquad (644)$$

for the electrostrictive effect. By working with materials of low donor density, where the penetration of the light is small compared with the width of the depletion layer, Raccah's group has amassed considerable evidence in favour of their model [187, 188].

By contrast, Hamnett and co-workers at Oxford have been more concerned with electrodes that possess significantly higher donor densities corresponding to those commonly encountered in electrochemical investigations [189]. In these, the electric field may vary very significantly over the penetration depth of the light and the thermal broadening is often comparable with the value of $\hbar\Omega$, especially in materials of low effective mass. To take account of the latter problem, the complete theory developed from eqn. (638) must be used and to take account of the inhomogeneous electric field [200], the depletion layer may be divided up into thin layers, in each of which the electric field is constant. The overall reflection coefficient may then be calculated. This approach is shown in Figs. 110 and 111; the experimental electroreflectance spectra for n-GaAs are shown as a function of the applied

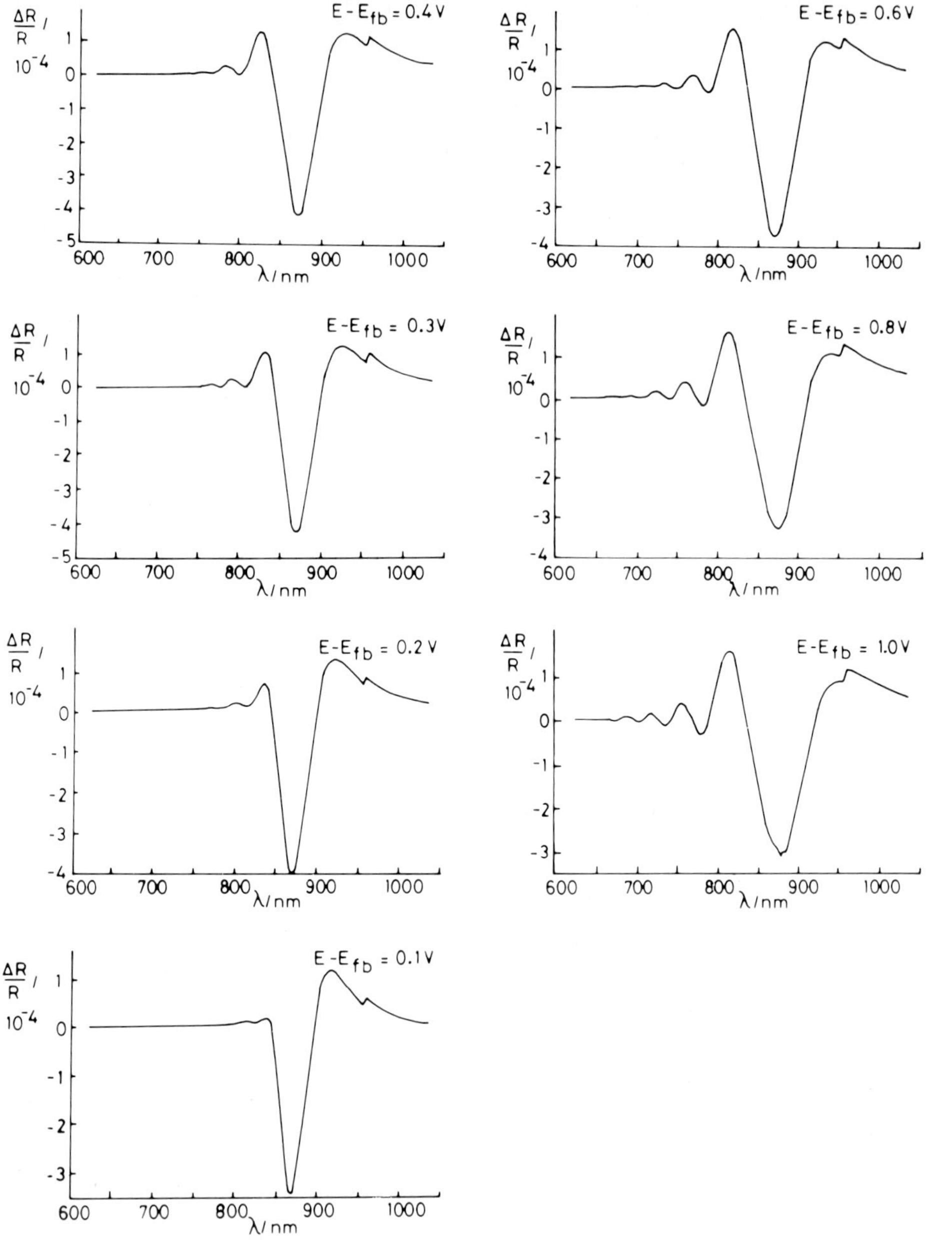

Fig. 110. Theoretical best fits to the data of Fig. 111 using the full model. The parameters used were $\Gamma = 40\,\mathrm{meV}$ and $E_g = 1.414\,\mathrm{eV}$. The small feature at 960 nm is an artifact of the program and can be ignored.

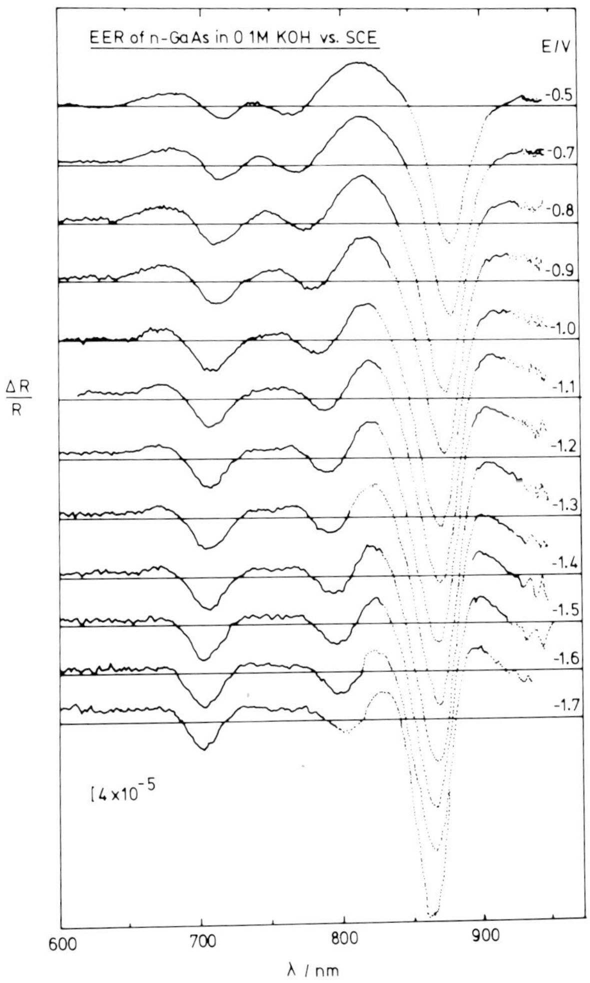

Fig. 111. Experimental electroreflectance spectra for n-GaAs in 0.5 M H_2SO_4 at various potentials. The flat-band potential is estimated to be close to -1.8 V vs. SCE and the modulation potential is 16 mV square wave.

d.c. potential in Fig. 111, the modulation potential being 16 mV. It can be seen that there are small but definite changes in the shape of the spectrum as the d.c. potential is moved away from flat-band potential (which is believed to be in the neighbourhood of 1.8 V). The theoretically generated spectra using the complete theory are shown in Fig. 110. It is easily seen that, qualitatively, the trends observed experimentally are reproduced theoretically. At a quantitative level, a careful analysis revealed that not all the d.c. potential was being dropped across the depletion layer. In fact, only about 70% appears to be dropped in this way, the remainder being accommodated across the Helmholtz layer. This is in good agreement with the detailed investigations reported in Sect. 3 [54, 56]. Similar investigations have been carried out for p-GaAs and, again, there is good qualitative agreement between theory and experiment, at least at potentials well removed from flat-band potential. However, at potentials within ca. 0.7 V of flat-band, the experimental spectrum becomes much less sensitive to applied d.c. potential and a substantial ER spectrum is obtained even at the apparent flat-band potential itself. The fact that the spectra can be modelled so successfully in the potential region above 0.7 V gives us confidence that the effects observed below 0.7 V are real; in other words, there are very slow surface states below 0.7 V that effectively pin the semiconductor at this potential.

This result does not stand in isolation. Peter et al. [150] have recently reported that transient data on p-GaP, using the theory of section 5, also show evidence that the potential distribution is far from the ideal portrayed in Sect. 3, and recent rotating ring–disc studies by Kelly and Notten [190] have also lent credence to this idea. If these results are indeed correct, it is evident that much of the current theory of semiconductors will have to be applied with more critical care than has been done so far. It is clear that much remains to be done.

7.8 DIFFERENTIAL REFLECTANCE STUDIES

This technique differs from electroreflectance in that the change in reflectivity is caused not by varying the electric field but by a change in surface composition [191]. The technique has been used primarily in semiconductor electrochemistry to study reductive or oxidative decomposition and its initial advantage over ellipsometry was that data could be gathered much more rapidly, permitting reflectivity measurements simultaneously with such electrochemical techniques as cyclic voltammetry. However, the advent of fast rotating-analyser ellipsometers has made this latter a significantly more competitive tool and it is likely that the next few years will see considerable progress in this area.

A typical experimental set-up for differential reflectance is shown in Fig. 112. To ensure that slow variations in the power of the lamp are adequately monitored, a rotating silvered chopper is used that either reflects the light

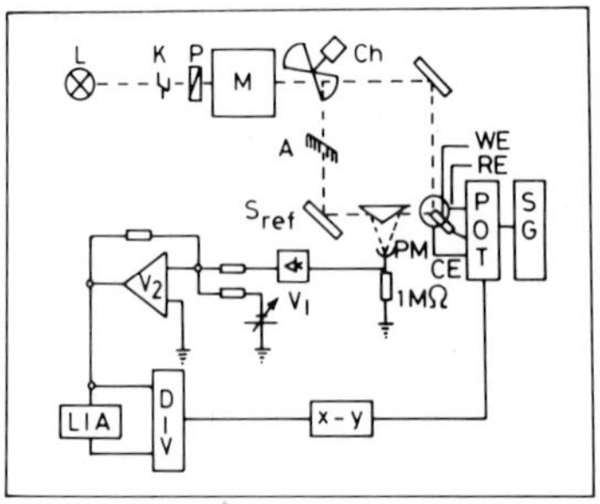

Fig. 112. Block diagram of the experimental set-up for differential reflectance. L, light source; K, order filter; P, polariser; M, monochromator; Ch, chopper; A, variable light attenuator; S_{ref}, reference mirror; PM, photomultiplier; V_1, V_2, amplifiers; LIA, lock-in amplifier; DIV, divider.

through an appropriate attenuator directly on to the photomultiplier tube or allows it to pass through to the sample. For thin films, and linearly polarised light, it may be shown that the change in reflectance on film formation is given by

$$\left(\frac{\Delta R}{R}\right)_{\perp} = \frac{8\pi n_1 d \cos\phi_1}{\lambda} \operatorname{Im}\left\{\frac{\hat{\varepsilon}_2 - \hat{\varepsilon}_3}{\varepsilon_1 - \varepsilon_3}\right\} \tag{645}$$

$$\left(\frac{\Delta R}{R}\right)_{\parallel} = \frac{8\pi n_1 d \cos\phi_1}{\lambda} \operatorname{Im}\left\{\left(\frac{\hat{\varepsilon}_2 - \tilde{\varepsilon}_3}{\varepsilon_1 - \hat{\varepsilon}_3}\right) \times \left[\frac{1 - (\varepsilon_1/\hat{\varepsilon}_2\hat{\varepsilon}_3)(\hat{\varepsilon}_2 + \hat{\varepsilon}_3)\sin^2\phi_1}{1 - (1/\hat{\varepsilon}_3)(\varepsilon_1 + \hat{\varepsilon}_3)\sin^2\phi_1}\right]\right\} \tag{646}$$

where ϕ_1 is the angle of incidence, d is the mean film thickness, λ the vacuum wavelength of the incident light, $\varepsilon_1 \equiv n_1^2$ the dielectric constant of the electrolyte, $\tilde{\varepsilon}_2 = \varepsilon_2' \quad i\varepsilon_2''$, that of the film and $\tilde{\varepsilon}_3$ that of the substrate, and the subscripts $\perp$ and $\parallel$ refer to the polarisation *direction* of the light being perpendicular or parallel, respectively, to the *plane of reflection*. The expected magnitude of the effect may be calculated for p-CdTe with a monolayer of Cd metal; at $\lambda = 5000$ Å, $d = 3$ Å, $\varepsilon_1 = 1.78$, $\hat{\varepsilon}_2 = (-12.1, 3.53)$, $\hat{\varepsilon}_3 = (6.77, 4.07)$ and $\phi_1 = 45°$, then $(\Delta R/R)_{\perp} = 2.6\%$ and $(\Delta R/R)_{\parallel} = 7.0\%$. These reflectivity changes are quite easy to measure and give some impression of the sensitivity of the technique.

An example of the use of the technique to obtain information during a cyclic voltammagram is shown in Fig. 113 for n-ZnO [192]. As the crystal was cycled to very negative potentials, a large cathodic current was found due both to hydrogen evolution and Zn-metal deposition from the reductive decomposition of the ZnO. Re-oxidation of the Zn metal is seen as an anodic peak in the CV, but it is very difficult, in the presence of both cathodic and anodic currents, to form a clear impression from the CV of the amount of Zn

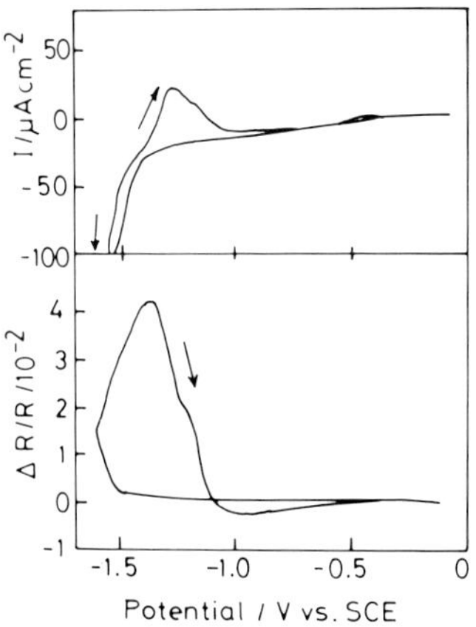

Fig. 113. CV of n-ZnO in 1 M KCl at a scan rate of $20\,\mathrm{mV\,s^{-1}}$ with accompanying reflectivity changes at 500 nm. Note that the large change in $\Delta R/R$ is due to zinc deposition.

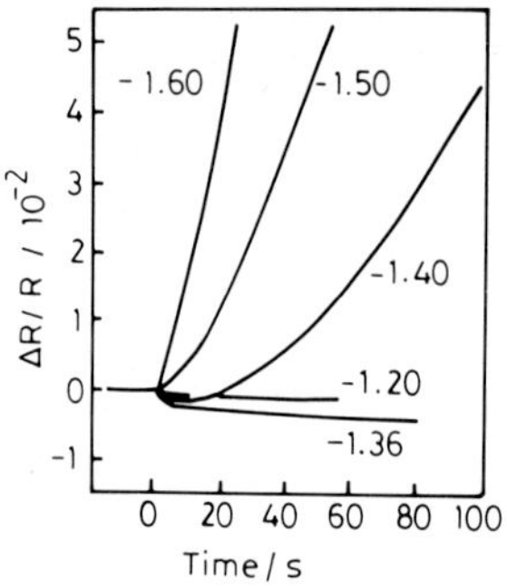

Fig. 114. Reflectivity changes as a function of time following potential steps from − 0.3 V to the potential indicated.

deposited. The simultaneous reflectivity changes are also seen in Fig. 113 and they clearly mirror the CV. The magnitude of ($\Delta R/R$) suggests 2–3 monolayers of Zn have formed. It is interesting to note that, if the CV is replaced by potential stepping to cathodic regions, the deposition of zinc has a very sharp onset between − 1.36 and − 1.40 V vs. SCE, as shown in Fig. 114.

References

1 J.B. Goodenough and A. Hamnett, in O. Madelung (Ed.), Non-tetrahedrally Bonded Semiconductors, Landolt-Bornstein Neue Serie Vol. XVII/g, Springer-Verlag, Berlin, 1984.
2 R.A. Smith, Semiconductors, Cambridge University Press, London, 1979.
3 C. Kittel, Introduction to Solid State Physics, Wiley, New York, 6th edn., 1986.
4 A. Hamnett, in A.K. Cheetham and P. Day (Eds.), Solid State Chemistry, Oxford University Press, Oxford, 1986.
5 J.B. Goodenough, Prog. Solid State Chem., 5 (1971) 145.
6 N.F. Mott and E.A. Davis, Electronic Processes in Non-Crystalline Materials, Clarendon Press, Oxford, 1979.
7 V.A. Myamlin and Yu.V. Pleskov, Semiconductor Electrochemistry, Plenum Press, New York, 1967.
8 Yu.V. Pleskov, in J.O'M. Bockris, B.E. Conway and E. Yeager (Eds.), Comprehensive Treatise of Electrochemistry, Vol. 1, Plenum Press, New York, 1980.
9 M. Green, J. Chem. Phys., 31 (1959) 200.
10 J.F. DeWald, in B.N. Hannay (Ed.), Semiconductors, Reinhold, New York, 1959.
11 H. Gerischer, in H. Eyring (Ed.), Physical Chemistry. An Advanced Treatise, Vol. 9, Academic Press, New York, 1970.
12 A. Many, Y. Goldstein and N.B. Grover, Semiconductor Surfaces, North-Holland, Amsterdam, 1965.
13 D.R. Frankl, Electrical Properties of Semiconductor surfaces, Pergamon Press, Oxford, 1967.
14 Yu.Ya. Gurevich and Yu.V. Pleskov, Usp. Khim., 52 (1983) 563.
15 S.R. Morrison, Electrochemistry at Semiconductor and Oxidised-metal Electrodes, Plenum Press, New York, 1980.
16 C. Hinnen, C. Nguyen van Huong, A. Rousseau and J.P. Dalbera, J. Electroanal. Chem., 95 (1979) 131.

17 R.H. Kingston and S.F. Neustadtler, J. Appl. Phys., 26 (1955) 718.
18 R. Seiwatz and M. Green, J. Appl. Phys., 29 (1958) 1034.
19 H. Gerischer, A. Mauerer and W. Mindt, Surf. Sci., 4 (1966) 431.
20 S. Trassatti, in J. O'M. Bockris, B.E. Conway and E. Yeager, Comprehensive Treatise of Physical Chemistry, Vol. I, Plenum Press, New York, 1980.
21 T.W. Healy, D.E. Yates, L.R. White and D. Chan, J. Electroanal. Chem., 80 (1977) 57.
22 L. Bousse and P. Bergveld, J. Electroanal. Chem., 152 (1983) 25.
23 L. Bousse, N.F. de Rooij and P. Bergveld, IEEE Trans. Electron Devices, ED-30 (1983) 1263.
24 M.A. Butler, J. Electrochem. Soc., 125 (1978) 228.
25 R.K.S. El Wakked and H.A. Risk, J. Phys. Chem., 61 (1957) 494.
26 G.A. Park, Chem. Rev., 65 (1965) 177.
27 S. Mukai, Y. Yokoyama, T. Wakamatsu and N. Narazaki, Shuiyokai-Shi, 17 (1970) 49.
28 K.C. Ray and S. Khan, Ind. J. Chem., 13 (1975) 577.
29 Y.G. Berube and P.L. DeBruyn, J. Colloid Interface Sci., 27 (1968) 305.
30 R.J. Atkinson, A.M. Posner and J.P. Quirk, J. Phys. Chem., 71 (1967) 550.
31 I. Iwasaki, S.R.B. Cooke and Y.S. Kim, Trans. AIME, 223 (1962) 113.
32 L. Blok and P.L. DeBruyn, J. Colloid Interface Sci., 32 (1970) 518.
33 K. Emerson and W.M. Gravea, J. Inorg. Nucl. Chem., 11 (1959) 309.
34 H.R. Kruyt, Colloid Sci., 1 (1952) 231.
35 M. Robinson, J.A. Pask and D.W. Furstenan, J. Am. Ceram. Soc., 47 (1964) 516.
36 G.A. Kokarev, V.A. Kolesnikov, A.F. Gubin and A.A. Korobanov, Elektrokhimiya, 18 (1982) 466.
37 A.H. Nethercot, Phys. Rev. Lett., 33 (1974) 1088.
38 R.T. Poole, D.R. Williams, J.D. Riley, J.G. Jenkins, J. Liesegang and R.C.G. Leckey, Chem. Phys. Lett., 36 (1975) 401.
39 D.S. Ginley and M.A. Butler, J. Electrochem. Soc., 125 (1978) 1968.
40 H. Gerischer, J. Electroanal. Chem., 150 (1983) 553.
41 R.M. Noyes, J. Am. Chem. Soc., 84 (1962) 513.
42 S. Trassatti, Adv. Electrochem. Electrochem. Eng., 10 (1977) 213.
43 F. Lohmann, Z. Naturforsch., 221 (1967) 843.
44 R. Gomer and T. Tryson, J. Chem. Phys., 66 (1977) 4413.
45 F. El Halonami and A. Desschanvres, Mater. Res. Bull., 17 (1982) 1045.
46 B. Reichmann and C.E. Byvik, J. Electrochem Soc., 128 (1981) 2601.
47 A.J. Bard, A.B. Bocarsly, F.R. Fan, E.G. Walton and M.S. Wrighton, J. Am. Chem. Soc., 102 (1980) 3671.
48 F.R. Fan and A.J. Bard, J. Am. Chem. Soc., 102 (1980) 3677.
49 A.B. Bocarsly, D.C. Bookbinder, R.M. Dominey, N.S. Lewis and M.S. Wrighton, J. Am. Chem. Soc., 102 (1980) 3683.
50 J.N. Chazalviel and T.B. Truong, J. Electroanal. Chem., 114 (1980) 299.
51 J. Bardeen, Phys. Rev., 71 (1947) 717.
52 A.M. Cowley and S.M. Sze, J. Appl. Phys., 36 (1965) 3212.
53 S. Kurtin, T.C. McGill and C.A. Mead, Phys. Rev. Lett., 22 (1969) 1433.
54 G. Horowitz, P. Allongue and H. Cachet, J. Electrochem. Soc., 131 (1984) 2563.
55 W.E. Spicer, P.W. Chyl, P.R. Skeath, C.Y. Su and I. Lindau, J. Vac. Sci. Technol., 16 (1979) 1422.
56 R.L. van Meirhaeghe, F. Cardon and W.P. Gomes, J. Electrochem. Soc., 188 (1985) 287.
57 W. Schmickler, Ber. Bunsenges. Phys. Chem., 82 (1978) 477.
58 W.P. Gomes and F. Cardon, Prog. Surf. Sci., 12 (1982) 155.
59 M.P. Dare-Edwards, A. Hamnett and P.R. Trevellick, J. Chem. Soc. Faraday Trans. 1, 79 (1983) 2111.
60 H.S. Jarrett, J. Electroanal. Chem., 150 (1983) 629.
61 J. Rosenthal and B. Westerby, J. Electrochem. Soc., 129 (1982) 2147.
62 M.P. Dare-Edwards, D. Phil. Thesis, Oxford University, 1982.

63 W.H. Laflere, R.L. van Meirhaeghe, F. Cardon and W. Gomes, Surf. Sci., 59 (1976) 401.
64 E.C. Dutoit, R.L. van Meirhaeghe, F. Cardon and W.P. Gomes, Ber. Bunsenges. Phys. Chem., 79 (1975) 1206.
65 M.P. Dare-Edwards, J.B. Goodenough, A. Hamnett and P.R. Trevellick, J. Chem. Soc. Faraday Trans. 1, 79 (1983) 2111.
66 M.P. Dare-Edwards and A. Hamnett, unpublished work.
67 K. Uosaki and H. Kita, J. Electrochem. Soc., 130 (1983) 895.
68 R. de Gruyse, W.P. Gomes, F. Cardon and J. Vennik, J. Electrochem. Soc., 125 (1975) 711.
69 J.F. McCann and S.P.S. Badwal, J. Electrochem. Soc., 129 (1982) 551.
70 W. Sirpala and M. Tomkiewicz, J. Electrochem. Soc., 128 (1981) 1240.
71 P. Allongue, H. Cachet and G. Horowitz, J. Electrochem. Soc., 130 (1983) 2352.
72 P. Zoltowski, J.Electroanal. Chem., 178 (1984) 11.
73 E.C. Dutoit, R.L. van Meirhaeghe, F. Cardon and W.P. Gomes, Ber. Bunsenges. Phys. Chem., 79 (1975) 1206.
74 R.U. E. 't Lam, J. Schoonman and G. Blasse, Ber. Bunsenges. Phys. Chem., 85 (1981) 592.
75 R.U.E. 't Lam, L.H.J.M. Janssen and J. Schoonman, Ber. Bunsenges. Phys. Chem., 88 (1984) 163.
76 G. Nogami, J. Electrochem. Soc., 129 (1982) 2219.
77 J.N. Chazalviel and T.B. Truong, J. Electroanal. Chem., 114 (1980) 299.
78 L. Bousse and P. Bergveld, J. Electroanal. Chem., 152 (1983) 25.
79 J. Schoonman, K. Vos and G. Blasse, J. Electrochem. Soc., 128 (1981) 1154.
80 R.U.E. 't Lam, L.H.J.M. Janssen and J. Schoonman, Ber. Bunsenges. Phys. Chem., 88 (1984) 163.
81 R. Memming, Philips Res. Rep., 19 (1964) 323.
82 F. Berz, J. Phys. Chem. Solids, 23 (1962) 1795.
83 C.G.B. Garrett and W.H. Brattain, Phys. Rev., 99 (1955) 376.
84 C.D. Jaeger, H. Gerischer and W. Kautek, Ber. Bunsenges. Phys. Chem., 86 (1982) 20.
85 J. Ulstrup, Charge Transfer Processes in Condensed Media, Springer Verlag, Berlin, 1979.
86 H. Gerischer, Surf. Sci., 18 (1969) 97.
87 E.A. Efimov and I.G. Erusalimchik, Zh. Fiz. Khim., 32 (1958) 1967.
88 R.A.L. Vanden Berghe, F. Cardon and W.P. Gomes, Surf. Sci., 39 (1973) 368.
89 S.R. Morrison, Surf. Sci., 15 (1969) 363.
90 V.A. Tyagai, Eletrokhimiya, 1 (1965) 377.
91 D.L. Ullman, J. Electrochem. Soc., 128 (1981) 1269.
92 W.H. Brattain and C.G.B. Garrett, Bell Syst. Tech. J., 34 (1955) 129.
93 W.H. Brattain and C.G.B. Garrett, Physica, 20 (1954) 885.
94 W.H. Brattain and C.G.B. Garrett, Phys. Rev., 94 (1954) 750.
95 E.A. Efimov and I.G. Erusalemchik, Zh. Fiz. Khim., 32 (1958) 1103.
96 E.A. Efimov and I.G. Erusalemchik, Dokl. Akad. Nauk SSSR, 128 (1959) 124.
97 E.A. Efimov and I.G. Erusalemchik, Zh. Fiz. Khim., 35 (1961) 543.
98 E.A. Efimov and I.G. Erusalemchik, Zh. Fiz. Khim., 36 (1962) 1791.
99 J.B. Flynn, J. Electrochem. Soc., 105 (1958) 715.
100 E.A. Efimov and I.G. Erusalemchik, Zh. Fiz. Khim., 35 (1961) 384.
101 E.A. Efimov and I.G. Erusalemchik, Dokl. Akad. Nauk SSSR, 130 (1960) 353.
102 E.A. Efimov, I.G. Erusalemchik and G.P. Sokolova, Zh. Fiz. Khim., 38 (1964) 2178.
103 R. Memming and G. Schwandt, Surf. Sci., 5 (1966) 97.
104 R. Memming and G. Schwandt, Angew. Chem. Int. Ed. Engl., 6 (1967) 851.
105 E.N. Paleolog, K.S. Korokhova and N.D. Tomaskov, Dokl. Akad. Nauk SSSR, 133 (1960) 170.
106 W. Mehl and F. Lohmann, Ber. Bunsenges. Phys. Chem., 71 (1967) 1055.
107 J. Vandermolen, W.P. Gomes and F. Cardon, J. Electrochem. Soc., 127 (1980) 324.
108 P. Salvador and C. Gutierrez, J. Electrochem. Soc., 131 (1984) 326.
109 V.A. Tyagai and G.Ya. Kolbasov, Surf. Sci., 28 (1971) 423.

110 P. Salvador and C. Gutierrez, Surf. Sci., 124 (1983) 398.
111 W. Schmickler, J. Electroanal. Chem., 137 (1982) 189.
112 H. Gerischer and F. Beck, Z. Phys. Chem., 24 (1960) 378.
113 H. Gerischer and I. Wallem-Battes, Z. Phys. Chem., 64 (1969) 187.
114 F. Beck and H. Gerischer, Z. Elektrochem., 63 (1959) 943.
115 R. Memming and F. Mōllers, Ber. Bunsenges. Phys. Chem., 76 (1972) 610.
116 P. Janietz, R. Woiche and R. Landsberg, J. Electroanal. Chem., 112 (1980) 63.
117 H. Gerischer and H. Mindt, Electrochim. Acta, 13 (1968) 1329.
118 R. Memming and F. Mollers, Ber. Bunsenges. Phys. Chem., 77 (1973) 960.
119 J.W. Cowley, C.B. Drake, G.D. Mahan and J.J. Tiemann, Phys. Rev., 150 (1966) 466.
120 K. Kobayashi, X. Aikawa and M. Sukigara, J. Electroanal. Chem., 134 (1982) 11.
121 W. Schmickler, Bunsenges. Phys. Chem., 82 (1978) 477.
122 J. Ulstrup, J. Chem. Phys., 63 (1975) 4358.
123 N.R. Kestner, J. Logan and J. Jortner, J. Phys. Chem., 78 (1974) 2148.
124 D. Elliott, D.L. Zellmed and H.A. Laitinen, J. Electrochem. Soc., 117 (1970) 1343.
125 B. Pettinger, H.R. Schoppel and H. Gerischer, Ber. Bunsenges. Phys. Chem., 78 (1974) 450.
126 R. Memming and F. Mōllers, Ber. Bunsenges. Phys. Chem., 76 (1972) 475.
127 V.G. Levich, Adv. Electrochem. Electrochem. Eng., 4 (1966) 249.
128 K.I.W Frese, J. Phys. Chem., 85 (1981) 3911.
129 R.A.L. Vanden Berghe, F. Cardon and W.P. Gomes, Surf. Sci., 39 (1973) 368.
130 B. Pettinger, H.-R. Schoppel, T. Yokoyama and H. Gerischer, Ber. Bunsenges. Phys. Chem., 78 (1974) 1024.
131 W.J. Albery and P.N. Bartlett, J. Electrochem. Soc., 130 (1983) 1699.
132 W.W. Gaertner, Phys. Rev., 116 (1959) 84.
133 H.H. Streckert and A.B. Ellis, J. Phys. Chem., 86 (1982) 4921.
134 V.A. Tyagai, Russ. J. Phys. Chem., 38 (1965) 1335.
135 R.A. Wilson, J. Appl. Phys. 48 (1977) 4292.
136 H. Gerischer, J. Electroanal. Chem., 150 (1983) 553.
137 R. Memming, Surf. Sci., 1 (1964) 88.
138 A. Stevenson and R. Keyes, Physica, 20 (1954) 1041.
139 V.A. Tyagai and Yu.V. Pleskov, Fiz. Tverd. Tela, 4 (1962) 343.
140 H.U. Harten, Z. Naturforsch. Teil A, 16 (1961) 459.
141 H.U. Harten, J. Phys. Chem. Solids, 14 (1960) 220.
142 W.J. Albery, P.N. Bartlett, M.P. Dare-Edwards and A. Hamnett, J. Electrochem. Soc., 128 (1981) 1492.
143 M. Abramovich and I. Stegun, Handbook of Mathematical Functions, Dover, New York, 1970, pp. 504 et seq.
144 H. Margenau and G.M. Murphy, The Mathematics of Physics and Chemistry, Vol. I, Van Nostrand, Princeton, 1944.
145 P. Lemasson, A. Etcheberry and J. Gautron, Electrochim. Acta, 27 (1982) 607.
146 J. Reichman, Appl. Phys. Lett., 36 (1980) 574.
147 M.P. Dare-Edwards, A. Hamnett and P.R. Trevellick, J. Chem. Soc. Faraday Trans. 1, 79 (1983) 2027.
148 B. Wolf, K.D. Schultze and W. Lorenz, Z. Phys. Chem. (Leipzig), 263 (1982) 1258.
149 A. Hamnett, in A. Harriman and M.A. West (Eds.), Photogeneration of Hydrogen, Academic Press, New York, 1982.
150 L.M. Peter, J. Li and R. Peat, J. Electroanal. Chem., 165 (1984) 29.
151 J.J. Kelley and R. Memming, J. Electrochem. Soc., 129 (1982) 730.
152 W.J. Albery and P.N. Bartlett, J. Electrochem. Soc., 129 (1982) 2254.
153 H. Gerischer, J. Electroanal. Chem., 150 (1983) 553.
154 G. Bin-Daar, M.P. Dare-Edwards, J.B. Goodenough and A. Hamnett, J. Chem. Soc. Faraday Trans. 1, 79 (1983) 1199.
155 F. Cardon, W.P. Gomes, F. Vander Kerchove, D. Vanmaekelbergh and F. Van Overmeise, Faraday Discuss. Chem. Soc., 70 (1980) 153.
156 D. Vanmaekelberg, W.P. Gomes and F. Cardon, J. Electrochem. Soc., 129 (1982) 546.

157 D. Vanmaekelbergh, W.P. Gomes and F. Cardon, J. Chem. Soc. Faraday Trans. 1, 79 (1983) 1391.
158 K.W. Frese, M.J. Madou and S.R. Morrison, J. Phys. Chem., 84 (1980) 3172.
159 K.W. Frese, M.J. Madou and S.R. Morrison, J. Electrochem. Soc., 128 (1981) 1527.
160 R. Tenne, N. Muller, Y. Mirovsky and D. Landa, J. Electrochem. Soc., 130 (1983) 852.
161 J.-N. Chazalviel, J. Electrochem. Soc., 127 (1980) 1822.
162 J. Li, R. Peat and L.M. Peter, J. Electroanal. Chem., 165 (1984) 41.
163 K.H. Beckmann and R. Memming, J. Electrochem. Soc., 116 (1969) 368.
164 R. Haak, C. Ogden and D. Tench, J. Electrochem. Soc., 129 (1982) 891.
165 R. Haak and D. Tench, J. Electrochem. Soc., 131 (1984) 275.
166 R. Haak and D. Tench, J. Electrochem. Soc., 131 (1984) 1442.
167 Y. Nakato, A. Tsumura and H. Tsubomura, Bull. Chem. Soc. Jpn., 55 (1982) 3390.
168 K. Uosaki and H. Kita, J. Phys. Chem., 88 (1984) 4197.
169 B. Pettinger, H.-R. Schoppel and H. Gerischer, Ber. Bunsenges. Phys. Chem., 80 (1976) 849.
170 N.S. Lewis, J. Electrochem. Soc., 131 (1984) 2496.
171 H.C. Gatos, Mater. Sci. Res., 7 (1974) 195.
172 W. Kautek and H. Gerischer, Electrochim. Acta, 27 (1982) 355.
173 C.D. Jaeger, H. Gerischer and W. Kautek, Ber. Bunsenges. Phys. Chem., 86 (1982) 20.
174 P.A. Allongue and H. Cachet, J. Electrochem. Soc., 132 (1985) 45.
175 J.E.A.M. Van der Meerakker, J.J. Kelly and P.H.L. Notten, J. Electrochem. Soc., 132 (1985) 638.
176 H. Gobrecht and O. Meinhart, Ber. Bunsenges. Phys. Chem., 67 (1963) 151.
177 K. Chandrasakaram, M. Weichold, F. Gutmann and J.O'M. Bockris, Electrochim. Acta, 30 (1985) 961.
178 B.L. Wheeler, G. Nagasubramanian and A.J. Bard, J. Electrochem. Soc., 131 (1984) 1038.
179 Z. Harzion, N. Croitoron and S. Gottesfeld, J. Electrochem. Soc., 128 (1981) 551.
180 R.H. Wilson, T. Sakata, T. Kawai and K. Hashimoto, J. Electrochem. Soc., 132 (1985) 1082.
181 S. Gottesfeld and S.W. Feldberg, J. Electroanal. Chem., 146 (1983) 47.
182 Z. Harzion, D. Happert, S. Gottesfeld and N. Croitoron, J. Electroanal. Chem., 150 (1983) 571.
183 M. Cardona, Modulation Spectroscopy, Academic Press, New York, 1969.
184 R.K. Willardson and A.C. Beer (Eds.), Semiconductors and Semimetals, Vol. 9, Academic Press, New York, 1972. B.O. Seraphin (Ed.), Optical Properties of Solids. New Developments, North-Holland, Amsterdam, 1976.
185 D.E. Aspnes, Surf. Sci., 37 (1973) 418.
186 D.E. Aspnes, P. Handler and D.F. Blossey, Phys. Rev., 166 (1968) 921.
187 P.M. Raccah, J.W. Garland, Z. Zhang, U. Lee, D.Z. Xue, L.L. Abels, S. Ugur and W. Wilinsky, Phys. Rev. Lett., 55 (1984) 1958.
188 P.M. Raccah, J.W. Garland, Z. Zhang, L.L. Abels, S. Ugur, S. Mioc and M. Brown, Phys. Rev. Lett., 55 (1985) 1323.
189 A. Hamnett, in R.G. Compton and A. Hamnett (Eds.), Comprehensive Chemical Kinetics, Vol. 29, Elsevier, Amsterdam, to be published.
190 J.J. Kelly and P.H.L. Notten, Electrochim. Acta, 29 (1984) 589.
191 J.D.E. McIntyre and D.E. Aspnes, Surf. Sci., 24 (1971) 417.
192 D.M. Kolb and H. Gerischer, Electrochim. Acta, 18 (1973) 987.
193 J. Li and L.M. Peter, J. Electroanal. Chem., 193 (1985) 27.
194 J. Li and L.M. Peter, J. Electroanal. Chem., 199 (1986) 1.
195 R. Peat and L.M. Peter, J. Electroanal. Chem., 209 (1986) 307.
196 R. Peat and L.M. Peter, Ber. Bunsenges. Phys. Chem., 91 (1987) 381.
197 G. Nagasubramanian, B.L. Wheeler and A.J. Bard, J. Electrochem. Soc., 130 (1983) 1680.
198 K. Uosaki and H. Kita, Ber. Bunsenges. Phys. Chem., 91 (1987) 447.
199 S. Gottesfeld, Ber. Bunsenges. Phys. Chem., 91 (1987) 362.
200 L.M. Abrantes, R. Peat, L.M. Peter and A. Hamnett, Ber. Bunsenges. Phys. Chem., 91 (1987) 369.

Chapter 3

Reactions at Metal Oxide Electrodes

EUGENE J.M. O'SULLIVAN and ERNESTO J. CALVO

1. Introduction

The object of the present chapter is to review the status of interfacial reactions, mainly Faradaic, which have been studied at the metal oxide–electrolyte boundary. Not much attention will be given to the intrinsic properties of oxides, except where relevant to the discussion of Faradaic reactions. A comprehensive discussion on the properties of oxide electrodes can be found, for example, in refs. 1–3.

Reactions at oxide electrodes have attracted the attention of electrochemists for both fundamental and technological reasons. The semiconducting properties of many oxides, for example, can be advantageously used to test current theories on electron transfer at electrochemical interfaces. Unlike a metal, where the number density of carriers is fairly constant and the energy can be varied over a wide range by means of the electrode potential, in the case of semiconducting oxides, the number of electrical carriers on the surface may be varied over a narrow energy range, although the span of available energies is more limited. Metallic oxides are employed, often as coatings or thin films, as electrode materials for a range of electrocatalytic reactions, such as oxygen evolution and reduction, chlorine evolution (chlor-alkali industry) and organic electrosynthesis. In many instances, the chemistry of the different cations at the oxide surface (active catalytic centers, or reactive groups) may be exploited to enhance the kinetics of electrochemical reactions.

Many heterogeneous reactions extensively studied in catalysis can be extended to the electrochemical environment, thereby increasing the selectivity, with fine control of reaction rate via the electrode potential. The conversion of various forms of energy into chemical energy can be achieved in electrosynthesis or generation of electricity. Practical interest in reactions at oxide electrodes comes from a very wide range of technologies: dimensionally stable anodes for the chlor-alkali industry, electrosynthesis, photoelectrochemical energy conversions including splitting of water, battery electrodes, electrochromic devices, and corrosion processes where oxhydroxy passivation layers enhance the resistance of a metal to a corrosive environment.

Electrode reactions may be regarded as a switch for electrical charge to flow between two phases of different electrical conductivity: electrode and electrolyte. Oxide electrodes attract a special interest in this respect because charge transfer across the electrochemical interface may occur by either ion or electron transfer, or both simultaneously. Oxide electrodes are often not chemically inert since their constituent cations and oxygen ions may undergo chemical and electrochemical reactions in the oxide–solution interfacial region. When cations can change their oxidation state in the oxide, variations of chemical composition in the oxide electrode are possible. These intraphase transformations, which may occur simultaneously with interfacial (Faradaic) electrode reactions, may alter the electronic properties of the oxide electrode surface and affect the latter reactions. The electrochemical stability of oxide electrodes has very important consequences in the performance of technologically important electrodes and lack of stability may result from, for example, photocorrosion, loss of electrocatalytic activity, and wear.

Oxide electrodes exhibit a wide range of electronic properties from nearly insulating materials (SiO_2) to typical semiconductors (TiO_2, ZnO_2, and SnO_2), to nearly metallic (PbO_2 and RuO_2), and even the surface electronic properties may change with electrode potential and electrolyte composition.

2. Types of oxide electrode

The ideal oxide electrode for a fundamental kinetic study is an oxide single crystal where the surface exposed to the electrolyte is well defined geometrically, chemically, and morphologically. However, most of the times it is not possible to employ a single crystal oxide electrode because the material is not stable enough to prepare a single crystal, or because the oxide exhibits variable stoichiometry. Also, most commonly used oxide electrodes in practical applications have very large surface areas and hence the electrochemically active surface and the reaction plane are ill-defined in these systems. Electrochemical reactions at single crystal NiO, RuO_2, and TiO_2 electrodes, along with some other oxides, have been studied. The possibility of polymorphism and non-stoichiometry in single and mixed oxides (more than one cation) enlarges the range and variety of properties observed.

Oxide materials which are attractive because of their catalytic activity are often employed in the form of finely divided powders of considerable surface area. The history of the material and the preparation technique employed are important aspects to be considered when electrokinetic data are compared. Oxide powders can be hot pressed into sintered pellets, supported, or impregnated, on to carbons of high specific area, and bonded with Teflon or other inert material into composite electrodes. Microporosity of the system may produce an ill-defined surface zone flooded by electrolyte with imprecise ratio of real surface area to geometric cross-section. Sputtered oxide layers and precicipated hydroxides on inert substrates have been studied in fundamental work as well as in technological applications such as electrochromic devices and battery oxides. In these systems, different degrees of hydration can be obtained, which can signficantly affect their electrocatalytic activity.

Passive layers on metals tend to be oxhydroxide multilayer films with variable composition in the direction normal to the underlying metal and variable degree of hydration. Although they show some evidence of semiconducting behavior, e.g. exhibiting photo-potentials, photocurrents, and inhibition and rectification in simple electron transfer reactions with redox couples in the electrolyte, the behavior of such thin systems (ca. 2–10 nm) is strongly influenced by the underlying metal and several models have been proposed to describe these films. Probably the most important system from a practical point of view, and the most extensively explored, is the passive layer on iron for which models such as the duplex layer model, i.e. $Fe/Fe_3O_4/Fe_2O_3$ [4], along with a model involving continuously varying Fe(II) stoichiometry [5] have been proposed. Hydroxide layers deposited electrochemically on to inert metal electrodes, i.e. sandwich-type systems, e.g. $Au/Au_2O_3/FeOOH$, and diodes, e.g. $Au/TiO_2/Au$, have attracted considerable interest in recent years [6]. They are often employed as model systems in the study of electron transfer reactions in order to gain insight into the properties of systems such as passive films.

The technologically important dimensionally stable anodes (DSA) are thermally prepared mixed oxide films supported on an inert substrate, usually Ti, which contain RuO_2 as the catalytically active component. These anodes exhibit high performance in the industrial generation of chlorine.

Finally, optically transparent electrodes comprised of SnO_2 or In_2O_3 thin films on glass, quartz, or plastic are widely used to study electrochemical reactions under illumination in solution. Spectral studies of reactants, intermediates, or products can be also performed to gain some molecular insight into electrode kinetics.

3. The metal oxide–electrolyte interface

The reaction site at oxide electrodes, the oxide–electrolyte interface, differs from the metal–electrolyte interface in several respects and its structure and properties are of the utmost importance for the understanding of reaction kinetics at oxide electrodes. Most of the information available on the properties of the interfacial region in oxides comes from colloid chemistry, i.e. electrokinetic, or zeta, potentials and surface titration curves. Several models developed by Lyklema, Berube and De Bruyn, and Levine and Smith to explain these experimental results have been reviewed elsewhere [7–9].

The most important differences between the oxide–electrolyte and the metal–solution interfaces are (i) the surface charge is higher than predicted by the Gouy–Chapman theory and is strongly dependent on electrolyte pH and (ii) more symmetrical capacity–potential curves are observed in the case of oxides, with much higher values of capacity being observed than in the case of, for example, Hg or AgI. Acid–base titrations yield plots of surface charge vs. pH from which the point of zero charge (pzc) is obtained, that is the pH at which the oxide surface has no net charge.

The surface of an oxide is amphoteric with surface groups which become protonated in acid solutions and negatively charged in basic solutions

$$MOH + H^{+} \longrightarrow MOH_2^{+} \tag{1}$$

$$MOH \longrightarrow MO^{-} + H^{+} \tag{2}$$

giving rise to a reversible double layer. Due to fast protonation of these surface groups, protons and hydroxyl ions are potential-determining ions with nearly Nernstian dependence on electrolyte pH being exhibited by many oxides. Theoretical models suggest that the Nernstian response is only an approximate solution to the free energy—pH equation and that the change in surface charge with electrolyte pH results in a change of surface potential that opposes further change. It has been suggested that the first water layer adjacent to an oxide electrode is more structured than water dipoled at metal electrodes due to appreciable hydrogen bonding by water to the oxide surface [10]. The recent work relating to hydrous oxides prepared by electrochemical methods deserves mention, since the potential–pH response of these films is close to 1.5 RT/F [11,12]. These films will be discussed in Sect. 5.4.

At oxide semiconductor electrode–electrolyte interfaces, with no contribution from surface states, the electrical potential drop exhibits three components: the potential drop across the space-charge region, ϕ_{sc}, across the Helmholtz layer, ϕ_H, and across the diffuse double layer, ϕ_d, the latter becoming negligible in concentrated electrolytes

$$\phi_{\text{oxide–electrolyte}} = \phi_H + \phi_{sc} \tag{3}$$

If the contribution of a depletion layer in the semiconducting oxide to the interfacial potential is not negligible, the Mott–Schottky relationship holds between the interfacial capacitance and the electrode potential [13]. For an n-type oxide

$$C^{-2} = \frac{2}{\varepsilon\varepsilon_o N_D}\left(E - E_{FB} - \frac{kT}{e}\right) \tag{4}$$

where ε and ε_o are the dielectric constants of the oxide and vacuum, respectively, e is the unity electrical charge, and N_D is the concentration of donors in the oxide. The extrapolation $C^{-2} \rightarrow 0$ yields the flat band potential of the oxide, E_{FB}, at which the potential drop across the space-charge region is zero and all the potential difference applies at the Helmholtz layer. Since the surface potential is fixed at a constant pH and since, in many oxides, it varies with a nearly Nernstian dependence, the flatband potential follows the same pH dependence for a number of oxide electrodes, e.g. ZnO [14], TiO_2 [15], and SnO_2 [16]. The potential drop across the Helmholtz layer in oxide electrodes reflects the chemical nature of the oxide surface. This is illustrated in Fig. 1; the E_{FB} values for iron oxides of different bulk properties follow the same behavior within the experimental scatter.

An important consequence of the pH dependence of the interfacial potential at oxide electrodes is the observation of pseudo-orders for protons and hydroxyl ions in simple redox reactions. This will be discussed in Sect. 5.3.

Since anions and cations adsorb at oxide electrodes positive and negative to the pzc, respectively, electrostatic work terms (double layer corrections) should contribute to the activation free energy barrier for adsorbed electroactive ions depending on the position of the reaction site. Not much attention has been paid to this phenomenon yet. Trasatti and co-workers

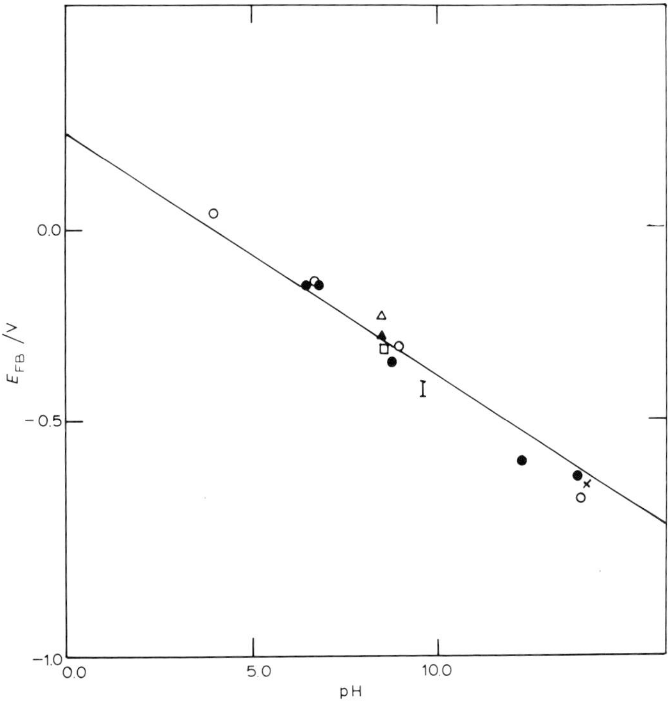

Fig. 1. Plot of flatband potential vs. pH for different iron(III) oxides in contact with aqueous electrolytes obtained by capacitance–potential measurements. ○, Thermal α-Fe_2O_3 [24]; ●, single crystal α-Fe_2O_3 [25]; △, ▲, thermal α-Fe_2O_3 [26, 27]; □, passive iron [28]; I, passive iron [29]; ×, passive iron [30].

have shown that, for several oxide electrodes of technical relevance, e.g. RuO_2 [17], Co_3O_4 [18,19], IrO_2 [20,21], RuO_2–IrO_2 [22], and $NiCo_2O_4$ [21, 23], the structural, chemical, and electrical properties of the oxide are strongly affected by the preparation temperature and method.

4. Thermodynamic aspects of metal oxide electrodes

An oxide electrode in equilibrium with the aqueous solution is a multicomponent system with metal, oxygen, and hydrogen ions in both places. At constant water activity, the system can be considered as quasi-binary [31]. Oxide electrodes can be regarded simultaneously as oxygen electrodes and as metal electrodes [31–33]. The electrode potential of MO_n in contact with aqueous electrolyte can be expressed with respect to oxygen by

$$E^{\circ}_{O}(n) \;=\; E^{s}_{O} + \frac{RT}{F}\ln p^{1/4}_{O_2} a_{(H^+)} \tag{5}$$

with E^{s}_{O} the standard potential of the oxygen electrode, O_2/H_2O. The oxide electrode potential can be expressed with respect to metal ions by

$$E^{\circ}_{M}(n) \;=\; E^{s}_{M} + \frac{RT}{zF}\ln \frac{a_{(M^{z+})}}{a_{(M)}} \tag{6}$$

with E^{s}_{M} the standard potential for the metal electrode M^{z+}/M. When the oxide electrode is at equilibrium with metal ions and oxygen in solution, $E^{\circ}_{M}(n)$ and $E^{\circ}_{M}(n)$ are equal and the solubility product, $L_z(n)$, depends on the oxygen partial pressure [31]. The oxygen pressure controlling the electrode potential is [34]

$$p_{O_2} \;=\; \exp\{(4F/RT)[E^{\circ}_{O}(n) \;-\; E^{s}_{O}]\} \tag{7}$$

Aspects of the thermodynamic pH dependence of the oxide electrode potential with x in non-stoichiometric oxides MO_2H_x have been discussed by Pohl and Atlung [35].

5. Ion and electron transfer reactions at metal oxide electrodes

5.1 ION TRANSFER REACTIONS

Electrical current flows at the oxide–electrolyte interface at a rate fixed by electrochemical kinetics; charge can be transferred across this interface either by ions or electrons. While electrons are transferred between the electrode and the electrolyte only if acceptor/donor redox species are available in solution, solvated ions can be transferred to or from the solution, changing the composition of the adjacent phases. Ion transfer reactions (ITR) at the oxide–electrolyte interface are consecutive multistep processes with at least one ionic charge transfer step, structural relaxation at the oxide, ligand exchange processes, transport phenomena in the oxide and the electrolyte, and so on. Exchange of metal cations, oxygen ions, and protons must be considered at oxide electrodes in contact with aqueous electrolytes. Cationic and anionic electrode reactions are linked together by the interfacial potential $\phi_{2,3}$, but otherwise they may be taken as statistically independent processes in the same sense that Wagner and Traud considered the corrosion reactions establishing a mixed potential [36]. However, Valverde and Wagner [37] considered that the release of a surface cation and a neighboring OH^- at a kink site are coupled processes.

The free energy barrier for the flow of ionic charge across the oxide electrode–electrolyte interface has an electrical contribution and consequently the reaction rate can be formally described by Butler–Volmer-type equations [38]. The cation current density corresponding to the process

$$ {}_2M^{z+} = {}_3M^{z+} \tag{8}$$

can be written [31]

$$\begin{aligned} j_c &= j_c^+ - j_c^- \\ &= k_c^+\, {}_2a_{(M^{z+})}\, {}_3a_X^r \exp[(\alpha_c F/RT)\phi_{2,3}] \\ &\quad - k_c^-\, {}_3a_{(M^{z+})}\, {}_3a_X^{r'} \exp[-(-z-\alpha_c)F\phi_{2,3}/RT] \end{aligned} \tag{9}$$

Here α is the effective cationic transfer coefficient, X is the complexing agent, and r and r' are reaction orders; the reaction order in cations is assumed to be one. The subscript indices in front of the chemical symbols and activity terms (a) denote electrolyte (3) and oxide (2), while (1) is reserved for the underlying metal. Another cationic reaction at the oxide–solution interface is the transfer of hydrogen ions

$$ {}_2H^+ = {}_3H^+ \tag{10}$$

The anion current density corresponding to transfer of oxygen ions between the oxide electrode and the aqueous solution is

$$ {}_3H_2O = {}_2O^{2-} + 2\,{}_3H^+ \tag{11}$$

is given by

$$\begin{aligned} j_a &= j_a^+ - j_a^- \\ &= k_a^+\, {}_3a^y_{(HO^-)} \exp(\alpha_a F\phi_{2,3}/RT) - k_a^-\, {}_2a_{(O^{2-})}\, {}_3a^{(2-y)}_{(HO^-)} \\ &\quad \exp[-(2-\alpha_a)F\phi_{2,3}/RT] \end{aligned} \tag{12}$$

where α_a is the effective anionic transfer coefficient and y is the reaction order with respect to HO^-. Hydroxide ions may also be involved in the complexation of cations at the oxide surface

$$ {}_2M(OH)_l^{(z-1)+} + m\,HO^- \longrightarrow {}_3M(OH)_{(l+m)}^{(z-1-m)+} \quad (13) $$

and for $X = HO^-$, $r = l + m$. Under steady state conditions, the net current at the interface is given by the sum of cation and anion partial ionic currents. Deposition and dissolution of the oxide requires that both cations and anions flow in the same direction while the corresponding electrical currents are opposite [31]

$$ {}_3H_2O + {}_2M^{z+} \longrightarrow {}_3HO^- + {}_3M^{(z+1)+} \quad (14) $$

Dissolution–deposition Tafel plots of the partial ionic current densities as a function of the oxide electrode potential are schematically represented in Fig. 2 [31,32].

The steady state potential E_{ss} of the oxide electrode at a particular composition MO_n lies between the formal equilibrium potentials with respect to oxygen, $E_O^o(n)$, and metal, $E_M^o(n)$, and its value is determined by the electroneutrality condition at a particular total current density

$$ (j = j_a - j_c) : \frac{j_a}{j_c} = \frac{-z}{2n} \quad (15) $$

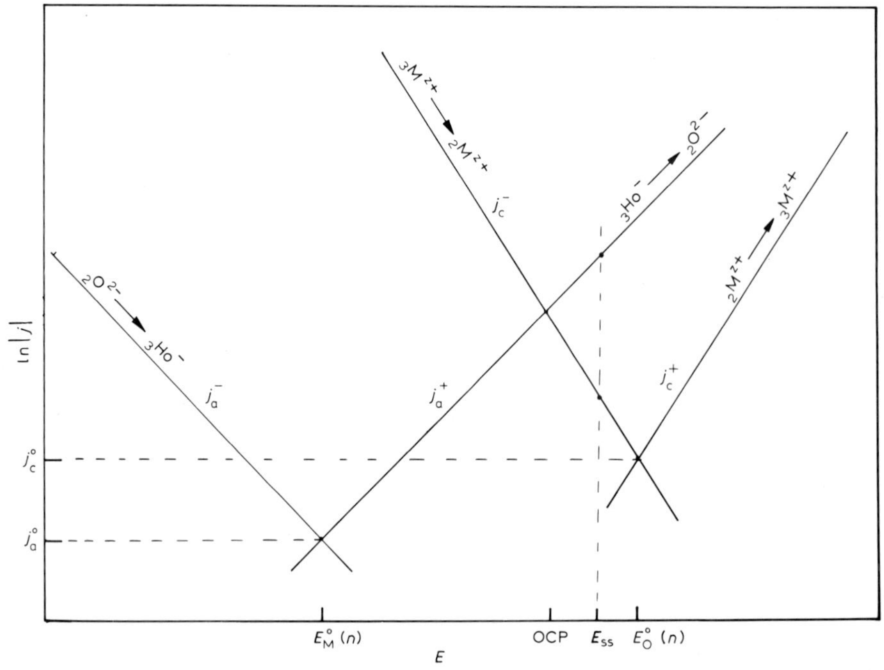

Fig. 2. Schematic Tafel plots for ITR for steady-state anodic oxide deposition between the equilibrium potentials with respect to metal E_M^o (n) and with oxygen E_O^o (n).

where the constant ratio of the ionic partial current densities is set by the oxide composition (n). Since $E^{\circ}_{\mathrm{M}}(n) > E^{\circ}_{\mathrm{O}}(n)$ in Fig. 2, deposition of the oxide would take place between the two equilibrium potentials for the metal and oxygen reactions ($j_c^- > j_c^+$ and $j_a^+ > j_a > j_a^-$). At the open circuit potential (OCP), steady state, currentless deposition (or dissolution) will occur at the rate $j_a^+ = j_c^-$ (or $j_a^- = j_c^+$) and $j_{\mathrm{total}} = 0$. It is interesting to note the case where cations of different oxidation states are transferred across the interface in parallel steady state ITRs. This the case of the passivating film on iron where the rate of reaction

$$ {}_2\mathrm{Fe}^{3+} = {}_3\mathrm{Fe}^{3+} \tag{16}$$

defines the corrosion rate of iron through the passive layer. The rate of ITR does not depend on electrode potential over a wide potential range (passivity range) because of the constant potential drop at the oxide–electrolyte interface [38], but it is proportional to the activities of SO_4^{2-} and Cl^- ions in solution and has a reaction order of zero in HO^- [42]. On this basis, the mechanism

$$ {}_2\mathrm{Fe}^{3+} + {}_3\mathrm{SO}_4^{2-} \longrightarrow {}_{\mathrm{ad}}\mathrm{FeSO}_4^+ \tag{17}$$

$$ {}_{\mathrm{ad}}\mathrm{FeSO}_4^+ \longrightarrow {}_3\mathrm{FeSO}_4^+ \qquad \text{rds} \tag{18}$$

$$ {}_3\mathrm{FeSO}_4^+ \longrightarrow {}_3\mathrm{Fe}^{3+} + {}_3\mathrm{SO}_4^{2-} \tag{19}$$

has been proposed in sulphate solutions [42]. On the other hand, the rate of iron(II) transfer

$$ {}_2\mathrm{Fe}^{2+} = {}_3\mathrm{Fe}^{2+} \tag{20}$$

decreases exponentially on increasing the electrode potential [41]; both cation and anion ITR in this system are statistically independent [31].

Ion transfer reactions have been studied in detail at a number of oxide electrodes, e.g. TiO_2 [43], Al_2O_3 [44], Fe_2O_3 [39–42], PbO_2 [43–45], MnO_2 [46], and $Ni(OH)_2$ [31]. On the basis of kinetic data summarized in Table 1, the mechanisms

TABLE 1

Kinetic parameters for various ion transfer reactions [see eqns. (9) and (12)]

	TiO_2 [a]	Al_2O_3 [b]
α_c	2.5	1.2
α_a	2	2
r	2 (acid)	1 (acid)
r	3 (alkaline)	2 (alkaline)
y	2	2

[a] Ref. 43.
[b] Ref. 44.

$$_2TiO^{2+} \;=\; _3TiO^{2+} \qquad \text{at pH} < 6 \qquad (21)$$

$$_2TiO^{2+} + {}_3OH^- \;=\; _3TiOOH^+ \qquad \text{at pH} > 7 \qquad (22)$$

and

$$_2AlOH^{2+} \;=\; _3AlOH^{2+} \qquad \text{at pH} < 9 \qquad (23)$$

$$_2AlOH^{2+} + {}_3OH^- \;=\; _3AlO^+ + H_2O \qquad \text{at pH} > 10 \qquad (24)$$

have been proposed for TiO_2 and Al_2O_3 where small changes at the oxide surface occur.

In the non-steady state, changes of stoichiometry in the bulk or at the oxide surface can be detected by comparison of transient total and partial ionic currents [32]. Because of the stability of the surface charge at oxide electrodes at a given pH, oxidation of oxide surface cations under applied potential would produce simultaneous injection of protons into the solution or uptake of hydroxide ions by the surface, resulting in ionic transient currents [10]. It has also been observed that, after the applied potential is removed from the oxide electrode, the surface composition equilibrates slowly with the electrolyte, and proton (or hydroxide ion) fluxes across the Helmholtz layer can be detected with the rotating ring disk electrode in the potentiometric-pH mode [47]. This pseudo-capacitive process would also result in a drift of the electrode potential, but its interpretation may be difficult if the relative relaxation of the potential distribution in the oxide space charge and across the Helmholtz double layer is not known [48].

5.2 KINETICS OF OXIDE DISSOLUTION

Oxide dissolution in aqueous electrolytes involves transfer of metal and oxygen ions to the solution. Since O^{2-} ions cannot be transferred into the solution, protonation must precede the ion transfer reaction, which leads to strongly pH-dependent dissolution rates [37]. The cation and oxygen transfer reactions may be regarded as statistically independent with

$$_2O^{2-} + {}_3H_2O \longrightarrow {}_2HO^- + {}_3HO^- \qquad (25)$$

followed by

$$_2HO^- \;=\; _3HO^- \qquad (26)$$

or, alternatively, complexing of the metal cation by an adjacent HO^- ion at an active site

$$_2M(OH)_l^{(z-1)+} \;=\; _3M(OH)_r^{(z-r)+} \qquad (27)$$

Oxide dissolution has been reviewed by Diggle [49]. According to the type of ion transfer rate-determining step, the different mechanisms can be classified into a chemical

$$MO + 2\,H^+ \longrightarrow M^{2+} + H_2O \qquad (28)$$

or an electrochemical type involving both ion and electron transfer. The latter mechanism may be oxidative

$$\tfrac{1}{2}\,Cr_2O_3 + 5\,HO^- \longrightarrow CrO_4^{2-} + \tfrac{5}{2}\,H_2O + 3\,e \tag{29}$$

or reductive

$$Fe_2O_3 + 3\,H^+ + 2\,e \longrightarrow 2\,Fe^{2+} + 3\,HO^- \tag{30}$$

The rates of dissolution of oxides have been measured by a number of authors, including Engell [50, 51], Vermilyea [52] and Valverde and Wagner [37]. Engell [50] has examined the dependence of the rate of FeO and Fe_3O_4 dissolution on electrode potential and Valverde [53] has investigated the dependence of the rate of dissolution of various transition metal oxides on the solution redox potential by using redox couples. The dissolution kinetics of mosaic, lithiated NiO single crystals has been studied by Lee et al. [54] in acid solutions and found to be strongly dependent on potential, the dissolution rates increasing with anodic polarization of the electrode. The rate of dissolution was determined by the rate of cation or anion transfer, depending on the electrode potential. Vermilyea [52] pointed out, however, that since the potential difference at the Helmholtz layer is constant at the oxide–electrolyte interface, the applied potential driving the oxide dissolution reaction would apply in the space-charge region. Valverde and Wagner [37] considered the factors affecting the rate of metal oxide dissolution, e.g. temperature, hydrogen ion concentration, non-stoichiometry, redox potential of the solution, complexing substances in the electrolyte, blocking of active kinks, photodissolution, and distribution of crystallographic planes. These authors also considered that transfer of cations across the oxide–electrolyte interface would proceed according to the multistep surface sequence

$$M^{z+}\ (\text{kink}) \longrightarrow M^{z+}\ (\text{ledge}) \longrightarrow M^{z+}\ (\text{ad}) \longrightarrow M^{z+}\ (\text{aq})$$

The rate of metal ion transfer from the oxide electrode to the electrolyte can be enhanced by complexing substances in the solution which can adsorb at active sites and weaken the M–O bonds. However, it is known that certain organic complexing ions can slow down the rate of dissolution by adsorbing and blocking active sites at the surface [37].

5.3 ELECTRON TRANSFER REACTIONS

A large number of metal oxides exhibit semiconductivity, e.g. ZnO, SnO_2, TiO_2, NiO, CuO, α-Fe_2O_3, and a few show nearly metallic properties, e.g. RuO_2, PbO_2, $LaNiO_3$, while many other oxides range in an intermediate category. The kinetics of redox electrode reactions, of which the simplest cases are electron transfer reactions (ETR), at oxide electrodes are affected by the density of charge carriers at the oxide surface and the potential distribution at the oxide–electrolyte interface. The relative position of the energy levels, both in the redox solution and in the oxide, determine the probability of electron transfer at the oxide electrodes. The positions of the band edges at the interface of the oxide with the electrolyte are relatively fixed by the acid–base properties of the oxide material and the Fermi level in the electrode and the electrolyte is adjusted at equilibrium by bending of the energy bands with respect to the bulk oxide. As a result, there is a blocking effect for charge to flow under anodic polarization at an n-type oxide semiconductor, while p-type oxide semiconductors exhibit a similar effect in the cathodic direction. In the ideal case where surface states are absent, changes in electrode–electrolyte potential difference occur across the oxide space-charge layer only, the potential drop in the Helmholtz double layer remaining constant because of fast hydroxylation at the oxide surface. Since the positions of the bands at the surface shift with electrolyte pH due to fast protonation, a reaction order in protons or hydroxide ions is commonly observed, even for redox reactions with pH-independent kinetics on metals. Since electron transfer reactions between an electrode and aceptor/donor electronic states in the electrolyte are isoenergetic processes, simple outer sphere electron transfer reactions can be used as probes of surface electronic levels at oxide electrodes. Current–potential curves are determined by the behavior of minority electrical carriers in the oxide electrode and therefore it is possible to estimate the type of carrier (hole or electron) in the semiconductor oxide, the potential distribution, involvement of surface states mediating the electron transfer, and so on.

It is very important that the probing redox system is stable and that no adsorption or chemisorption occurs at the oxide surface, since this may alter the potential distribution at the oxide–electrolyte interface. Low ionic charge on the species of the redox system is also preferred whenever possible to minimize double layer electrostatic effects. The hexacyanoferrate system generally involves weak interaction with the electrode and is stable over a wide pH range of 4–12, while its standard potential is essentially constant in that pH range. Although this couple has large ionic charge and presents uneven protonation of reduced and oxidized species at low pH values, it is probably the most widely used in this type of study. Other couples like Fe^{3+}/Fe^{2+} and Ce^{4+}/Ce^{3+} have been used but are less stable in electrolytes of higher pH. The open circuit potential which most oxide electrodes attain in contact with the redox system in aqueous solutions is very close to the equilibrium potential of the redox couple at a platinum or gold electrode in the same solution. This property has been verified for many couples in contact with a variety of oxide electrodes such as NiO, RuO_2, SnO_2, Fe_2O_3, $La_{(1-x)}Sr_xCoO_3$, $LaNiO_3$, and passive films on metals and it implies that the oxide electrode and the redox system are in electronic equilibrium.

Reduction of $Fe(CN)_6^{3-}$ ions at n-type semiconductor ZnO electrodes involves the conduction band with reaction orders 1 and 0 with respect to ferricyanide and ferrocyanide ions, respectively [55]. The electrode reaction can be described by a simple model of direct electron transfer from the conduction band with no surface states being involved. The potential distribution at the oxide–electrolyte interface has been obtained from capacitance–potential measurements and it consists of a potential drop at the space-charge region in the ZnO with constant energy levels at the surface at constant electrolyte pH. The flatband potential shifts 0.054 V per pH unit [14, 55, 56]. From the slope of linear log j vs. potential plots, $\alpha = 1$ has been obtained and plots of log j are also linear with band bending (as calculated from Mott–Schottky data and the applied potential), in good agreement with theoretical predictions of electron transfer at semiconductors [57].

Unlike the simple behavior of typical semiconductor oxides such as ZnO, many oxides used as electrode materials involve surface states due to the existence of transition metal cations at their surfaces with redox properties or involve tunnelling across the space-charge region for electron transfer to or from solution species. Other oxides may be considered hopping-type semiconductors with a narrow conduction band or even amorphous semiconductors with short-range periodicitiy. Memming and Mollers [16,58] studied the charge transfer kinetics of several redox couples in aqueous solutions such as Fe^{3+}/Fe^{2+}, Ce^{3+}/Ce^{4+}, and $Fe(CN)_6^{3-}/Fe(CN)_6^{4-}$ on n-type semiconducting SnO_2 electrodes of different carrier densities. On SnO_2, most redox processes occur via the conduction band by tunnelling through the space-charge region and variations in electrode potential are manifested only across the space-charge region. The redox potential, the potential distribution at the oxide–electrolyte interface, and the 60 mV per pH unit shift of the flatband potential were taken into account to explain the observed rates of electron transfer reactions at this oxide electrode. Slow kinetics were observed for the anodic oxidation reactions and the reaction order in protons for the hexacyanoferrate system was associated with the difference between flatband and redox potentials, which increases at higher pH values due to the relative positions of the energy levels. Memming and Mollers [16] also suggested the possiblity of obtaining absolute values of the reorganization energy of the redox couple in solution from experimental current–potential curves at SnO_2 electrodes of different carrier densities and capacitance–potential data.

Noufi et al. [59] investigated several redox couples at n-type TiO_2 single crystal electrodes in aqueous solutions to probe the electron transfer process at this oxide semiconductor–electrolyte interface, the formal equilibrium potentials of the couples spanning a wide range. Reduction currents positive to the flatband potential were observed, but most reduction reactions occurred close to the flatband potential. It was concluded that participation of surface states occurred in the electron transfer process from current–potential experimental curves ($\alpha_c < 1$); pure conduction band and tunnelling mechanisms were considered unlikely. All couples exhibited slower kinetics on TiO_2 than on platinum and cathodic peak potentials were observed to shift negative by 54 mV per pH unit, in agreement with the energy levels at the oxide semiconductor-electrolyte interface from capacitance measurements at different pH values. Danzfuss and Stimming [60] investigated the rate of ETR of Fe^{3+}/Fe^{2+} on amorphous $Fe_xTi_{(1-x)}O_y$ electrodes; anodic inhibition was found with appreciable currents for $x \geq 0.5$ and $\alpha < 0.1$. n-Type semiconducting properties were apparently exhibited by these mixed oxides since a cathodic Tafel slope was observed and blocking of the reaction in the anodic direction was attributed to the depletion layer at the oxide surface acting as a barrier for ETR. Higher activity in the cathodic direction has been observed with the iron doping and was explained by a shift in the flatband potential with composition. According to Fredlein and Bard [61], cyclic voltammograms of various electroactive species in acetonitrile have shown evidence of surface electronic states mediating ETR on n-type Fe_2O_3 at 0.9 and 1.8 eV below the conduction band edge. Reaction order studies (dark currents) suggested that reduction of $Fe(CN)_6^{3-}$ on α-Fe_2O_3 proceeds by a sequential two-step process from electrons in the conduction band via surface states [62].

Studies on lithiated NiO (p-type semiconductor) in aqueous solutions where Ni^{3+} concentration and the electrical conductivity are enhanced by Li doping of the bulk oxide, have shown asymmetry in the current–potential curves corrected for mass transport with the rotating-disk electrode. Conductivity in the system is explained by a narrow-band polaron model, the charge carrier being a hole in the Ni 3*d* level [63]. Yeager and co-workers [63,64] reported that the kinetics of the $Fe(CN)_6^{3-}/Fe(CN)_6^{4-}$ redox couple on NiO(Li) strongly depend on the pH of the electrolyte; the rate of reaction increases as the pH of the electrolyte increases. If all the potential drop occurs in the space-charge region of the p-type oxide, the cathodic current should reach a limiting value, which should correspond to the exchange current density in the absence of surface states. In acidic solutions, the reduction reaction was strongly inhibited and a potential-independent limiting current equal to the exchange current density j_o was apparent, as expected for a p-type semiconductor. The results are consistent with Mott–Schottky behavior of the NiO(Li)/aqueous solution interface, which indicates that a substantial potential drop operates across the oxide space-charge region [65–68]. At low anodic overpotentials, some inhibition is still observed at pH 2.9, but at higher oxidizing overpotentials, a 120 mV decade^{-1} Tafel slope indicates that changes in electrode potential occur across the Helmholtz double layer only and $\alpha_a = 0.5$. Convective diffusion anodic limiting currents were eventually reached at high overpotentials. In electrolytes of higher pH, the cathodic limiting current and the exchange current density extrapolated from the anodic Tafel line to the formal equilibrium potential were observed to increase. At pH < 6.5, the rate of hexacyanoferrate(II) oxidation was found to be on the fringe between pure diffusion and mixed kinetic–diffusion while in alkaline solutions (pH $= 12$), the reaction became purely diffusion-controlled and it was impossible to get electrokinetic information from the polarization data. Other cathodic reactions, such as reduction of Fe^{3+} ions and quinone, were found to be inhibited [64], but the analysis of the electrokinetic data led to the conclusion that specific adsorption of the reactants at the NiO electrode possibly occurred. The apparent standard rate coefficients, k_o, were observed to be two orders of magnitude lower than the corresponding rate coefficients on platinum in all cases and the difference was considered not to be due to any double layer effect since positively and negatively charged redox couples showed the same inhibition trend. For the hexacyanoferrate system, the apparent standard rate coefficient, k_o, on NiO(Li) at 25°C was $1.9 \times 10^{-3}\,cm\,s^{-1}$ in 0.5 M potassium sulfate, compared with the value of $1.3 \times 10^{-1}\,cm\,s^{-1}$ for platinum obtained by Randles and Sommerton [69] by an a.c. method. For the anodic oxidation of hexaquoiron(II) on the same electrode, Tench and Yeager [64] found $k_o = 8.3 \times 10^{-5}\,cm\,s^{-1}$ in comparison with $3 \times 10^{-3}\,cm\,s^{-1}$ for platinum in 1 M sulfuric acid which was reported by Gerischer [70]. Slower electrokinetics on NiO than on platinum at all pH values reflects the semiconducting behavior of the oxide electrode: (a) the density of charge carriers at the surface is controlled by the potential drop in the space-charge region, and by the surface redox transformations and (b) the rate of electron transfer to electronic levels in solution is modified due to the NiO surface contributing signficantly to the free energy of activation of the charge transfer process [64]. Yoneyama and Tamura [71] also examined several redox reactions at NiO electrodes in sulfuric acid solutions as a function of lithium doping level; enhanced anodic currents and cathodic saturation were reported. The electrical conductivity of the oxide electrode and the exchange current density for $Fe(CN)_6^{3-}/Fe(CN)_6^{4-}$, Fe^{3+}/Fe^{2+}, and Ce^{4+}/Ce^{3+} redox couples increased with the charge carrier density of the lithiated oxide.

Brenet and co-workers [72,73] studied the kinetics of the hexacyanoferrate redox system on polycrystalline spinel mixed ozides prepared by thermal decomposition on platinum grids: $Cu_{1-x}Mn_{3-x}O_4$ ($x = 0$ and 1.2), $NiMn_2O_4$, and $AlMn_2O_4$ at pH = 6. Partial inhibition of the cathodic reaction was observed with $Cu_{(1-x)}Mn_{(3-x)}O_4$ and $NiMn_2O_4$. Anodic convective–diffusion limiting currents were found in both cases, with faster kinetics for $CuMn_2O_4$ than for $NiMn_2O_4$, with $\alpha_a = 0.54$ and 0.14, respectively. The oxide with highest resistivity, $AlMn_2O_4$, exhibited currents two orders of magnitude lower. These authors observed that the reaction order was unity only for equimolar ratios of the hexacyanoferrate couple both at $CuMn_2O_4$ spinel oxide [73] and $La_{0.8}Sr_{0.2}CoO_3$ perovskite oxide [74,75] and attributed the phenomenon to adsorption or chemisorption of redox components on to the oxide electrode surface.

Perovskite, $LaCoO_3$, is a semiconductor with a band gap of only 0.3 eV which can easily be made either a p- or n-type semiconductor; the mobility of holes however is an order of magnitude higher than that of electrons. Conductivity can be further increased with strontium doping, which replaces La sites in the lattice and has a maximum at about $x = 0.5$ [76], but is high enough for $x > 0.2$ for use as an electrode material. Brenet and co-workers [73] and van Buren et al. [77] studied the electrode kinetics of the hexacyanoferrate couple on $La_{1-x}Sr_xCoO_3$ employing the rotating disk electrode technique. Inhibition of both cathodic and anodic processes was reported at $LaCoO_3$, although symmetric current–potential curves sensitive to rotation frequency were observed but with much slower kinetics than on the doped oxide. A limiting current (not diffusion-controlled) was observed for hexacyanoferrate reduction by Brenet and co-workers [73] for $La_{0.8}Sr_{0.2}CoO_3$ in 0.1 M NaOH solutions, while the anodic limiting current was observed to follow the convective–diffusion Levich equation; electrokinetic data are summarized in Table 2.

On the basis of electrode kinetic data obtained in 1 M NaOH for oxides in the range $0.1 < x < 0.5$, van Buren et al. [77] concluded that the solid state electronic properties of these mixed oxies have no observable effect on the electron transfer kinetics and the oxides can be considered as pseudo-metallic from an electrochemical point of view. There are, however, several observations that make this conclusion questionable: (a) Characterization data for the oxide electrode surfaces were not presented. In particular, the electrochemical real surface area (capacity, or BET) of the electrodes, and therefore comparison of apparent rate coefficients, are uncertain. (b) The

TABLE 2

Kinetic data for the hexacyanoferrate system at various oxide electrodes

x	k_o (cm s^{-1})	C_{NaOH^a} (mol dm^{-3})
$La_{(1-x)}Sr_sCoO_3$ in 1 M NaOH		
0.1	4.6×10^{-2}	0.65
0.2	3.1×10^{-2}	0.65
0.2	4.1×10^{-3}	0.49
0.3	3.1×10^{-2}	0.55
0.4	3.6×10^{-2}	0.65
0.5	3.1×10^{-2}	0.58
Pt	7.3×10^{-2}	0.50
$Cu_xMn_{(3-x)}O_4$ in 0.5 M K_2SO_4[b]		
1.2	3×10^{-3}	0.5
Na_xWO_3 in 0.5 M H_2SO_4[c]		
0.86	6.5×10^{-4}	0.48
0.75	6.2×10^{-4}	0.48
0.60	5.2×10^{-4}	0.48
0.53	8.0×10^{-5}	0.48
Pt	2.0×10^{-2}	0.50

[a] Ref. 77.
[b] Ref. 75.
[c] Ref. 91.

analysis of resistivity data reported by Obayashi and Kudo [76] shows that the resistivity varies by less than an order of magnitude in the range of compositions studied and much less in the range $0.2 < x < 0.6$. (c) Electrokinetic data obtained with the rotating disk electrode technique lacks accuracy when the apparent electrochemical rate coefficient is of comparable magnitude to the mass transport coefficient k_m.

It has been shown [78] that, for fast electrode kinetics at a hydrodynamic electrode

$$\ln [k_m/k_o] + \eta F/RT = \ln [\{(j_o/j) - 1\} - (D_o/D_r)^{2/3}\{(j_r/j) - 1\}x \exp (\eta F/RT)] \quad (31)$$

and for the rotating disk electrode

$$k_m = 1.55 D^{2/3} \nu^{-1/6} \omega^{1/2} \quad (32)$$

with D is the diffusion coefficient, ν the kinematic viscosity, and ω the rotation frequency (Hz). In eqn. (31), j_o and j_r are the convective–diffusion limiting currents. If $k_o \sim k_m$, then the right-hand side of eqn. (31) will involve differences between small numbers and kinetic data cannot be derived accurately. Furthermore, in the limit when $k_o >> k_m$, eqn. (31) becomes

$$\frac{(j_o - j)}{(j - j_r)} = \left(\frac{D_o}{D_r}\right)^{2/3} \exp (\eta F/RT) \quad (33)$$

and the surface concentrations of oxidized and reduced species follow the Nernstian equilibrium as is demonstrated in Fig. 3 for the hexacyanoferrate system at $La_{0.8}Sr_{0.2}CoO_3$ and $LaNiO_3$ with $D_o \sim D_r$.

$LaNiO_3$ can be regarded as a model oxide electrode since it shows bulk metallic conductivity with a narrow σ^* band [79] arising from the Ni^{+3}–O^{2-}–Ni^{+3} arrays between d(Ni) and p(O) electrons of neighboring lattice cations and anions which are not completely orthogonal in the ABO_3 perovskite structure. Non-stoichiometry in the La–Ni perovskite mixed oxides system arises from the co-existence of structurally related intergrowth phases [80]. These extended defects cannot be detected by X-ray examination which reveals only the average structural features over approximately 10^{18} unit cells. Some of these phases in the $La_{(n+1)}Ni_nO_{(3n+1)}$ system have been prepared ($n = 1,2,3$) and characterized [81]. Similar intergrowth phases have been postulated in other related perovskite oxide systems like $Sr_nTi_nO_{(3n+1)}$ [82] and $La_{(n+1)}Co_nO_{(3n+1)}$ [83] to accommodate non-stoichiometry. The extreme cases in the $La_{(n+1)}Ni_nO_{(3n+1)}$ homologous series are $LaNiO_3$ and La_2NiO_4, which nominally contain Ni(III) and Ni(II) ions, respectively. Unlike $LaNiO_3$, La_2NiO_4 is an n-type semiconductor with $\varrho = 0.5\,\Omega\,\text{cm}$ [84]. All these oxide materials show very similar surface redox electrochemistry in alkaline solutions, which is associated with the transition metal and this has been taken as indicative of a common surface

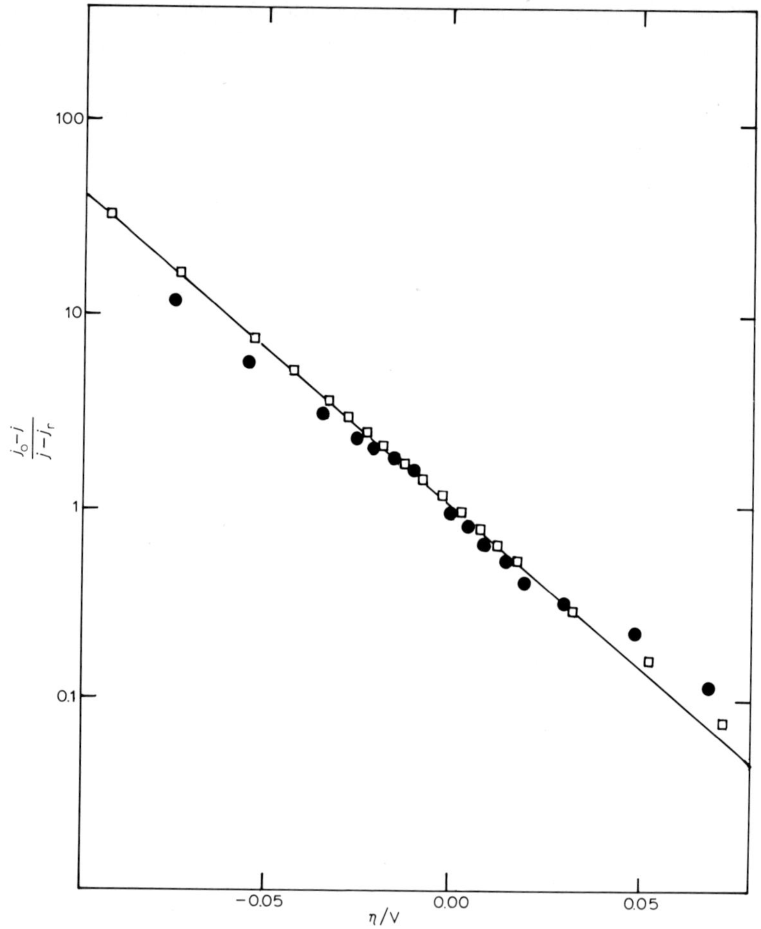

Fig. 3. Plot of $(j_o - j)/(j - j_r)$ vs. η according to eqn. (33) for 1 mM $Fe(CN)_6^{-3}/Fe(CN)_6^{-4}$ in 1 M NaOH. □, $LaNiO_3$; ●, $La_{0.5}Sr_{0.5}CoO_3$ electrodes.

modification of the oxides in contact with the aqueous electrolytes [48]. Every time the electrode potential is changed both in the absence and presence of a redox species in solution, a transient current is observed with this oxide materials as shown in Fig. 4. This phenomenon is not only characteristic of perovskite oxides but also of many other oxide electrodes and generally involves a long time scale which is associated with the large surface area (interfacial capacity) of the oxide and slow surface redox transformations. When redox species are present in solution, the measured current, j, is the sum of the Faradaic, j_f, and the surface pseudo-capacitive, j_s, current components.

$$j = j_f + j_s \tag{34}$$

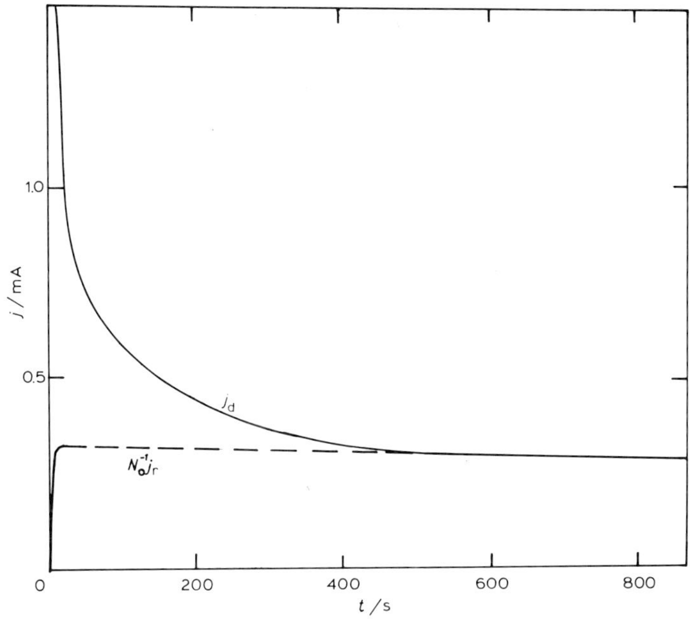

Fig. 4. Current transients at disk (j_d) and ring electrodes (j_r) for a potential step at $LaNiO_3$ disc electrode in 1 mM $Fe(CN)_6^{3-}$, 1 mM NaOH solution. Oxidation of $Fe(CN)_6^{4-}$ at the ring electrode under convective diffusion conditions [48].

If the ring electrode of the rotating ring–disc system monitors the flux of the oxidized or reduced components of the redox couple, then deconvolution of both Faradaic and pseudo-capacitive currents may be done [85]

$$j_f = N_o j_r \tag{35}$$

where N_o is the collection efficiency of the RRDE system. In Fig. 4, it can be seen that, while the $LaNiO_3$ disk current decays slowly, the contribution due to reduction of $Fe(CN)_6^{3-}$ species is constant from the beginning of the disk transient. Use of this technique may prove very useful in studies of oxide electrodes where electrochemical transformations at the surface may alter the electronic properties and therefore the kinetics of Faradaic reactions.

Cyclic voltammetry of $LaNiO_3$ in the presence of the hexacyanoferrate system and deconvolution of Faradaic and surface processes by means of the RRDE are depicted in Fig. 5. Steady-state results obtained for the hexacyanoferrate redox couple at $LaNiO_3$ in alkaline solutions were similar to those reported for $La_{0.5}Sr_{0.5}CoO_3$ with very fast kinetics, comparable with the reaction on platinum electrodes, and convective–diffusional limiting currents which obey the Levich equation are observed close to the equilibrium potential (Fig. 5).

In the potential range where a higher valence surface oxide is formed and before the onset of oxygen evolution (Fig. 5), the redox couple has such fast

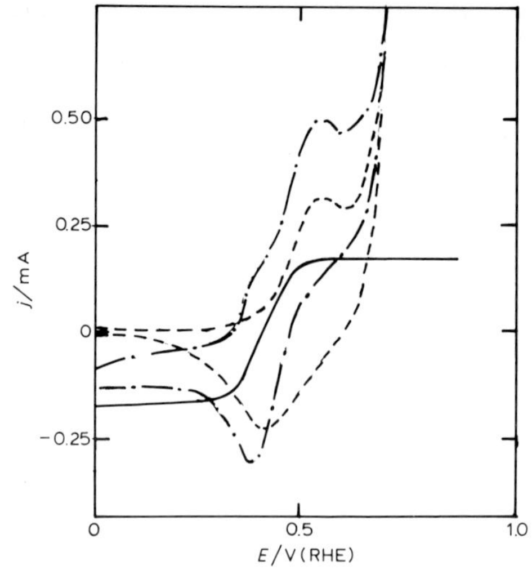

Fig. 5. Voltammetry of $LaNiO_3$ under same conditions as Fig. 3 (–●–●–), deconvolution of surface transformations (– – –), and extracted interfacial hexaxcyanoferrate system kinetics (———) with the RRDE system.

kinetics on $LaNiO_3$ that the Nernstian equilibrium for the redox system at the surface is observed and no kinetic data can reliably be obtained with the rotating disk technique at accessible rotation frequencies, as seen in Fig. 3. Impedance or transient techniques are more appropriate for the measurement of fast electron transfer kinetics ($k'_o > 10^{-2}\,cm^{-2}\,s^{-1}$), but such techniques are unsuitable for large surface area electrodes which also undergo electrochemical transformations.

Electrochemical reduction of the $LaNiO_3$ surface at potentials negative to 0 V vs. NHE changes the electronic properties of the oxide electrode. Matsumoto et al. [86] have interpreted the increase of electrode resistance in the potential range where reduction of Ni^{3+} to Ni^{2+} is apparent in terms of a decrease in the interaction between Ni^{3+} and O^{2-} in the Ni^{+3}–O–Ni^{+3} arrays. The reduction of the Fe(III)–EDTA system was studied in the potential range where the reduction of $LaNiO_3$ occurs in boric–borate buffer of pH = 8 [48]. The oxidation of Fe(II)–EDTA is slower than on platinum, but eventually the convective–diffusion limiting current is reached; however, the cathodic process is hindered on $LaNiO_3$ with a strong memory effect exhibited by the oxide electrode surface. If the electrode was previously polarized at −0.5 V (SCE), then lower current densities were observed at less negative potentials. Figure 6 shows the variation of the steady-state current ratio for $LaNiO_3$ with reference to platinum for the Fe(EDTA) redox system as a function of electrode potential.

A strong inhibition of the electrode kinetics on $LaNiO_3$ relative to platinum can be seen as the surface of the oxide is reduced, which also results in an increase of electrode resistance [86]. This behavior has been interpreted

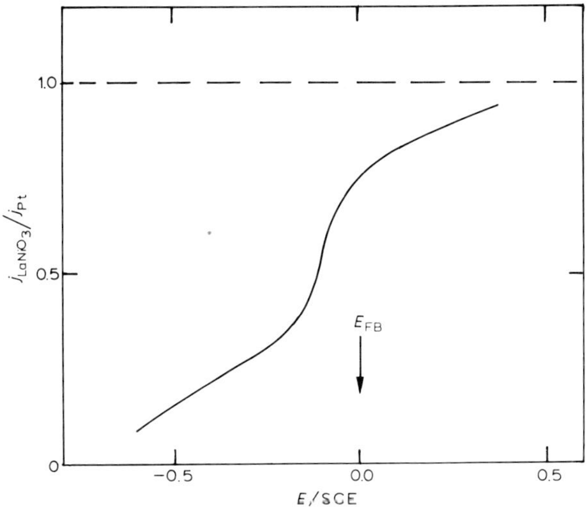

Fig. 6. Ratio of steady-state current densities at $LaNiO_3$ and platinum for the Fe(EDTA) redox system in boric–borate buffer of pH 8.4 as a function of electrode potential [48].

in terms of a transformation of the surface layer of $La_{(n+1)}Ni_nO_3$ oxides to give a poorly conducting, hydrated surface layer upon reduction of the oxide surface. The reduced surface hydrous layer apparently acts by partially screening the redox reaction from changes in the electrode–electrolyte applied potential and the hydrous layer is evidently thick enough such that direct elastic electron tunnelling from the bulk oxide to solution species is not operative.

According to Bockris and Otagawa [87], perovskite oxides of first row transition metals show p-type semiconducting surface properties as deduced from linear Mott–Schottky plots with negative slope for the oxides in contact with aqueous electrolyte and Hall effect measurements in dry oxide pellets. The interpretation of Bockris and Otagawa [87] that the surface of an n-type semiconducting oxide like $LaNiO_3$ with negligible band gap has p-type character is somewhat unusual. Although $LaNiO_3$ has metallic conductivity in the bulk (n-type), the shape of C^{-2} vs. E plots in alkaline solutions over a range of 0.4 V [87] is similar to that for p-type NiO(Li) in acid solutions [65–68]. Linear Mott–Schottky plots indicated that almost all changes of applied potential were manifested as a change in the potential drop across the space-charge region of the oxide, with negligible potential variation across the Helmholtz layer. At potentials above the flatband potential derived by extrapolation of C^{-2} vs. E plots obtained by Otagawa et al. [87], the electrokinetic behavior of hexacyanoferrate indicates that almost all the applied potential drop occurs across the Helmholtz layer. It has been suggested that NiO segregation at the surface of $LaNiO_3$ occurs as the perovskite oxide is reduced [88]. It is possible that the modified surface hydrous layer which has lost the chemical and electrical properties of

$LaNiO_3$ shows two-dimensional semiconducting character which would be responsible for the C^{-2} vs. E plots. Similar results have been reported for the reduction of $Fe(CN)_6^{3-}$ on PbO_2 in 0.5 M K_2SO_4 solutions by Lovrecek et. al [89]. While diffusion-controlled anodic oxidation currents on PbO_2 are the same as on platinum, at potentials below 0.175 V vs. SCE where reduction of the oxide surface becomes noticeable, cathodic currents fall to very small values and the j–E curve shows a maximum.

The kinetics of several redox couples have been investigated at sodium tungsten bronzes: $Fe(oxalate)^{3-}/Fe(oxalate)^{4-}$, $F(CN)_6^{3-}/Fe(CN)_6^{4-}$, $W(CN)_8^{3-}/W(CN)_8^{4-}$, and Fe^{3+}/Fe^{2+} in acid solutions [90,91]. Simple ETR at sodium tungsten bronzes, Na_xWO_3, with the perovskite structure are fast and are influenced by the sodium bulk content of the electrode as can be seen in Table 2. Unfortunately, the kinetic pattern is not simple because the variation of ETR rate coefficients with sodium content is not the same for each couple [91]. A qualitative interpretation of the ETR kinetic results has been attempted in terms of the density of electronic states at the Fermi level of the oxide electrode [90].

In other work, Trasatti and co-workers investigated the kinetics of ETR on RuO_2 films (essentially metallic conductors) thermally deposited on tantalum and platinum and reported reversible behavior for the iron hexaquo redox system [92].

5.3.1 Electron transfer reactions at surface films and passive layers

Surface layers on metals obtained by thermal or anodic oxidation and by chemical or electrochemical deposition strongly influence the rate of electron transfer at such substrates. For instance, inhibition of the standard rate coefficient of the $Fe(CN)_6^{3-}/Fe(CN)_6^{4-}$ system on nickel subjected to varying degrees of thermal oxidation has been reported [93]. Different types of surface film have been investigated, e.g. passive layers consisting of hydrous amorphous hydroxide films on the corresponding metals with varying stoichiometry, deposition layers on inert substrates, e.g. Υ-FeOOH/Pt [94,95], sandwich oxides, i.e. $Au/Au_2O_3/\Upsilon$-FeOOH [97], and diodes, i.e. $Ti/TiO_2/Au$ [96]. Electron transfer reactions on surface layers and passive films exhibit different kinetics from those observed on the underlying metal. Electronic equilibrium with the redox solution is observed since the formal equilibrium potential is established. Strong inhibition of ETR is observed on these layers and, generally, the exchange current densities are lower than on bare metals by several orders of magnitude and the current–potential curves are commonly asymmetric ($\alpha_a + \alpha_c \neq 1$). Besides the influence of variables such as concentration of the redox system in solution, electrode potential and temperature, the rate of ETR on surface films may also depend on (a) the thickness of the film for thin layers, (b) the pH of the electrolyte, even for redox couples with rates of ETR independent of pH on bare metals, (c) the structure of the surface layer, degree of hydration, ageing, and so on, (d) the nature of the underlying metal and electronic equilibrium at the metal/film/electrolyte interface, and (e) the potential range of the redox couple. The study of a large number of systems [98–105] has shown that several ETR mechanisms at surface layers and passive films are possible and that electron transfer at these interfaces generally proceeds by combination of more than one of these mechanisms.

Several theoretical models have been developed in recent years to explain the experimental observations [102–106]. The most important ETR mechanisms are as follows.

(i) Direct elastic tunnelling of electrons from the underlying metal to redox states in solution for very thin films. The film represents a barrier and its thickness determines the tunnelling probability. The electron transverses the barrier in a single step with no loss of energy. The exchange current densities strongly decrease with film thickness and with the average barrier height but transfer coefficients remain constant.

(ii) For thicker surface films (ca. 2–10 nm), the rate of ETR becomes independent of thickness and participation of electron terms or bands in the surface layer must be taken into consideration. With the exception of a few oxide layers such as Cu_2O and NiO, most passive layers behave like n-type semiconductors with donor terms in oxygen vacancies and the dominant contribution to ETR is through the conduction band. The exchange current density and transfer coefficient are determined by the electronic properties of the film which is regarded as exhibiting semiconducting properties in quantitative calculations.

(iii) Resonance tunnelling with short-lived intermediate states in the film if impurity levels are available in the film. The tunnelling probability of an electron is enhanced if it can be scattered by localized terms of similar energy within the barrier. ETR of an Fe^{3+}/Fe^{2+} couple on platinum-doped, passive titanium is an example of this mechanism [101].

5.4 PROTON INSERTION REACTIONS IN ELECTROCHEMICALLY FORMED, ELECTROCHROMIC OXIDE FILMS

Metal oxides which undergo proton insertion reactions find extensive application in batteries and are currently being investigated as potential electrochromic materials. The properties of battery oxides, e.g. manganese dioxide [107–110] and nickel [111–114] have been extensively reviewed in the literature and will therefore not be discussed here. Rather, the properties of electrochemically grown, electrochromic oxide films will be described since this is a relatively new and interesting field.

Metal oxide films which change color in an electrolyte with change in applied potential [11, 115–123, 127, 128, 137] have attracted a lot of attention in the past 15 years or so because of their potential application in electrochromic displays. Tungsten trioxide was the first oxide to receive significant attention in this regard [115–119] and, later, Ir oxide films [11, 120, 121, 127,

128] proved more interesting from the viewpoint of fundamental electrochemical properties, e.g. potential–pH response, and will therefore be preferentially discussed here.

5.4.1 Iridium oxide

The anodic behavior [121, 122, 130–134] of iridium in aqueous solution differs from that of platinum in that cyclic voltammograms (Fig. 7) display very little hysteresis between the oxidation and reduction processes in the oxide region and that the peaks in this region increase readily in acid solution under continuous potential cycling. Earlier interpretations of this behavior were in terms of reversible chemisorption of oxygen species on the iridium surface. For example, Böld and Breiter [130] interpreted the enhancement of charge in the oxide region in terms of an expanded-lattice model: they attributed the increase to the expansion of the surface layer of the metal lattice accompanied by penetration of oxygen into the expanded metal lattice. However, in view of the unaltered charge capacity of the hydrogen region on multicycling, this model is clearly inadequate. In a subsequent investigation, Otten and Visscher [132, 134], proposed a pit model for the oxide region on the basis of both voltammetric and ellipsometric measurements. However, like the expanded-lattice model, of which the pit model may be considered a special case (reaction occurring at active sites on the electrode surface), this model did not explain the invariance of the hydrogen region of the voltammogram with potential cycling. Kurnikov et al. [135] observed oxygen coverage up to 20 monolayers and concluded

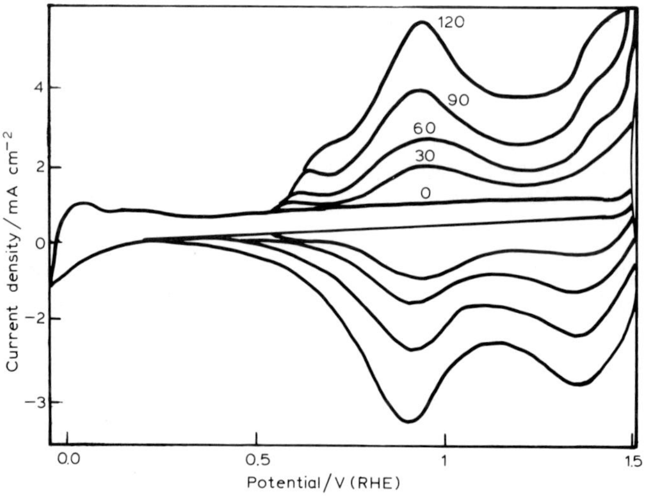

Fig. 7. Cyclic voltammograms [− 0.05 to 1.50 V(RHE), 50 mV s^{-1}] for Ir in 1 M H_2SO_4 at 25°C after being subjected to various numbers of potential sweeps [− 0.05 to 1.50 V(RHE) at ca. 3 V s^{-1}]. The numbers refer to the number of potential sweeps [11].

that the symmetrical anodic and cathodic peaks of the voltammogram resulted from the reversible formation and reduction of a phase oxide rather than of a chemisorbed film.

Detailed investigations carried out by Rand and Woods [122] and by Burke and co-workers [121, 136] confirmed the irreversible formation of a multilayer phase oxide on the iridium surface during potential cycling. Ellipsometry [123] and electron microscope [138] studies clearly show the presence of this oxide film, which can reach thicknesses of several microns. Rand and Woods [122] and Burke and co-workers [126, 136] recognized that the main anodic and cathodic peaks at ca. 0.97 V (RHE) (Fig. 7) were associated with stoichiometry changes within the oxide layers. The reaction giving rise to this stoichiometry change will be discussed in some detail later. The electrochromic effect arises because the reduced form of the Ir oxide films [i.e. when the films are poised at potentials below the main redox peak occurring at ca. 0.97 V (RHE) (Fig. 7)] is transparent to visible radiation, while the oxidized form is strongly absorbing, and exhibits a deep blue color [120, 121].

The importance of potential sweep limits on the processes involved in thick oxide growth at Ir was demonstrated by Burke and co-workers [121, 136]. The optimum range at moderate cycling frequencies (1–10 Hz) in acid involves a lower limit of about 0.025 V and an upper limit of 1.55 V (RHE). According to these authors, exceeding this upper limit results in dissolution of the oxide film, while exceeding the lower limit of 0.025 V reduces the rate of oxide formation.

The probable mechanism of film growth may be described as follows [11, 121, 136, 139]. In the initial stages of oxidation of the Ir metal surface, the metal surface becomes covered with a layer of hydroxy species. This surface layer is transformed ultimately, probably via place-exchange reactions [140], to an oxide layer which is essentially anhydrous and of the order of a monolayer or two in thickness. The oxide film at this point may be considered to be highly passivating, as shown, for example, by the absence of multilayer oxide formation on holding the electrode potential at anodic potentials, e.g. 1.5 V (RHE). (However, this may be also due to the highly conducting nature of the film at the high anodic potentials which disallows ionic transport necessary for film growth). During the cathodic potential sweep, the oxide is subjected to partial reduction only and the remnant of the previously compact oxide is left in a disrupted, active state. In the subsequent anodization period, oxidation of displaced Ir metal atoms, e.g. those left after reduction of the previous compact layer, and of some of the underlying metal occurs. The small amount of unreduced oxide material which remains at the completion of each reduction period accumulates to give the multilayer oxide film. This outer, multilayer film is quite hydrous and stable to reduction compared with the inner, compact layer. McIntyre et al. [141] estimated a mean density of 2.0 g cm^{-3} for these films, compared with 11.68 g cm^{-3} for bulk crystalline IrO_2. Because of the disperse nature of the multilayer film, growth of this layer continues at a decreasing rate on the underlying metal surface. Dissolution–precipitation processes do not seem to play a significant role in the formation of the film, as indicated by, for example, the lack of dependence of rotation frequency on the rate of growth of the multilayer film at an Ir rotating disk electrode [139].

Perhaps the most interesting fundamental electrochemical property of the hydrous Ir films is the potential–pH behavior of the pair of reversible peaks at 0.97 V (RHE). The reaction originally suggested by Buckley and Burke [121] to account for these peaks was of the form

$$IrO_2 + H^+ + e = IrOOH \tag{36}$$

McIntrye et al. [141] obtained compelling evidence for an Ir(III)/Ir(IV) transition using, for example, combined gravimetric and coulometric measurments. The theoretical reversible potential for eqn. (36) shows a pH dependence of 59 mV per pH unit. However, the peak potential values of the reversible voltammetric peaks for cycled Ir have been found to exhibit a potential–pH dependence of close to 88.5 mV per pH unit [i.e. 1.5 (2.303 RT/F)] [11, 12, 142, 143]. Burke and Whelan [12, 142] asssociated this behavior with increasing acidity of the oxyspecies due to increasing degree of oxidation of the central metal ion. These authors suggested that the oxidized state of the hydrous layers might be an aggregate of the octahedral species $H_2Ir(OH)_6$, existing in the ionized form $Ir(OH)_6^{2-}$ or possibly $[IrO_2(OH)_2 \cdot 2H_2O]^{2-}$. Similarly, the reduced form was regarded as an aggregate of an $H_3Ir_2(OH)_9$ species, again existing in the ionized form $Ir_2(OH)_9^{3-}$, or possibly $[Ir_2O_3(OH)_3 \cdot 3H_2O]^{3-}$. Thus, a more complete form of eqn. (36), which appears to be in agreement with the potential–pH data, is [12]

$$2\,\{[Ir(OH)_6]^{2-} \cdot 2H_f^+\} + 3\,H_s^+ + 2\,e = [Ir_2(OH)_9]^{3-} \cdot 3H_f^+ + 3\,H_2O \tag{37}$$

or

$$2\,\{[IrO_2(OH)_2 \cdot 2H_2O]^{2-} \cdot 2H_f^+\} + 2\,H_s^+ + 2\,e = [Ir_2O_3(OH)_3 \cdot 3H_2O]^{3-} \cdot 3H_f^+ + 3\,H_2O \tag{38}$$

Delocalized H^+ counterions are denoted with a subscript f, while H^+ species which transfer between tbe film and bulk solution during the redox reaction are identified by the subscripts s. Thus, for each electron injected into the film there is a simultaneous transfer of one proton, i.e. H_s^+, from the solution bulk into the hydrous oxide material, while at the same time there is a transfer locally of 1.5 protons into the ligand sphere of the central metal ion for each electron added to the latter. Proton transport is likely to occur via a Grotthus-type mechanism in these films and is much more likely than OH^- movement as suggested by other authors [144].

In general, the hydrous oxide layer is regarded as a porous, cross-linked gel or polymer, the polymer strands bearing a negative charge with the counterion present in the aqueous phase permeating the disperse oxide layer [12]. It is evident that the principal difference between the hydrous and anhydrous oxides is that the open, disperse structure of the former enables interaction to occur between most of the Ir oxyspecies in the film, not just those at the external surface as in the case of the anhydrous material, and solution species, especially H^+ and OH^- ions. In support of the Ir species depicted in eqn. (37), Burke and Whelan [12] point out that iridic acid, $H_2Ir(OH)_6$, is analogous to platinic acid, $H_2Pt(OH)_6$, for which not only are structural details available, but a number of salts, $M_2Pt(OH)_6$ (M = Li, Na and K), are also known [145]. Further, hydrous oxide layers formed on platinum, which reduce directly to the metal in acid electrolyte [146], have been shown to exhibit the same type of potential–pH behavior observed for the hydrous Ir films. However, in the absence of spectroscopic evidence, eqns. (37) and (38) have to be regarded as tentative assignments.

Yuen et al. [143] recently studied the pH dependence of the open-circuit potential and voltammetric peaks of anodic Ir oxide films (AIROF's) and sputtered Ir oxide films (SIROFs). These workers observed a 85 mV per pH unit variation of the main redox peaks for the AIROF films, and a 59 mV per pH unit variation in the case of the SIROFS. An electronic density-of-states distribution constructed for the AIROF films was found to be different in acid and base, which may indicate a pH-driven structural transformation for these films, which is reflected in a super-Nernstian pH response. On the other hand, a less structured electron density-of-states distribution, which is essentially pH-independent, appears to be applicable in the case of the SIROF layers.

Other aspects of electrochemically grown Ir oxide films are the question of conductivity changes and the invariance of the charge in the hydrogen adsorption/desorption and double layer regions of the voltammogram (Fig. 7). Glarum and Marshall [147] have reported that the a.c. conductivity of the Ir films increases about four orders of magnitude on going from the reduced state ($\rho < 10^{-2}$ $ohm^{-1} cm^{-1}$) to the oxidized state. Results obtained by Gottesfeld [148] in a study of the $Fe(CN)_6^{4-}/Fe(CN)_6^{3-}$ redox couple at the hydrous Ir films showed total inhibition of the Faradaic reactions at the reduced forms of the films and are in agreement with such a change in conductivity. Relatively simple band structure considerations [139,149] involving the Ir(III) and Ir(IV) oxidation states as treated, for example, by Mattheis [151] and Goodenough [152], possibly coupled with a lack of non-stoichiometry in the fully reduced [i.e. essentially Ir(III)] layer [142], have been invoked to explain the difference in conductivities between the oxidized and reduced forms of the Ir layers. The monolayer type, compact oxide formation/reduction processes, along with the hydrogen adsorption/desorption and double layer charging processes occur as usual at the Ir surface beneath the hydrous oxide layer (Fig. 7) [150]. These processes can occur because of the large amount of electrolyte available throughout the disperse oxide layer.

5.4.2 Other oxides

To date, tungsten trioxide is the most widely studied material from the viewpoint of practical electrochromic properties. Besides its potential use in displays [116], WO_3 is also being investigated as a potential electrochromic coating for windows for controlled radiant energy transfer in buildings [153]. WO_3 films have been formed by a number of methods, e.g. evaporation [117–119,154], anodization [155–157], and thermal methods [157]. Burke et al. [158] have shown that such films can also be grown by potential cycling in acid electrolyte. The change in optical properties with potential is generally associated with a reaction of the form

$$WO_{(3-y)} + x\,M^+ + x\,e \rightleftharpoons M_xWO_{(3-y)} \qquad (39)$$

where $0 < x \leq 0.5$ and $y \leq 0.03$ [116], and M^+ is H^+ or, for example, Li^+, Na^+ in non-aqueous or solid electrolytes. For a number of reasons, including improved stability, much of the research on WO_3 films is centered on the use of Li^+ as an insertion ion in electrolytes other than aqueous systems.

Hydrous oxides films, many of them with electrochromic properties, have been grown on a number of other metals using potential cycling methods, for example Rh [126] and Ni [159] in alkaline solution. Film formation seems to occur via a similar mechanism to Ir on the various metals under potential cycling conditions. Further, the films exhibit the same type of potential–pH response [150,160] as that described above in the case of Ir.

6. Oxygen electrode

The oxygen electrode is of major importance to energy conversion, storage, and conservation. Consequently, it has been the subject of vigorous investigation over the last fifty years or so. Oxygen electroreduction is employed in fuel cell systems and in metal–air batteries and will probably be used as the cathodic process in chloralkali cells. Oxygen electrogeneration is involved in water electrolysis and other industrial electrolytic processes, as in the recharging of metal–air cells. The oxygen electrode is a highly irreversible system in aqueous electrolytes and this results in substantial energy losses in electrochemical cells involving oxygen electrodes.

The oxygen electrode reaction with accompanying standard potentials at 25°C may be represented as

$$O_2 + 4\,H^+ + 4\,e^- \rightleftharpoons 2\,H_2O \qquad E^0 = 1.229\text{ V (NHE)} \qquad (40)$$

in acid solutions and

$$O_2 + 2\,H_2O + 4\,e^- \rightleftharpoons 4\,OH^- \qquad E^0 = 0.401\text{ V (NHE)} \qquad (41)$$

in alkaline solution. The reversible potential for this system has a pH dependence of about 59 mV per pH unit and a dependence of approximately 14.8 mV per tenfold change of O_2 pressure. Tarasevich et al. [161] have discussed the component reactions and the thermodynamic data of the oxygen electrode system in detail.

The reversible oxygen electrode potential is exceedingly difficult to attain experimentally. The difficulty in obtaining the reversible potential stems from the irreversibility of the oxygen electrode reaction system and its extremely low exchange current density (e.g. $j_o \sim 10^{-10}$ A cm^{-2}). Other side reactions, even though possessing slow kinetics, may compete with eqns. (40) and (41) in establishing the rest potential of the system. Usually, the rest potentials of even highly active platinum electrodes in ultra-pure aqueous acid or alkaline solutions for p_{O_2} atm and ambient temperature is only a little in excess of 1.0 V (NHE). The reversible potential has been most notably reported by Bockris and Huq [162], Hoare [163], Watanabe and Devanathan [164], and Burshtein et al. [165] These workers generally found it necessary to employ strongly oxidized platinum electrodes, either by heating in oxygen or contacting with concentrated HNO_3, coupled with careful purification of the electrolyte. Various theories have been advanced to explain the nature of the commonly observed, irreversible rest potential [i.e. $E \sim 1.0$–1.1 V (NHE)] at noble metal electrodes especially platinum in purified, O_2-saturated aqueous solution where impurities are not considered to play an important role. These theories are based on the concept of either (a) the presence of an oxide film on the metal surface [166,167] or (b) that the observed potential is in actuality a mixed potential [168]. In the mixed potential approach, the 4-electron reduction of O_2 occurs in conjunction with one or more anodic reactions, with the rest potential having a value between the equilibrium potentials of these reactions. Reactions such as oxidation of hydrogen peroxide accumulated in solution due to 2-electron reduction of O_2 [169] and oxidation of principally organic impurities from the solution [162,170], have been suggested for the complementary anodic reactions of the mixed (rest) potential at noble metal electrodes in O_2-saturated solution. There are now indications that the rest potential of platinum in aqueous acid and alkaline solution is a mixed potential involving platinum dissolution as the principal complementary anodic reaction. There is considerable evidence that platinum undergoes dissolution in acid and alkaline solutions of potentials ~ 1.0 V (RHE) [171, 172].

The ability to chemisorb oxygen dissociatively would seem to be a necessary property of an electrocatalyst in order for it to exhibit the reversible potential of the oxygen electrode in an oxygen-saturated electrolyte. This concept has been a motivating factor in the work by Tseung and co-workers [173–175] on perovskite oxides and doped nickel oxides for oxygen electrocatalysis. Tseung's group have argued that magnetic considerations are important in catalysis on oxides. The oxygen molecule is paramagnetic and has two unpaired electrons with parallel spins in antibonding π orbitals. Thus, electrons transferred at the time of O–O bond rupture from the catalyst surface to the adsorbed oxygen species must necessarily be of parallel spin. According to this view, ferromagnetic or paramagnetic oxides should be good catalysts for establishing the reversible oxygen electrode potential.

Tseung and Bevan [173] have reported observing the reversible potential for the oxygen electrode at the perovskite $La_{0.5}Sr_{0.5}CoO_3$ in 45% KOH at 25°C, which is a temperature above the Neel point for this oxide. However, the $La_{0.5}Sr_{0.5}CoO_3$ electrode had to be immersed in 75% KOH overnight at 220°C before the desired potential was achieved. The O_2 partial pressure dependence followed the Nernst equation for a 4-electron process. This result has been questioned and not reproduced by van Buren et al. [176]. The latter authors suggested that a non-equilibrium phenomenon involving slow diffusion of oxygen ions inside the perovskite electrode was responsible for the "reversible" behavior observed by Tseung and Bevan [173]. However, Calvo et al. [48] have disputed the conclusions of van Buren et al. regarding the involvement of slow lattice oxygen ion diffusion at ambient temperatures (see Sect. 6.2). Calvo et al. [48] suggested that the oxide aqueous solution interfacial properties determined the electrochemistry of the perovskite rather than bulk features of the catalysts.

Tseung et al. [174, 175] also studied the open circuit behavior of lithiated nickel oxide electrodes. Nickel oxide undergoes a transition from its normal antiferromagnetic state to a paramagnetic state at the Neel point, T_N, at 523 K [177]. However, it is known that the Neel point can vary significantly with intrinsic surface area, the T_N value decreasing with increase of surface area, i.e. decreasing particle size [178]. Tseung et al. [174,175] observed potentials corresponding to the reversible O_2 electrode potential at NiO at temperatures above the Neel point. Below the Neel point, where the NiO becomes antiferromagnetic, the potentials observed were appreciably lower than the theoretical reversible potentials. However, in view of the variation of the Neel point with particle size [178,179], the correlation of Tseung et al.'s electrochemical results with magnetic properties have to be treated with some caution. This approach remains an important contribution to oxide electrocatalysis, however.

The marked irreversibility of the oxygen evolution and reduction reactions in aqueous solutions has imposed severe limitations on the mechanistic information which can be obtained for both reactions. In general, at the current densities normally employed for kinetic studies, the current–potential data are insensitive to the back reaction, which normally occurs early on in the multi-step reaction sequence. Further, the reduction and oxidation processes are usually studied only at widely separated potentials. Thus, the surface conditions, whether in the case of metals or bulk oxides, probably differ sufficiently such that the reduction and oxidation pathways may not be complementary. The situation is complicated further by the large number of possible pathways for both reactions.

During oxygen reduction on most electrode surfaces, hydrogen peroxide is generated, with exhange current densities for the O_2–peroxide couple which are relatively large compared to that of the O_2–water reduction reaction. Therefore, a considerable amount of information has been obtained regarding the pathways involved in the O_2–hydrogen peroxide reaction.

6.1 OXYGEN EVOLUTION

Much of the published work on the electrocatalysis of oxygen evolution deals with the use of metallic electrodes. Metal surfaces, however, always contain an oxide film under conditions of oxygen evolution. Such oxide films begin to form at potentials below that of the reversible O_2 electrode potential. The films undergo further transformation, e.g. the so-called place exchange process discussed by Conway and co-workers [140] in the case of platinum, and generally complicate the study of the kinetics of the oxygen evolution reaction by the introduction of time-dependent effects in the data. It is not possible in a short review such as this to go into the details of oxygen evolution on metallic electrodes. A number of review articles may be referred to which summarize the kinetic data available for oxygen evolution on metals [161,168,180,181].

An increasing amount of attention is being given to oxides as possible anodes for oxygen evolution because of the importance of this reaction in water electrolysis. In this connection, numerous studies have been carried out on noble metal oxides, spinel and perovskite type oxides, and other oxides such as lead and manganese dioxide. Kinetic parameters for the oxygen evolution reaction at a variety of single oxides and mixed oxides are shown in Table 3.

In general, a good electrocatalyst should exhibit a low Tafel slope value, b, but a high exchange current density, j_o. It is difficult, however, to compare apparent j_o values for oxide electrocatalysts due to differences in electrode preparation between research groups, which results in, for example, variable or unknown intrinsic surface area values. RuO_2 and $NiCo_2O_4$ are generally considered to be amongst the most active electrocatalysts for oxygen evolution, with both of these materials exhibiting b values of ca. 40 mV decade^{-1}. However, most oxide catalysts to date which exhibit fairly good activity for oxygen evolution tend to undergo some dissolution in practical water electrolysis conditions, e.g. 30% KOH at 80°C. Besides the oxide electrodes prepared by common methods such as thermal decomposition of films on to inert substrates and ceramic or pressed pellet, other types of oxides have recently attracted fundamental interest as electrocatalysts, such as reactively sputtered films (IrO_x) [182, 183] and electrochemically grown hydrous films on Ir [123] and Rh [184].

TABLE 3

Kinetic parameters for the oxygen evolution reaction on various oxides

Oxide	Electrode[a]	Electrolyte[b]	Tafel slope (mV $decade^{-1}$)	Reaction order[c]	Ref.
RuO_2	TD	Acid	35–40 (low η)	−1	193, 200, 228–230
	TD	KOH	40–67	1	227
	TD	NaOH (80°C)	41		202
	SC	Acid	75–105		237
IrO_2	TD	H_2SO_4	50–56	−1	241
	TD	KOH	40	2	241
	S	H_2SO_4	40		182, 183
	E	H_2SO_4	50		123, 249
RhO_x	TD	KOH (80°C)	67		242
	TD	NaOH	50	1.2	257
	E	NaOH	50	1.2	184, 257
PtO_2	TD	KOH	59 (low η)	2	252
			114 (high η)	1	252
		H_2SO_4	80°C		255, 256
PdO_x	TD	KOH (80°C)	67		242
	TD	H_2SO_4			445
β-MnO_2	C	H_2SO_4	110	0.1	315
	C	KOH	110	0.9	315
β-MnO_2 (+ α-Mn_2O_3)	TD	H_2SO_4	110	0.1	214
	TD	KOH	110	1	214

β-PbO_2	E	H_2SO_4	120–160		321, 323, 324
α-PbO_2	E	H_2SO_4	70		321
NiO	TD	KOH (80°C)	43–71 (low η)		242, 334
NiO_x	E	Base	40–50 (low η)		329, 331, 332
NiM_2O_4	TD	KOH (85°C)	40 (low η)		271, 272
(M = La, Pr, Nd)					
			120 (high η)		271, 272
Co_3O_4	TD	KOH	42–48	1.3	309
Co_3O_4	TB	KOH	60		296
Co_3O_4 (10% Li)	TB	KOH	60	1.76	296
$NiCo_2O_4$	TB	KOH	33–48 (low η)		297, 299, 301, 303
			80–120 (high η)		297, 299, 303
MFe_2O_4	TD	KOH	110–115	~1	312
(M = Ni, Mg, Co)					
$Li_{0.5}Fe_{2.5}O_4$	TD		110	1.18	312
$Ni_xFe_{(3-x)}O_4$	TB	KOH	38–50		314
$La_{(1-x)}Sr_xMnO_3$	C	KOH	130–140	1	277
$La_{0.7}Pb_{0.3}MnO_3$	SC	KOH	95	0.91	276
$La_{(1-x)}M_xMnO_3$	C	NaOH	125–130	0.6–0.65	265
(M = Sr, Ca, K)					
$SrFeO_3$	C	KOH	65–70	2	274
	C	H_2SO_4	65–70	0	274
$La_{(1-x)}Sr_xFeO_3$	C	NaOH	110–130	0.75–0.82	265
$La_{(1-x)}Sr_xFe_{(1-y)}Co_yO_3$	C	KOH	45–80		278
$La_{0.2}Sr_{0.8}Fe_{0.2}Co_{0.8}O_3$	PJS	NaOH	50		263
$La_{(1-x)}Sr_xCoO_3$	C	KOH	59–74	1.8	275, 281
$La_{(1-x)}Sr_xCoO_3$	C	NaOH	64–70	0.70–1.0	265
$La_{(1-x)}Ba_xCoO_3$	C	KOH	57–60		281

TABLE 3 (Continued)

Oxide	Electrode[a]	Electrolyte[b]	Tafel slope (mV $decade^{-1}$)	Reaction order[c]	Ref.
$La_{(1-x)}M_xMnO_3$ (M = Sr, Ca, K)	C	NaOH	125–130	0.6–0.65	265
$LaNiO_3$	C	NaOH	40–43	0.95	265, 289
$A_2[B_{(2-x)}A_x]O_{(7-y)}$ (A = Pb, Bi)	TB	KOH (≤ 110°C)			335
$Bi_2Ru_2O_7$	TD		40	1.12	337
$MCoO_2$ (M = Pt, Pd)	S	NaOH	50		422
$PdRhO_2$	S	NaOH	50		422

[a]TD = thermal decomposition; C = ceramic; TB = Teflon-bonded; RS = reactive sputtering; E = electrochemical decomposition or formation; SC = single crystal; PJS = plasma jet spray.
[b]Experiments performed at room temperature unless otherwise indicated.
[c]Reaction order is with respect to H^+ in acid, OH^- in base.

6.1.1 Noble metal oxides

The thermally prepared oxides of the so-called rarer platinum metals are among the best electrocatalysts known for the oxygen gas evolution reaction from aqueous systems. Of these oxides, RuO_2 exhibits the highest catalytic activity (at least in relatively short term tests) and has been investigated in most detail. Much of the published work on RuO_2 has been stimulated by the success of RuO_2-based anodes in chlor-alkali cells.

(*a*) *Ruthenium dioxide.*

A number of fairly current reviews are available which deal with the preparation and the physicochemical and electrochemical properties of RuO_2 electrodes [185–187]. The following is a summary of some of the more salient features of this important electrocatalyst material. RuO_2, which adopts the rutile structure, is the most stable of the oxides of ruthenium [188]. It develops appreciable volatility at temperatures above ca. 800°C [189] and can transform to ruthenium metal plus oxygen as well as volatile RuO_4 and RuO_3 at higher temperatures [190,191]. Ruthenium dioxide single crystals exhibit essentially metallic conductivity, the value of the conductivity being 2.5×10^4 ohm^{-1} cm^{-1} at room temperature.

(i) Preparation and physicochemical properties. Ruthenium dioxide is generally investigated as a thin film, either alone or as a binary mixture with, for example, TiO_2 and SnO_2, usually supported on Ta or Ti substrates. The films are generally formed by thermal decomposition of $RuCl_3 \cdot nH_2O$ (where $n = 1$–3) salt at temperatures from about 350°C upwards. The basic preparation method [192, 193] consists of dissolving the the Ru salt, or mixture of salts, in a suitable solvent, e.g. isopropanol, and then spreading it on the support. After drying at 50–100°C, the layer is usually briefly fired at the selected temperature, after which several more layers are applied in a similar manner until the desired loading of catalyst is achieved. The final firing may be carried out for several hours. The final thickness of films prepared in this manner is typically 2–3 μm, which corresponds to about 2 mg RuO_2 cm^{-2} or 10^{-5} mol RuO_2 cm^{-2}.

Substantial amounts of chloride are present in the fixed ruthenium oxide-based films, e.g. about 4% for a preparation temperature of 400°C [194, 195]. The chlorine content has been observed to decrease slightly on going to the external surface as indicated by, for example, secondary ion mass spectrometry [196]. The exact location of chlorine in the bulk lattice is somewhat unclear at present. Oxygen content has been found to increase sharply over the last few monolayers at the external surface, as shown by SIMS [196] and XPS measurements for powders [197] and films [198]. There is now evidence from several groups that suggests the existence of some RuO_3 in the surface regions of ruthenium dioxide electrodes [196–200].

The basic rutile structure of RuO_2 is exhibited by the thermally prepared films of RuO_2. The films are microcrystalline in nature, and X-ray diffraction analysis data was found to be dependent on firing temperature [195], possibly as a result of change in crystallite size. The general morphology of RuO_2 films is dependent to a considerable extent on preparation procedure and films may range from highly microcracked [194] to being relatively compact [195]. BET measurements indicate that the real surface area of the films increases linearly with oxide loading [193] for films supported on etched Ti and, as would be expected, decreases with increasing firing temperature. Generally, BET areas of the RuO_2 films are about 175–250 times the geometric surface area.

There is some disagreement over the nature of the electrical properties of RuO_2 films. Earlier measurements [194] on pressed pellets yielded conductivity values about 3 orders of magnitude (10 $ohm^{-1}\,cm^{-1}$) less than that of the single crystal value. The temperature dependence of the conductivity suggested that the RuO_2 (pressed-pellet form) possessed semiconducting properties. Later studies by Lodi et al. [195] on fairly compact RuO_2 films yielded conductivity values of the same order of magnitude as for single crystal RuO_2 and suggested that the electrical conductivity was metallic type. The electronic states of RuO_2 crystals have been recently studied by photoemission spectroscopy using synchroton radiation [201].

Technologically, RuO_2 is used in a mixed oxide mode, generally in association with TiO_2. Although RuO_2 has also been investigated as a binary mixture with other oxides as well, e.g. ZrO_2 [202], most of the basic research has focussed on the $Ti_{(1-x)}Ru_xO_2$ ($0 \leq x \leq 1$) system. The mixed oxide films are prepared by essentially the same procedure as that described for the pure RuO_2 films above. There is disagreement in the literature as to the extent of solid solution formation between RuO_2 and TiO_2 [203, 204]. From the viewpoint of morphology, the RuO_2/TiO_2 films are qualitatively similar to the pure RuO_2 films. Between 0 and 70% TiO_2, the resistivity of RuO_2/TiO_2 mixtures is essentially unchanged and metallic-like, whereas above 70% TiO_2, a large increase in resistivity is observed [204, 205].

(ii) Electrochemical properties. A typical voltammogram is shown in Fig. 8 for RuO_2 in 1.0 mol dm^{-3} NaOH. Oxygen gas evolution is seen to commence at ca. 1.45 V (RHE).

There has been considerable discussion regarding the extent of the contribution of bulk redox processes, with associated proton diffusion, e.g. along grain boundaries, to the non-Faradaic currents observed in voltammograms of RuO_2 electrodes. Burke's group [193, 206, 208] have generally maintained that the voltammetric charge is due to surface processes only, on the basis of detailed BET surface area and electrochemical analysis. There are strong indications [209] that the pair of peaks at ca. 1.05 V (RHE) is associated with the interconversion of RuO_2 and RuO_4^{2-} (ruthenate), with the pair of reversible peaks at ca. 1.38 V (RHE) involving the RuO_4^{2-}/RuO_4^- (perruthenate) couple, as shown in eqns. (42) and (43), respectively:

$$RuO_4^{2-} + 2\,H_2O + 2\,e = RuO_2 + 4\,OH^- \quad E = 0.9 - 1.15\ V\ (RHE) \tag{42}$$

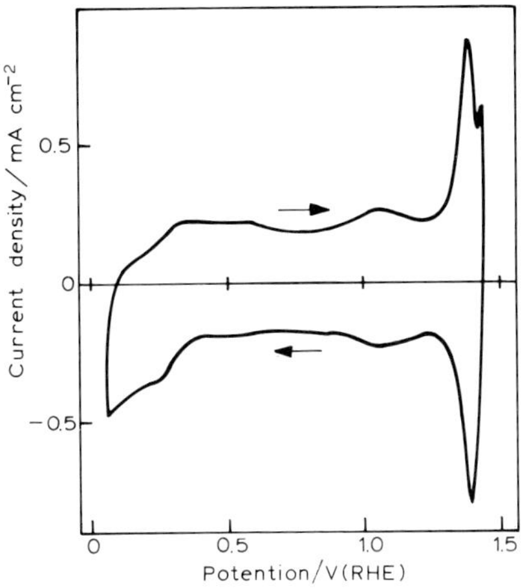

Fig. 8. Voltammogram for RuO_2 supported on Ti in 1 M NaOH at 25°C. Sweep rate = $10\,mV\,s^{-1}$; oxide preparation temperature = 400°C.

$$RuO_4^- + e = RuO_4^{2-} \qquad E = 1.35 - 1.45\ V\ (RHE) \qquad (43)$$

Besides concluding that these redox processes are essentially confined to the surface region of the anhydrous oxide on the basis of BET surface area and cyclic voltammetric charge correlation studies [193, 208], cyclic voltammetry was proposed as a technique for estimating the real surface area of RuO_2 electrodes [206]. However, it has been maintained [207] that not all surface ruthenium species participate in the redox processes. It has been estimated that the ruthenium species which react according to eqn. (43) corresponds to only about 12% of the number of apparent surface species [206, 207]. Qualitatively similar voltammetric features have been observed in the case of RuO_2/TiO_2 mixed oxide electrodes in base [209].

Doblhofer et al. [210] have investigated the electrochemical behavior of RuO_2 supported on Ti in $1.0\,mol\,dm^{-3}$ H_2SO_4. These authors concluded that the oxidation state of surface ruthenium atoms changes from +2 to +6 in the potential range −0.2 to 1.4 V (RHE). In general, the electrochemical studies of ruthenium species on platinum in solution by Lam et al. [211] support the assignments of Burke's group [212] and Doblhofer et al. [210] for the oxidation state changes of ruthenium on thermally prepared RuO_2 surfaces.

An attempt was made by Doblhofer et al. [210] to separate surface from bulk charging processes for thermally prepared RuO_2 using the potential step technique. These authors [210] concluded that some bulk diffusion was involved, presumably involving protons, and estimated a diffusion coefficient of $10^{-19}\,cm^2\,s^{-1}$. Weston and Steele [213] deduced a diffusion coefficient value for protons in porous powder electrodes of RuO_2 which is approximately similar to the value of Doblhofer et al. [210]. Iwakura and co-workers [214], on the other hand, employed cyclic voltammetry in deducing a much higher value of the diffusion coefficient, 10^{-8} to 10^{-10} $cm^2 s$. It is interesting to note that Li is incorporated into RuO_2 from non-aqueous electrolytes [215–217] and a diffusion coefficient of ca. 10^{-11} $cm^2 s^{-1}$ has been estimated for Li diffusion in the RuO_2 host structure [216].

In general, the discrepancies between the results of various groups for thermally prepared RuO_2 electrodes may be due to the variability in preparation from laboratory to laboratory and the absence of a sharp cut-off point between internal surface and bulk electrode material, or grain boundary and defect regions. The direct correlation of BET surface area with voltammetric charge [206, 218] is attractive and simple. However, there may still be uncertainties in this approach, e.g. sweep rate dependence and the lack of definite confirmation of the redox reactions contributing to the voltammetric charge. Further, there may be charge associated with electrochemical reactions occurring in regions of the RuO_2 layers not accessed in gas adsorption experiments. There will probably continue to be more discussion on this topic in the literature.

A rest potential of 0.95 V (RHE) has been reported for RuO_2 films throughout the pH range of 0–14, i.e. the rest potential exhibits a pH dependence of 59 mV per pH unit [161, 219]. Safonova et al. [220] observed that the isoelectric potential shift, or electrode potential at constant total surface charge, showed a pH dependence of 59 mV per pH unit in the range 0.3–0.9 V (RHE). Burke et al. [207, 209], on the other hand, have reported that the voltammetric peak potentials for RuO_2, i.e. ca. 0.5 V (RHE) and 1.1 V (RHE) in 1.0 mol dm^{-3} NaOH (Fig. 8), exhibit a pH dependence of 75 mV per pH unit rather than the Nernstian 59 mV value. Burke and Healy [207] used the electro-oxidation of benzaldehyde as a probe reaction in studying the pH dependence of the voltammetric peak which is believed to correspond to eqn. (42) (Fig. 9). The oxidation of benzaldehyde coincides with the Ru(IV) → - Ru(VI) transition. Burke and Healy [207] suggested that surface oxycomplexes may be formed such that eqn. (42) may best be written in the form

$$RuO_2 + 2.5\ OH^- = RuO_2(OH)_{2.5}^{0.5-} + 2\ e \tag{44}$$

The species on the right of eqn. (44), though speculative, may be part of a polymeric or interlinked system on the surface of otherwise anhydrous RuO_2. The existence of oxyruthenium species in solution has been suggested by other workers [211, 221]. A somewhat greater pH dependence of 88 mV per pH unit has been observed for redox transitions in the case of hydrous, electrochemically grown oxide films [222, 223] as discussed in Sect. 5.4. The apparent discrepancy between the data of Burke et al. [207, 209] and the data of Trasatti and co-workers [185, 219] and Safonova et al. [220] is difficult to rationalize. Different experimental approaches were employed by the different groups, however. It does not seem likely that kinetic effects contribute significantly to the data of Burke et al. [207, 209] which was derived from linear potential sweep measurements at relatively slow sweeps. The pH dependence of the electrode potential of the RuO_2 films is likely to remain a topic of discussion for some time.

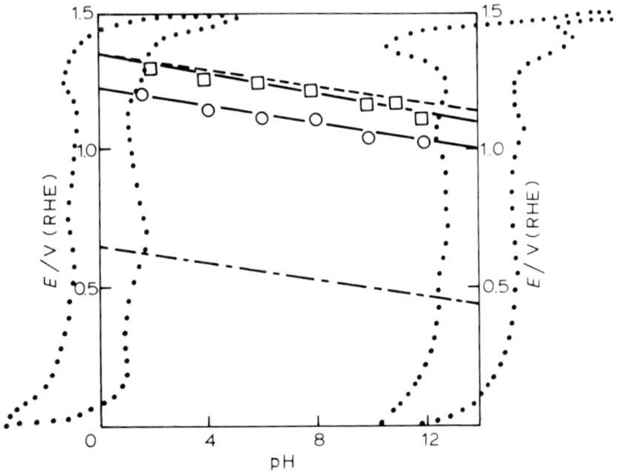

Fig. 9. Effect of solution pH on the redox behavior of an RuO_2-aqueous solution interface. (-·-·-), From work with an RuO_2/TiO_2 film [209]. The upper lines are for the Ru(IV)/Ru(VI) transition as obtained from voltammograms for pure RuO_2 (on Ti) in a range of borate (pH > 8) and phthalate (pH < 8) buffer solutions: ○ and □ refer to the cathodic and anodic peak maxima respectively. (---), Variation of the half-wave potential for benzaldehyde oxidation (0.1 mol dm^{-3}, scan rate = 1.5 mV s^{-1}) on pure RuO_2 in buffered 10% *t*-butanol in water mixtures. The voltammograms outlined on the left and right for RuO_2 in acid and base, respectively [207].

(iii) Oxygen evolution. RuO_2 exhibits a very good electrocatalytic activity for the O_2 gas evolution in both acid and base. Although Ru metal is considered to be an even better catalyst for oxygen evolution, it undergoes rapid dissolution under the conditions of O_2 evolution [212, 224–226].

Oxygen evolution on RuO_2 and RuO_2/TiO_2 mixed oxide electrodes is characterized by a relatively low Tafel slope, 40 mV $decade^{-1}$, at low current densities. For plots of log j vs. E, an increase in slope is observed in these plots at high current densities. The deviation from Tafel linearity in this current density region has been related to, for example, uncorrected iR drops [227], i.e. within the film or between the oxide and Ti substrate.

Not surprisingly, the method of preparation of the RuO_2 influences the current density (i.e. with respect to geometric area) for oxygen evolution: the current density decreases with increasing firing temperature [193]. This can be readily explained in terms of a surface area effect. As discussed above, the surface area of the films and the voltammetric charge decrease with increasing firing temperature. Oxygen evolution current densities (geometric area) increase with oxide loading. The log j vs. E plot of oxygen evolution for different RuO_2 loadings (Fig. 10) shows that the current densities estimated using BET surface areas are essentially independent of the weight of RuO_2, despite a substantial variation in catalyst loading. Lodi et al. [228], however, have suggested that effects such as degree of compactness or the defect nature of the RuO_2 films can influence electrochemical parameters, e.g. the Tafel slope, for O_2 evolution.

The mechanism of O_2 evolution on RuO_2 and RuO_2/TiO_2 electrodes in acid electrolyte has been addressed by a number of different groups [193, 200, 210,

228–230]. The proposed mechanisms range from a classical type, with reaction occurring at relatively unspecified active sites on the oxide surface, to those involving formation and decomposition of higher oxides as an integral part of the oxygen evolution pathway. Doblhofer et al. [210] suggested that a possible sequence of reactions leading to O_2 evolution might be

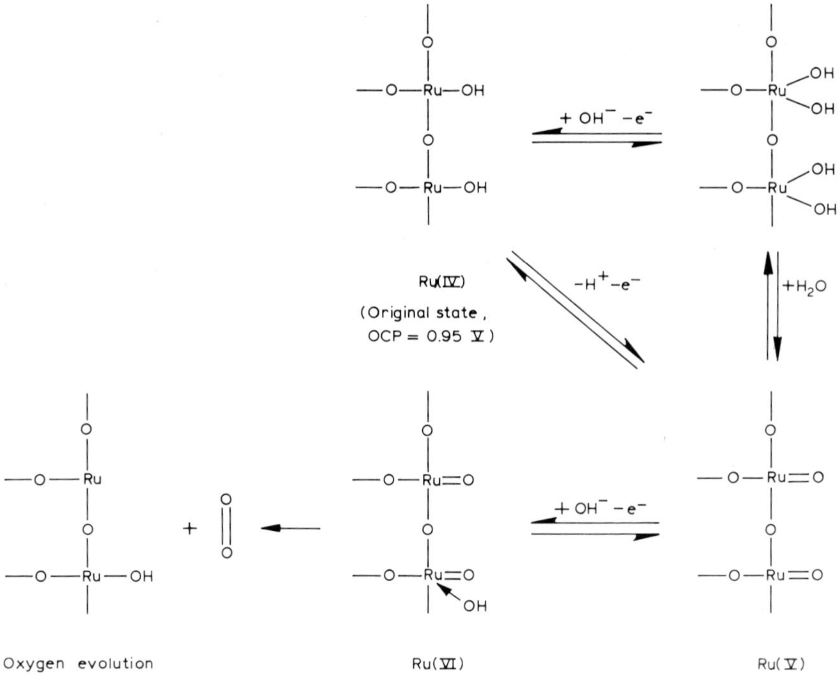

Ru(IV)
(Original state,
OCP = 0.95 V)
+ OH⁻ −e⁻
−H⁺−e⁻
+H₂O
+ OH⁻ −e⁻
Oxygen evolution
Ru(VI)
Ru(V)

(45)

Burke et al. [193] also suggested that the decomposition of electrochemically generated unstable RuO_3 was an intrinsic part of the O_2 evolution pathway on RuO_2 anodes. Kötz et al. [200, 231] have employed XPS to study the oxidation states of ruthenium at ruthenium metal and ruthenium dioxide electrodes which were polarized at high potentials. Electrodes were transferred farily rapidly to the spectrometer after completion of the electrochemical measurements. Kötz et al. [200, 231] confirmed the existence of a Ru(VI) species on the surface of thermally prepared RuO_2. The general scheme shown in Fig. 11 was proposed for oxygen evolution and corrosion at Ru and RuO_2 electrodes. RuO_4 was proposed as a common intermediate [200, 231, 232] in both oxygen gas evolution and ruthenium dissolution. The Ru(VI) oxy species was considered to be much less stable in the case of Ru metal anodes, which develop a thick hydrous oxide layer, than in the case of thermally prepared RuO_2, hence the enhanced tendency of Ru metal anodes for dissolution. The mechanism proposed by Kötz et al. [200, 231] for oxygen evolution on RuO_2 has some attractive features. Ring–disk and optical measurements indicate that formation of solution-phase RuO_4 and oxygen evolution commence at the same potentials [232], which supports, to

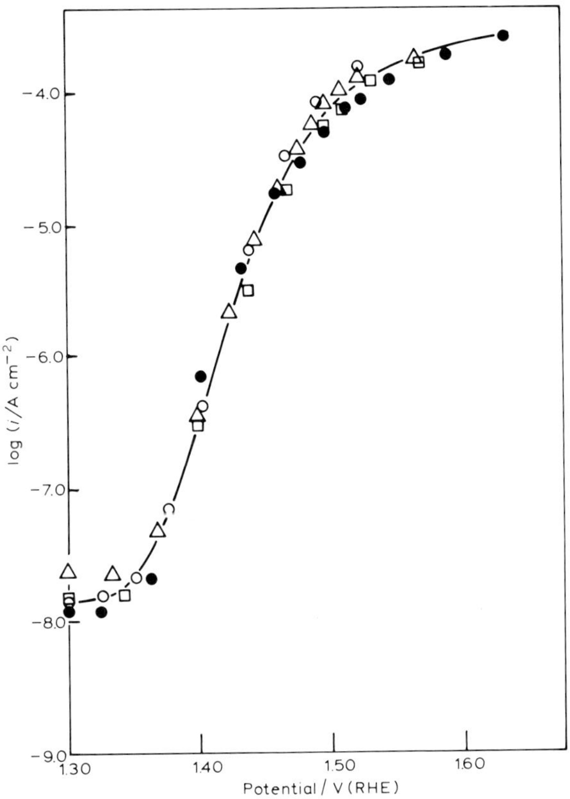

Fig. 10. Tafel plot for oxygen evolution in 1.0 mol dm^{-3} H_2SO_4 at 25°C on RuO_2-coated Ti foils: current densities were estimated on the basis of real (i.e. BET) surface areas. Ratio of real area to geometric area: ○, 350; △, 1200; □, 1800; ●, 2900 [193].

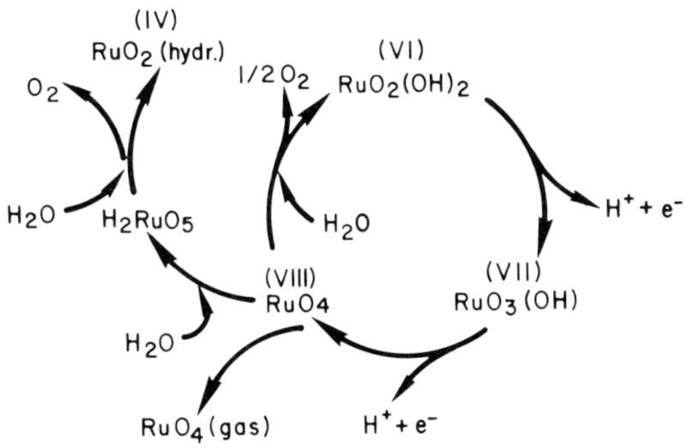

Fig. 11. Model for the oxygen evolution and corrosion reactions on Ru and RuO_2 electrodes [200].

some extent, the view that the pathways of ruthenium dissolution and O_2 evolution involve a common intermediate. It is interesting to note that conversion of RuO_2 to gaseous RuO_4 can be a significant side reaction in the cleavage of water by Ce^{4+} to give O_2 gas with hydrated RuO_2 as catalyst [233, 234]. However, Mills et al. [234] found that dehydration of the RuO_2 by mild heat treatment resulted in a stable, active catalyst for O_2 generation from the Ce^{4+}/H_2O system.

The most widely quoted mechanism for oxygen evolution at RuO_2 electrodes in alkaline solution is that of O'Grady et al. [227], viz.

$$M^{+z} + OH^- \underset{k_{-1}}{\overset{k_1}{\rightleftharpoons}} (M-OH)^{+z} + e \qquad (46)$$

$$(M-OH)^{+z} \xrightarrow{k_2} (M-OH)^{z+1} + e \qquad (47)$$

with reaction (47) as the rate-determining step and followed by subsequent processes yielding O_2 and regenerating the site M^{z+}, e.g.

$$\begin{array}{c}(M-OH)^{z+1}\\+\\(M-OH)^{z+1}\end{array} \quad \begin{array}{c}OH^-\\\\OH^-\end{array} \rightarrow \begin{array}{c}M\text{-}\text{-}O\text{-}\text{-}H\text{-}\text{-}\text{-}OH\\|\\|\\M\text{-}\text{-}O\text{-}\text{-}H\text{-}\text{-}\text{-}OH\end{array} \quad \begin{array}{c}M^{+z}\\+\\M^{+z}\end{array} \quad \begin{array}{c}O\\\\O\end{array} \quad \begin{array}{c}HOH\\+\\HOH\end{array} \qquad (48)$$

This mechanism is in agreement with the experimentally observed Tafel slope of 40 mV decade^{-1} [assuming a symmetry factor $\beta = 1/2$ for reaction (47)] and a reaction order of 1 with respect to OH^-. The species M was suggested to be a hydrated surface Ru complex [227].

Burke et al. have, for a number of years, promoted the idea of direct involvement of higher oxidation states of Ru in anodic reactions such as O_2 evolution [193, 202] and organic oxidations [235]. Burke and McCarthy [202] recently advanced the following variation of reactions (46)–(48) for oxygen evolution on RuO_2/ZrO_2 electrodes, viz.

$$\text{RuO}^-(\text{OH})_3 + OH^- \underset{k_{-1}}{\overset{k_1}{\rightleftharpoons}} \text{RuOO}^-(\text{OH})_2 + H_2O + e^- \qquad (49)$$

$$\text{RuOO}^-(\text{OH})_2 \xrightarrow{k_2} \text{RuO}_2(\text{OH}) + e^- \qquad (50)$$

$$\text{RuO}_2(\text{OH})_2 \rightleftharpoons \text{RuO}_2 + 2H^+ + O_2 + 2e^- \qquad (51)$$

$$+ H_2O + OH^-$$

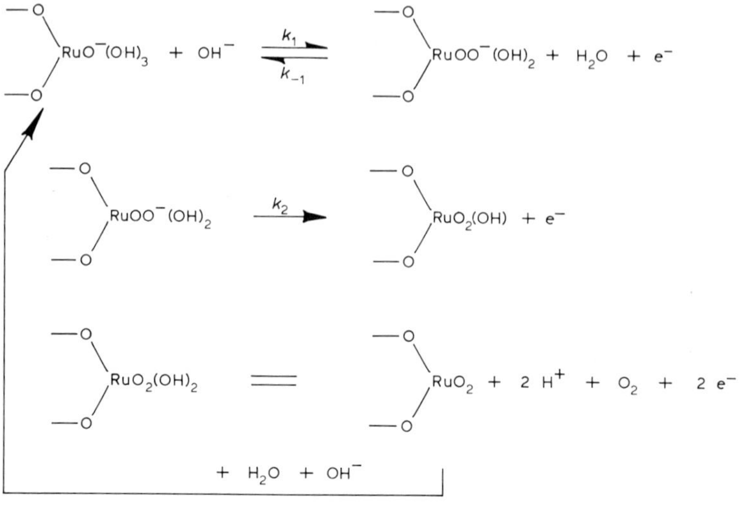

There are good indications that surface oxyruthenium species in base can at least attain the somewhat unstable Ru(VII) state at a potential just prior to the onset of oxygen evolution [212, 235]. Further discussion of the mechanistic aspects of oxygen evolution on RuO_2-based electrode systems is available in a relatively recent paper by Krishtalik [236]. In general, the concept of involvement of higher oxidation states of the catalyst metal ions in the mechanism of oxygen evolution at transition metal oxide electrodes in a mediator-type role is interesting and is probably valid. Further experimental verification of the involvement of such oxidation states, however, in anodic processes is necessary.

RuO_2 single crystals have been subjected to a limited amount of electrochemical investigation to date [237–240]. Hepel et al. [238] observed significant differences in electrochemical behavior for the hydrogen absorption region for various single crystal faces in H_2SO_4 electrolyte. For a number of different faces, viz. (110), (001), (111), and (101), a correlation was found between the hydrogen adsorption/desorption regions of the cyclic voltammograms and the composition and structure of the idealized surface of the rutile structure of RuO_2 [238]. Shafer et al. [237] observed that the Tafel slope for O_2 evolution in acid solution was, for the most part, independent of the crystal face, although some face specificity seemed to be exhibited by the exchange current density. The Tafel slope for oxygen evolution was higher, e.g. 80 mV decade^{-1} for (101) and (111), for all faces studied [237] than that which is usually reported for the thermally prepared RuO_2 films, i.e. 40 mV decade^{-1}.

(b) Iridium oxide.

Very little electrochemical study has been carried out on thermally prepared films of noble metal oxides other than RuO_2. A number of reports [241–243] are available regarding the electrocatalytic activity of thermally prepared IrO_2 for oxygen gas evolution. IrO_2 is a somewhat poorer electrocatalyst than RuO_2 for oxygen evolution although it may be less susceptible to corrosion than RuO_2. Tafel slopes of 56 mV decade^{-1} in acid solution [241] and 40 [241] and 50 mV decade^{-1} [242] in base have been observed for O_2 evolution on IrO_2. Iwakura et al. [241] proposed that the mechanism of O_2 evolution in acid solution involves chemical reaction of absorbed OH species as a rate-determining step. The same authors [241] suggested that the slow step for oxygen evolution in alkaline solution at IrO_2 is the second electron transfer step involving absorbed OH species, assuming a Langmuirian adsorption conditions.

IrO_2 and RuO_2 are mutually soluble in one another and consequently the IrO_2–RuO_2 binary oxide system [244] along with the ternary systems (Ir, Ru, Ta)O_x [245] and (Ir, Ru, Sn)O_x [246] have been investigated as catalysts for oxygen evolution. Enhanced corrosion resistance is claimed for RuO_2 in such binary and ternary oxide systems for oxygen evolution in acidic electrolytes including solid polymer electrolyte [244].

Beni and co-workers [182, 183] found that reactively sputtered (in pure oxygen) IrO_x films exhibited good catalytic activity for oxygen evolution in acid as well as good corrosion resistance. A sputtered IrO_x films of thickness ~3200 Å maintained a steady-state current density of about 75 mA cm^{-2} (geometric area) for 18 days after initially exhibiting a higher current in 0.5 M H_2SO_4 at 25°C without undergoing detectable corrosion [183]. At low overpotentials, the enhancement in oxygen evolution current was about 2 orders of magnitude greater than that of bare Ir metal. The enhanced corrosion resistance of the sputtered IrO_x layers relative to the electrochemically formed hydrous Ir oxide layers, which were discussed in Sect. 5.4, may be ascribed to the essentially anhydrous nature of the sputtered layers. It may well be, as speculated by Hackwood et al. [183], that the sputtered IrO_x films owe their relatively high catalytic activity to their amorphous, high-free-energy structure. Hackwood et al. [247], using such techniques as differential thermal analysis, evolved gas analysis and X-ray diffraction, observed that sputtered IrO_x films underwent an irreversible exothermic transition at about 300°C with ~1 eV energy release, which was associated with an amorphous-to-crystalline transition. The amorphous state was nonmetallic, whereas metallic conductivity was observed in the crystalline state. Further, the sputtered films lost their electrochromic properties and exhibited a considerable reduction in electrocatalytic activity upon transformation to the crystalline state [247]. Recently IrO_x films [248] prepared by d.c. sputtering in an argon–oxygen mixture were observed to be more electrochemically active as regards the redox process governing the electrochromic response of the films than the films prepared in a pure oxygen environment [182]. However, there is no report to date of any oxygen evolution studies carried out on these films, although is may be speculated that the corrosion resistance under oxygen evolution conditions of these films is likely to be less than that of the films sputtered in pure oxygen. In general, a more detailed study of the electrochemical properties and real surface area (e.g. BET) of reactively sputtered IrO_x films, including other catalytically active oxides, would be interesting.

The class of hydrous Ir oxide films has received a tremendous amount of interest in the last decade, primarily because of their electrochromic properties [11, 120]. These oxide films, were discussed in Sect. 5.4. The films are extremely disperse and hydrous in nature, and undergo fairly rapid dissolution at potentials above ~1.6 V (RHE). The oxygen evolution reaction has been investigated at these films in the potential region where the hydrous films are stable. Gottesfeld and Srinivasan [123] and Frazer and Woods [249] observed that the rate of oxygen evolution at the Ir–solution interface increased appreciably (by up to an order of magnitude) upon growing a hydrous oxide by potential multicycling. Frazer and Woods [249] observed that the oxygen evolution current density (based on geometric area) increased linearly with amount of oxide on the Ir surface, as measured by cyclic voltammetry. This linear variation of oxygen evolution current density with oxide thickness could be simplistically understood as follows: the voltammetric charge (which is associated with a bulk redox reaction) is proportional to the mass of the film which for a microporous film is essentially linearly related to real surface area. However, in view of the apparent highly disperse nature of the hydrous oxide layers [12, 150], a clearly defined oxide electrode solution interface may not exist. These iridium hydrous oxide layers may substantially consist of polymeric oxide chains which contain a fairly high concentration of coordinated OH and H_2O species [12]. Thus, oxygen gas evolution may best be regarded as occurring at active sites not necessarily located at a well-defined oxide–solution interface, but rather at hydrated polymer chains and other hydrated, microdisperse oxide regions.

Gottesfeld and Srinivasan [123] observed a lower Tafel slope for oxygen evolution and hydrous Ir oxide, i.e. 50 mV decade^{-1}, than for hydrous-oxide-free Ir, for which they reported a slope of 90 mV decade^{-1}. This decrease of Tafel slope for oxygen evolution on growing hydrous films will be discussed later in conjunction with the behavior of hydrous rhodium oxide films. Kötz et al. [250] using XPS data postulated a mechanistic scheme for oxygen evolution and oxide dissolution at hydrous iridium oxide films which is the same type as that shown in Fig. 11 for for ruthenium oxide films [200]. The XPS data of these authors [250] indicates that IrO_3 is formed at high anodic potentials [e.g. 1.25 V (SCE) in H_2SO_4]. Accordingly, Kötz et al. [250] proposed a mechanism for O_2 evolution at the Ir oxide films which involves the cyclic formation and decomposition of IrO_3. It was suggested that the latter species may also undergo dissolution into the electrolyte as IrO_4^{2-} ion. According to the Kötz et al. model, the IrO_3 species in the case of the hydrous films grown by potential cycling are more unstable and possibly are present in greater density than in the case of thermally prepared or reactively sputtered (especially in pure oxygen [183]) IrO_x films.

(c) *Other noble metal oxides.*

Very little published work is available on the catalytic activity of the remaining noble metal oxides. O'Grady et al. [243] have compared the activity of a range of thermally prepared noble metal oxide catalysts for oxygen evolution (Fig. 12). The Tafel slopes at low and moderate overpotentials are fairly linear and range from 40 mV decade^{-1} for RuO_2 to somewhat greater values for the poorer catalysts. The order of catalytic activity between the thermally prepared oxides at low overpotentials is Ru > Ir > Rh > Pd > Pt. Differences in the true surface areas of these oxides, however, may account for some of the differences between the polarization curves. Miles et al. [242] surveyed the activity of a range of oxide catalysts in 30% KOH at 80°C. However, this work did not clearly establish the relative activities of the various noble metal oxides other than to show that RuO_2 was the better catalyst overall. Cipris and Pouli [251] observed that the activity of Rh in mixed oxide electrodes was significantly less that that of similar

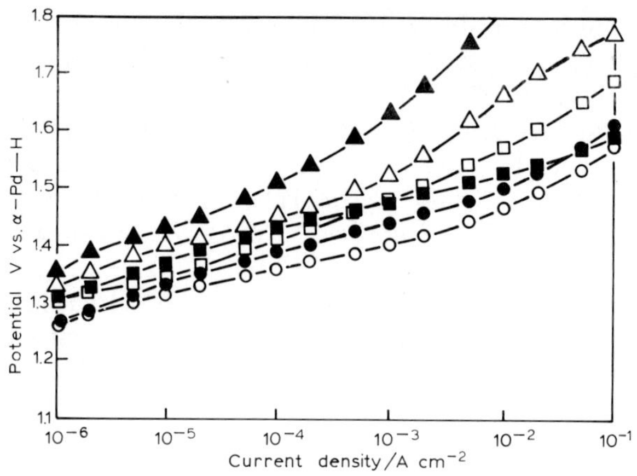

Fig. 12. Polarization curves of various metal oxides on Ti substrate electrodes in O_2-saturated 4 M KOH obtained by pseudo-steady-state galvanostatic method (3 min./point). Curves recorded from low to high currents; T = 22°C [243]. ▲, Fe; △, Pr; ■, Pd; □, Rh; ●, Ir; ○, Ru.

electrodes containing Ru or Ir. In general, the data of O'Grady et al. [243] seems to represent a good approximation of the relative catalytic activities of the various noble metal oxides for oxygen evolution.

Iwakura et al. [252] investigated oxygen evolution in alkaline solution on thermally prepared platinum oxide, which was characterized as PtO_2. A Tafel slope of 59 mV decade^{-1} was observed at low overpotentials and a slope of 114 mV decade^{-1} at high overpotentials. Following evaluation of the usual mechanistic indicators, Iwakura et al. [252] proposed a mechanism for oxygen evolution which has as its rate-limiting step the deprotonation of adsorbed OH species by the OH^- ion. The authors noted that the oxygen evolution currents on the thermally produced PtO_2 were stable and reproducible, in contrast to the usual behavior at platinum metal anodes. It is well established that the rate of oxygen evolution at platinum anodes decreases considerably with the extent of oxidation (at the monolayer level) of the platinum surface [253, 254]. On the other hand, it has also been reported that multilayer oxide films formed at high positive potentials [2.1–2.5 V (RHE)] at platinum give rise to enhanced rates of oxygen evolution and a lowering of the Tafel slope for this reaction [255, 256]. These multilayer oxide films on platinum have recently been reinvestigated by Burke and Roche [124] and it has been found that such films can be readily grown using relatively fast potential cycling techniques [e.g. 100 V s^{-1}, 0.42–1.9 V (RHE) in base] at rates that exceed those previously reported for potentiostatic growth conditions [125]. The films are evidently quite hydrous in nature.

Relatively little information is available on the electrocatalytic activity of thermally prepared rhodium oxide for oxygen evolution, this oxide having been investigated for the most part in conjunction with other oxide catalysts [242, 243]. A Tafel slope of ca. 50 mV decade^{-1} has been observed at low overpotentials for oxygen evolution on thermally prepared Rh_2O_3 supported on Ti in acid and base [257].

Of greater interest from a fundamental electrocatalytic viewpoint is the hydrous, multilayer oxide film which can be grown on Rh in base by the potential cycling technique [126]. This film is analogous to the one described earlier in this review paper for Ir (Sect. 5.4). The rhodium film, which exhibits electrochromic activity [137], is resistant to dissolution up to potentials of 1.7–1.8 V (RHE) and hence is somewhat more interesting as a model hydrous oxide system for studying oxygen than the corresponding Ir films, which undergo dissolution above about 1.6 V (RHE). An enhancement in oxygen evolution rate of up to three orders of magnitude at low overpotentials has been observed upon growing a multilayer oxide film at a Rh surface in base [184]. A further feature of these results was the observation that the Tafel slope at low overpotential decreased gradually from about 75 mV decade^{-1}, for an electrode free of any multilayer film, to a limiting value in the region 47–50 mV decade^{-1} for a surface covered initially with a substantial thickness of oxide (Fig. 13) [257]. From about 1.57 to 1.76 V (RHE), a Tafel slope of ca. 120 mV decade^{-1} was observed, while above 1.76 V (RHE), oxide dissolution became significant. It may also be noted that the potential at which the change over from the first to the second linear Tafel region occurred decreased from a value of ca. 1.64 V (RHE) for multilayer oxide free Rh to a limiting value of ca. 1.57 V (RHE) for a Rh electrode with appreciable coverage of multilayer oxide. The gradual lowering of the Tafel slope value with increase in oxide thickness suggests that the hydrous oxide formation occurs initially by an "island" mechanism, the lower slope being reached when the rhodium surface is essentially completely covered with this oxide. Alternatively, the change in slope could be due to a gradual increase in the extent of hydration of a film of uniform thickness. The higher Tafel slope in the case of the uncycled surface suggests weaker hydroxyl coordination of the intermediate at the largely anhydrous compact oxide on this surface, so that the rate of oxygen evolution is not determined exclusively by a reaction of an intermediate as seems to be the case on the hydrated film. The observation that the change in Tafel slope between the first and second regions occurs at lower potentials in the case of thicker films (Fig. 13), suggests that limiting coverage of reaction intermediates is observed at lower potentials in the case of the hydrated films. This is readily understandable on the basis that increased intermediate stability (as reflected in the decrease of the Tafel slope) leads to saturation coverage at progressively lower potentials as oxide charge capacity, or thickness, develops until a constant 50 mV decade^{-1} slope is observed.

With regard to other catalysts, little is known about the activity of palladium oxide catalysts for oxygen gas evolution. As discussed earlier in this section, it seems to be a poorer catalyst than Ru, Ir, or Rh oxides. A report by Razina and Gur'eva [258] suggests that PdO has relatively good

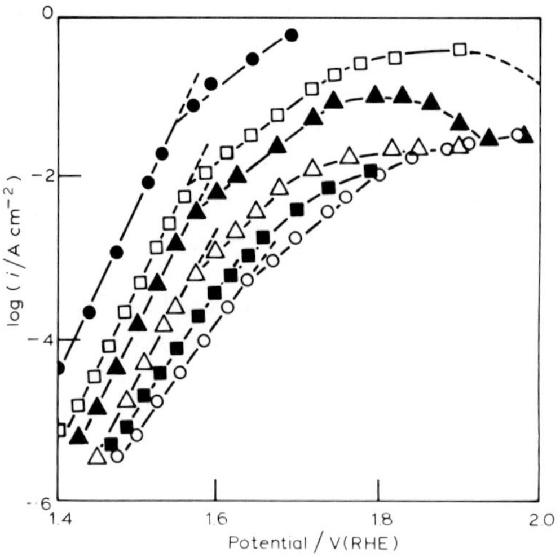

Fig. 13. Tafel plots for oxygen evolution on multicycled Rh in 1.0 mol dm^{-3} NaOH at 25°C. Number of oxide growth cycles (0–1.55 V, 3.0 V s^{-1}): ○, 0; ■, 20; △, 120; ▲, 600; □, 2100; ●, 4500 [257].

activity for oxygen evolution as well as good corrosion resistance in 1 M H_2SO_4 electrolyte.

6.1.2 Perovskite-type oxides

The perovskite series of oxides have been subjected to a considerable amount of investigation over the past 15 or so years from the standpoint of their activity as catalysts for oxygen electrode reactions. The basic perovskite oxide structure may be represented as ABO_3, where A is a large cation such as a lanthanide element and B is a transition metal cation. The ABO_3 composition can be modified, in many cases, by partial substitution of either the A ion with another similar type to give $A_{(1-x)}A'_xBO_3$. Similarly, B can be be partially substitued to give $AB_{(1-x)}B'_xO_3$. Such substitution leads to modification of the resistivity and magnetism properties of the oxides, which may have a significant effect on their catalytic activity [259]. Although the ideal structure of perovskite oxides is cubic, deviations from this structure occur due to variations in ionic radii of the constituent ions.

Perovskite oxides are generally prepared by thermal decomposition of a mixture of the metal salts or by the high-temperature solid state reaction of appropriate oxides. In general, perovskite oxides prepared by standard thermal decomposition methods or by solid state reaction exhibit fairly low real surface area which is usually of the order of a few square meters per gram. Therefore, a variety of other techniques have been employed to produce higher surface area and more homogeneous oxides, e.g. freeze drying, [260–262], precipitation [88], and plasma jet spraying [263]. To prepare electrodes, oxide powders are often pressed into pellets and sintered or pressed with the aid of a binder. Some aspects of the preparation of perovskite oxides have been described by Tamura et al. [264] and will not be discussed further here. The lattice constant of perovksite oxides is large enough such that the overlap of *d*-orbitals of the transition metal cation B with the *p*-orbitals of the oxygen anion to a large extent determines the electrical conduction properties, the latter also being dependent on the number of *d*-electrons exhibited by the transition metal cation [152]. Perovksite oxides exhibit semiconducting properties, e.g. $LaCoO_3$, p-type, 10^{-1}–$1\,\Omega\,cm$ [265]; $La_{(1-x)}Sr_x$-FeO_3, p-type, 10^{-2}–$1\,\Omega\,cm$ [266]; or metallic conductivity, e.g. $LaNiO_3$, $\varrho \sim 10^{-3}\,\Omega\,cm$ [265]. Electrical properties may change significantly, however, with the value of x within a particular perovskite oxide family. For example, in the case of sintered pellets of $La_{(1-x)}Sr_xCoO_3$ [176, 267, 268], at low values of x these oxides are p-type semiconductors with a relatively high acceptor concentration (10^{21}–$10^{22}\,cm^{-3}$) and with a band gap of 0.29 eV. For $x > 0.2$, the conductivity is transformed into a semimetallic or metallic type. p-Type semiconducting perovskite oxides should be more suitable than n-type oxides as anodes for oxygen evolution, since the potential drop in the space charge layer will be negligible for p-type oxides under anodic polarization conditions.

The kinetics of oxygen evolution have been investigated at a variety of perovskite oxides, mainly in alkaline solution. Notwithstanding the work of Bockris and co-workers [269] on the electrocatalytic activity of the perovskite analog oxide Na_xWO_3 for oxygen reduction, the first report of a study of the electrocatalytic activity of perovskite oxides was by Meadowcroft [270] for oxygen reduction on $La_{(1-x)}Sr_xCoO_3$.

Fiori et al. carried out an investigation of the kinetics of oxygen evolution on the perovskite-like oxide class NiM_2O_4 [271] where M = La, Pr and Nd as well as nickel and cobalt and lanthanum mixed oxides in base [272, 273]. Little difference in the activity for oxygen evolution was found on varying M from La to Nd to Pr in the NiM_2O_4 series, which is in accord with the idea that Ni is the electrocatalytically active agent in these oxides. Two Tafel slope regions were observed for all electrodes studied [271, 272], i.e. 40 mV decade^{-1} at low and 120 mV decade^{-1} at high overpotentials. Kinetic studies of the oxygen evolution reaction on NiM_2O_4 oxides suggested that the mechanism of oxygen evolution involved second electron transfer from adsorbed OH species. These authors [271] rather surprisingly concluded that the change in Tafel slope from 40 to 120 mV decade^{-1} occurred due to a change over from low coverage of adsorbed intermediates (assuming Langmuirian conditions) at low overpotentials to high coverage at high overpotentials. The more standard explanation for such a break in the Tafel slope would be that the first electron transfer step becomes rate-determining at high overpotentials. In general, quite good electrocatalytic activity was exhibited by the NiM_2O_4 mixed Co and Ni and La oxide catalysts studied by Fiori et al. [271–273], the highest activity being exhibited by an oxide of nominal composition $Ni_{0.2}Co_{0.8}LaO_3$ [272]. However, some degradation of these oxide catalysts occurs under prolonged oxygen evolution polarization conditions which may be due to a number of factors, e.g. La dissolution [273].

Matsumoto et al. [274–278] investigated the kinetics of oxygen evolution at a number of perovskite type oxides e,g., $SrFeO_3$ [274], $La_{(1-x)}Sr_xCoO_3$ [278], and $La_{(1-x)}Sr_xMnO_3$ [277]. An important aspect of these authors' work was the correlation of the electrocatalytic activity of the oxide catalysts studied with the extent of σ^* antibonding orbital formation involved in the M–O–M moiety in the lattice [278–280]. δ^* band formation occurs as a result of overlap between an e_g orbital of the transition metal ion with an sp_σ orbital of oxygen. In oxygen evolution from base, for example, δ^* band formation was considered to enhance electron transfer from OH^- via the e_g orbital to the δ^* band of the oxide. A rapid rate of electron transfer in the primary discharge step means that rate control may shift to a subsequent step, resulting in more effective electrocatalysis. The ability to generate a relatively high oxidation state, or positive charge density in the transition metal ion, was also considered desirable in order to enhance the rate of chemical reaction steps in the oxygen evolution process. Based on standard kinetic analysis, Matsumoto et al. [274, 275, 277, 278] proposed the general mechanism

$$S + OH^- \longrightarrow SOH + e \tag{52}$$

$$SOH + OH^- \longrightarrow SO^- + H_2O \tag{53}$$

$$SO^- \longrightarrow SO + e \tag{54}$$

$$2\,SO \longrightarrow 2\,S + O_2 \tag{55}$$

for oxygen evolution at perovskite oxides in alkaline solution where S represents the transition metal ion on the electrode surface. The rate-controlling step in reactions (52)–(55) depended on the oxide catalyst studied in the grouping $SrFeO_3$, $LaFeO_3$, $SrNiO_3$, $LaNiO_3$ and the substituted oxides $La_{(1-x)}Sr_xFe_{(1-y)}Ni_yO_3$ and $La_{(1-x)}Sr_xFe_{(1-y)}Co_yO_3$. In general, catalytic activity for oxygen evolution increased with increases in the values of x and y for $La_{(1-x)}Sr_xFe_{(1-y)}Ni_yO_3$ [266] and $La_{(1-x)}Sr_xFe_{(1-y)}Co_yO_3$ [275, 278] perovskite series. The work of Matsumoto et al. constitutes an interesting study, both theoretically and experimentally, of the electrocatalytic activity of perovskite oxides. However, information is lacking on the stability behavior of their catalysts at high current densities and conditions relevant to water electrolysis.

Kobussen et al. [281–285] have carried out a detailed study of the oxygen evolution reaction on $La_{0.5}Ba_{0.5}CoO_3$ principally in strongly alkaline solution. Following reaction order studies [282], impedance measurements [283, 286], and overpotential decay behavior studies [287], a modified Krasil'shchikov [288] mechanism was proposed for oxygen evolution on this oxide catalyst. The first three steps of this mechanism are similar to eqns. (52)–(55) written as equilibrium processes. An alternative to step reaction (54) was also suggested to include successive stages of surface oxidation, which are followed by a rate-determining step involving formation of a perhydroxyl species.

$$O^-_{ads} + 2\,MeOOH \rightleftharpoons MeO_2 + H_2O + e \tag{54a}$$

$$2\,MeO_2 + H_2O \rightleftharpoons 2\,MeOOH + O_{ads} \tag{54b}$$

$$O_{ad} + OH^- \longrightarrow O_2H_{ads} \quad \text{rds} \tag{56}$$

$$O_2H^-_{ads} + OH^- \xrightarrow{\text{fast}} O_2 + H_2O + 2\,e \tag{57}$$

Deactivation, or "passivation", behavior has been found to be a feature of the oxygen evolution process on these electrodes [284]. While apparently not related to oxide dissolution, the loss of activity for oxygen evolution is evidently related to the formation of a hydrated Co-rich oxide multilayer film over a significant portion of the $La_{0.5}B_{0.5}CoO_3$ electrode surface [284]. Lowering of the electrode potential restored the initial activity for oxygen evolution. Further study of this deactivation process in warranted since similar deactivation processes may occur at other oxide catalysts.

Bockris and co-workers [87, 289–291] have conducted a systematic study of the activity of a range of perovskite oxide catalysts for oxygen evolution in base with special emphasis on correlating activity with electronic properties. A feature of these authors' work was the coupling of results obtained from solid state-type measurements, e.g. Hall parameters and magnetic susceptibilities, with electrochemical measurements. Oxide catalysts studied by Bockris and co-workers included $LaNiO_3$, $LaCoO_3$, $La_{(1-x)}Sr_xCoO_x$, $LaMnO_3$, and $La_{(1-x)}Sr_xMnO_3$. It was found that the kinetics of oxygen evolution did not depend on semiconductor-type properties. In general, the kinetics increased with decrease of the magnetic moment, with decrease of the enthalpy of formation of transition metal hydroxides, and with increase in the number of *d*-electrons in the transition metal ion. Bockris and Otagawa [290] proposed a mechanism for oxygen evolution which involves hydrogen peroxide as an intermediate in the process.

$$M^z + OH^- \rightleftharpoons M^z\text{—}OH + e \tag{58}$$

$$M^z\text{—}OH + OH^- \longrightarrow M^zH_2O_2 + e \quad \text{rds} \tag{59}$$

These reactions are followed by ones which involve the catalytic decomposition of physisorbed H_2O_2 to yield O_2. An important feature of Bockris and Otagawa's work [87, 290] was a reasonably linear correlation of current density (intrinsic surface area) for oxygen evolution at an overpotential of 0.3 V vs. M–OH bond strength for hydroxides $M(III)(OH)_3$ (Fig.14). The electrocatalytic activity for oxygen evolution is seen to decrease as the M–OH bond strength decreases. This suggests that the kinetics of oxygen evolution on perovskites are determined by the rate of reaction of OH

References pp. 347–360

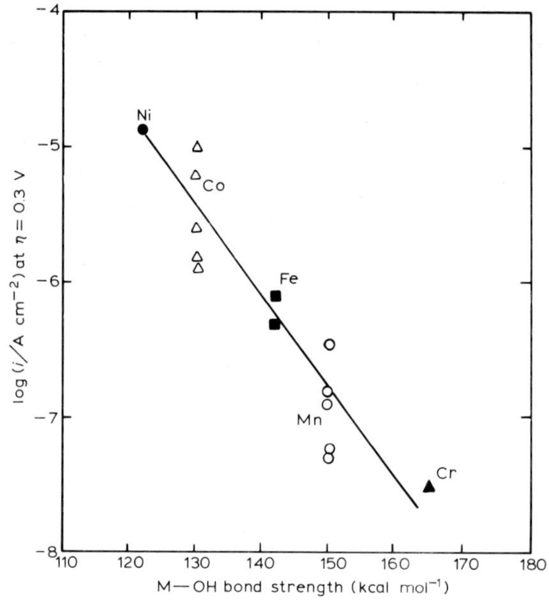

Fig. 14. Current density (based on real surface area) for oxygen evolution on perovskites at an overpotential of 0.3 V vs. M–OH bond strength. The transition metal ions (M) in perovskites are indicated with different symbols [290].

intermediates rather than adsorption of OH intermediates. Thus, the data in Fig. 14 suggest that breaking of the M–OH bond is involved in the rate-determining step [reaction (59)]. With regard to reactions (28) and (59), the contention that hydrogen peroxide is a concrete intermediate in the oxygen evolution process, and indeed present in a physisorbed state for a limited time, seems rather tenuous in view of the instability of peroxide at the potentials involved in oxygen evolution. In general, of the various perovskite catalysts investigated by Bockris et al., $LaNiO_3$ showed reasonably high activity, exhibiting a stable performance of 100 mA cm^{-2} at an overpotential of ca. 300 mV for over 70 h [289, 290].

6.1.3 Spinel-type oxides

The spinel group of metal oxides have the general composition MM'_2O_4 where M and M′ are metal cations. In the spinel structure, one of the cations is located in an octahedral site, the other in a tetrahedral site. The oxygen anions are arranged in an f.c.c. packing arrangement. A variety of methods are available for preparation of spinel oxides. These include high-temperature solid-state reaction, co-precipitation of hydroxides of the corresponding metals followed by thermal decomposition, freeze-drying [260], and the preparation of spinel catalysts on carriers such as dispersed carbon [292, 293] and nickel screen [294, 295]. A problem with spinel oxides which are of interest for electrocatalytic studies is their relatively low electrical conductivity. This has resulted in the use of conducting carriers as well as the use of promoters. Rasiyah and Tseung [296] observed, for example, that the conductivity of Co_3O_4 increased by up to 4 orders of magnitude on being doped with Li.

A variety of spinel oxides, including $NiCo_2O_4$, Co_3O_4, and their substituted oxides have been investigated as catalysts for oxygen gas evolution, principally in alkaline solution. $NiCo_2O_4$ has been subjected to extensive investigation in this regard [297–301]. Srinivasan and co-workers [301] observed a Tafel slope of 40 mV decade^{-1} in strong base at 80°C, suggesting that a step following the first electron transfer step is rate-determining. The data of Jasem and Tseung [297] indicates a Tafel slope of about 120 mV decade^{-1} at high overpotentials in 5 N KOH at 25°C, which is consistent with the discharge of OH^- being rate-determining. Efremov and Tarasevich [302] observed a somewhat similar trend in Tafel slope behavior for oxygen evolution on $NiCoO_4$.

In a relatively recent study, Rasiyah et al. [300, 303] suggested that the evolution of oxygen proceeds via the formation of higher valence oxides of Ni or Co on the surface of the $NiCo_2O_4$

$$MO + OH^- \longrightarrow MOOH + e \tag{60}$$

$$MOOH + OH^- \longrightarrow MO_2 + H_2O + e \tag{61}$$

$$2\,MO_2 \longrightarrow 2\,MO + O_2 \tag{62}$$

Here, M in MO is a divalent cobalt or nickel cation. When MO_2 is formed on the surface, M is in the tetravalent state. In this scheme, after formation in eqn. (62), MO is re-oxidized via steps (60) and (61). Results obtained by Hibbert [304] from ^{18}O-enriched KOH electrolyte strongly support the above mechanism. Hibbert's experiments show that oxygen atoms from the electrolyte are incorporated in a $NiCo_2O_4$ electrode in a relatively unstable form, possibly as a short-lived surface compound which subsequently decomposes releasing O_2 gas. Tseung and co-workers [297, 305, 306] maintain that, prior to oxygen evolution on metal oxide surfaces, the potential of the lower metal oxide/higher metal oxide couple of the oxide system must be reached before oxygen evolution may take place. Thus, according to this approach, the final step in the oxygen process on catalytically active oxide catalysts is the decomposition of unstable, higher valent metal oxide.

The oxygen gas evolution reaction has also been investigated to some extent at Co_3O_4 anodes, especially in alkaline solutions. In acid solutions, dissolution of the oxide is observed along with oxygen evolution [307]. A short Tafel slope region, i.e. ca. 60 mV decade^{-1}, was observed at low overpotentials for oxygen evolution in $HClO_4$ electrolyte, which was principally related to the effects of the potential drop in the space charge region of the Co_3O_3 [307]. Enhanced dissolution was observed after attaining potentials at which formation of the relatively unstable CoO_2 becomes thermodynamically stable [$E° = 1.48$ V (NHE)]. Although this oxide can dissolve with concomitant liberation of oxygen, it was estimated that the contribution of this process to the overall current did not exceed more than 1–2%, so that oxygen was liberated chiefly by the discharge of water. Shalaginov et al. [308] found that the activity for oxygen evolution increased as the preparation temperature of Co_3O_4 films on Ti was lowered from 450 to 300°C. This phenomenon was related to the influence of defects on the electrocatalytic activity of the Co_3O_4 electrodes. At a fixed potential, a linear relation was observed between the electrocatalytic activity and the concentration of defects, i.e. cation vacancies, which had a slope of 0.85–0.95. Roughness factors, estimated from BET measurements, were taken into account in this work [308].

Iwakura et al. [309] have studied the oxygen evolution reaction at Co_3O_4 on a range of different substrates in 1 M KOH at 30°C. The polarization characteristics of the Co_3O_4 films were found to be dependent on the nature of the substrate. This was especially the case with iron substrates, where iron species were found to dope the Co_3O_4 layer during electrode preparation, which resulted in relatively high electrocatalytic activity being observed for the Fe/Co_3O_4 electrodes [309]. Tafel slopes of ca. 60 mV decade^{-1} were observed for oxygen evolution on the Co_3O_4 films on substrates such as Ti, Co, Ni, Nb, and Ta, while average slopes of ca. 45 mV decade^{-1} were observed for Co_3O_4 films on Fe substrates and for Co_3O_4 films on inert substrates doped with Ru, Ir, and Rh species. A reaction order of 1 with respect to OH^- was observed for the oxygen evolution reaction on the Fe/Co_3O_4 electrodes [309]. For the low Tafel slope electrodes, the mechanism of oxygen evolution was considered to be similar to that proposed by O'Grady et al. [227] for oxygen evolution on Ti/RuO_2 electrodes.

Rasiyah and Tseung [296] observed that the kinetics of oxygen evolution were enhanced on doping Co_3O_4 with Li in KOH electrolyte. The bulk concentration of Co^{3+} ions increased with increase of Li, while the resistivity decreased with increase in Li doping. A Tafel slope of 60 mV decade^{-1} was observed on both doping and undoped Co_3O_4 electrodes which were of the Teflon-bonded type, the Co_3O_4 having been prepared by the freeze-drying method. The mechanism proposed for oxygen evolution is similar to that proposed by Tseung and co-workers [300, 303] for $NiCo_2O_4$.

The spinels $NiCo_2O_4$ and Li-doped Co_3O_4 are, in general, good catalysts for the oxygen evolution process in alkaline solution, not only because of their high electrocatalytic activity but also because of their relatively good resistance to dissolution. In relatively long term tests in 45% KOH at 85°C, $NiCo_2O_4$ and 10% Li-doped Co_3O_4 electrodes exhibited current densities of 1 A cm^{-2} at 1.60 (DHE) over a period of ca. 6000 h with only about 0.04 V increase in overvoltage in this period [306, 310]. A comparison of various methods for preparing electrodes of spinel oxides containing cobalt has been recently carried out by a number of groups [295, 311].

In other work on spinels, Iwakura et al. [312, 313] observed that the overpotential at a fixed current density on spinel-type ferrite film electrodes of MFe_2O_4 (M = Mg, Cu, Ni, Co, and Mn) on Fe substrates decreased linearly with increase in effective Bohr magneton (μB). In contrast, Orehotsky et al. [314] did not observe any dependence of the transfer coefficient on the room-temperature saturation magnetism of the $Ni_xFe_{(3-x)}O_4$ spinel series for oxygen evolution in base. Further work is obviously required in the correlation of electrocatalytic properties with magnetic properties.

6.1.4 Other oxides

A number of other oxides have been investigated as electrocatalysts for oxygen evolution, e.g. lead dioxide, manganese dioxide, and nickel oxide.

(*a*) *Manganese dioxide.*

MnO_2 has only moderate activity for oxygen evolution and has been found to exhibit a Tafel slope of about 120 mV decade^{-1} in acid and base [315–318]. This slope value is in accordance with the discharge of H_2O or OH^- being rate-determining in acid or base. Platinum [315] and RuO_2^- or Pt-coated Ti [319] are more suitable as supports for thermally prepared MnO_2 layers because of the ohmic drop which develops at the Ti/MnO_2 boundary. Electrodes made from tablets of so-called massive MnO_2 were found by Morita et al. [214] to give maximum activity for oxygen evolution on thermal pretreatment at 480°C. This pretreatment effects the partial transformation of β-MnO_2 to α-Mn_2O_3 which effectively sets up a solid state redox couple. These authors suggested that Mn^{3+} ions on the surface are the active sites for oxygen evolution, which, on being oxidized to the Mn^{4+} state, mediate the evolution of oxygen in a cyclic manner. Treatment of the massive MnO_2 above 480°C led to formation of mainly Mn_2O_3, which resulted in a substantial decrease in electrocatalytic activity, probably due to the relatively high electrical resistivity of the Mn_2O_3. Despite its rather low electrocatalytic activity for oxygen evolution, when added to the RuO_2/TiO_2 and IrO_2/TiO_2 binary oxide systems, MnO_2 is reported to decrease the overpotential for oxygen evolution at these mixed oxide catalysts [251]. This surprising result implies the existence of a synergistic catalytic effect; however, such catalysts need a more careful characterization of the relative surface areas and surface distribution of the atoms.

(*b*) *Lead dioxide.*

Lead dioxide has been the subject of study as an anode material from the early days of electrocatalysis due, in large part, to its importance in the lead-acid battery. Its good corrosion resistance at high anodic potentials has also resulted in its use in a number of other electrochemical processes, e.g. organic synthesis (see Sect. 8). Aspects of the anodic behavior of PbO_2 have been relatively recently reviewed by Randle and Kuhn [320]. In acid solution, β-PbO_2 has been shown to exhibit a Tafel slope of ca. 120 mV decade^{-1}

for oxygen evolution, while the non-rutile form, α-PbO_2 exhibits a slope of ca. 70 mV decade^{-1} [321]. In alkaline solution, both types of electrode exhibited Tafel slopes of 120 mV decade^{-1} [322, 323]. The Tafel slope of 120 mV decade^{-1} indicates that the first step discharge reaction is rate-determining. The higher activity of the α-PbO_2 (Tafel slope $\sim$70 mV decade^{-1}) is difficult to explain, although geometric surface factors, i.e. small spacing, have been cited as a possible reason [324]. As mentioned earlier in the case of MnO_2, electrodes consisting of thermally prepared films of PbO_2 on Ti substrates are complicated by ohmic drops at the PbO_2/Ti interfacial region [320]. However, good success has been achieved with Ag- [325] and Au-coated [326] Ti substrates for the PbO_2 layers.

(c) *Nickel oxide.*

Because of the importance of Ni metal as an anode material in practical water electrolysis cells [227, 328], the oxygen gas evolution has been extensively investigated at this metal in alkaline solutions [329–331]. Oxygen evolution occurs on an extensively oxidized Ni surface, where the oxidation state of most of the Ni ions is three. While possessing moderately good catalytic activity for O_2 evolution, Ni anodes exhibit a marked increase in overpotential with time. This decay in activity is thought to be related to the gradual conversion of Ni^{3+} to Ni^{4+} in the oxide layer [329]. However, activity is restored on potentiostating electrodes at 1.5 V (RHE). The Tafel slope observed at oxidized Ni anodes tends to be dependent on the pretreatment given to the electrodes. Lu and Srinavasan [329] observed, in the case of Ni electrodes pre-anodized or rejuvenated at 1.5 V (RHE), that plots of log i vs. E exhibit only one linear region of slope 40 mV decade^{-1}. On the other hand, dual Tafel regions were observed on Ni pretreated at 1.8–2.0 V (RHE), i.e. 40 mV decade^{-1} at low η and 170 mV at high η. Dual Tafel slope regions at Ni anodes have also been observed by a number of other authors, e.g. 50 mV decade^{-1} at low overpotentials [331,332] and 100 mV decade^{-1} at high overpotentials [332]. A variety of mechanisms have been suggested for O_2 evolution at Ni anodes. One of the most well known is due to Krasil'shchikov [333], which involves the decomposition of a higher Ni oxide as the rate-determining step; a modification of this mechanism was also suggested by Lu and Srinivasan [329].

Miles et al. [330] studied the kinetics of oxygen evolution of Ni anodes at temperatures up to 264°C. The exchange current density for oxygen evolution was found to increase by more than three orders of magnitude on going from 80 to 264°C. The transfer coefficient for oxygen evolution was observed to change from 0.7 at 80–208°C, to 3.3 at 264°C. A change in reaction mechanism was considered to occur close to the Neel temperature (250°C) for nickel oxide. Recombination of adsorbed oxygen atoms was suggested to be rate-determining at 264°C.

Only a few investigations of O_2 evolution on thermally prepared Ni oxide films have been carried out [242, 334]. Miles et al. [242] observed an average

Tafel slope of ca. 70 mV $decade^{-1}$ for O_2 evolution in alkaline solution at 80°C. Botejue Nadesan and Tseung [334] observed three apparent Tafel slope regions on NiO, the lowest slope value being ca. 43 mV $decade^{-1}$, the highest ca. 132 mV $decade^{-1}$. A considerable advancement in catalytic activity was observed on doping NiO with Li [334]. Such doping increases the concentration of Ni^{3+} ions in the nickel oxide electrode and a substantial lowering of oxide resistivity is observed. On the Li-doped NiO electrodes, Tafel slopes of ca. 44 and 73 mV $decade^{-1}$ were observed at low and high current densities, respectively. In their mechanistic analysis of O_2 evolution at nickel oxide, Botejue Nadesan and Tseung [334] stressed the importance of Ni^{3+} ions as active sites at low current densities on undoped-NiO and Li-doped NiO, while Ni^{3+} sites were kinetically important also at high current densities on the Li-doped catalysts.

(*d*) *Pyrochlore-type oxides.*

A limited amount of investigation has been carried out on the electrocatalytic properties for oxygen evolution of metallic oxides exhibiting the pyrochlore structure, i.e. $A_2[B_{(2-x)}A_x]O_{(7-y)}$, where A is usually Pb or Bi, B is usually Ru, $O \leqslant x \geqslant 1$ and $O \leqslant y \geqslant 0.5$ [335–337]. Horowitz et al. [335, 336] have reported that bismuth and lead ruthenates ($Bi_2Ru_2O_7$, $Pb_2Ru_2O_7$) displayed significantly lower overpotentials than any other electrocatalyst for oxygen evolution in alkaline solution at temperatures between 25 and 110°C (Fig. 15) and that these materials were also among the best electroca-

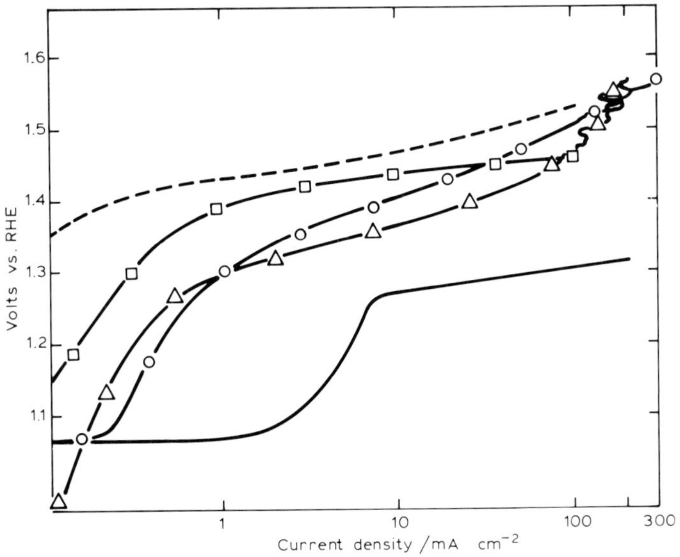

Fig. 15. O_2 evolution on various catalysts in 3 M KOH at 75°C. Catalyst loadings were: 60 mg cm^{-2} for the pyrochlore; 33 mg cm^{-2} for the Pt black; and 62 mg cm^{-2} for the RuO_2. The catalyst loading for $NiCoO_4$ spinel was not recorded. ———, Pb $[Ru_{1.67}Pb_{0.33}]O_{(7\text{-}y)}$ (85 $m^2 g^{-1}$); ○, Pt black (24 $m^2 g^{-1}$); △, RuO_2 (4 $m^2 g^{-1}$); □, $NiCo_2O_4$; – – – –, Ni sheet.

talysts for oxygen reduction (see also Edgell et al. [10]. The catalysts used by Horowitz et al. were prepared by the alkaline solution synthesis technique, which yielded materials with surface areas in the range 50–200 $m^2 g^{-1}$ and with metallic or near metallic conductivity. The authors observed that electrocatalytic activity was maintained at a reasonable level in both oxygen evolution and reduction tests, which were conducted for periods of 500–1800 h, although evidence of finite catalyst solubility on the electrolyte was obtained. The good catalytic activity exhibited by these catalysts for both reduction and evolution of oxygen marks them as potentially interesting bifunctional mode catalysts. However, their stability needs to be further evaluated. Iwakura et al. [337] observed that the electrocatalytic activity of thermally prepared $Bi_2Ru_2O_7$ films supported on Pt for oxygen evolution in base was greatly enhanced by electrochemical pretreatment, such as potential cycling and pre-anodization. This improvement was explained in terms of an intrinsic enhancement of the catalytic activity of the surface, possibility due to the formation of higher-valent Ru species rather than an increase in surface area. Perhaps generation of active Ru groups is involved here due to restructuring and concomitant hydration of the surface. It may be speculated, however, that such an enhancement should also be accompanied by an increased tendency of the surface Ru species to undergo disssolution. Iwakura et al. [337] observed a Tafel slope of about 40 mV $decade^{-1}$ for oxygen evolution at bismuth ruthenate electrodes and a reaction order of close to 1. This kinetic information is similar to that previously reported by O'Grady et al. [227] for oxygen evolution in base and it was assumed that the same reaction mechanism applied in both cases, i.e. eqns. (46)–(48).

6.2 OXYGEN REDUCTION

Metal oxide electrodes have attracted attention as candidate materials for the oxygen electroreduction reaction as well as for the oxygen evolution reaction. Besides the constituent anions normally present at the oxide surface, i.e. OH^- and O^{2-}, which are natural intermediates for both oxygen evolution and reduction, other stable oxygen species like O^- and O_2^- have also been found at oxide surfaces [338]. It has been suggested that non-stoichiometry at the oxide electrode surface plays an important role in oxygen evolution [340] and reduction [341]. Participation of surface oxygen in oxygen evolution on $NiCo_2O_4$ has been indicated by isotopic experiments [304]. Metal oxides are widely used as heterogeneous oxidation catalysts with molecular oxygen interacting with surface cations [338, 339]. These cations at the surface of transition metal oxides have *d*-electron orbitals directed towards the electrolyte which can interact specifically with species in solution, including molecular oxygen. Dissociative adsorption of oxygen with cleavage of the O–O bond is known to occur on the surface of certain oxides; this interaction would avoid the peroxide path at the oxide electrode during oxygen reduction.

Metal oxides may be considered close to ideal oxygen electrocatalysts if both oxygen evolution and reduction occur at low overpotentials with minimal surface redox transformation and good long term stability of the oxide in the electrolyte. This concept is particularly relevant in the development of bifunctional oxygen electrodes. It has been suggested that changes of oxygen stoichiometry at the surface of oxide electrodes must be operative for an efficient catalysis of the oxygen electrode

$$MO_{(n-x)} \underset{\text{cathodic}}{\longleftarrow} MO \underset{\text{anodic}}{\longrightarrow} MO_{(n+y)}$$

Before discussing the electrokinetics of oxygen reduction at specific oxide electrodes, it seems appropriate to review the modes at which molecular oxygen interacts with oxide surfaces at least in the gas phase. Bielanski and Haber [338] have classified transition metal oxides into three categories depending on the degree of interaction of molecular oxygen with the oxide surface (in the absence of water dipoles).

(1) Oxides with cations that increase the degree of oxidation and supply electrons to the oxygen molecule, yielding highly reduced species like O^- or O^{2-}; i.e. p-type oxides as NiO, MnO, CoO, and Co_3O_4.

(2) Oxides adsorbing oxygen as superoxide radical ion O^-; i.e. non-stoichiometric, n-type oxides as ZnO and TiO_2.

(3) Oxides which essentially do not adsorb oxygen, such as MoO_3 and WO_3.

Yeager [342] has reviewed the different modes of interaction of oxygen molecules with transition metal cations known in inorganic coordination chemistry both in the solid state and in solution. According to this author, bonding of oxygen involves π and π^* orbitals of oxygen interacting with d orbitals of the transition metal with partial electron transfer. Figure 16(a) shows the three configurations known as the Griffith model [343], the Bridge, or side-on, model, and the Pauling, or end-on, model [344]. Which one applies to a particular oxide surface will depend on the electronic configuration of metallic cations, geometric factors (i.e. metal cation spacing at the surface), magnetic restrictions, and so on.

A detailed model for the oxygen reduction reaction at semiconductor oxide electrodes has been developed by Presnov and Trunov [341, 345, 346] based on concepts of coordination chemistry and local interaction of surface cation d-electrons at the oxide surface with HO^-, H_2O, and O_2 acceptor species in solution. The oxygen reduction reaction is assumed to take place at active sites associated with cations at the oxide surface in a higher oxidation state. These cations would act as donor–acceptor reduction (DAR) sites, with acceptor character with respect to the solid by capture of electrons and donor electronic properties with respect to species in solution. At the surface, the long-range oxide structure is lost and short-range coordination by hydroxide ions and water molecules in three octahedral positions may occur [Fig. 16(b)]. One hydroxide ion can compensate coulombically for the excess charge on surface M^{2+} cations with two coordinated water mole-

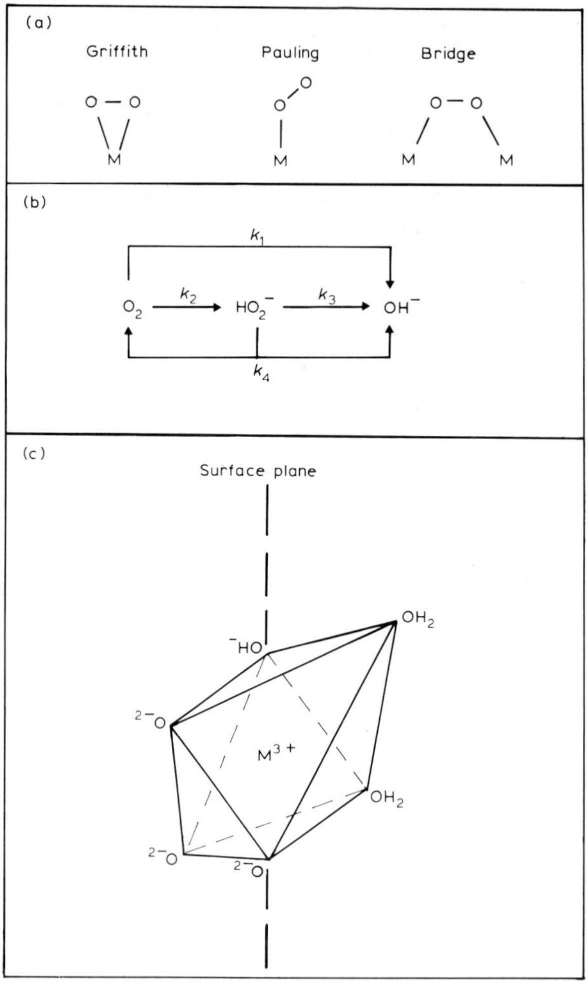

Fig. 16. (a) Adsorption modes of molecular oxygen [342]. (b) Different oxygen reduction pathways. (c) Coordination sphere of a surface cation at the oxide–electrolyte interface [345].

cules; on M^{3+} sites, an excess positive charge of one half still remains after coordination of a hydroxide ion. The asymmetric crystal field created at the surface produces surface electronic states by a Jahn Teller effect. These surface states can capture electrons from the bulk oxide to form excited cationic states [M^{3+} + e]. Because of its electron affinity, water can further react at these surface states and become coordinated to the surface. Simultaneously, oxygen vacancies at the surface provide sites of adsorption for oxygen containing species like HO^-, H_2O, and O_2. The interaction of oxygen with surface oxide cations can be envisaged as a substitution of water molecules in the coordination sphere of surface M^{3+} cations by molecular oxygen. This molecule is polarized by the cation and undergoes electron transfer to a π^* orbital. According to the model, the second electron transfer to an adsorbed oxygen molecule forms hydroperoxide ion which may follow one of two routes.

(1) Desorption due to the excess negative charge on the cationic site if it is coordinated to the surface cation via only one oxygen atom. This path would lead to hydroperoxide in solution.

(2) If oxygen is coordinated to the surface cation as a bidentate ligand via two oxygen atoms substituting two adjacent water molecules, four-electron reduction to hydroxyl ion may occur with subsequent desorption, and with no solution phase peroxide being generated.

The model of Presnov and Trunov [341,345] is based on the transition state theory and modern ideas on electron transfer at electrodes, and allows a quantitative calculation of the rate of oxygen electroreduction based on

(1) electrophysical and structural properties of the oxide surface;
(2) electronic structure of transition metal cations;
(3) coordination chemistry of the oxide surface;
(4) ion–dipole and dipole–dipole interactions at the oxide–electrolyte interface.

The model has been successful in predicting the effective electrocatalysts of cobalt-containing oxides and spinels.

6.2.1 Nickel and cobalt oxides

Yohe et al. [347] reported that the electroreduction of oxygen on lithiated NiO single crystals (mosaic ⟨100⟩ face) in alkaline solutions is very inhibited due to the potential distribution in the space charge region of the p-type oxide as shown by impedance measurements and reactions involving simple redox couples (Sect. 5.3). Lithium-doped polycrystalline NiO produced by high temperature oxidation of nickel exhibits similar behavior with production of peroxide as reduction product [342]. The open circuit potential of nickel oxide in oxygen-saturated alkaline solutions corresponds to the equilibrium potential of the O_2/H_2O_2 couple, which is more negative than the flat-band potential derived from capacitance measurements. Yeager and co-workers [347], using ^{18}O labelling techniques, have demonstrated that cleavage of the O–O bond does not occur during the oxidation of peroxide to oxygen and presumably the reverse reaction on nickel oxide. Yoneyama et al. [348] reported that the electroreduction of oxygen proceeds on single crystal NiO(Li) electrodes of ⟨100⟩ and ⟨110⟩ planes ($N_d = 5.86 \times 10^{21}\,cm^{-3}$) with 120 mV decade^{-1} Tafel slope and exchange current densities in the range $10^{-9}\,A\,cm^{-2}$ in 3.4 M KOH. Although nickel oxide is not a good electrocatalyst for the reduction of oxygen at room temperature, however at higher temperatures the rate of reaction in concentrated alkaline solutions rapidly increases above 150–170°C.

Tseung et al. [349] studied this reaction at NiO doped with lithium and chromium and reported that the activity increases with the content of Ni^{3+} in the oxide, which is promoted by Li doping. It was also observed that a steep increase of conductivity and catalytic activity occurred above 220°C [175]; at this point a break was observed in the Arrhenius plot. Tseung et al. have assumed that, above the Neel point (onset of disorder of magnetic domains), oxygen chemisorption changes from non-dissociative to dissociative in analogy with the gas-phase chemisorption of oxygen at nickel oxides [350]. A change in the adsorption mode would occur from "end on" at a single active site leading to peroxide to "side on" at two adjacent surface nickel cations, which favors the 4-electron reduction path (Fig. 16(a)]. Unfortunately, several experimental factors like large surface area, porosity, ohmic drops, and ill-defined mass transport conditions handicapped the mechanistic interpretation of these results. Tarasevich et al. [161] have pointed out that the thermal homogeneous reaction

$$\tfrac{3}{2}\,O_2 + 2\,OH^- \longrightarrow 2\,O_2^- + H_2O \tag{63}$$

above 170°C followed by electroreduction of superoxide

$$O_2^- + 2\,H_2O + 3\,e \longrightarrow 4\,OH^- \tag{64}$$

may explain the acceleration of oxygen reduction on nickel oxide at elevated temperatures.

The reduction of oxygen has been studied on nickel metal electrodes with reference to the degree of surface oxidation (both thermally and electrochemically); the reaction slows down with increase of surface oxidation [351–353]. Rotating ring–disk electrode studies have revealed that peroxide is formed as a stable intermediate of oxygen reduction, the yield of which increases with oxidation, and its further reduction, or catalytic decomposition, is inhibited by the oxidized surface. According to diagnostic criteria obtained with the RRDE, the reaction proceeds on oxidized nickel by two parallel mechanistic paths: (a) direct reduction to HO^- ions without generation of solution phase peroxide and (b) via peroxide, which can desorb and be detected in solution [354].

Savy [355] reported results for the electroreduction of oxygen on lithium-doped cobalt oxide obtained by thermal oxidation of Co–Li alloys. According to this author's results, the rate of steady-state reduction in alkaline electrolyte shows a maximum for 0.15 at. % Li. Russian researchers [356–359] have reported high catalytic activity for oxygen reduction on oxidized cobalt as being due to the presence of Co_3O_4 spinel. Fabjan et al. [360] studied the reaction on passive cobalt in 1 M NaOH, and reported Tafel slopes in the range 75–80 mV decade^{-1} and a fractional order of 0.5 in oxygen. According to these authors, the reaction shows higher activity on oxidized cobalt than on oxidized nickel.

6.2.2 Perovskite-type oxides

Perovskite-type oxides are well known oxidation catalysts in the gas phase [259]. In 1970, Meadowcroft suggested that $LaCoO_3$ doped with strontium was a less expensive alternative to platinum for air cathodes in alkaline solutions. Tseung and Bevan [361] reported that $La_{0.5}Sr_{0.5}CoO_3$ exhibits the reversible oxygen electrode potential in alkaline oxygenated solution. This unusual result with PTFE-bonded porous electrodes was explained on the basis of dissociative adsorption of O_2 on the perovskite mixed oxide [362,363]. However, the results of Tseung and Bevan could not be reproduced by other investigators [176, 364, 365] who observed long equilibration times, especially under cathodic polarization. van Buren et al. [176] concluded that the reversible behavior reported by Tseung and Bevan could result from unstable data observations. The slow time response of these electrodes has been discussed in Sect. 5.3 in connection with simple electron transfer reations. Kudo et al. [34,366] have established that $Ln_{(1-x)}M_xM'O_3$ (Ln = lanthanide element, M = alkaline earth and M′ = transition metal element) undergo simultaneous reduction of the oxide with oxygen reduction at cathodic potentials. These authors assumed that the process involves reduction of the bulk oxide and that oxygen ion vacancies are the mobile species at room temperature, thus regarding the material as a mixed electronic–ionic conductor. They developed a galvanostatic method to measure the diffusion constant [34] and reported values in the range 10^{-11} to $10^{-14}\,cm^2\,s^{-1}$ at 25°C. van Buren et al. [367,368], based on the same assumption that oxygen vacancies are mobile species and using a potentiostatic method which takes into account the oxide particle's microscopic size and current transients, reported a diffusion coefficient $5 \times 10^{-15}\,cm^2\,s^{-1}$ for $La_{0.5}Sr_{0.5}CoO_3$ in contact with 5 M KOH at 25°C. The model is based on the experimental observations:

(1) bulk oxygen deficiency (vacancies) as Sr doping increases the formation of Co^{4+};

(2) thermogravimetry shows oxygen deficiency at high temperature;

(3) decrease of electrical conductivity with partial pressure of oxygen, though changes are not detectable below 100°C; and

(4) dependence of the open circuit potential (OCP) of the oxide electrode in contact with aqueous electrolyte on the oxide preparation temperature (degree of non-stoichiometry as indicated in, for example, ref. 340).

However, neither Kudo et al. [34] nor van Buren et al. [367,368] have proved conclusively that bulk diffusion in the oxides was the cause for their transient measurements at room temperature nor have they given direct evidence of oxygen vacancies in the oxides to be the mobile species. Matsumoto and Sato [369] studied the reduction of oxygen on $La_{0.7}Pb_{0.3}MnO_3$ single crystal and demonstrated that the reaction does not occur with involvement of diffusion of oxygen vacancies in the lattice, but strictly as a Faradaic interfacial reaction. More recently, the investigation of diffusion profiles carried out at Imperial College (London) with ^{18}O-equilibrated perovskite sintered pellets have shown that oxygen self-diffusion values were below 10^{-16} to $10^{-17}\,cm^2\,s^{-1}$ in the temperature range 100–400°C. Furthermore, potentiostatic transients of $LaNiO_3$ electrodes were similar to those reported by van Buren et al. for the La–Co perovskite and could be fitted by several kinetic models [48]. Hibbert and Tseung [370] have made a correlation between the activation energy of electrochemical oxygen reduction on $La_{0.5}Sr_{0.5}CoO_3$ ($54.4\,kJ\,mol^{-1}$) with that of the gas-phase oxygen homomolecular exchange on the oxide ($66.1\,kJ\,mol^{-1}$).

$$^{18}O_2 + {}^{16}O_2 \rightleftharpoons 2\,{}^{16}O^{18}O \tag{65}$$

This correlation, together with the fractional reaction order of 0.5 in oxygen and the OCP, were taken by the authors [370] as evidence of dissociative adsorption of oxygen molecules on the La–Co perovskite oxide. This type of adsorption was considered to be related to the magnetic properties of the oxide. Tamura et al. [264] have placed more emphasis on the electronic properties of the perovskite oxide electrodes than on the magnetic behavior. In 1975, they reported that $LaNiO_3$ was an attractive material as an oxygen cathode in alkaline solutions with higher catalytic activity than platinum [371]. However, it has been shown that the activity reported was not referred to the true electrochemical area of the electrode [48].

In a series of papers, Matsumoto et al. [86, 280, 371–374] reported the results of extensive studies of the electrocatalysis of oxygen reduction on $LaNiO_3$ and related perovskite oxide electrodes from the viewpoint of their bulk solid state structural and electronic properties. The main conclusions of their studies are

(1) Distortion of the $La_{(1-x)}Ln_xNiO_3$ (Ln = Nd, Sm) lattice with increasing the ionic radius of the Ln element causes a decrease in the catalytic activity, which has been related to the decrease in the orbital overlapping in the O–Ni–O arrays;

(2) Reduction of $LaNiO_3$ oxide occurs in the same potential range where oxygen reduction takes place. The reduction of the solid oxide surface results in a large increase of the electrode resistance at cathodic polarizations and loss of electrocatalytic activity; and

(3) The influence of the oxide resistivity on oxygen electrocatalytic activity was studied in a series of oxides: $La_{(1-x)}Sr_xMnO_3$ and $LaNi_{(1-x)}M_xO_3$ (M = Fe, Co and V) of similar BET surface areas. It was found that j_o increased with the degree of Sr substitution and decreased with nickel substitution.

Electrocatalytic activity, as reflected in j_o, decreased with increase in the oxide resistivity and a relation between j_o and the resistivity has been derived theoretically

$$j_o = -\frac{1}{n}\ln \varrho + C \tag{66}$$

with n and C related to the electron mobility and its effective mass in the solid according to $u = C(m^*)^{-n}$ on the basis of the theoretical derivation for the exchange current density by Dogonadze and Chizmadzev [375]. Based on a study of a large number of metallic perovskite-related oxide electrodes, Matsumoto and co-workers [264] developed the δ^* band theory of oxygen electrocatalysts. They concluded that the relatively high conductivity of these oxides and their electrocatalytic activity was related to the presence of a partially filled σ^* band in the oxide. This idea was based on the model of Presnov and Trunov [341, 345] and considered that e_g orbitals of the metal cation at the surface (at the end of a σ^* conducting chain) would have a favorable interaction with π^* orbitals of molecular oxygen or water.

Calvo et al. [81] reported the behavior of mixed oxides in the $La_{(n+1)}Ni_nO_{(3n+1)}$ homologous series: e.g. $La_4Ni_3O_{10}$, $La_3Ni_2O_7$, and La_2NiO_4. Oxide powders of these compositions were carefully characterized and placed in the shallow disk cavity of a rotating single–disk electrode in order to study the surface redox electrochemistry and steady-state oxygen electroreduction in a controlled manner. These studies suggested that all the oxides exhibit similar behavior which is almost independent of the bulk composition. This observation is important as previous interpretations of electrocatalysis on perovskite oxides have stressed the importance of bulk electronic structure, the presence of Ni^{2+} or Ni^{3+} cations, or the mobility of oxygen vacancies. Furthermore, the same pattern of behavior for steady-state oxygen electroreduction is exhibited at low and high overpotentials for all perovskite oxides studied: $La_{(n+1)}Ni_nO_{(3n+1)}$, $La_{0.5}Sr_{0.5}MnO_3$, $La_{0.5}Sr_{0.5}CoO_3$, and $LaNiO_3$, the latter in both hot-pressed sintered and in composite electrodes with hydrophobic plastic binders. A common pattern for all Ni, Fe, Co and Mn perovskite oxides has been observed: a Tafel region (ca. 40 mV decade^{-1}) at low overpotentials and a region of almost potential-independent rate of oxygen reduction at high overpotentials. The specific electrocatalytic activity, however, was observed to depend strongly on the electrochemical surface area of the oxide electrode in contact with the electrolyte, decreasing in the order $LaNiO_3$ (10.6 m^2 g^{-1} BET area) > $La_{0.7}Sr_{0.3}MnO_3$ (26.6 m^2 g^{-1}) > $La_{0.5}Sr_{0.5}CoO_3$ (40.4 m^2 g^{-1}) as shown in Fig. 17.

Bronoel et al. [376] have compared the electrocatalytic activity of several perovskite ferrites, $La_{(1-x)}Sr_xFeO_{(3-y)}$ with $0.1 < x < 0.95$ and $0.06 < y < 0.27$, as a function of oxygen vacancy content in the bulk oxide. Although reduction of the oxide material was reported to take place simultaneously with the Faradaic reduction of oxygen, it could be demonstrated that the activity for oxygen reduction increased with y [376]. Since the authors have not presented a characterization of the oxide surfaces, particularly with respect to the electrochemically active area, in each case their conclusions are at question, because both the bulk stoichiometry (content of oxygen vacancies) and the oxide particle size (surface area) strongly depend upon the thermal conditions under which the oxides were prepared. Takeda et al. [377] examined total substitution of La in the ABO_3 perovskite structure of oxides of composition $SrFeO_{3-\delta}$ and $SrCoO_{3-\delta}$ as oxygen bifunctional electrodes. Both types of oxide showed high electrocatalytic activity for the oxygen electrode reactions, and $SrFeO_{3-\delta}$ had the highest activity for $0.24 < \delta < 0.29$. A detailed analysis of the effect of surface area in perovskite oxides on the activity for oxygen reduction has shown that materials

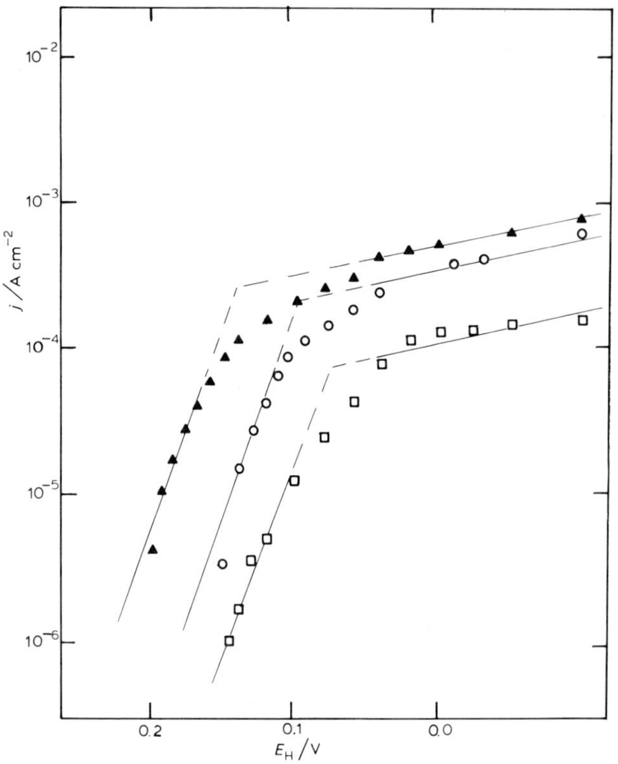

Fig. 17. Steady-state curves of oxygen reduction on perovskite oxide electrodes after mass transport correction using the RDE. ▲, $LaNiO_3$; ○, $La_{0.7}Sr_{0.3}MnO_3$; □, $La_{0.5}Sr_{0.5}CoO_3$ [81].

with the same composition and structure displayed a large range of activities depending on the electrochemical surface area as measured by BET and interfacial capacity [81]. In $LaNiO_3$, the perovskite structure stabilises Ni(III) cations in the bulk oxide lattice, although this is no longer valid at the surface, particularly at the oxide–aqueous interface. The surface composition of perovskite oxides in contact with alkaline solutions evidently depends more on the electrode potential and solution composition than on the bulk oxide composition.

An important feature of oxygen reduction on perovskite oxides is that hydrogen peroxide is not detected in significant quantities in the solution by using the rotating ring–disk electrode (RRDE) for either of the La–Ni, La–Co, and La–Mn perovskite oxides [48]. In the case of $LaNiO_3$, Matsumoto et al. [374] have related the small yield of peroxide (less than 3%) to NiO segregated as a small impurity since the reduction of oxygen proceeds by the peroxide path on nickel oxide. Yeager and co-workers [378] reported that hydrogen peroxide was an intermediate in the reduction of oxygen on $La_{0.5}Sr_{0.5}CoO_3$ perovskite oxide supported on carbon in a porous thin layer. However, the carbon support was primarily responsible for the yield of peroxide in solution in that case. Wang and Yeager [379] examined the catalytic decomposition of hydrogen peroxide on $La_{(1-x)}Sr_xCoO_3$ and reported that the reaction is first order with respect to peroxide, and that the activity increased with strontium doping of the perovskite, showing a maximum at a preparation temperature of 600°C.

The study of the electrochemical behavior of peroxide has indicated two possible routes of elimination at the $LaNiO_3$–aqueous interface [81]: electrochemical dismutation at low overvoltage and electroreduction at high overpotentials. Table 4 shows a comparison of electrokinetic data for several perovskite oxides.

A common electrokinetic pattern for the reduction of oxygen in alkaline solutions on all transition metal perovskite oxides is apparent from the analysis of the data in Table 4. Exchange current densities have not been included because no reliable data exist for true electrochemical surface area for these oxide electrodes and since the number density of active sites, and hence the catalytic activity, strongly depends on the true surface area, a fair comparison of reported data cannot be made. Studies with the rotating ring–disk electrode (RRDE) at Ni, Co, and Mn perovskite oxide electrodes have indicated that the reduction of oxygen proceeds with negligible yield of peroxide in solution and this important feature is related to the perovskite structure. With the exception of $La_{0.7}Pb_{0.3}MnO_3$, in all cases a Tafel region (b_c = 40–50 mV decade^{-1}) has been observed at low overpotentials. At higher overpotentials, an almost potential-independent current below the con-

TABLE 4

Kinetic data for the reduction of oxygen on various perovskite oxide electrodes

Electrode	Solution	T (°C)	b_c (mV decade^{-1})	Ref.
$LaNiO_3$	1 M NaOH	25	47	376
$LaNiO_3$	1 M NaOH	25	47	48
$LaNi_{(1-x)}Fe_xO_3$ ($0 < x < 0.5$)	1 M NaOH		47	280
$LaNi_{(1-x)}Co_xO_3$ ($0 < x < 0.3$)	1 M NaOH	25	47	280
$La_{0.5}Sr_{0.5}CoO_3$	45 wt.% KOH	25–80		372
	1 M NaOH	25	40	48
$Nd_{0.5}Sr_{0.5}CoO_3$	45 wt.% KOH	25–80		372
$La_{(1-x)}Sr_xMnO_3$	1 M NaOH	25	47	281
$La_{0.7}Pb_{0.3}MnO_3$	1 M NaOH	25	80	371
$La_{(n+1)}Ni_nO_{(3n+1)}$ (n = 1,2,3,)	1 M NaOH	25	40	81
$La_{0.7}Sr_{0.3}MnO_3$	1 M NaOH	25	40	81
$La_{0.5}Sr_{0.5}MnO_3$	1 M NaOH		0.100	382
$La_{0.5}Ca_{0.5}MnO_3$	1 M NaOH		115	382
$La_{0.5}Ba_{0.5}MnO_3$	1 M NaOH		115	382

vective–diffusion limit was apparent for all perovskites. This limiting current, however, was sensitive to the oxygen partial pressure (first order), surface area, and composition of the oxide. Raj et al. [380] have reported studies of oxygen reduction and evolution on manganates $La_{0.5}Sr_{0.5}MnO_3$, $La_{0.5}Ca_{0.5}MnO_3$, and $La_{0.5}Ba_{0.5}MnO_3$ with the rotating ring–disk electrode technique in alkaline solutions. The cyclic voltammetry of these materials show similarities with that of other oxides with the perovskite structure and the La–Mn oxides were reported to be more stable to electrochemical reduction. Convective–reduction limiting currents for the reduction of oxygen were reported to follow the Levich equation at high cathodic overpotentials. From ring–disk studies, the authors [380] conclude that the four-electron reduction to hydroxide ion is operative, with no appreciable peroxide detected in solution. Electrokinetic data for these oxides are compared in Table 4 where it can be seen that the values of the cathodic Tafel slope are higher than those reported by other authors. Hibbert and Tseung [370] reported a reaction order 0.5 in oxygen on La–Co porous PTFE-bonded oxides and, based on this evidence, the authors assumed that oxygen dissociatively adsorbs on the oxide. However, both $LaCoO_3$ and $LaNiO_3$ adsorb oxygen in the gas phase in the low-temperature range as molecular species with kinetics dependent on the transition metal [381]. The results of Hibbert and Tseung [370] may possibly be attributed to the porous nature of their electrodes. Karlsson [382] studied $LaNiO_3$ as an oxygen cathode in alkaline solutions and reported that low-temperature preparation of the oxide enhances the heterogeneous reduction with surface segregation of NiO and $La(OH)_3$. In another report, Karlsson [88] reported the performance of various TFE-bonded perovskite oxides as air cathodes. Some of these oxides which contained high concentrations of Mn^{4+} gave acceptable performance.

For oxygen reduction on $LaNiO_3$, Matsumoto and co-workers proposed the mechanism

$$O_2 \longrightarrow O_2(\mathrm{ads}) \qquad \mathrm{rds} \tag{67}$$

$$O_2(\mathrm{ads}) + e \longrightarrow O_2^- \tag{68}$$

$$O_2^-(\mathrm{ads}) + H_2O \longrightarrow O_2H(\mathrm{ads}) + OH^- \qquad \mathrm{rds} \tag{69}$$

$$O_2H(\mathrm{ads}) + e \longrightarrow O_2H^-(\mathrm{ads}) \tag{70}$$

followed by rapid elimination of O_2H^- (ads) by an oxygen vacancy at the oxide surface as proposed for spinel oxides like Co_2NiO_4 [341, 345]. This mechanism, which is in good agreement with the model of oxygen reduction electrocatalysts on oxides by Trunov and Presnov [345], would be applicable to all perovskite oxides exhibiting similar kinetic parameters. The transition metal at the surface would be expected to be the oxygen adsorption site. The limiting current at cathodic potentials has been associated by Matsumoto and co-workers [264] to the slow adsorption of molecular oxygen becoming the rate-determining step [step (67) in the mechanistic sequence].

A similar cathodic limiting current has also been observed for the electroreduction of peroxide on $LaNiO_3$ (Fig. 18) [48] and this behavior occurs at potentials where the reduction of the solid surface takes place changing the potential distribution at the oxide–electrolyte interface. This change of surface properties is quite similar to the behavior of NiO [347] under cathodic polarization and is also reflected in the inhibition of electron transfer to or from redox couples in solution [81] and capacitance Mott–Schottky type plots [87, 290, 291] of these interfaces.

Bockris et al. [87, 290, 291] have recently reported results of a comprehensive program of surface characterization of a large number of perovskite oxide electrodes in oxygen evolution investigations. Anodic and cathodic oxygen reactions were studied in detail as a function of the solid-state surface properties of these materials. Capacity–potential curves were analysed in terms of the Mott–Schottky treatment and indicated that the potential distribution in the oxide corresponds to a depletion of electrons at the oxide electrode surface in the potential region where oxygen reduction

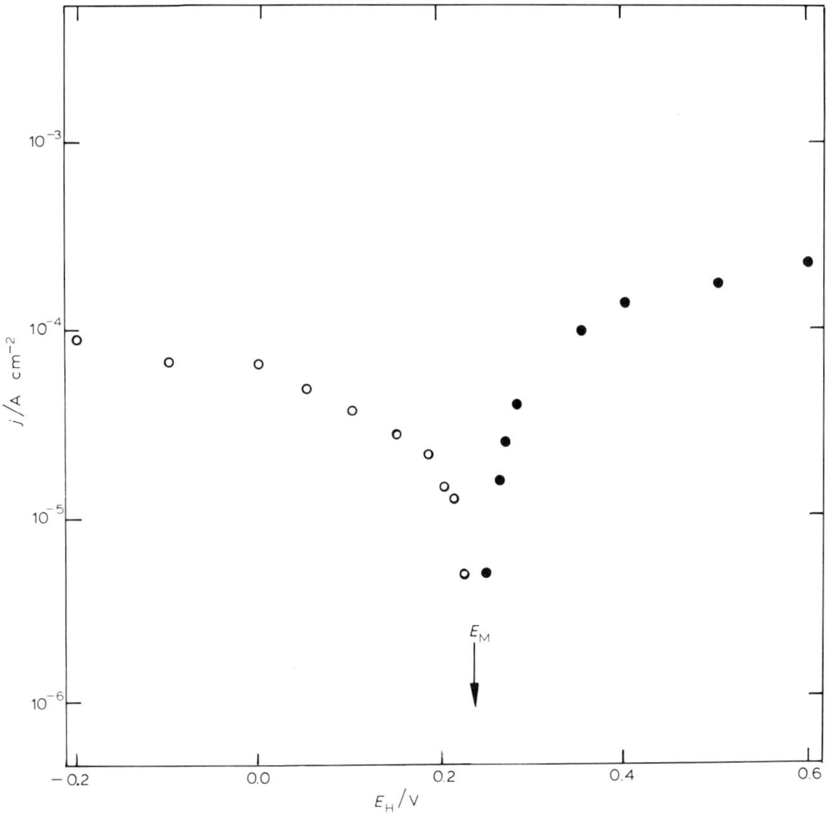

Fig. 18. Steady-state log j–E curves for H_2O_2 reduction and oxidation at a $LaNiO_3$ ($1.2\,m^2\,g^{-1}$) rotating disk electrode in 0.1 M KOH at 25°C after correction for mass transport in solution. H_2O_2 concentration = 1 mM [48].

proceeds at a commensurable rate. Bockris et al. [87, 290, 291] have interpreted this as indicative of p-type semiconducting behavior, even for materials with metallic conductivity in the bulk, like $LaNiO_3$. It should be stressed that, on these oxides, the surface chemical composition and the electronic properties change with electrode potential [81]. For instance, in the potential range where the C^{-2} vs. E plot is linear, reduction of the surface is observed with very slow kinetics. Vondrak and Dolezal [383] failed to find any photoelectrochemical effect or evidence of a Mott–Schottky barrier at the $La_{0.5}Sr_{0.5}CoO_3$ oxide–electrolyte interface at pH 5 under anodic polarization, but did observe a rectifying effect at the oxide–metal contact. Oxygen evolution and reduction takes place on perovskite oxides at surfaces of different composition and potential distribution. According to the values of pzc reported by Bockris et al. [87, 290, 291] for the La–Ni, La–Co, and La–Mn perovskite oxides, oxygen reduction proceeds in alkaline solutions at oxide electrode surfaces with excess negative charge ($> 100\,\mu C\,cm^{-2}$ true area) where $M{-}O^-$ groups predominate at the surface.

Sodium tungsten bronzes, Na_xWO_3, with the perovskite-like structure [384–386] have been investigated as oxygen cathode materials in acid solutions, but they have to be promoted by traces of platinum on the surface [386–389].

In general, although surface protonation of perovskite oxides in contact with aqueous electrolytes and consequent build up of a potential barrier at the oxide surface under cathodic polarization represents a serious drawback, perovskite oxide electrodes are good candidate electrocatalyst materials in molten carbonate [390] and high-temperature solid electrolytes [391].

6.2.3 Spinel-type oxides

A number of oxides with the spinel structure which exhibit acceptable conductivity have been reported to be good oxygen electrocatalysts for oxygen reduction and evolution. The effect of the spinel structure on oxygen reduction kinetics has been discussed by Davidson [392] who compared several mixed oxides with the corresponding spinels. A comparison of values in Table 5 shows that, in all cases, the spinel oxides have better performance than the simple mixtures of the corresponding single oxides. Tarasevich and Efremov [393] discussed the increase of electrical conductivity, carrier concentration, carrier mobility, and electrochemical activity for the oxygen reduction reaction in the Ni–Co system for both single oxides and spinel systems. In line with the opinion of Bockris et al. [87, 290, 291] for perovskite oxide electrodes, Tarasevich and Efremov [393] suggested that a common approach to the study of both the oxygen evolution and reduction reactions needs to be made for a proper understanding of the oxygen electrode on spinel oxides. The importance of the geometric factor in the electrocatalysis of oxygen reduction has been considered by Yeager [342] who suggested that "side on," bridge-type adsorption of oxygen [see Fig. 16(a)] may occur at

TABLE 5

Current density ($mA\,cm^{-2}$) at 75 V (RHE) for oxygen reduction on various mixed oxides and spinels in alkaline solutions

Oxide	Current density	Oxide	Current density
$NiO-Fe_2O_3$	19.5	$CuO-Cr_2O_3$	23.4
$NiFe_2O_4$	67.5	$CuCr_2O_4$	47.0
$NiO-Cr_2O_3$	18.2	$CoO-Cr_2O_3$	25.2
$NiCr_2O_4$	55.5	$CoCr_2O_4$	38.5
$CoO-Fe_2O_3$	25.2	$CuO-Fe_2O_3$	18.5
$CoFe_2O_4$	58.6	$CuFe_2O_4$	29.5
$CoO-Al_2O_3$	22.0	$NiO-Al_2O_3$	23.8
$CoAl_2O_4$	53.0	$NiAl_2O_4$	28.4
$CuO-Al_2O_3$	22.0		
$CuAl_2O_4$	26.0		

oxide spinels with the proper lattice spacing, plus partially filled d_{xz} or d_{yz} orbitals at surface cations to form bonds with sp^2 orbitals of molecular oxygen.

Oxygen reduction on $NiCo_2O_4$ has been studied by several authors [352, 353, 394–398]. In this mixed spinel oxide, Co^{2+} cations are tetrahedrally coordinated, and Ni^{2+} and Co^{3+} cations octahedrally coordinated in a cubic structure of close-packed oxygen anions, which is probably best represented [395] as $Co^{2+}[Ni^{2+}Co^{3+}]O_3^{2-}O^-$ where [ ] denotes an octahedral site in the lattice. Tseung and Yeung [396] reported high catalytic activity on Teflon-bonded $NiCo_2O_4$ and attributed this to the presence of loosely bound O^- species on the surface. They proposed a double catalytic effect of the spinel oxide electrode, i.e. as a cathode for the oxygen reduction reaction and as a catalyst for the decomposition of peroxide according to

$$O_2 + H_2O + 2\,e \;=\; HO_2^- + OH^- \tag{71}$$

and

$$2\,HO_2^- \;=\; O_2 + 2\,HO^- \tag{72}$$

the latter reaction occurring inside the porous structure of the composite electrode. It is well established that the catalytic activity for peroxide decomposition is poor on NiO [353, 398]. Trunov et al. [397] correlated the electrophysical and electrocatalytic properties of the semiconducting oxides NiO, $NiCo_2O_4$, and Co_3O_4 for the reduction of oxygen and concluded that the reaction proceeds by the peroxide path on NiO, but that on spinel oxides, oxygen is reduced to hydroxide ion with less than 1–3% peroxide yield in solution according to results obtained with the rotating ring–disk electrode [353, 399]. This difference in catalytic behavior for oxides of comparable electrophysical properties (concentration and mobility of carriers) may be explained by an intrinsic structural factor and the presence of high effective charge on cobalt cations at the oxide surface. Singh et al. [301] reported that

$NiCo_2O_4$ is not stable in the alkaline environment in the cathodic conditions under which oxygen reduction proceeds. King and Tseung [394] found that the catalytic activity decreases considerably within a few hours of beginning electrolysis, and X-ray diffraction of the reduced electrode material showed transformation of the original spinel into inactive CoO, Co_2O_3, NiO, and NiOOH. The reduction of the oxide electrode materials simultaneously with the electroreduction of oxygen seems to be a general phenomenon, characteristic not only of spinels but also of perovskites and other oxides as shown by Bronoel et al. [376].

Since its introduction by Frumkin and Nekrasov in 1959 [400], the rotating ring–disk electrode technique has been extensively used in mechanistic studies of the oxygen reduction reaction. Damjanovic et al. [401] developed some twenty years ago a diagnostic test to distinguish between two possible parallel pathways for the oxygen reduction reaction: a direct four-electron reduction to water or hydroxide ion and two sequential two-electron consecutive steps with soluble, stable peroxide intermediate detected at the ring electrode downstream in the solution [Fig. 16(b)]. The diagnostic test is based on the comparison of the total flux of electrons at the disk, j_d, and the outcoming flux of peroxide, j_r, as a function of rotation frequency

$$\frac{N_o j_d}{j_r} = J + S\omega^{-1/2} \tag{73}$$

where j_d and j_r are the disk and ring currents, respectively, and N_o the collection efficiency of the system. This constant depends only on the electrode geometry and can be calculated from the radii of the disk and the ring electrodes [85]. It is interesting to note that J and S are solely related to the diffusion and rate coefficients; ω is the rotation frequency of the electrode. Wroblowa et al. [402] considered the adsorption–desorption equilibrium of intermediate peroxide at the surface of the disk electrode in eqn. (73) and introduced a new diagnostic linear plot (J vs. S) at different potentials. For this plot to be linear, both direct and sequential paths should have the same potential dependence. No unambiguous mechanism interpretation, however, can be made since several mechanisms can accommodate the same experimental data as shown by Zurilla et al. [403]. Strongly adsorbed peroxide, for instance, may be involved in the four-electron reduction, but should not desorb into the solution before being reduced, as otherwise the sequential path would apply. It is not possible, therefore, to distinguish between the four-electron mechanism through adsorbed peroxide and direct reduction with cleavage of the O–O bond before the formation of the intermediate.

Cobaltite spinels, MCo_2O_4, deposited on carbon supports have been extensively studied with the rotating ring–disk electrode by the Russian school. Ring–disk analysis has shown that both direct (k_1) and sequential (k_2) paths are operative for oxygen reduction, with the largest k_1/k_2 ratio (ca. 10) for $NiCo_2O_4$. It has been suggested that the promotion of the activity for the reduction of oxygen to hydroxide with respect to carbon was due to the increased catalytic activity for peroxide decomposition. The activity decreases in the order $MnCo_2O_4$ < $MgCo_2O_4$ < Co_3O_4 < < C [404]. The reduction of oxygen on a series of manganite-type, mixed spinel oxides of general formula $A_xB_yMn_{3-(x+y)}O_4$ (where A = Zn, Cr, Al; B = Ni, Cu) with $x > 0$ and $y < 1$, has been reported by Nguyen Cong et al. [405–407]. The reaction was assumed to occur at Mn^{4+} active surface sites with Mn^{3+} contributing to the electronic transport in the bulk of the oxide. A redox catalytic mechanism has been suggested

$$\mathrm{Mn(IV)} + \mathrm{e} = \mathrm{Mn(III)} \tag{74}$$

$$\mathrm{Mn(III)} + \mathrm{O_2} = \mathrm{Mn(IV)} + \mathrm{O_2^-} \tag{75}$$

The main features of oxygen electrocatalysis on spinel oxide cathodes as presented by Tarasevich and Efremov [393] are as follows. According to the model of Presnov and Trunov [341,345], surface cobalt cations in Ni–Co oxides are seen as active "donor–acceptor" centers with respect to electrons in the bulk oxide and oxygen species in solution. Unlike the view of Tseung et al. [394–396], who considered the rate of oxygen adsorption as limiting the overall kinetics, Tarasevich suggested that the formation of chemisorbed O_2^- (ads) and O^- (ads) is rate-determining according to the mechanism

$$\mathrm{M} + \mathrm{O_2} = \mathrm{MO_2} \tag{76}$$

$$\mathrm{MO_2} + \mathrm{e} = \mathrm{MO_2^-} \tag{77}$$

$$\mathrm{MO_2^-} + \mathrm{H_2O} = \mathrm{MO_2H} + \mathrm{OH^-} \tag{78}$$

where M represents a metal cation at the oxide surface, followed by cleavage of the O–O bond or formation of peroxide. In the low overpotential range, eqn. (77), via a barrierless mechanism [78], or eqn. (78) are rate determining. At higher polarizations, eqn. (77) occurring at M^{3+} sites would be the limiting kinetic process. A linear correlation has been reported from the gas phase which would depend on the nature of the surface M^{3+} cations.

The absence of peroxide in solution as a soluble reduction intermediate on Ni–Co spinel oxides is explained by assuming the occurrence of dissociative adsorption of oxygen on the basis of oxygen homomolecular exchange results [393]. Other authors hold the view that the role of the catalyst oxide is to decompose in an efficient manner the intermediate peroxide formed [355]. The elucidation of the reaction order with respect to hydroxide ion is complicated by the different coordinating environments of M^{2+} and M^{3+} cations if the model of Presnov and Trunov is followed [341, 345]. Schiffrin [408] has pointed out that the electrochemical reaction order in hydroxide ions for oxygen reactions at oxide electrodes deserves a careful analysis since this kinetic parameter may not only reflect the existence of a protonation step at or before the rate-determining step, but its value may also arise from the acid–base properties of the oxide surface as discussed in Sect. 3. Hydroxylation of the surface groups, in particular those species at the end of a conduction chain", may result in pseudo-orders in hydroxide ions. Bagotsky and Shumilova [353] have studied Ni–Co oxides at thermally, or electrochemically, oxidized metals or alloys below 300°C and reported an optimum activity for oxides with the spinel structure Co_3O_4. Negligible production of peroxide was found on $NiCo_2O_4$ and the authors suggested that this may be explained by the occurrence of dissociative adsorption of oxygen. Other authors [409–412] have suggested that the reaction proceeds through the intermediate formation of hydrogen peroxide and that the promoting effect of spinel catalysts is to accelerate the decomposition of the intermediate.

The catalytic decomposition of hydrogen peroxide as catalysed by oxide surfaces has been extensively studied by several authors [413–415]. The activity on spinel cobaltite oxides supported on carbon has been reported [409–412, 416–419] to follow the order

$$NiCo_2O_4 > CuCo_2O_4 > Co_3O_4 > MnCo_2O_4 > MgCo_2O_4$$

Yeager and co-workers [379, 419] studied the kinetics of peroxide decomposition on dispersed oxide powders by measuring changes in the convective–diffusion limiting current for peroxide oxidation at a rotating gold electrode immersed in the liquid dispersion. The current decay was observed to be proportional to the peroxide concentration decay due to the catalytic decomposition process. On perovskite [379] and spinel oxides [419], it follows first-order kinetics described by the equation

$$\ln k = -S \ln [OH_2^-] + J \tag{79}$$

with $S \sim 1$. The authors proposed a mechanism of hydroperoxy ion dismutation to account for the observed kinetics, with the simultaneous occurrence of

$$HO_2^- + HO^- = O_2 + H_2O + 2\,e \tag{80}$$

and

$$HO_2^- + H_2O + 2\,e = 3\,HO^- \tag{81}$$

at local cells at the oxide surface and resulting in the establishment of a mixed potential, E_M, at the oxide electrode. Since the exchange current density for eqn. (80) is generally much higher than that for eqn. (81), E_m was found to be close to the standard potential for the OH_2^-/O_2 couple. By combining eqns. (80) and (81), one obtains the overall peroxide decomposition reaction [eqn. (72)].

Tarasevich et al. [417, 420] employed the rotating disk electrode with an oxide disk electrode to study the electrochemical reactions of peroxide in conjunction with a gasometric method, by means of which the rate of the peroxide decomposition via a purely chemical pathway [eqn. (72)] could be followed independently. The authors compared the rate of gas evolution and peroxide electroreduction and oxidation, respectively, as a function of electrode potential and attributed the difference of these rates to chemical decomposition of the peroxide [eqn. (72)]. As already mentioned, the so-called "chemical decomposition" pathway has an electrochemical aspect with the oxidation and reduction of peroxide taking place simultaneously at the electrode surface at E_m. This heterogeneous dismutation must be distinguished, however, from the well-known free-radical chemical decomposition of peroxide [421]. The rate of chemical peroxide decomposition on Co_3O_4 is reported to be potential-independent except in a narrow potential range close to E_m, where the Co^{3+}/Co^{2+} ratio strongly depends on potential [417]. At high anodic and cathodic polarizations, the direct electrochemical transformation of peroxide takes over with less gasometric activity being observed. Tarasevich et al. [417] assumed that adsorption of peroxide is the first step of both chemical and electrochemical peroxide decomposition, and takes place at surface metal cations (active sites) according to

$$(Co^{3+} + e) + HO_2^- \longrightarrow (Co^{3+})(HO_2^{2-}) \tag{82}$$

$$(Co^{3+}) + HO_2^- \longrightarrow (Co^{3+}/ + e)(HO_2) \tag{83}$$

with subsequent oxidation and reduction of the surface complexes either in local cells or through involvement of a "conducting chain" in the bulk oxide. On manganese oxides, MnO_x, it has been suggested that peroxide catalytic decomposition proceeds via adsorbed peroxide and involves consecutive oxidation–reduction of the Mn^{4+}/Mn^{3+} and Mn^{3+}/Mn^{2+} couples [408].

6.2.4 Other ternary oxides

Oxygen electrocatalysis on thin films of metallic oxide electrodes with Delafossite structure was studied by Carcia et al. [422]. The layered oxides $PtCoO_2$, $PdCoO_2$, $PdRhO_2$, and $PdCrO_2$ with alternating layers of transition metal and platinum atoms are reported to exhibit metallic conductivity with $\varrho \simeq 10^{-4}\ \Omega$ cm. The catalytic activity was related to Pt and Pd atoms and was observed to be almost indedependent of the transition metal atom; $PtCoO_2$ exhibited the highest activity, which approached that of platinum metal for the same intrinsic area. The mechanism of oxygen reduction in alkaline solutions is probably similar to that observed on Pt and Pd.

Goodenough and co-workers [10] made a detailed study of the solid state chemistry and electrochemistry of ruthenates of general formula $Bi_{(2-2x)}Pb_{2x}Ru_2O_{(7-y)}$ with the pyrochlore structure and reported that the electroreduction of oxygen proceeds at low overpotentials according to the electrokinetic equation

$$j = nFkp_{O_2}(C_{HO^-})^{-q} \exp(-F\eta/RT) \tag{84}$$

with $q \simeq 0.5$–0.7. At higher overpotentials, the current–potential curve exhibited litle potential dependence and also became independent of hydroxide ion concentration. This behavior is similar to that observed on $LaNiO_3$ perovskite oxide [81]. The authors considered that metallic oxides should be intrinsically less active as oxygen electrocatalysts than noble metals because water is more structured at the oxide surface due to surface hydroxylation and hydrogen bonding. As a result, they postulated that molecular oxygen can penetrate only at the outer Helmholtz plane to undergo one-electron reduction to superoxide by an outer sphere-type mechanism as suggested by Sawyer et al. [423, 424]

$$O_2(OHP) + e \longrightarrow O_2^- \tag{85}$$

with subsequent displacement of surface coordinated HO^- by the resulting HO_2 [Fig. 16(c)]. The upper limit of superoxide concentration in solution is set by the disproportionation reaction

$$2\,O_2^- + H_2O \rightleftharpoons O_2 + HO_2^- + HO^- \tag{86}$$

with pK = -7.51 and a second-order rate coefficient for the forward process of $14.4\,M^{-1}s^{-1}$ [409–412]. The electrochemical reduction of superoxide is a heterogeneous reaction competing with eqn. (86)

$$O_2^- + HOH + e \longrightarrow HO_2^- + HO^- \tag{87}$$

The kinetics of eqn. (85) should be independent of the electrode surface it it proceeds by an outer sphere mechanism.

Sen et al. [425], however, have reported a variation of three orders of magnitude for the first-order rate coefficient for the reduction of oxygen to peroxide on several substrates where the first electron transfer step is rate-determining. On the other hand, it has been shown that oxygen reduction proceeds by an outer sphere mechanism in alkaline solutions on stress annealed pyrolytic graphite (SAPG), the surface of which lacks electrocatalytic groups [426, 427]. The reaction kinetics of this process are, however, very inhibited by at least two orders of magnitude with respect to more active carbon surfaces. It should be stressed that Trunov and Presnov [341, 345] have considered replacement of coordinated water rather than hydroxide ion by molecular oxygen. Yeager [342] has argued that displacement of water by molecular oxygen is unfavourable unless the oxygen adduct has a pronounced dipolar character, i.e. $M\text{–}O\text{–}O^{\delta-}$. Nevertheless, Goodenough and co-workers [10] have made an important contribution by stressing the importance of considering the surface chemistry of oxides in mechanistic considerations of faradaic reactions at their surfaces.

6.2.5 Single oxides, surface and passive layers

O'Grady et al. [227] studied the reduction and evolution of oxygen on RuO_x on titanium and reported that the cathodic reaction involves the formation of peroxide in solution. From $j^{-1/2}$ vs. $\omega^{-1/2}$ plots obtained using rotating disk electrodes, the authors concluded that the reaction O_2/H_2O_2 was in equilibrium at the surface [227]. Miles et al. [428] also studied the oxygen electrode reactions on several metal oxides (Ir, Ru, Pd, and Rd) coated on titanium substrates in 30 wt. % KOH at 80°C. The electrodes were less active for oxygen reduction than for anodic evolution of oxygen, with Tafel slopes $\sim 40\,\text{mV decade}^{-1}$ being observed for Ru, Rh, and Pd oxides and $60\,\text{mV decade}^{-1}$ for Ir and Pt oxides with respect to oxygen reduction.

On single-crystal n-type TiO_2 (rutile), the reaction proceeds through a surface species at potentials positive to the flatband potential by nearly four electrons with a small percentage via peroxide [429]. A shift ($-RT/F$) of the oxygen reduction wave with pH was associated with the variation in the potential drop at the oxide–solution interface, since simple redox couples showed a similar shift. Peroxide intermediate has been found on polycrystalline anodic films on titanium [430]. Gratzel and co-workers [431] have suggested that oxygen-photo-uptake probably occurs at TiO_2 anodes, yielding a surface μ-peroxo complex. Salvador and Gutierrez [432] investigated the role of surface states in the oxygen reactions on titanium oxide.

Matsuki et al. [419] studied the reduction of oxygen on MnO_x supported on RB carbon with the rotating ring–disk electrode. Diagnostic plots suggested that the series mechanism $O_2 \rightarrow HO_2^- \rightarrow HO^-$ is probably operative. Matsuki and Kamada [433] have recently reported that γ-MnO_2 has greater activity than γ-MnOOH below -0.1 V (NHE) and that the oxides supported on graphite produced solution-phase peroxide which decomposes on γ-MnO_2 according to the results of their RRDE experiments. Rao et al. [434] reported for oxygen reduction in alkaline solutions on MnO_2 a dissociative type adsorption mechanism on the basis of a $140\,\text{mV decade}^{-1}$ Tafel slope and reaction orders 0.5 and 0 in oxygen and hydroxide, respectively.

The mechanisms of oxygen electroreduction on iron oxides have more than academic interest since oxygen reduction is the cathodic reaction in the corrosion process of metals like steel in aqueous solutions [435], a reaction that usually occurs at an oxidized surface. In this connection, Stratmann et al. [436,437] have recently suggested a model for the complex atmospheric corrosion of iron. The kinetics of oxygen reduction on α-Fe_2O_3 hot pressed pellets, which were doped with NiO (0.2–0.4 mol %) to increase electrical conductivity [438], have been studied in alkaline solutions by Kamalova and Razina [439]. Thermal treatment of the oxides at 1200°C produced oxygen-deficient nickel ferrites according to the reaction

$$NiO + \alpha\text{-}Fe_2O_3 \longrightarrow Ni^{2+}Fe_{2-2\delta}Fe^{3+}Fe^{2+}_{2\delta}O^{2-}_{4-\delta} + \tfrac{1}{2}\delta\, O_2 \qquad (88)$$

Materials exhibiting high conductivity were obtained by quenching at room temperature (0.05–$0.43\ \Omega^{-1}\,\text{cm}^{-1}$) and dissociative adsorption of oxygen was postulated to account for the fractional reaction order of 0.5 in oxygen. The kinetics of the reaction on ceramic oxide containing 0.4 mol % NiO was studied in detail and a reaction order of 0.5 both in oxygen and hydroxide, 100–$150\,\text{mV decade}^{-1}$ Tafel slope, and an activation energy in the range of 26–$30\,\text{kJ mol}^{-1}$ were observed in the range 18–52°C. The reduction of oxygen was suggested to proceed by a similar mechanism to that on Fe–Ti mixed oxide.

McAlpine and Fredlein [440] studied the electroreduction of oxygen on TiO_2-doped α-Fe_2O_3 where titanium also promotes Fe^{2+}, which induces an increase of conductivity in the oxide by thermal activated hopping ($Fe^{2+}/Fe^{3+} + e$). These authors suggested that a small amount of hydrogen absorption upon electroreduction of the oxide occurs at potentials where the oxygen reduction reaction takes place. This cathodic reaction proceeds at potentials negative to the flatband potential where surface degeneration of electronic states occurs and exhibits steady-state Tafel slopes of $-2RT/F$. Positive to the flatband potential, a depletion layer exists at the surface and cathodic processes involving charge-transfer rate-determining steps should exhibit a $-RT/F$ slope in the absence of contribution from surface states. McAlpine and Fredlein [440] concluded that the reduction of oxygen on TiO_2-doped α-Fe_2O_3 would proceed with

$$O_2 + H_2O + e \longrightarrow HO_2(ads) + HO^- \tag{89}$$

as the rate-determining step with titanium donors mediating the charge transfer. The comparison in Fig. 19 between TiO_2-doped α-Fe_2O_3 and the iron passive layer suggests that the iron active centers catalyze the reaction.

Takei and Laitinen [442,443] reported that iron oxide coated on SnO_2/Si electrodes by spraying of ferric chloride solution at 450°C promoted the four-electron reduction of oxygen, which otherwise would proceed by the two-electron peroxide path on the SnO_2/Si substrate. The study of both hydrogen peroxide and oxygen reduction reactions on the iron oxide-coated electrode have indicated that, because the former occurs at more positive potentials, the reduction of oxygen is seen as a successive reaction $O_2/HO_2^-/HO^-$ in a four-electron single reduction wave. Figure 19 shows that this is also the case on passive iron in alkaline solutions at high overpotentials. As shown in Fig. 19, the steady-state j–E oxygen reduction curves show two well-defined Tafel slopes both greater than RT/F at high and low overpotentials, respectively. An intermediate region where the current is much less sensitive to the electrode potential is observed positive to the flatband potential. McAlpine and Fredlein [440] have suggested that the change in slope ($\delta \log i/\delta E$) below E_{FB} could be related to the semiconducting properties of the oxide electrode. Comparison with results of oxygen reduction on passive iron under similar conditions in Fig. 19 indicate that on TiO_2-doped α-Fe_2O_3 and the passive layer on iron the reduction of oxygen should have very similar electrode kinetics since the potential drop at the oxide–solution interface has been shown to vary with pH in both cases (Sect. 3, Fig. 1). On passive iron electrodes, however, other cathodic reactions, such as the reduction of hydrogen peroxide in the same potential range, do not show a change in the Tafel behavior and thus the change in Tafel slope may be related to a feature of the catalytic cathodic oxygen electroreduction reduction reaction rather than the semiconducting properties of the electrode (Fig. 19).

Danzfuss and Stimming [60] reported that oxygen reduction proceeds at

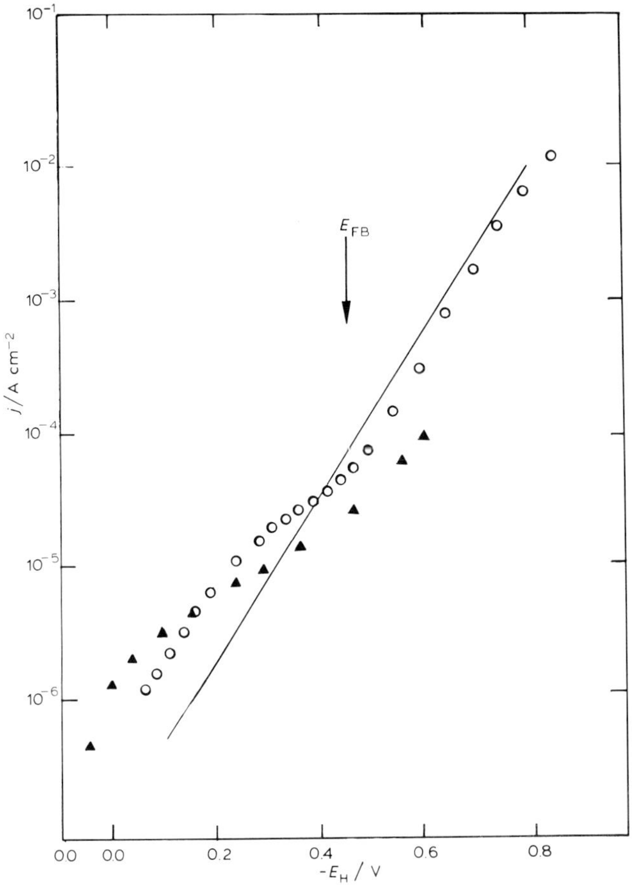

Fig. 19. Steady-state log *j*–*E* curves for oxygen reduction at iron oxide electrodes in 1 M NaOH at 25°C. ▲, TiO_2-doped α-Fe_2O_3; ○, passive iron electrode; ——, steady-state polarization curve of peroxide reduction on passive iron electrode under similar conditions. Currents were corrected for mass transport effects with the RDE. E_{FB} is indicated by arrow according to data in Fig. 1 [440].

$Fe_xTi_{(1-x)}O_y$ electrodes by a two-electron reaction leading to peroxide. The activity decreases appreciably in the case of an iron content greater than $x = 0.027$, while the reaction becomes readily observed on nearly pure TiO_2.

Fabjan et al.. [360] reported that the reduction of oxygen on passive iron in alkaline solutions proceeds with a Tafel slope of 140–160 mV decade^{-1} and reaction orders of 0.5 and −1 for O_2 and HO^-, respectively. Similar results were reported by Calvo and Schiffrin [444] for this system. Rotating ring–disk electrode studies indicated the formation of a small amount of peroxide in the course of an almost 4-electron reduction of oxygen. Calvo and Schiffrin [444] also studied peroxide electroreduction on passive iron in alkaline solution. Unlike McAlpine et al. [440], who reported a reaction order of zero with respect to hydroxyl ions at TiO_2-doped α-Fe_2O_3, Calvo and Schiffrin

[441] found a reaction order of -1 with respect to HO^- on passive iron. A similar reaction order and Tafel slope of 150 mV decade^{-1} were also found in the reduction of hydroperoxide and periodate ions under similar conditions on passive iron in alkaline solution [441] and these kinetic parameters have been related to the properties of the oxide electrode: density of states and potential distribution at the interface and their variation with pH. Mediation by surface Fe^{2+} cations formed by reduction of the passive film, viz.

$$(Fe^{3+}_{ox}\,OH^-) + H_2O + e = (Fe^{3+}_{ox}\,OH_2) + OH^- \tag{90}$$

has been assumed to explain the kinetics of the different cathodic reactions at the iron passive layer in alkaline solutions. While Fe_2O_3 and related iron(III) oxides are very poor catalysts for the oxygen reduction reaction, Fe_3O_4 inverse spinel shows better performance [445]. Ogura and Tanaka [446] reported that oxygen reduction at a cation-deficient-oxide, passive iron electrode proceeds by the 4-electron pathway after partial reduction of the oxide to Fe_3O_4. Reduction of the oxide prior to the cathodic reduction of oxygen seems to be a general kinetic feature of the reaction at iron electrodes.

The cathodic reduction of oxygen on catalytically active substrates is very inhibited by deposited hydrated oxide layers. Such is the case of $FeOOH \cdot xH_2O$ (sandwich oxide electrode) on gold at pH 7.2 [94] and $Ni(OH)_2$ deposited on platinum [447]. This inhibition is probably related to the electronic properties of the surface layer, since simple outer sphere couples are highly inhibited also (Sect. 5.3).

Passive layers on copper and its alloys, which have technological relevance in corrosion, have been investigated in studies of oxygen reduction. Balakrishnan and Venkatesan [448] reported a two-step reduction of oxygen via peroxide on copper(I) surface oxides in neutral solutions.

7. Chlorine evolution

The chlorine–chloride reaction viz.

$$Cl_2 + 2\,e \rightleftharpoons 2\,Cl^- \tag{91}$$

is fairly reversible with a standard reversible potential of 1.358 V (NHE). Practical reversible electrodes based on eqn. (91) can be established at a number of electrode surfaces, e.g. platinized Pt and RuO_2 [449, 450] electrodes in an analogous manner to reversible hydrogen electrodes. Reaction (91) exhibits current densities of 10^{-4} to 10^{-3} A cm^{-2} (real area) at noble metal electrodes [450, 451].

Although the thermodynamic potential for the oxygen electrode reaction [i.e. 1.229 V (RHE)] is lower than that of the chlorine electrode, in practice the kinetics of the chlorine electrode reactions are substantially faster than those of the oxygen electrode reactions. Thus, chlorine tends to evolve preferentially in the electrolysis of aqueous Cl^--containing solutions. Since the chlorine evolution reaction usually exhibits a lower Tafel slope than the oxygen evolution reaction, quite high efficiencies of Cl_2 evolution over O_2 evolution are observed at high current densities in relatively concentrated chloride solutions. With reasonable care, coulombic efficiencies close to 100% can be achieved for chlorine evolution and, commercially, the efficiency of chlorine production from brine is greater than 98%.

Because of the relatively high anodic potentials involved, chlorine evolution takes place on an oxide film on most metal substrates. In the case of noble metals, e.g. Pt, such films are in the range of monolayer or sub-monolayer, dimensions. Therefore, it is often the properties of the oxide rather than those of the metal which determine the kinetics of chlorine evolution. The influence of the oxide film on Pt on the kinetics of Cl_2 evolution has been investigated by Conway and Novak who studied chlorine evolution in aqueous [452] and non-aqueous [453] (i.e. trifluoroacetic acid) electrolytes. Such monolayer-type oxide films will not be discussed here. Rather, attention will be directed to thicker films, such as those formed by thermal decomposition methods, and hydrous oxide films formed by potential multicycling techniques. A number of relatively recent reviews have appeared on the chlorine evolution reaction: Novak et al. [454], a general review of fundamental and applied aspects; Trasatti and Lodi [186], pertaining to oxide catalysts only; Trasatti and O'Grady [185], pertaining to RuO_2 catalysts only.

7.1 RUTHENIUM DIOXIDE

Thermally prepared, mixed oxide DSA anodes containing RuO_2 as the catalytically active component have been employed with considerable success in chlor-alkali cells [455]. Despite its success in anodes for chlorine evolution, RuO_2 is an appreciably poorer catalyst than platinum when comparisons of Cl_2 evolution behavior are carried on a real area basis [456]. However, it is well etablished that Pt electrodes exhibit passivation phenomena at moderately high current densities [451, 452, 457]. Furthermore, noble emtal catalysts are known to degrade due to contact with Hg in amalgam cells [458]. Thus RuO_2-based anodes are good catalysts for practical Cl_2 evolution cells for a number of reasons, e.g. they do not exhibit passivation behavior or undergo significant dissolution and the absence of Hg wetting effects and wear due to gas bubbles [459].

Owing to the success of RuO_2-based DSA electrodes in the chlor-alkali industry, a significant amount of study has been carried out on the kinetics and mechanism of chlorine evolution at RuO_2-based electrodes over the past 15 years or so. A considerable body of experimental data has therefore been accumulated regarding the chlorine evolution reaction at RuO_2 electrodes, which includes E vs. $\log j$ plots, reaction order determinations, pH dependence studies, impedance measurements, and XPS studies of electrodes before and after Cl_2 evolution experiments. However, despite the amount of research that has been conducted on the chlorine evolution reaction at RuO_2, it seems premature to state that the kinetics and mechanism of this reaction are understood. There is considerable disagreement between the results of the various workers in the field. Space limitations do not permit a detailed review and discussion in this article of the entire body of literature associated with the Cl_2 evolution reaction at RuO_2 electrodes. Therefore, only the salient aspects of this topic will be discussed here.

Prior to discussing the kinetics and mechanism of chlorine evolution at RuO_2, a few general observations regarding the reported results for this reaction will be made. Chlorine is evolved in practical chlor-alkali cells at about pH = 2–3 and with NaCl concentration of about 300 g dm^{-3} (i.e. ~5 M) [460], and many authors have employed such electrolytes in studies of the chlorine evolution reaction. Most authors quote Tafel slopes of 30–40 mV $decade^{-1}$ for chlorine evolution at RuO_2 [450, 461–466]. However, the Tafel slope has been reported by other authors to vary from as low as 20 mV $decade^{-1}$ [467, 468] to even as high as 108 mV $decade^{-1}$ [449]. Differences in the mode of electron preparation cannot be relied upon to account completely for such a wide spread in Tafel slope. As has been pointed out by Yorodetskii and co-workers [469, 470], the Tafel slope obtained in concentrated chloride solutions, i.e. under conditions similar to those prevailing in commercial chlorine cells, is sometimes less than $2.3\,RT/2F$ (i.e. slope ~36.5 mV $decade^{-1}$ at 90°C). This theoretical value represents the lowest possible slope which can be obtained under conditions where the electrode reaction rate is limited by an electrochemical or subsequent diffusion step. Losev [471] has recently confirmed that anomalously low Tafel slopes can arise in fast electrochemical gas evolution reactions due to rate-limiting transport of gas away from the electrode surface. Thus, in experiments conducted in concentrated chloride solutions at high temperatures, where the solubility of chlorine is appreciably reduced, Tafel slope data may not be suitable for kinetic analysis. The effects of supersaturation of chlorine at the electrode surface have been the subject of some attention by Kadija [472] and by Müller et al. [473, 474]. The latter authors [473, 474] observed that such supersaturation decreases, e.g. by increasing the temperature, the roughness of the electrode, the KCl concentration of the solution and, of course, by stirring. The pH of the solution in the pH range 1–3 had no effect on the supersaturation.

In general, the Tafel slope value has been found to be independent of Cl^- ion concentration, although a decrease in slope with increase in Cl^- concentration has been reported in at least one instance [475]. The effect of electrolyte pH on the kinetics of chlorine evolution has generally not been considered in detail in the literature and it has been suggested by Denton et al. [465] that there is no pH effect in the case of RuO_2/TiO_2 electrodes. If, as will be discussed later, surface oxyruthenium complexes play a role in the mechanism of chlorine evolution, then a pH effect should be evident in the kinetics of chlorine evolution. Arikado et al. [476] and Erenburgh et al. [477] have reported such a pH effect. The reaction order with respect to the Cl^- ion is generally found to be about 1 [461, 465, 478], although a value of 2 has been reported by Arikado et al. [476].

The following two mechanisms have been mostly considered for chlorine evolution at RuO_2, viz.

Pathway I

$$Cl^- + S \longrightarrow S-Cl + e \tag{92}$$

$$2\,S-Cl \longrightarrow 2\,S + Cl_2 \tag{93}$$

and

Pathway II

$$Cl^- + S \longrightarrow S-Cl + e \tag{92}$$

$$S-Cl + Cl^- \longrightarrow M + Cl_2 + e \tag{94}$$

These mechanisms are analogous to the so-called Volmer–Tafel and Volmer–Heyrovsky mechanisms of the H_2 evolution reaction, respectively [469]. In pathway I, with eqn. (93) rate-limiting, the theoretical Tafel slope is 30 mV decade^{-1} and the reaction order is 2 with respect to Cl^- ion at low overpotentials and 0 at high overpotentials. In the case of pathway II, with eqn. (94) rate-limiting, the theoretical Tafel slope is 40 mV decade^{-1} and the reaction order is 3/2, assuming the transfer coefficient, α_a, to be 1/2. In this simple approach to the mechanism of the chlorine evolution reaction at RuO_2, probably the only conclusion which can be made is that the discharge of Cl^- ion [reaction (92)] is probably rate-limiting at high overpotentials. It has been suggested by some authors [456, 457, 462], principally on the basis of Tafel slope analysis only, that the Volmer–Tafel type mechanism applies at low overpotentials at RuO_2, while others [451, 476] argue in favor of the Volmer–Heyrovsky type mechanism.

Erenburg et al. [461, 478, 479] have carried out a fairly in depth study of the mechanism of chlorine evolution and reduction at RuO_2-based electrodes. Representative curves for chlorine evolution and reduction at RuO_2 are shown in Fig. 20, for which the anodic slope is 40 mV decade^{-1} and the cathodic slope is ca. 120 mV decade^{-1}. The extrapolated cathodic and anodic currents coincide. The electrolytes employed by these workers were mostly 1.5 M HCl + 2.5 M NaCl, and measurements were carried out using a stationary electrode of either pure RuO_2 or 35% RuO_2 + TiO_2. Initially, these authors suggested that chlorine evolution at mixed RuO_2/TiO_2 occurred via a rather complicated mechanism involving two parallel processes, recombination and electrochemical desorption, with an equilibrial discharge stage. Erenburg et al. [461, 478] later reappraised this mechanism and proposed instead

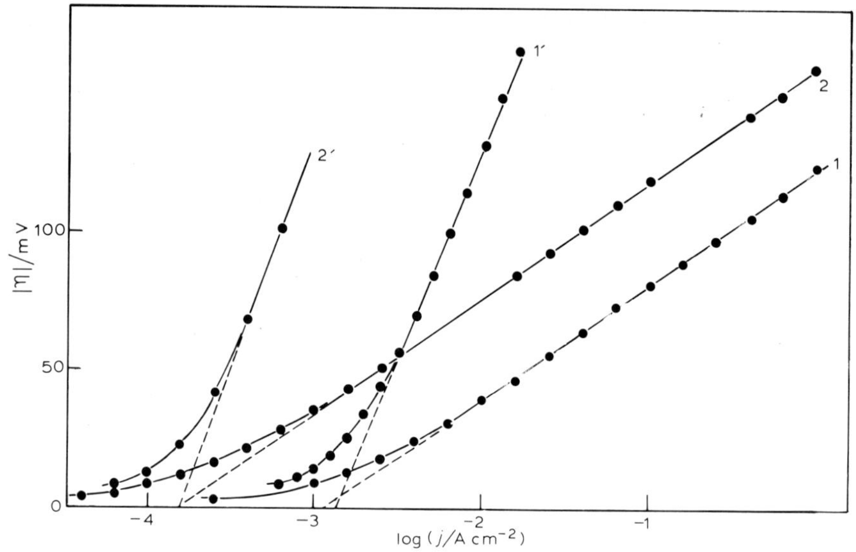

Fig. 20. Anodic (1,2) and cathodic (1′, 2′) polarization curves for 100% RuO_2. Heat treatment temperature 480°C. Solution composition: 1.5 M HCl + 2.5 M NaCl; p_{Cl_2}: 1,1′ = 1 atm; 2,2′ = 0.057 atm; temperature = 30°C [461].

$$S + Cl^- \longrightarrow S - Cl + e \qquad (92)$$

$$S - Cl \longrightarrow S - Cl^+ + e \qquad (95)$$

$$S - Cl^+ + Cl^- \longrightarrow Cl_2 \qquad (96)$$

A low coverage of chloride intermediate is implied by this mechanism. With eqn. (95) rate-limiting, the Tafel slope takes on the value $b = 2.303RT/(1 + \alpha_a)$ where α_a is the transfer coefficient. The experimental mechanistic indicators obtained by these authors included: a Tafel slope of about 30 mV decade^{-1} at 30% RuO_2/TiO_2 electrodes, which increased to 40 mV decade^{-1} at higher η, and 40 mV decade^{-1} only for pure RuO_2; reaction orders with respect to Cl_2 and Cl^- of 0 and 1, respectively; and an apparent stoichiometric number of about 1. The most marked feature of this mechanism is the formation of Cl^+ species in the rate-determining step [eqn. (95)]. In view of the instability of Cl^+ as a strong Lewis acid in water, the involvement of this intermediate in the Cl_2 evolution reaction as outlined in eqns. (95) and (96) is speculative [454]. Erenburgh et al. [479] have speculated that Cl^+ may actually exist as HClO, i.e. eqn (95) occurs with participation of water. A Tafel slope of 40 mV decade^{-1} is predicted by eqn. (95). To account for the lower slope of 30 mV decade^{-1} observed at 30% RuO_2/TiO_2 at low η values, a barrierless (i.e. $\alpha_a = 1$), second-electron transfer pathway was invoked [479]. On increasing the overpotential, a transition from barrierless to regular (i.e. $\alpha = 0.5$) second electron transfer rate control was considered to occur. These two reactions were considered to occur at two different active sites on the surface of the RuO_2/TiO_2 electrode, the natures of which are, however, obscure [461]. In general, the supposition in this work that $\theta << 1$ for chloride intermediates on the RuO_2 surface is open to question. Augustinski et al. [198], for example, have observed the presence of a considerable amount of chloride on the surfaces of RuO_2 and RuO_2/TiO_2 surfaces using XPS after electrolysis in 4 M NaCl. Two types of chlorine species were distinguished, one being chloride ions, the other absorbed atomic chlorine, or possibly ClO^-. Substantial incorporation of chlorine into RuO_2/TiO_2 layers has also been noted by Yorodetskii et al. [482] on the basis of Auger measurements. More recently, Krishtalik [236] discussed the possibility that valence changes at Ru active sites may be an intrinsic part of the chlorine evolution mechanism which was outlined in eqns. (92), (95), and (96). In further work on the mechanism of the chlorine evolution reaction at RuO_2/TiO_2 electrodes, Erenburgh et al. [177] observed that the kinetics of chlorine evolution were strongly dependent on pH, the apparent reaction order with respect to H^+ increasing with increasing acidity. A mechanism for chlorine evolution was proposed in which a fast electrochemical step involving the elimination of a proton preceded a slower Cl^- discharge step. It was suggested that hydrated Ru oxide groups may play the role of active sites in the chlorine evolution process.

Arikado et al. [476] studied the kinetics of Cl_2 evolution at several oxide electrodes with special emphasis on pH effects. Chlorine evolution kinetics were found to increase with increase in acidity of the electrolyte. The reaction order with respect to Cl^- was determined from the slope of a plot of exchange current vs. rest potential at constant p_{Cl_2}, using a method of calculation suggested by Enyo and Yokoyama [480] and found to be somewhat greater than 2. This value of the reaction order combined with a Tafel slope of 40 mV decade^{-1} led Arikado et al. [476] to conclude that the Volmer–Heyrovsky mechanism was operative for chlorine evolution at RuO_2. They concluded that Ru(III) cations were the active sites for Cl^- adsorption on the RuO_2 surface. In Fig. 21 is shown a schematic representation of the model proposed by Arikado et al. [476] for the interaction of chlorine species with RuO_2. These authors invoked arguments based on the *d*-electron configurations of RuO_2 and IrO_2 (partially filled t_{2g} and empty e_g levels) in support of the Volmer–Heyrovsky mechanism for Cl_2 evolution at these oxides. The Ru–Cl bond was regarded as being quite strong and short due to the absence of an electron in the e_g orbital of Ru. Consequently, such coordinated Cl atoms are unable to interact sufficiently [Fig. 21(b)] to give Cl_2 evolution by a combination mechanism. Thus Cl^- attack on Ru–Cl is favored to give Cl_2 molecules.

Inai et al. [483] suggested that an activated complex with a pentagonal bipyramid-type structure was formed in the transition state in the Volmer–Heyrovksy-type mechanism for chlorine evolution at RuO_2. A theoretical activation energy for this reaction was calculated by using the difference in the crystal field stabilization energy between the initial and transition

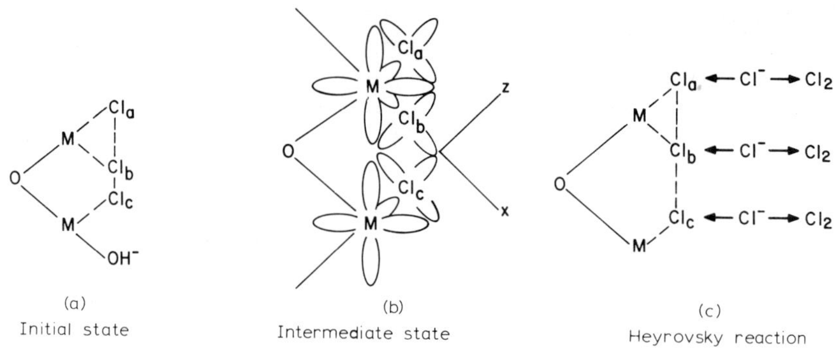

Fig. 21. Schematic methods of the surface of the rutile-type oxide electrode during chlorine evolution. Models shown for the xz plane [476].

states. The theoretical activation energy was estimated to be 9.4 kcal mol^{-1}, which is in fair agreement with experimentally determined values of 9 [483] and 8.5 [462].

The possibility of involvement of oxygenated species in the intermediate stages of Cl_2 evolution have also been discussed by, e.g. Burke and O'Neill [481] and Augustinski et al. [198]. Burke and O'Neill [481] proposed a modified Volmer–Heyrovsky scheme for RuO_2

$$O_{ads} + Cl^- = OCl_{ads} + e \tag{97}$$

$$OCl_{ads} + Cl^- = O_{ads} + Cl_2 + e \tag{98}$$

The O_{ads} species could actually be RuO_3. Augustinski et al. [198] concluded that RuO_3, as well as two different chloride species, were present on the surfaces of RuO_2 and RuO_2/TiO_2 electrodes which had been used as anodes for chlorine evolution. These authors suggested that Ru(VI) species may play a role in the chlorine evolution process and concluded that reaction sequences such as

$$2\,Cl^- + RuO_3 + 2\,H_3O^+ \longrightarrow (RuO_2)\cdot 2Cl_{ads} + 3\,H_2O \tag{99}$$

$$RuO_2 + 3\,H_2O \longrightarrow RuO_3 + 2\,H_3O^+ + 2\,e \tag{100}$$

warranted consideration. Since XPS could not distinguish between the presence of absorbed chlorine atoms or ClO^- ions on the ruthenium dioxide electrode surface, the reactions

$$Cl^- + RuO_3 \longrightarrow (RuO_2)ClO^-_{ads} \tag{101}$$

followed by eqn. (100) and then

$$ClO^-_{ads} + Cl^-_{ads}\ (\text{or } Cl^-_{aq}) + 2\,H_3O^+ \longrightarrow Cl_2 + 3\,H_2O \tag{102}$$

are also possible. Harrison and co-workers have applied impedance methods to the study of chlorine evolution at Pt [457] and RuO_2/TiO_2 electrodes [456,

457, 465, 484]. This group has suggested that HOCL may be involved in the chlorine evolution process, viz.

$$Cl_2 + H_2O \rightleftharpoons HOCl + H^+ + Cl^- \qquad (103)$$

$$H^+ + HOCl + 2\,e \rightleftharpoons Cl^- + H_2O \qquad (104)$$

In a recent study, Harrison et al. [485] used steady-state j–E and $Z(\omega)$–E data to characterize the chlorine evolution reaction at RuO_2/TiO_2 electrodes using a simple redox reaction description of the chlorine evolution process with HOCl and Cl^- as reactant and product, respectively. The impedance potential data were analyzed by the equivalent circuit method parameter curves such as C_{dl}–E and R_{ct}–E. It has been suggested by the authors [485] that this type of parametric analysis of impedance data can be useful for comparison of the activity of various types of electrodes.

The activity of RuO_2/TiO_2 electrodes for Cl_2 evolution decreases only a little on decreasing the RuO_2 content from 100 down to 20–30% [205], at which point a rapid fall off with further decrease in RuO_2 content occurs. This critical composition is apparently associated with the appearance of another phase, anatase, in the oxide layer [204, 205]. A few studies have been carried out on the effect of preparation temperature on the activity of RuO_2 and RuO_2/TiO_2 electrodes. Kalinovskii et al. [486] observed in the case of pure RuO_2 that the potential of constant current increased only slightly as the preparation temperature was increased from 350 to 600°C. Burke and O'Neill [481], on the other hand, have reported a slight increase in Cl_2 evolution activity with increase in electrode preparation temperature for RuO_2 in the range 350–500°C. In the case of RuO_2/TiO_2 catalysts, it has been shown that the catalytic activity for chlorine evolution remains virtually unchanged as electrode preparation temperature is increased [487]. A slight decrease in activity was observed by Jannsen et al. [462] for RuO_2/TiO_2 with increase in preparation temperature. In general, it may be concluded that the catalytic activity for chlorine evolution exhibited by RuO_2 based electrodes does not exhibit a major dependence on the preparation temperature. This indicates that internal surface area is not a critical factor in the chlorine evolution process at practical current densities. However, Ardizzone et al. [466] observed differences in the kinetics of chlorine evolution between "cracked", i.e. highly defective, layers and "compact", i.e. low-defect, layers of pure RuO_2. A Tafel slope of 30 mV decade^{-1} was oberved for the former type of electrode, while a slope of 40 mV decade^{-1} was found for the latter type.

7.2 MANGANESE DIOXIDE

Because of its inexpensive nature and widespread use as a cathode in batteries, MnO_2 has been extensively studied from the viewpoint of its physicochemical properties and its cathodic behavior. It has also attracted attention as a potential anode material. However, as discussed earlier in this chapter, it is a rather poor electrocatalyst for oxygen evolution. MnO_2 is a fairly effective catalyst for chlorine evolution [316, 488]. The behavior of thermally prepared MnO_2 film electrodes is substrate-dependent, however. An ohmic drop develops at the MnO_2–substrate interface in the case of a Ti substrate, which has been overcome by the use of thin Pt or RuO_2 layers between the MnO_2 and Ti [315, 316]. Arikado et al. [476] have reported Tafel slope values of 30 mV decade^{-1} in the case of chlorine evolution at Pt-support/MnO_2 electrodes, and concluded that the Volmer–Tafel-type mechanism [eqns. (92) and (93)] was probably operative for this catalyst. The thermally prepared MnO_2 electrode system was observed to give fairly stable overpotentials for chlorine evolution over a period of about 20 h (for Ti/RuO_2/MnO_2 electrodes) [316].

In order to rule out substrate effects, so-called "massive" MnO_2 electrodes have been used in studies of chlorine evolution by Morita et al. [214]. These electrodes essentially consist of pressed pellets of thermally prepared MnO_2 (β form) which were subsequently pore filled with polystyrene resin. Rather surprisingly, quite high overpotentials were observed for chlorine evolution at the massive electrodes, possibly related to ohmic loss in the electrodes [214]. An intriguing observation by Morita et al. [214] was that the polarization curves for chlorine and oxygen evolution at massive MnO_2 electrodes were apparently superimposable. This led these authors to suggest that the rate-determining step in oxygen evolution was similar to that for chlorine evolution. It was proposed that Mn(III) is oxidized to Mn(IV) on the surface, which then mediates the oxidation of Cl^- ion to generate Cl_2 in the case of the chlorine evolution reaction [214]. Maximum catalytic activity was observed at a preparation temperature of 485°C, which is a mixture of β-MnO_2 and α-Mn_2O_3. Above 485°C, mainly inactive α-Mn_2O_3 results which is quite resistive [214].

The electrocatalytic properties of massive MnO_2 doped with small amounts of noble metals have also been investigated [488, 489]. A considerable decrease in chlorine evolution overpotential was observed with Pd as dopant, while a somewhat smaller decrease was observed for Ru, Ir, and Pt. Rather surprisingly, a negligible effect was exhibited on doping with Rh. The Tafel slope was observed to vary from 33 to 57 mV decade^{-1} on going from Pd to Ir, while a value of 45 mV decade^{-1} was noted for the Pt doped catalyst. A similar trend in catalytic activity was observed in the case of catalysts prepared by prolonged soaking in solutions containing the corresponding noble metal chlorides [214]. Clearly, noble metal sites on the surface of the bulk-doped MnO_2 act as the active sites for chlorine evolution, rather than a modification of bulk oxide properties being responsible for the enhanced catalysis. In the case of the Pd-doped MnO_2 catalyst, Morita et al. [490] observed a second-order relation between the rate of chlorine evolution and the number of Pd sites at the electrode surface. It was concluded that the Pd sites were highly dispersed on the surface.

7.3 SPINEL-TYPE OXIDES

The investigation of spinel oxides as catalysts for chlorine evolution has focussed mainly on cobaltic oxides, especially Co_3O_4 [491–495]. It has been generally observed that the overpotentials exhibited by Co_3O_4 layers on Ti strongly depend on the surface pretreatment of the Ti support [496, 497]. Boggio et al. [494] noted that Co_3O_4 could not be readily employed as catalyst if prepared at temperatures higher than about 400°C on Ti substrates because of the formation of an intermediate TiO_2 layer between the Co_3O_4 and Ti. Even in the case of electrodes prepared at lower temperatures, the insulating oxide layer builds up at the Co_3O_4–Ti interface due to the non-protective nature of the Co_3O_4 layer [494, 497]. The use of an intermediate layer of RuO_2 between the Co_3O_4 layer and the Ti substrate has been found to minimize the effect of tho insulating layer [494, 497].

The preparation [498] and electrocatalytic properties of Co_3O_4 for chlorine evolution [494] have recently been described in detail by Trasatti and co-workers. These authors [498] observed formation of single phase, pure Co_3O_4 at $T \leq 200°C$ from the nitrate salt and at $T \geq 400°C$ from the chloride salt. Quite good activity was exhibited by the Co_3O_4 electrodes for Cl_2 evolution, which was only a little less than that of RuO_2 [494]. The long-term stability of Co_3O_4 under conditions of Cl_2 evolution has not yet been studied, however. The catalytic activity of Co_3O_4 electrodes was found to increase with decraesing temperature below 300°C, although the stability of the Co_3O_4 became poorer. Above about 400°C, the higher stability of the Co_3O_4 layer was offset by decreased catalytic activity due to crystallization and sintering. A Tafel slope of 40 mV decade^{-1} was found for chlorine evolution at Co_3O_4 along with reaction order of 1 with respect to Cl^- ion. An interesting result in this work was the observation that chlorine evolution rate decreased with increasing acidity, a reaction order of -1 with respect to H^+ being observed. This result indicates involvement of oxy/hydroxy species in the process of Cl_2 evolution. The following mechanism was tentatively suggested [494].

$$S\text{–}OH_2^+ + Cl^- \rightleftharpoons S\text{–}OH_2Cl \tag{105}$$

$$S\text{–}OH_2Cl \rightleftharpoons S\text{–}OHCl + H^+ + e \tag{106}$$

$$S\text{–}OHCl \longrightarrow S\text{–}(OHCl)^+ + e \qquad \text{rds} \tag{107}$$

$$S\text{–}(OHCl)^+ + Cl^- \rightleftharpoons S\text{–}OH_2^+ + Cl_2 \tag{108}$$

Equation (105) simply represents the electrostatic adsorption of Cl^- ions on to positively charged sites (point of zero charge of $Co_3O_4 \sim 7.5$ [499]).

Mostkova et al. [500] studied the kinetics of chlorine evolution at $NiCo_2O_4$ and observed Tafel slopes of 40 and 150 mV decade^{-1} for the anodic and cathodic reactions, respectively. Reaction orders close to 1 with respect to

Cl^- and OH^- were observed for the anodic reaction. The pH dependence suggests that water participates directly in the chlorine reaction. Accordingly, the mechanism

$$Co + H_2O \rightleftharpoons CoOH + H^+ + e \tag{109}$$

$$CoOH + Cl^- \longrightarrow CoOHCl + e \tag{110}$$

$$CoOHCl + Cl^- + H^+ \rightleftharpoons Co + Cl_2 + H_2O \tag{111}$$

was suggested for the chlorine reaction at $Ni_2Co_2O_4$ [501] with eqn. (110) as the rate-determining step. Since the pH dependence implies that the first step occurs with detachment of a proton, eqn. (110) could also be written in terms of a valence state change of cobalt ion, while allowing for hydration of the surface oxide, viz.

$$Co^{z+}OH^- \rightleftharpoons Co^{(z+1)+}O^{2-} + H^+ + e^- \tag{110}$$

The stability of $NiCo_2O_4$ in long term chlorine generation remains to be clarified, however.

In studies of other cobalt spinels, Hazelrigg and Caldwell [501] observed reasonably high catalytic activity for spinels of general formula $M_xCo_{(3-x)}O_4$ where $0 < x < 1$ and M is Cu, Mg, or Zn. In the case of the Zn spinel, an initial overvoltage of 30 mV was observed in 5.1 mol dm^{-3} NaCl at 70°C and 77 mA cm^{-2}, while an overvoltage of 50 mV was observed after 425 days in a small diaphragm cell.

Fe_3O_4 has been subjected to investigation as a possible anode material since the early part of this century and has been reviewed by Hayes and Kuhn [502]. Fe_3O_4 is not an attractive electrode material, however, due mainly to poor electrical conductivity and an inherent brittleness. Magnetite anodes prepared in the conventional DSA manner by thermal decomposition on Ti substrates suffer from the development of a resistive TiO_2 layer at the interface of the Fe_3O_4 and Ti [502], although the effects of this layer can be minimized by employing an interlayer of Pt. Very few kinetic studies of chlorine evolution at Fe_3O_4 electrodes have been reported in the literature [450, 502, 503]. Matsumura et al. [503] observed a Tafel slope of about 70 mV $decade^{-1}$ for chlorine evolution on magnetite in 5.3 mol dm^{-3} NaCl at 25°C. Tafel slopes in the range of 90 mV $decade^{-1}$ were observed by Hayes and Kuhn [502]. The latter authors suggested that the catalytic activity of magnetite electrodes should be improved by the incorporation of small amounts of either precious metals or their oxides.

In a study of a series of iron-based spinels, Iwakura et al. [313] did not observe a correlation of chlorine evolution overvoltage with the magnetic properties of the oxide. In contrast, a dependence of overpotential on magnetic properties was observed for oxygen evolution.

7.4 OTHER OXIDES

Thermally prepared films of SnO_2 exhibit semiconducting properties and are poor catalysts for chlorine evolution due to a significant potential drop in the space charge layer [504, 505]. However, addition of Sb to the SnO_2 causes quasi-metallic conduction to be exhibited by the SnO_2 [506]. The addition of 10% Sb to the SnO_2 results in electrodes which exhibit the reversible potential in Cl_2-saturated electrolyte, while a Tafel slope of approximately 40 mV decade^{-1} is obtained for the chlorine evolution reaction [506]. Further, the addition of RuO_2 to Sb-doped SnO_2 results in a major enhancement in the kinetics of chlorine evolution, which is thought to result from a synergistic effect, due possibly to spill over effects, between the Ru and Sb dopants [507].

An oxide which has received attention as a chlorine evolution catalyst is PbO_2 [508–510]; however the activities of both the α- and β-forms are poor in relation to RuO_2TiO_2 electrodes.

Conway and co-workers [511, 512] studied the kinetics of chlorine evolution on the hydrous oxide film formed on Ir by potential multicycling. An increase of up to 180 in Cl_2 evolution rate was observed (Fig. 22), the apparent exchange current density increasing almost linearly with oxide thickness. This enhancement in Cl_2 evolution rate at the Ir–solution interface was mainly attributed to an increase in microscopic surface area with

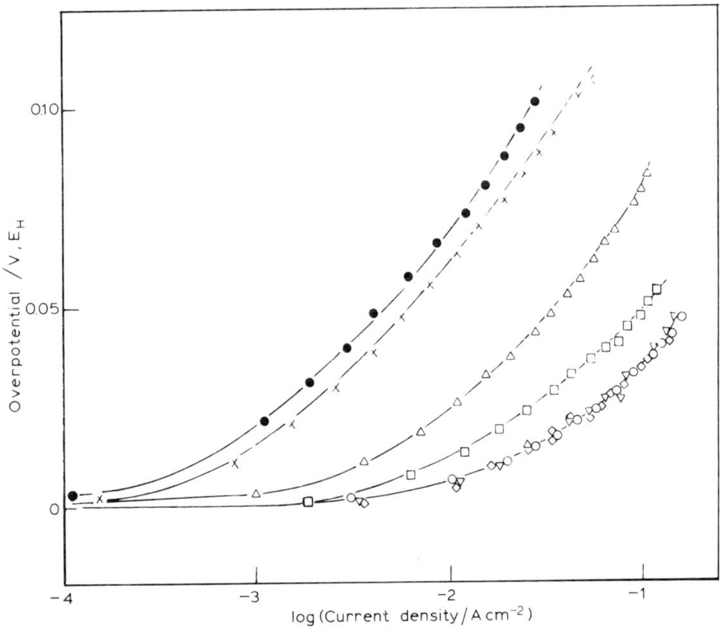

Fig. 22. Steady-state log j vs. overpotential relations for anodic Cl_2 evolution on Ir at 298 K in 1 M aqueous HCl + HCl to pH 2: charge enhancement factor (CEF) values were ●, 1; ×, 2; △, 9; □, 30; ▽, 70; ◇, 132; ○, 182 [512].

increasing oxide growth. A Tafel slope of about 60 mV decade^{-1} was observed for Cl_2 evolution at the hydrous films for all oxide thicknesses studied [512]. Vukovic et al. [513] have shown that potential cycling induces an altered state of surface oxidation at Ru at the monolayer level which causes an enhancement of ca. 30 in the rate of chlorine evolution. On the other hand, upon growing a thick oxide film on Ru in the manner described for Ir, the rate of Cl_2 evolution did not increase with increase in oxide film thickness [513]. This lack of enhancement may be due to the compact layer nature of the multilayer oxide film at Ru in contrast to the disperse, hydrous films grown at Ir [488].

8. Electro-organic reactions

Organic electrochemistry continues to gain in momentum. A number of recent reviews have appeared dealing with selected aspects of this burgeoning field. These include: kinetics, mechanisms, and prospects for commercial development by Rudd and Conway [514]; problem areas in modern organic electrochemistry by Fioshin and Tomilov [515]; electrocatalytic aspects by Wendt [516]; electro-organic reactions with and at anodic oxide films by Conway [517]; solvent and electrolyte systems in electro-organic chemistry by Miller [518]; a review of electro-organic processes presently practised throughout the world by Baizer [519]; and, a survey of different reaction types with some HMO-theory based theoretical background by Köster and Wendt [520].

The historical development of electro-organic chemistry is well documented by several authors, e.g. in refs. 514 and 521–527, and therefore will not be repeated here. Much of the pioneering work in the field was carried out at Pt, i.e. Pt covered with an oxide film of monolayer dimensions in the case of anodic reactions. Electro-organic reactions at Pt have been analyzed in considerable detail by Conway [517] and will not be discussed here. Rather, attention will be focussed on oxide electrocatalysts and metal anodes covered with oxide films of multilayer dimensions, e.g. Ni and Pb. However, before commencing with a discussion of such oxide catalysts, some important factors in electroorganic chemistry will be briefly reviewed.

8.1 PRELIMINARY REMARKS

In aqueous electrolytes, simple exchange of one or two electrons and protons is observed at many electrode surfaces. In anodic reactions, however, radical intermediate formation may also occur, as for example in the Kolbe reaction [525]. The order in which electron and proton exchange occur, whether accompanied by chemical reactions or not, can often be characterized by standard electrochemical analysis. Radical ion formation is mostly involved in electro-organic reactions in aprotic non-aqueous media. In many cases, electron paramagnetic resonance spectroscopy can be employed to study such radicals [521].

The basic principles of electrocatalysis have never received their due attention in the field of electro-organic chemistry. In those reactions where the nature of the electrode material plays a key role, the organic molecule tends to adsorb strongly on the electrode surface, generally in a dissociative manner. The resultant adsorbed organic moieties subsequently undergo electrochemical steps of, for example, hydrogenation in case of reduction reactions or oxidation by adsorbed oxygen species of the anodic film in the case of anodic oxidation. The electrochemical transformation in many of these classes of reactions may be regarded as indirect, since passage of electric current merely serves to maintain the hydrogen layer (cathodic reactions) or the anodic oxide film (anodic reactions). The electro-oxidation of molecules of fuel cell interest, e.g. CH_3OH and HCOOH, is characteristic of this class of reaction [528].

Another important type of electro-organic process is that in which reaction takes place at a bulk phase oxide layer on metallic electrode surfaces. Such oxide layers can mediate the oxidation of organic molecules, as in the case of alcohols and amines at oxidized Ni [529], and passage of current may be regarded as serving to maintain the oxide layer.

8.2 LEAD DIOXIDE

Lead dioxide has received considerable attention as an electrode material for electro-organic processes. This is due to its high overpotential for oxygen evolution, its relatively good stability under anodic conditions, and high electrical conductivity. Lead dioxide can exist in two modifications [530, 531]: α-PbO_2 has an orthorhombic structure, while β-PbO_2 has a tetragonal rutile structure. Under normal laboratory conditions, β-PbO_2 is the most stable polymorph and is generally prepared by electrodeposition on an inert metal substrate, e.g. platinum. It is the modification that will be discussed in the present paper along with anodic oxide films formed at Pb electrodes.

A review of the early literature reveals that a variety of functional group oxidations has been studied at PbO_2 anodes [532–534]. Fichter and Grizard [534] have shown that toluenesulphonamide is oxidized to benyzoylsulphonamide, while the conversion of $PhCH_3CN$ to PhCHO has also been demonstrated [532, 533], *p*-nitrotoluene is oxidized to *p*-nitrobenzoic acid [535], while *p*-toluic acid is also converted to terephthalic acid in high yield [536].

Clarke et al. [537] have studied the oxidation of benzene, toluene, and anisole at anodic PbO_2 films using the rotating disk electrode technique. Product analysis revealed that partial oxidation of benzene to benzoquinone occurred at PbO_2 electrodes with a current efficiency approaching 100% at potentials prior to oxygen evolution. These authors [537] observed that fragmentation occurred in the oxygen evolution region yielding maleic acid and CO_2. The reaction order in the low potential region for benzene and toluene was observed to be 2, while a zero-order pH dependence was also observed in this region. A mechanism of arene oxidation was proposed which involves electrochemical formation of PbO_2 followed by a chemical reaction of this with the organic compound. That PbO_2 should act to mediate the oxidation of organic compounds in a heterogeneous manner at an anode surface is quite tenable. PbO_2 is a relatively strong oxidizing agent as shown by the fact that PbO_2 in suspension has been found to give quantitative oxidation of toluene to CO_2 under certain conditions [538]. The work of Michler and Pattinson [539] of almost a century ago, in which higher yields of tetramethylbenzidine and tetramethyldiaminodiphenylmethane from dimethylaniline were observed at a PbO_2 anode compared with direct chemical oxidation with PbO_2, suggests that the electrochemical–chemical type mechanism favored by Clarke et al. [537] may be an oversimplification. It is interesting to note that a number of authors have confirmed the technical feasibility of the electrolysis of benzene-emulsions in sulfuric acid at PbO_2 for the production of benzoquinone [540, 541].

Harrison and Mayne [542] recently investigated the electro-oxidation of a number of aromatic compounds, e.g. *p-tert*-butyltoluene and *p-tert*-butylbenzyl alcohol, at a PbO_2 film on a Pb rotating disk electrode with techniques which included a.c. impedance. In contrast, to Clarke et al.'s [537] electrochemcial–chemical mechanism for oxidation of simple aromatics like benzene, Harrison and Mayne [542] suggested that the electro-oxidation of their more complicated aromatic compounds proceeded through a E.E. (or E.C.E.) type mechanism, the slow step being a 1*e* transfer step. Tissot et al. [543] have studied the electro-oxidation of *p-tert*-butyltoluene to give *p-tert*-butylbenzaldehyde at Pb–Sb and Pb–Ag alloy anodes at 60°C in H_2SO_4 electrolyte. The Pb–Sb alloy (5 and 20%) anodes gave higher yields of the aldehyde product compared with Pb alone. On the basis of the observation that PbO_2 chemically oxidizes *p-tert*-butyltoluene to p-*tert*-butylbenzaldehyde and p-*tert*-butylbenzoic acid, albeit at slow rate at room temperature, Tissot et al. [543] concluded that a surface oxide species was probably involved in the oxidation process at PbO_2. The reason for the enhanced activity of the Pb–Sb alloys remains unclear.

Nilsson et al. [544] studied the anodic hydroxylation of monohydric phenols at a variety of anode materials, including PbO_2. In all cases studied, the hydroxy group entered the 4 position. Thus 4-substituted phenols gave 4-substituted 4-hydroxyclohexa-2, 5-dienones and phenols without substituents at the C-4 position gave *p*-benzoquinones. A mechanism involving hydrolysis of an anodically generated phenoxonium ion was suggested. Experimental evidence suggested that the phenoxonium ion results from chemical oxidation by anodically generated lead dioxide [544]. PbO_2 anodes were observed to have higher activity than carbon, platinum, and nickel anodes for the hydroxylation reactions. Similarly, Fioshin et al. [545] observed that PbO_2 exhibited the highest activity for oxidation of phenol to quinone when compared with anodes such as graphite, Pt, MnO_2, and magnetite.

Recently, Fleszar and Ploszynska [546] re-examined the kinetics of electro-oxidation of benzene and phenol at PbO_2 anodes. These authors concluded that the oxidation process does follow an E.C.E.-type mechanism, there not being enough evidence to indicate direct participation of PbO_2 in the oxidation of these molecules. They observed a linear dependence between oxidation rate and the rate of water electrolysis. Fleszar and Ploszynska [546] advanced the hypothesis that hydroxyl radicals formed on the anode surface caused direct hydroxylation of the benzene and phenol compounds as shown in the following scheme. In a rationalization of this mechanism, the authors invoked semiconductor-based arguments, viz.

$$PbO_2H^+ + H_2O_{ads} = PbO_2(OH)_{ads} + H^+ \tag{112}$$

The OH radicals either recombine into oxygen and water or react with adsorbed organic compounds. It may be noted that Levina et al. [547] have suggested that adsorbed OH radicals are responsible for phenol oxidation at platinum. Fleszar and Ploszynska's [546] results are somewhat at variance with those of Clarke et al. [537] in that the latter authors observed the highest efficiency for benzene oxidation at PbO_2 anodes at potentials prior to the onset of oxygen evolution.

It is possible that differences in experimental approach, e.g. manner of PbO_2 anode preparation, are responsible for some of the disagreement in the literature regarding the mechanism of oxidation of organic compounds at PbO_2. It may also be the case that chemical oxidation experiments carried out with PbO_2 suspensions are not directly related to oxidation processes conducted at PbO_2 anodes due to differences in the nature of the PbO_2, as well as intrinsic differences in the two types of experiment. Thus, an electro-oxidation mechanism involving direct chemical participation by a lead oxide or oxyhydroxide species on the surface may be prevalent for a number of organic processes. Further work is clearly needed to clarify the mechanism of organic oxidation at PbO_2 anodes, e.g. isotopic labelling experiments.

8.3 NICKEL OXIDE

A considerable amount of work has been published dealing with the electro-oxidation of organics at Ni anodes in aqueous base [548–552]. These reactions have generally been dehydrogenations, e.g. primary alcohols to aldehydes, secondary alcohols to ketones and primary amines to nitriles. The reactions occur on a relatively thick layer of oxide on the Ni anodes. Pletcher and co-workers [529, 548, 549] observed that most of the oxidizable compounds were found to oxidize at the same potential and this potential coincided with that at which the surface of the Ni became oxidized. A typical cyclic voltammogram recorded at Ni in dilute KOH in the presence and absence of *n*-propylamine is shown in Fig. 23. It can be seen that addition of *n*-propylamine results in an oxidation wave being observed which is

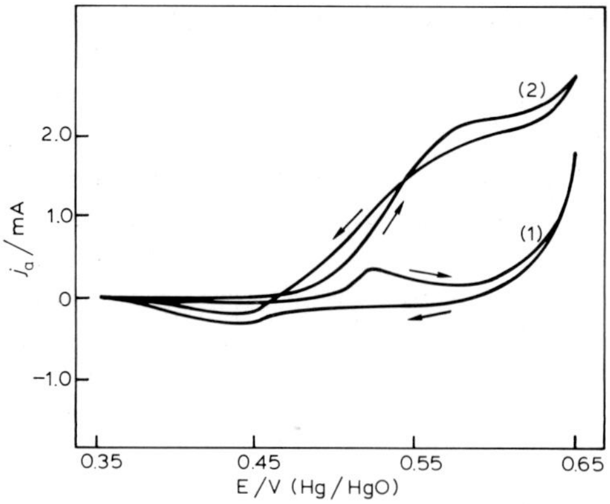

Fig. 23. Cyclic voltammograms for Ni in (1), 8.7×10^{-1} M KOH; and (2) $8.7\,10^{-1}$ M KOH + 8.7×10^{-1} M *n*-propylamine [529].

superimposed on the anodic peak due to the Ni surface. Limiting currents were observed for all compounds, which were variable and were well below that expected for a diffusion-controlled process. Pletcher and co-workers [529, 548, 549] concluded that the mechanism of oxidation of simple aliphatic alcohols and amines at Ni in base is

$$Ni(OH)_2 \rightleftharpoons NiOOH + H^+ + e \tag{113}$$

$$NiOOH + RCH_2OH \xrightarrow{\text{slow step}} Ni(OH)_2 + R\dot{C}HOH \tag{114}$$

$$R\dot{C}HOH \xrightarrow{\text{further steps}} \text{products} \tag{115}$$

The limiting currents observed, which were rotation-independent, were considered to be directly proportional to the rate constant for the slow chemical step. The above mechanism of Pletcher and co-workers [529, 548, 549] was confirmed by Robertson [553] for the case of high OH^- concentration. At low C_{OH} values, the reaction rate was observed to be limited by the diffusion of OH^- to the electrode surface [553]. Further, Robertson [553] observed a peak-shaped *j* vs. *E* plot in place of the limiting current observed by Pletcher et al. [548, 549] for oxidation of the organics. Vertes and Horanyi [554] have cautioned that the mechanism of Pletcher and co-workers [529, 548, 549] cannot be unquestionably accepted because of the possibility that eqn. (113) is not reversible enough.

The foregoing electro-oxidation processes at Ni anodes have the disadvantage that they can only be carried out at very low current densities, typically < 1% of the calculated diffusion-limiting current. In an effort to overcome this current density limitation, Manandhar and Pletcher [555] have recently prepared high surface area nickel electrodes by electrodeposition from a Watts bath on graphite or Cu substrates. Following anodic polarization of the coating in base, greatly enhanced current densities for alcohol oxidation relative to smooth Ni were observed with this type of electrode [555].

The foregoing discussion of organic oxidations at Ni anodes logically leads into the technologically important oxidation of diacetone-L-sorbose (DAS) to diaceto-2-keto-L-gulonic acid (DAG), which is in essence the oxidation of a primary alcohol to a carboxylic acid. This reaction is a step in the synthesis of vitamin C and is normally carried out via catalytic air or sodium hypochlorite oxidation. However, work which began in the early seventies has culminated in the development of a viable electrochemical process at Hoffman–LaRoche for the conversion of DAS to DAG at Ni anodes [550, 556–558]. A key factor in the success of this synthesis was the development of the so-called "Swiss-Roll" electrolysis cell [559, 560]. The primary structure of this cell is a sandwich obtained by laying the working and secondary electrode materials on top of each other, with a suitable insulating separator layer between them and another separator on top. The specific electrode area of the cell, i.e. the electrode area divided by the cell volume, exhibited a relatively high value of $50\,cm^{-1}$. The mechanism of oxidation of DAS to DAG at Ni anodes may well be similar to that outlined in eqns. (113)–(115), although the observations of Robertson et al. [558] suggest that Ni(III) species may not, in fact, be the active species since the oxidation of DAS occurs in the range where further oxidation of Ni(III) occurs, i.e. partially to Ni(IV).

8.4 OTHER OXIDES

Fleischmann et al. [549] studied the electro-oxidation of a series of amines and alcohols at Cu, Co, and Ag anodes in conjunction with the previously described work for Ni anodes in base. In cyclic voltammetry experiments, conducted at low to moderate sweep rates, organic oxidation waves were observed superimposed on the peaks associated with the surface transitions, Ni(II) → Ni(III), Co(II) → Co(III), Ag(I) → Ag(II), and Cu(II) → Cu(III). These observations are in accord with an electrogenerated higher oxide species chemically oxidizing the organic compound in a manner similar to eqns. (112)–(114). For alcohol oxidation, the rate constants decreased in the order $k_{Cu} > k_{Ni} > k_{Ag} > k_{Co}$. Fleischmann et al. [549] observed that the rate of anodic oxidations increases across the first row of the transition metals series. These authors observed that the products of their electrolysis experiments were essentially identical to those obtained in heterogeneous reactions with the corresponding bulk oxides.

Horowitz et al. [561] studied the oxidation of a number of organic compounds at high surface area lead ruthenate in aqueous alkali. The ruthenate used may be represented as $Pb_2(Ru_{(1-x)}Pb_x)_2O_{(7-\varepsilon)}$ where ε may have oxygen vacancies up to $\varepsilon = 1$. Electrode fabrication was carried out by mounting the ruthenate powder on gold screen using Teflon as binder. Horowitz et al. [561] observed that lead ruthenate catalyzes the electro-oxidation of primary alcohols to acids, secondary alcohols to ketones and the cleavage of ketones and olefins to carboxylic acids. Some of their results are summarized in Table 6. Besides exhibiting relatively high activity for the electro-oxidation reactions, the most significant attribute of the lead ruthenate catalyst was its ability to catalyze cleavage reactions cleanly (e.g. Table 6) [561]. It was tentatively suggested that the active moieties on the ruthenate catalyst surface were ruthenium groups which formed cyclic complexes with olefinic double bonds, leading to their cleavage.

Relatively few electro-organic processes have been studied at noble metal oxide, DSA-type, anodes [207, 562, 563]. Benzyl alcohol oxidation occurs at a much lower potential at RuO_2 electrodes than at Pd [207, 562]. Burke and Healy [207] observed that oxidation of benzaldehyde base commences at a potential of about 1.0 V (RHE), i.e. in the region where, according to eqn. (42), RuO_2 is converted to RuO_4^{2-} (Fig. 9). The onset of benzyl alcohol oxidation occurs at 1.4 V (RHE) in base where, according to eqn. (43), RuO_4^- is formed on the surface of the ruthenium dioxide electrodes. Further, Burke and Healy [235] observed that dissolved ruthenate (RuO_4^{2-}) oxidizes benzaldehyde but not benzyl alcohol. Perruthenate (RuO_4^-) was found to oxidize both organic compounds, the product in each case being benzoic acid. This work highlights the important role that electrochemically generated higher oxide species can play in electrocatalysis at oxide electrodes.

O'Sullivan and White [563] recently investigated formaldehyde oxidation at thermally prepared RuO_2 electrodes. Two waves were observed for this reaction in alkaline solution (Fig. 24), the first wave beginning at ~0.6 V

TABLE 6

Secondary alcohols, ketones cleaved

Feed	Product	Coulombic yield (%)		Electrons per mol
		Product	CO_3^- [a]	
$CH_3CH_2\underset{H}{\overset{OH}{C}}—CH_3$	$2\,CH_3COO^-$	67 76	8–25 13–25	8
$CH_3–CH_2\overset{O}{\overset{\|}{C}}CH_3$	$2\,CH_3COO^-$	81	14–24	6
$CH_3–\underset{OH}{CH}–\underset{OH}{CH}–CH_3$	$2\,CH_3COO^-$	84	3–4	6
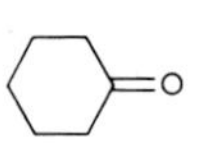	$CH_2–CH_2COO^-$ $\vert$ $CH_2–CH_2–COO^-$	87	1	6

[a] Range of values depends on electrons per mol of CO_3 assumed.

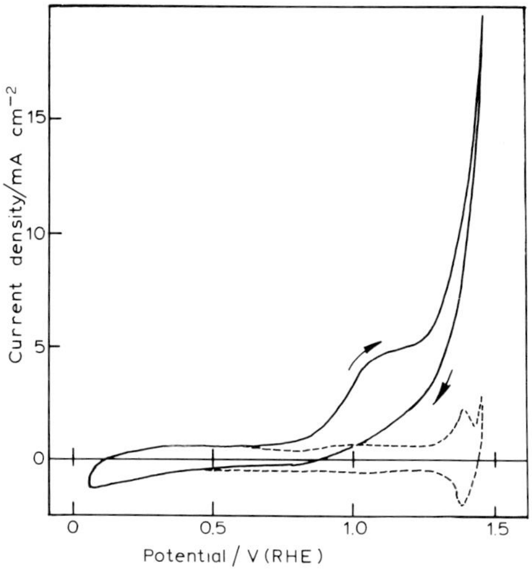

Fig. 24. Current–potential curves for a Teflon-bonded RuO_2 layer on a GC disk electrode in 1.0 mol dm^{-3} NaOH with (——) and without (– – –) 0.05 mol dm^{-3} HCHO. Sweep rate = 5 mV s^{-1}; rotation rate = 2500 rpm; T = 23°C; oxide preparation temperature = 400°C [563].

(RHE) and coinciding with the peak (see Fig. 8) associated with the Ru(IV)/Ru(VI) couple beginning at about 1.25 V (RHE) and the second coinciding with the peak usually associated with the Ru(VI)/Ru(VII) couple [209]. Formate was observed to be the product of HCHO oxidation in the first wave, while 4e oxidation of HCHO to carbonate was observed in the second wave. Despite its interesting catalytic features for the electro-oxidation of formaldehyde, RuO_2 is not as active as the noble metal catalysts Pt and Pd for this reaction, however.

9. Conclusions and recommendations

(1) Careful characterization of the oxide–electrolyte interface is needed: electrochemical area, surface structure, and electronic properties (potential distribution and density of electrical carriers). Chemical and electrochemically induced transformations of the oxide surface in contact with electrolyte can substantially modify the behavior of oxide electrodes. Extrapolation of gas/solid oxide results to oxide electrodes is not always valid since the oxide–electrolyte interface can strongly depend on electrolyte type and applied potential.

(2) There is increasing evidence that several oxides, especially electrochemically formed hydrous oxides, exhibit potential–pH behavior, as derived from cyclic voltammetry data, close to $1.5(2.303RT/F)$ rather than the more commonly accepted $2.303RT/F$. Such behavior has been associated, for example, with increasing acidity of the oxy species at the oxide surface accompanying the increase in degree of oxidation of the central metal ion [12, 142].

(3) In the case of catalytic reactions at oxide electrodes, the importance of the coordination chemistry of surface cations, presence of reactive groups, and ability to form unstable metal oxide species (e.g. in oxygen evolution and organic oxidations) is increasingly being recognized for such reactions in the literature. The results of Hibbert [304] for oxygen evolution at $NiCo_2O_4$ from ^{18}O-enriched electrolyte are notable in this regard.

(4) Interesting results have been obtained for oxygen evolution at reactively sputtered IrO_2 oxide electrodes in acid [182, 183, 247]. In the case of films sputtered in pure oxygen, these underwent an irreversible exothermic transition at ca. 300°C which was associated with an amorphous-to-crystalline transition. The films exhibited a considerable reduction in electrocatalytic activity, as well as a loss of electrochromic behavior, upon undergoing this transition. A complete characterization of this transition, e.g. intrinsic area change, change of degree of hydration, would be useful to gain insight into the factors governing the change in electrochemical properties. The study of sputtered and chemically vapor deposited films of other oxides would be worthwhile, e.g. of RuO_2.

(5) Of the various models which have been proposed for the oxygen electrode reactions, the model of Presnov and Trunov [341, 345] for oxygen reduction at semiconductor oxide electrodes deserves special mention. This model is based on concepts of coordination chemistry and local interaction of surface cation *d*-electrons with HO^-, H_2O, and O_2 acceptor species in solution.

(6) It is now recognized that anomalously low Tafel slopes can be observed for the chlorine evolution reaction due to rate-limiting transport of gas away from the electrode surface [471, 474], e.g. in concentrated chloride solutions at high temperatures.

(7) There is considerable evidence that the kinetics of chlorine evolution at a number of oxide electrodes are sensitive to electrolyte pH, e.g. at RuO_2 [477] and Co_3O_4 [494], the apparent reaction order with respect to H^+ increasing with increasing acidity [494]. This is an unexpected result from the viewpoint of thermodynamics and suggests involvement of surface cation oxy species in the chlorine evolution pathway. Notwithstanding the technological importance of RuO_2 based DSA electrodes in chlorine evolution, this pH effect needs to be further explored since it provides insight into the mechanism of chlorine evolution and may account, in part, for the discrepancies between the results of various workers in the field.

(8) Metal oxide electrodes have been relatively infrequently employed in electro-organic reactions and, even in those cases which have been moderately well studied, there are still some questions regarding the reaction mechanisms, e.g. whether a surface oxide species mediates the organic transformation or not in the case of oxidation reactions. The study of certain types of model organic compounds, e.g. alcohols and aldehydes, at metal oxide electrodes could lead to further insight into oxide electrocatalysis.

(9) Simple Faradaic electron transfer reactions at semiconducting oxides

and passive films are strongly influenced by many factors, e.g. the oxide electronic properties, electrolyte pH, potential range.

(10) There is a need to develop new types of oxide electrodes for reactions of technological importance with emphasis on both high electrocatalytic activity and stability. For example, pyrochlore-type oxides, e.g. lead or bismuth ruthenates, have shown excellent catalytic activity for the oxygen evolution and reduction reactions and should be further investigated to elucidate the reasons for such high activity. The long term stability of such ruthenate electrodes is questionable, however.

(11) In many instances for metal oxide electrodes, high electrocatalytic activity is accompanied by poor catalyst stability, i.e. mainly poor resistance to dissolution. It may be possible for certain oxide systems to establish some type of general relationship between catalytic activity and catalyst stability, e.g. based on extent of hydration, crystallization and so on.

References

1 J.W. Diggle (Ed.), Oxides and Oxide Films, Vol. 1, Dekker, New York, 1972.
2 J.W. Diggle (Ed.), Oxides and Oxide Films, Vol. 2, Dekker, New York, 1973.
3 S. Trasatti (Ed.), Electrodes of Conductive Metallic Oxides, Elsevier, New York, Part A, 1980, and Part B, 1981.
4 M.I. Nagayama and M. Cohen, J. Electrochem. Soc., 110 (1963) 670.
5 C. Wagner, Ber. Bunsenges. Phys. Chem., 77 (1973) 1090.
6 M.M. Lohrengel, P.K. Richter and J.W. Schultze, Ber. Bunsenges. Phys. Chem., 83 (1979) 490.
7 S.A. Ahmed, in J.W. Diggle (Ed.), Oxides and Oxide Films, Vol. 1, Dekker, New York, 1972, p. 319.
8 D.N. Furlong, D.E. Yates and T.W. Healy, in S. Trasatti (Ed.), Electrodes of Conductive Metallic Oxides, Part 8, Elsevier, New York, 1981, p. 367.
9 S. Levine and A.L. Smith, Discuss Faraday Soc., 52 (1971) 290.
10 R.G. Egdell, J.B. Goodenough, A. Hammet and C.C. Naish, J. Chem. Soc. Faraday Trans. 1, 79 (1983) 893.
11 L.D. Burke and R.A. Scannell, Platinum Met. Rev., 28 (1984) 56.
12 L.D. Burke and D.P. Whelan, J. Electroanal. Chem., 162 (1984) 121.
13 F. Cardon and W.P. Gomes, J. Phys. D, 11 (1978) L63.
14 V.F. Lohmann, Ber. Bunsenges. Phys. Chem., 70 (1966) 428.
15 E.C. Dutoi, F. Cardon and W.P. Gomes, Ber. Bunsenges. Phys. Chem., 76 (1972) 475.
16 R. Memming and F. Mollers, Ber. Bunsenges. Phys. Chem., 76 (1972) 475.
17 P. Siviglia, A. Daghetti and S. Trasatti, Colloids Surf., 7 (1983) 15.
18 C. Pirovano and S. Trasatti, J. Electroanal. Chem., 180 (1984) 171.
19 R. Caravaglia, C,M. Mari and S. Trasatti, Surf. Technol., 19 (1983) 197.
20 R. Caravaglia, C.M. Mari and S. Trasatti, Surf. Technol., 23 (1984) 41.
21 S. Ardizzone, D. Lettieri and S. Trasatti, J. Electroanal. Chem., 146 (1983) 431.
22 R. Vigano, J. Taraszewska, A. Daghetti and S. Trasatti, J. Electroanal. Chem., 182 (1985) 203.
23 A. Carugati, G. Lodi and S. Trasatti, J. Electroanal. Chem., 143 (1983) 419.
24 J.H. Kennedy and K.W. Frese, Jr., J. Electrochem Soc., 125 (1978) 723.
25 R.K. Quinn, R.D. Nasby and R.J. Baughman, Mater. Res. Bull., 11 (1976) 1011.
26 S.H. Wilhelm, K.S. Yun, L. Ballenger and N. Hackerman, J. Electrochem. Soc., 127 (1979) 419.

27 K.S. Yun, S.M. Wilhelm, S. Kapusta and N. Hackerman, J. Electrochem. Soc., 127 (1980) 85.
28 U. Stimming and J.W. Schultze, Ber. Bunsenges. Phys. Chem., 80 (1976) 1297.
29 D.J. Wheeler, B.D. Cahan, C.T. Chen and E. Yeager, in R.P. Frankenthal and J. Kruger (Eds.), The Passivity of Metals, The Electrochemical Society, Pennington, NJ, 1978, p. 546.
30 E.J. Calvo, unpublished data.
31 K.E. Heusler, Electrochim. Acta, 28 (1983) 439.
32 J.P. Pohl and H. Rickert, in S. Trasatti (Ed.), Electrodes of Conductive Metallic Oxides, Part A, Elsevier, New York, 1980, p. 183.
33 D.J.G. Ives, in D.J.G. Ives and G.J. Janz, (Eds.), Reference Electrodes Theory and Practice, Academic Press, New York, 1961, p. 322.
34 T. Kudo, H. Obayashi and T. Gejo, J. Electrochem. Soc., 122 (1975) 159.
35 J.P. Pohl and S. Atlung, Electrochim. Acta, 31 (1986) 391.
36 C. Wagner and W. Traud, S. Elektrochem., 44 (1938) 391.
37 N. Valverde and C. Wagner, Ber. Bunsenges. Phys. Chem., 80 (1976) 330.
38 K.J. Vetter, Electrochemical Kinetics, Academic Press, New York, 1967 (English translation).
39 K.J. Vetter and F. Gorn, Electrochim. Acta, 18 (1973) 321.
40 K.E. Heusler and L. Fischer, Werkst. Korros., 27 (1976) 697.
41 K.E. Heusler, Ber. Bunsenges. Phys. Chem., 72 (1968) 1197.
42 K.E. Heusler, in A.J. Bard (Ed.), Encylopedia of the Electrochemistry of the Elements, Vol. IXA, Dekker, New York, 1982, p. 229.
43 K.D. Allard and K. Heusler, J. Electroanal. Chem., 77 (1977) 35.
44 T. Valand and K.E. Heusler, J.Electroanal. Chem., 149 (1983) 71.
45 J.P. Pohl and J. Rickert, Z. Phys. Chem. N.F., 95 (1975) 59.
46 A. Grzegorzewski and K.E. Heusler. Proc. Int. Soc. Electrochem. M., Salamanca, 1985. p. 5040.
47 W.J. Albery and E.J. Calvo, J. Chem. Soc. Faraday Trans. 1, 79 (1983) 2583.
48 E.J. Calvo, J. Drennan, J.A. Kilver, W.J Albery and B.C.H. Steele, in J.D.E. McIntyre, M.J. Weaver and E.B. Yeager (Eds.), The Chemistry and Physics of Electrocatalysis, The Electrochemical Society, Pennington, NJ, 1984, p. 489.
49 J.W. Diggle, in J.W. Diggle (Ed.), Oxides and Oxide Films, Vol. 2, Dekker, New York, 1973, p. 281.
50 H.J. Engell, Z. Phys. Chem. N.F., 7 (1956) 158.
51 H.J. Engell, Z. Elektrochem., 60 (1956) 905.
52 D.A. Vermilyea, J. Electrochem. Soc., 113 (1966) 1067.
53 N. Valverde, Ber. Bunsenges. Phys. Chem., 80 (1976) 333.
54 C.H. Lee, A. Riga and E. Yeager, Off. Nav. Res. (U.S.) Tech. Rep. 40, Contract No. NOO14-67-A-O4O4-OOO6, Case Western Reserve University, 1975.
55 S.R. Morrison and T. Freund, Electrochim. Acta, 13 (1968) 1343.
56 R.A.L. Van den Berghe, F. Cardon and W.P. Gomes, Surf. Sci., 39 (1973) 368.
57 H. Gerischer, Adv. Electrochem. Eng., 1 (1961) 139.
58 F. Mollers and R. Memming, Ber. Bunsenges. Phys. Chem., 76 (1972) 469.
59 R.N. Noufi, P.A. Kohl, S.N. Frank and A.J. Bard, J. Electrochem. Soc., 125 (1978) 246.
60 B. Danzfuss and U. Stimming, J. Electroanal. Chem., 164 (1984) 89.
61 R.A. Fredlein and A.J. Bard, J. Electrochem. Soc., 126 (1979) 1892.
62 P. Iwanski, J.S. Curran, W. Gissler and R. Memming, J. Electrochem. Soc., 128 (1981) 2128.
63 D. Yohe, A. Riga, R. Greef and E. Yeager, Electrochim. Acta, 13 (1968) 1351.
64 D. M. Tench and E. Yeager, J. Electrochem. Soc., 121 (1974) 318.
65 T.O. Rouse and J.L. Weininger, J. Electrochem. Soc., 113 (1966) 186.
66 D. M. Tench and E. Yeager, J. Electrochem. Soc., 120 (1973) 164.
67 H. Yoneyama and H. Tamura, Bull. Chem. Soc. J., 45 (1972) 3048.
68 M.P. Dare-Edwards, J.B. Goodenough, A. Hammett and N.D. Nickolson, J. Chem. Soc. Faraday Trans. 2, 77 (1981) 643.

69 J.E.B. Randles and K.W. Sommerton, Trans. Faraday Soc., 48 (1952) 937.
70 H. Gerischer, Z. Elektrochem., 54 (1960) 366.
71 H. Yoneyama and H. Tamura, Bull. Chem. Soc. J., 43 (1970) 1603.
72 M. Beley, J. Brenet and P. Chartier, Ber. Bunsenges. Phys. Chem., 78 (1974) 455.
73 M. Beley, J. Brenet and P. Chartier, Ber. Bunsenges. Phys. Chem., 79 (1975) 317.
74 H. Nguyen Cong, J. Brenet and P. Chartier, Ber. Bunsenges, Phys. Chem., 79 (1975) 323.
75 M. Beley, J. Brenet and P. Chartier, Electrochim. Acta, 24 (1979) 1.
76 H. Obayashi and T. Kudo, Mater. Res. Bull., 13 (1978) 1409.
77 F.R. van Buren, G.H.J. Broers and T.G.M. Van den Belt, Ber. Bunsenges. Phys. Chem., 83 (1979) 82.
78 E.J. Calvo, in C.H. Bamford and R. Compton (Eds.), Comprehensive Chemical Kinetics, Vol. 26, Elsevier, Amsterdam, 1986, p. 1.
79 J.B. Goodenough and P.M. Raccah, J. Appl. Phys., 36 (1965) 1031.
80 J. Drennan, C.P. Tavares and B.C.H. Steele, Mater. Res. Bull., 17 (1982) 621.
81 E. Calvo, J. Drennan, B.C.H. Steele and W.J. Albery, in J. Goodenough, J. Jensen and M. Kleitz (Eds.), Solid State Protonic Conductors, Vol. 2, Odense University Press, Denmark, 1983, p. 289.
82 R.J.D. Tilley, J. Solid State Chem., 21 (1977) 293.
83 J.J. Janecek and G.P. Wirtz, J. Am. Ceram. Soc., 61 (1978) 242.
84 P. Ganguli and C.N.R. Rao, Mater. Res. Bull., 8 (1973) 405.
85 W.J. Albery and S. Bruckenstein, Trans, Faraday Soc., 62 (1966) 1920.
86 Y. Matsumoto, H. Yoneyama and H. Tamura, J. Electroanal. Chem., 80 (1977) 115.
87 J.O'M. Bockris and T. Otagawa, J. Phys. Chem., 87 (1983) 2960.
88 G. Karlsson, J. Power Sources, 10 (1983), 319.
89 B. Lovrecek, J. Sefaja and M. Shala, Z. Phys. Chem. (Leipzig), 262 (1981) 439.
90 M.V. Vojnovic and D.B. Sepa, J. Chem. Phys., 51 (1969) 5344.
91 M. Amjad and D. Pletcher, J. Electroanal. Chem., 59 (1975) 61.
92 D. Galizziolli, F. Tantardini and S. Trasatti, J. Appl. Electrochem., 5 (1975) 203.
93 R.V. Bucur and A. Bartes, Electrochim. Acta, 25 (1980) 879.
94 J.W. Schultze, S. Mohr and M.M. Lohrengel, J. Electroanal. Chem., 154 (1983) 57.
95 R. Babic and B. Lovrecek, Elektrokhimiya, 15 (1979) 16.
96 M.M. Lohrengel, P.K. Richter and J.W. Schultze, Ber. Bunsenges. Phys. Chem., 83 (1979) 495.
97 J.W. Schultze and C. Bartels, J. Electroanal. Chem., 150 (1983) 583
98 J.W. Schultze, in R.P. Frankenthal and J. Kruger (Eds.), The Passivity of Metals, The Electrochemical Society, Pennington, NJ, 1978, p. 82.
99 W.Schmickler and J.W. Schultze, in J.O'M. Bockris, B.E. Conway and R.E. White (Eds.), Modern Aspects of Electrochemistry, Vol. 17, Plenum Press, New York, 1986, p. 357.
100 K.E. Heusler and K.S. Yun, Electrochim, Acta, 22 (1977) 977.
101 J.W. Schultze and L. Elfenthal, J. Electroanal. Chem., 204 (1986) 153.
102 W. Schmickler, in M. Froment (Ed.), Passivity of Metals and Semiconductors, Elsevier, Amsterdam, 1983, p. 23.
103 W. Schmickler, J. Electroanal. Chem., 82 (1977) 65.
104 W. Schmickler, J. Electroanal. Chem., 83 (1977) 387.
105 W. Schmickler, J. Electroanal. Chem., 137 (1982) 189.
106 W. Schmickler and U. Stimming, Thin Solid Films, 75 (1981) 331.
107 B. Schumm Jr., H.M. Joseph and A. Kozawa (Eds.), Manganese Dioxide Symposium, Vol. 2, Tokyo, 1980; I.C. Sample Office, Box 6116, Cleveland, OH, 44101, U.S.A.
108 B.D. Desai and J.B. Fernandes and V.N.K. Dalal, J. Power Sources, 16 (1985) 1.
109 J.B. Fernandes, B.D. Desai and V.N.K. Dalal, J. Power Sources, 15 (1985) 209.
110 B. Schumm, Jr., R.L. Middaugh, M.P. Grother and J.C. Hunter (Eds.), Manganese Dioxide Electrode Theory and Practice for Electrochemical Applications, The Electrochemical Society, Pennington, NJ, 1985.
111 R.G. Gunther and S. Gross (Eds.), The Nickel Electrode, The Electrochemical Society, Pennington, NJ, 1982.

112 P. Oliva, J. Leonardi, J.F. Laurent, C. Delmas, J.J. Baconnier, M. Figlarz, F. Fievet and A. de Guibert, J. Power Sources, 8 (1982) 229.
113 R.T. Barton, P.J. Mitchell and N.A. Hampson, Surf. Coat. Technol., 28 (1986) 1.
114 G.W.D. Briggs, Chem. Soc. Spec. Period. Rep. Electrochem. 4 (1974) 33.
115 A.R. Kmetz and F.K. Von Willisen (Eds.), Non-Emissive Electrooptic Displays, Plenum Press, New York, 1976.
116 B.W. Faughnan and R.S. Crandall, in J.I. Pankove (Ed.), Topics in Applied Physics, Vol. 40, Springer-Verlag, Berlin, 1980, p. 181.
117 S.K. Deb, Philos. Mag., 27 (1973) 801.
118 B.W. Faughnan, R.S. Crandall and P.M. Heyman, RCA Rev., 36 (1975) 177.
119 M. Green and D. Richman, Thin Solid Films, 24 (1974) 545.
120 S. Gottesfeld, J.D.E. McIntyre, G. Beni and J.L. Shay, Appl. Phys. Lett., 33 (1978) 208.
121 D.N. Buckley and L.D. Burke, J. Chem. Soc. Faraday Trans. 1, 71 (1975) 1447.
122 D.A.J. Rand and R. Woods, J. Electroanal. Chem., 55 (1974) 375.
123 S. Gottesfeld and S. Srinivasan, J. Electroanal. Chem., 86 (1978) 89.
124 L.D. Burke and M.B.C. Roche, J. Electroanal. Chem., 164 (1984) 315.
125 S.D. James, J. Electrochem. Soc., 116 (1969) 1681.
126 L.D. Burke and E.J.M. O'Sullivan, J. Electroanl. Chem., 93 (1978) 11.
127 S. Gottesfeld and J.D.E. McIntyre, J. Electrochem. Soc., 126 (1979) 742.
128 J.L. Shay, L.M. Schiavone, R.W. Epworth and D.W. Taylor, J. Appl. Phys., 53 (1982) 6004.
129 L.D. Burke and E.J.M. O'Sullivan, unpublished results.
130 W. Böld and M. Breiter, Electrochim. Acta, 5 (1961) 169.
131 A. Capon and R. Parsons, J. Electroanal. Chem., 39 (1972) 275.
132 J.M. Otten and W. Visscher, J. Electroanal. Chem., 55 (1974) 1.
133 J.O. Zerbino, N.R. deTacconi and A.J. Arvia, J. Electrochem. Soc., 125 (1978) 1266.
134 J.M. Otten and W. Visscher, J. Electroanal. Chem., 55 (1974) 13.
135 B.D. Kurnikov, A.I. Zhurin, V.V. Chernyi, Yu.B. Vasil'ev and V.S. Bagotskii, Elektrokhimiya, 9 (1973) 833.
136 D.N. Buckley, L.D. Burke and K.J. Mulcahy, J. Chem. Soc. Faraday Trans. 1, 72 (1976) 1896.
137 S. Gottesfeld, J. Electrochem. Soc., 127 (1980) 272.
138 V. Birss, R. Myers, H. Angerstein-Kozlowska and B.E. Conway, J. Electrochem. Soc., 131 (1984) 1502.
139 B.E. Conway and J. Mozota, Electrochim. Acta, 28 (1983) 9.
140 H. Angerstein-Kozlowska, B.E. Conway and W.B.A. Sharp, J. Electroanal. Chem., 43 (1973) 9.
141 J.D.E. McIntyre, W.F. Peck, Jr. and S. Nakahara, J. Electrochem, Soc., 127 (1980) 1264.
142 L.D. Burke and D.P. Whelan, J. Electroanal. Chem., 124 (1981) 333.
143 M.F Yuen, I. Lauks and W.C. Dautremont-Smith, Solid State Ion., 11 (1983) 19.
144 G. Beni, C.E. Rice and J.L. Shay, J. Electrochem, Soc., 127 (1980) 1342.
145 H.G. Scott, Acta Crystallogr. B, 35 (1979) 3014.
146 L.D. Burke and M.C.B. Roche, J. Electroanal. Chem., 164 (1984) 315.
147 S.H. Glarum and J.H. Marshall, J. Electrochem. Soc., 127 (1980) 1467.
148 S. Gottesfeld, J. Electrochem. Soc., 127 (1980) 1922.
149 E.R. Kotz and H. Neff, Surf. Sci., 160 (1985) 517.
150 L.D. Burke and E.J.M. O'Sullivan, J. Electroanal. Chem., 117 (1981) 155.
151 L.F. Mattheiss, Phys. Rev. B, 13 (1976) 2433.
152 J.B. Goodenough, in H. Reiss, (Ed.), Progress in Solid State Chemistry, Vol. 5, Pergamon, Press, Oxford, 1971, p. 143.
153 J.S.E.M. Svensson and C.G. Granqvist, Sol. Energy Mater., 12 (1985) 391.
154 N. Yoskiike, M. Ayusawa and S. Kondo, J. Electrochem. Soc., 131 (1984) 2600.
155 A. Di Paola, F. Di Quarto and C. Sunseri, J Electrochem. Soc., 125 (1978) 1344.
156 B. Reichman and A.J. Bard, J. Electrochem. Soc., 126 (1979) 583.
157 B. Reichman and A.J. Bard, J. Electrochem. Soc., 126 (1979) 2133.
158 L.D. Burke, T.A.M. Twomey and D.P. Whelan, J. Electroanal. Chem., 107 (1980) 201.

159 L.D. Burke and D.P. Whelan, J. Electroanal. Chem., 109 (1980) 385.
160 L.D. Burke and T.A.M. Twomey, J. Electroanal. Chem., 167 (1984) 285.
161 M.R. Tarasevich, A. Sadkowski and E. Yeager, in E. Yeager, J.O'M. Bockris, B.E. Conway and S. Sarangapani (Eds.), Comprehensive Treatise of Electrochemistry, Vol. 7, Plenum Press, New York, 1983, p. 301.
162 J.O'M. Bockris and A.K.M.S. Huq, Proc. R. Soc. London Ser. A, 237 (1956) 277.
163 J.P. Hoare, J. Electrochem. Soc., 110 (1963) 1019.
164 N. Watanabe and M.A.V. Devanathan, J. Electrochem. Soc., 111 (1964) 615.
165 R.Kh. Burshtein, M.R. Tarasevich and V.A. Bogdanovskaya, Elektrokhimiya, 8 (1972) 1542.
166 R. Lorenz and H. Hauser, Z. Anorg. Allgem. Chem., 51 (1906) 615.
167 R. Lorenz, Z. Elektrochem., 14 (1908) 781.
168 J.P. Hoare, The Electrochemistry of Oxygen, Interscience, New York, 1968.
169 F.J. Brislee, Trans. Faraday Soc., 1 (1905) 65.
170 H. Wroblowa, M.L.B. Rao, A. Damjanovic and J.O'M. Bockris, J. Electroanal. Chem., 15 (1967) 139.
171 P. Bindra, S. Clouser and E. Yeager, J. Electrochem. Soc., 126 (1979) 1631.
172 E.I. Khrushcheva, V.S. Bagotskii and N.A. Shumilova, Elektrokhimiya, 16 (1980) 743.
173 A.C.C. Tseung and H.L. Bevan, J. Electroanal. Chem., 45 (1973) 429.
174 A.C.C. Tseung, B.S. Hobbs and A.D.S. Tantram, Electrochim. Acta, 15 (1970) 473.
175 H.L. Bevan and A.C.C. Tseung, Electrochim. Acta, 19 (1974) 201.
176 F.R. van Buren, G.H.J. Broers, C. Boesveld and A.J. Bouman, J. Electroanal. Chem., 87 (1978) 381.
177 M. Foex, C.R. Acad. Sci., 227 (1948) 193.
178 J.T. Richardson and W.O. Milligan, Phys. Rev., 102 (1956) 1289.
179 J. Cohen, K.M. Creer, R. Pantbenet and K.G. Srivastava, J. Phys. Soc. J., 17 (1962) 685.
180 L.D. Burke, in S. Trasatti (Ed.), Electrodes of Conductive Metallic Oxides, Part A, Elsevier, New York, 1980, p. 11.
181 A. Damjanovic, J.O'M. Bockris and B.E. Conway (Eds.), Modern Aspects of Electrochemistry, Vol. 5, Plenum Press, New York, 1969, p. 369.
182 G. Beni, L.M. Schiavore, J.L. Shay, W.C. Dautremont-Smith and B.S. Schneider, Nature (London), 282 (1979) 281.
183 S. Hackwood, L.M. Schiavone, W.C. Dautremont-Smith and G. Beni, J. Electrochem. Soc., 128 (1981) 2569.
184 L.D. Burke and E.J.M. O'Sullivan, J. Electroanal. Chem., 97 (1979) 123.
185 S. Trasatti and W.E. O'Grady, Adv. Electrochem. Electrochem. Eng., 12 (1981) 177.
186 S. Trasatti and G. Lodi, in S. Trasatti (Ed.), Electrodes of Conductive Metallic Oxides, Part A, Elsevier, New York, 1980, p. 521.
187 S. Trasatti and G. Lodi, in S. Trasatti (Ed.), Electrodes of Conductive Metallic Oxides, Part B, Elsevier, 1981, p. 301.
188 J.A. Rard, Chem. Rev., 85 (1985) 1.
189 S.R. Butler and J.L. Gillson, Mater. Res. Bull., 6 (1971) 81.
190 W.E. Bell and M. Tagami, J. Am. Chem. Soc., 67 (1963) 2432.
191 P.F. Campbell, M.H. Ortner and C.J. Anderson, Anal. Chem., 33 (1961) 58.
192 D. Galizzioli, F. Tantardini and S. Trasatti, J. Appl. Electrochem., 4 (1974) 57.
193 L.D. Burke, O.J. Murphy, J.F. O'Neill and S. Venkatesan, J. Chem. Soc. Faraday Trans. 1, 73 (1977) 1659.
194 S. Pizzini, G. Buzzanca, C. Mari, L. Rossi and S. Torchio, Mater. Res. Bull., 7 (1972) 449.
195 G. Lodi,. G. Bighi and C. deAsmundis, Mater. Chem., 1 (1976) 177.
196 C. Iwakura, H. Tada and H. Tamura, Electrochim. Acta, 22 (1977) 217.
197 K.S. Kim and N. Winograd, J. Catal., 35 (1974) 66.
198 J. Agustinski, L. Balsenc and J. Hinden, J. Electrochem. Soc., 125 (1978) 1093.
199 D.R. Rolison, K. Kuo, M. Umana, D. Brundage and R.W. Murray, J. Electrochem. Soc., 126 (1979) 407.
200 R. Kötz, H.J. Lewerenz and S. Stucki, J. Electrochem. Soc., 130 (1983) 825.

201 R.R. Daniels, G. Margaritondo, C.A. Georg and F. Levy, Phys. Rev. B, 29 (1984) 1813.
202 L.D. Burke and M. McCarthy, Electrochim. Acta, 29 (1984) 211.
203 F. Hine, M. Yasuda and T. Yoshida, J. Electrochem. Soc., 124 (1977) 500.
204 W.A. Gerrard and B.C.H. Steele, J. Appl. Electrochem., 8 (1978) 417.
205 E.K. Spasskaya, Yu. B. Makarychev, A.A. Yakovleva and L.M. Yakimenko, Elektrokhimiya, 13 (1977) 327.
206 L.D. Burke and O.J. Murphy, J. Electroanal. Chem., 96 (1979) 19.
207 L.D. Burke and J.F. Healy, J. Electroanal. Chem., 124 (1981) 327.
208 L.D. Burke, O.J. Murphy and J.F. O'Neill, J. Electroanal. Chem., 81 (1977) 391.
209 L.D. Burke and O.J. Murphy, J. Electroanal. Chem., 109 (1980) 199.
210 K. Doblhofer, M. Metikos, Z. Ogumi and H. Gerischer, Ber. Bunsenges. Phys. Chem., 82 (1978) 1046.
211 D.W. Lam, K.E. Johnson and D.G. Lee, J. Electrochem. Soc., 125 (1978) 1069.
212 L.D. Burke and D.P. Whelan, J. Electroanal. Chem., 103 (1979) 179.
213 J.E. Weston and B.C.H. Steele, J. Appl. Electrochem., 10 (1980) 49.
214 M. Morita, C. Iwakura and H. Tamura, Electrochim. Acta, 24 (1979) 357.
215 D.W. Murphy, F.J. DiSalvo, J.M. Carides and J.V. Waszcak, Mater. Res. Bull., 13 (1978) 1395.
216 M. Armand, F. Dalard, D. Deroo and C. Mouliom, Solid State Ion., 15 (1985) 205.
217 I.J. Davidson and J.E. Greedman, J. Solid State Chem., 51 (1984) 104.
218 D.V. Kokoulina, T.V. Ivanova, Yu.I. Krasovitskaya, Z.I. Kudryavtseva and L.I. Krishtalik, Elektrokhimiya, 13 (1977) 1511.
219 G. Lodi, G. Zucchini, A. DeBattisti, E. Siviere and S. Trasatti, Mater. Chem., 3 (1978) 179.
220 T.Y. Safonova, O.A. Petrii and E.A. Gudkova, Elektrokhimiya, 16 (1980) 1607.
221 A. Carrington and M.C.R. Symons, J. Chem. Soc., (1960) 284.
222 L.D. Burke and M. McRann, J. Electroanal. Chem., 125 (1981) 387.
223 L.D. Burke, M.E. Lyons, E.J.M. O'Sullivan and D.P. Whelan, J. Electroanal. Chem., 122 (1981) 403.
224 C. Iwakura, K. Hirao and H. Tamura, Electrochim. Acta, 22 (1977) 329.
225 M.H. Miles, E.A. Klaus, B.P. Gunn, J.R. Locker and S. Srinivasan, Electrochim. Acta, 23 (1978) 521.
226 M.M. Peckerskii, V.V. Gorodetskii, V.M. Pulina and V.V. Losev, Elektrokhimiya, 12 (1976) 1445.
227 W.E. O'Grady, C. Iwakura, J. Huang and E. Yeager, in M.W. Breiter (Ed.), Electrocatalysis, The Electrochemical Society, Princeton, NJ, 1974, p. 286.
228 G. Lodi, E. Sivieri, A. DeBattisti and S. Trasatti, J. Appl. Electrochem., 8 (1978) 135.
229 D.A. Denton, J.A. Harrison and R.I. Knowles, Electrochim. Acta, 26 (1981) 1197.
230 A. Carugati, G. Lodi and S. Trasatti, Mater. Chem., 6 (1981) 255.
231 R. Kötz, H.J. Lewerenz, P Brüsch and S. Stucki, J. Electroanal. Chem., 150 (1983) 209.
232 R. Kötz, S. Stucki, D. Scherson and D.M. Kolb, J. Electroanal. Chem., 172 (1984) 211.
233 J. Kiwi, M. Grätzel and G. Blondael, J. Chem. Soc. Dalton Trans., (1983) 2215.
234 A. Mills, C. Lawrence and R. Enos, J. Chem. Soc. Chem. Commun., (1984) 1436.
235 L.D. Burke and J.F. Healy, J. Chem. Soc. Dalton Trans., (1982) 1091.
236 L.I. Krishtalik, Electrochim. Acta, 26 (1981) 329.
237 M.W. Shafer, R.A. Figat, R. Johnson and R.A. Pollack, Extended Abstracts 30th ISE Meeting, Trondheim, 1979, p. 313.
238 T. Hepel, F.H. Pollak and W.E. O'Grady, J. Electrochem. Soc., 133 (1984) 2094.
239 T. Hepel, F.H. Pollak and W.E. O'Grady, J. Electrochem. Soc., 133 (1986) 69.
240 W.E. O'Grady, A.K. Goel, F.H. Pollak, H.L. Park and Y.S. Huang, J. Electroanal. Chem., 151 (1983) 295.
241 C. Iwakura, H. Tada and H. Tamura, Denki Kagaku, 45 (1977) 202.
242 M.H. Miles, Y.H. Huang and S. Srinivasan, J. Electrochem. Soc., 125 (1978) 1931.
243 W.E. O'Grady, C. Iwakura and E. Yeager, in Proc. Intersoc. Conf. Environ. Systems, The American Society of Mechanical Engineers, San Diego, 1976, p. 37.

244 S. Stucki and R. Müller, in T.N. Veziroglu, K. Fueki and T. Ohta (Eds.), Proc. 3rd World Hydrogen Energy Conf., Tokyo, 1980, Pergamon Press, Oxford, 1981, p. 1799.
245 R.S. Yeo, J. Orehotsky, W. Visscher and S. Srinivasan, J. Electrochem. Soc., 128 (1981) 1900.
246 R. Hutchings, K. Müller, R Kötz and S. Stucki, J. Mater. Sci., 19 (1984) 3987.
247 S. Hackwood, A.H. Dayem and G. Beni, Phys. Rev. B, 26 (1982) 471.
248 K.S. Kang and J.L. Shay, J. Electrochem. Soc., 130 (1983) 766.
249 E.J. Frazer and R. Woods, J. Electroanal. Chem., 102 (1979) 127.
250 R. Kötz, H. Neff and S. Stucki, J. Electrochem. Soc., 131 (1984) 72.
251 D. Cipris and D. Pouli, J. Electroanal. Chem., 73 (1976) 125.
252 C. Iwakura, K. Fukuda and H. Tamura, Electrochim. Acta, 21 (1976) 501.
253 A. Damjavovic, A.T. Ward and M. O'Jea, J. Electrochem. Soc., 121 (1974) 1186.
254 D. Gilroy, J. Electroanal. Chem., 83 (1977) 327.
255 O. Spalek and J. Balej, Collect. Czech. Chem. Commun., 38 (1973) 29.
256 S. Gottesfeld and S. Srinivasan, Electrochem. Soc. Spring Meeting, Extended Abstr., 76-1, 1976, p. 907.
257 E.J.M. O'Sullivan and L.D. Burke, in W.E. O'Grady, P.N. Ross, Jr. and F.G. Will (Eds.), Electrocatalysis, The Electrochemical Society, Pennington, NJ, 1982, p. 209.
258 N.F. Razina and L.N. Gur'cva, Izv. Akad. Nauk Kaz. SSR Ser. Khim., (1983) 21.
259 V.H. Voorhoeve, in J.J. Burton and R.L. Garten (Eds.), Advanced Materials in Catalysis, Academic Press, New York, 1977, p. 129.
260 A.C.C. Tseung and H.L. Bevan, J. Mater. Sci., 5 (1970) 604.
261 J. Kelly, D.B. Hibbert and A.C.C. Tseung, J. Mater. Sci., 13 (1978) 1053.
262 D.B. Hibbert and A.C.C. Tseung, J. Mater. Sci., 14 (1979) 2665.
263 Y. Matsumoto and E. Sato, Denki Kagaku, 51 (1983) 783.
264 H. Tamura, Y. Yoneyama and Y. Matsumoto, in S. Trasatti (Ed.), Electrodes of Conductive Metallic Oxides, Part A, Elsevier, New York, 1980, p. 261.
265 J.O'M. Bockris and T. Otagawa, J. Electrochem. Soc., 131 (1984) 290.
266 S. Yamada, Y. Matsumoto and E. Sato, Denki Kagaku, 49 (1981) 269.
267 G.H. Jonker, Philips Res. Rep., 24 (1969) 1.
268 F.R. van Buren, H.G.J. Broers and T.G.M. van den Belt, Ber. Bunsenges. Phys. Chem., 83 (1979) 82.
269 D.B. Sepa, A. Damjanovic and J.O'M. Bockris, Electrochim. Acta, 12 (1967) 746.
270 D.M. Meadowcroft, Nature (London), 226 (1970) 847.
271 G. Fiori, C. Mandelli, C.M. Mari and P.V. Scolari, in T.N. Veziroglu and W. Seifritz (Eds.), Hydrogen Energy System, Vol. 1, Pergamon Press, Oxford, 1978, p. 193.
272 G. Fiori and C.M. Mari, in T.N. Veziroglu, K. Fueki and T. Ohta (Eds.), Hydrogen Energy Progress, Vol. 1, Pergamon Press, Oxford, 1981, p. 165.
273 G. Fiori and C.M. Mari, Adv. Hydrogen Energy, 3 (1982) 291.
274 Y. Matsumoto, J. Kurimoto and E. Sato, J. Electroanal. Chem., 102 (1979) 77.
275 Y. Matsumoto, H. Manabe and E. Sato, J. Electrochem. Soc., 127 (1980) 811.
276 Y. Matsumoto and E. Sato, Electrochim. Acta, 25 (1980) 585.
277 Y. Matsumoto and E. Sato, Electrochim. Acta, 24 (1979) 421.
278 Y. Matsumoto, S. Yamada, T. Nishida and E. Sato, J. Electrochem. Soc., 127 (1980) 2360.
279 Y. Matsumoto, H. Yoneyama and H. Tamura, J. Electroanal. Chem., 79 (1977) 319.
280 Y. Matsumoto, H. Yoneyama and H. Tamura, J. Electroanal. Chem., 83 (1977) 237.
281 A.G.C. Kobussen, F.R. van Buren, T.G.M. van den Belt and H.J.A. Van Wees, J. Electroanal. Chem., 96 (1979) 123.
282 A.G.C. Kobussen and C.M.A.M. Mesters, J. Electroanal. Chem., 115 (1980) 131.
283 A.G.C. Kobussen and G.H.J. Broers, J. Electroanal. Chem., 126 (1981) 221.
284 A.G.C. Kobussen, H. Willems and G.H.J. Broers, J. Electroanal. Chem., 142 (1982) 85.
285 H. Willems, A.G.C. Kobussen and J.H.W. DeWitt, J. Electroanal. Chem., 194 (1985) 317.
286 A.G.C. Kobussen, J. Electroanal. Chem., 126 (1981) 199.
287 A.G.C. Kobussen, H. Willems and G.H.J. Broers, J. Electroanal. Chem., 142 (1982) 67.

288 A.N. Krasil'shchikov, Zh. Fiz. Khim., 37 (1963) 531.
289 T. Otagawa and J.O'M. Bockris, J. Electrochem. Soc., 129 (1982) 2391.
290 J.O'M. Bockris and T. Otagawa, J. Electrochem. Soc., 131 (1984) 290.
291 J.O'M. Bockris, T. Otagawa and V. Young, J. Electroanal. Chem., 150 (1983) 633.
292 M.R. Tarasevich, G.I. Zakharkin, A.M. Khutornoi, F.V. Makordei and V.I. Nikitin, Zh. Prikl. Khim., 49 (1976) 953.
293 B.N. Efremov, M.R. Tarasevich, G.I. Zhakharkin, F.Z. Sabirov and I.A. Kropp, Zh. Prikl. Khim., 51 (1978) 731.
294 A.C.C. Tseung and S. Jasem, Electrochim. Acta, 22 (1977) 31.
295 J.G.D. Haenen, W. Visscher and E. Barendrecht, J. Appl. Electrochem., 15 (1985) 29.
296 P. Rasiyah and A.C.C. Tseung, J. Electrochem. Soc., 130 (1983) 365.
297 S.M. Jasem and A.C.C. Tseung, J. Electrochem. Soc., 126 (1979) 1353.
298 V.S. Bagotzky, N.A. Shumilova and E.I. Khruscheva, Electrochim. Acta, 21 (1976) 919.
299 C.R. Davidson, G. Kissel and S. Srinivasan, J. Electroanal. Chem., 13 (1982) 129.
300 P. Rasiyah, A.C.C. Tseung and D.B. Hibbert, J. Electrochem. Soc., 129 (1982) 1724.
301 G. Singh, M.H. Miles and S. Srinivasan, in A.D. Franklin (Ed.), Electrocatalysis on Non-Metallic Surfaces, (N.B.S. Spec. Publ. No. 455), U.S. Government Printing Office, Washington, DC, 1976, p.289.
302 B.N. Efremov and M.R. Tarasevich, Elektrokhimiya, 17 (1981) 1672.
303 P. Rasiyah and A.C.C. Tseung, J. Electrochem. Soc., 130 (1983) 2384.
304 D.B. Hibbert, J. Chem. Soc. Chem. Commun., (1980) 202.
305 A.C.C. Tseung and S.M. Jasem., Electrochim. Acta, 22 (1976) 501.
306 P. Rasiyah and A.C.C. Tseung, J. Electrochem. Soc., 131 (1984) 803.
307 D.M. Shub, A.N. Chemodanov and V.V. Shalaginov, Elektrokhimiya, 14 (1978) 595.
308 V.V. Shalaginov, I.D. Belova, Yu.E. Roginskaya and D.M. Shub, Elektrokhimiya, 14 (1978) 1708.
309 C. Iwakura, A. Honji and H. Tamura, Electrochim. Acta, 26 (1981) 1319.
310 A.C.C. Tseung, P. Rasiyah, M.C.M. Mann and K.L.K. Yeung, Hydrogen as an Energy Vector, Commission of European Communities, 1978, p. 199.
311 H. Wendt, H. Hofmann and V. Plzak, in J.D.E. McIntyre, M.J. Weaver and E.B Yeager (Eds.), The Chemistry and Physics of Electrocatalysis, The Electrochemical Society, Pennington, NJ, 1984, p. 537.
312 C. Iwakura, M. Nishioka and H. Tamura, Nippon Kagaku Kaishi, (1982) 1136.
313 C. Iwakura, M. Nishioka and H. Tamura, Denki Kagaku, 49 (1981) 535; Chem. Abstr., 95 (1981) 158689c.
314 J. Orehotsky, H. Huang, C.R. Davidson and S. Srinivasan, J. Electroanal. Chem., 95 (1979) 233.
315 M. Morita, C. Iwakura and H. Tamura, Electrochim. Acta, 22 (1977) 325.
316 M. Morita, C. Iwakura and H. Tamura, Electrochim. Acta, 23 (1978) 331.
317 A.K. Gorbachev, E.E. Krech and V.I. Shmorgun, Elektrokhimiya, 13 (1977) 1046.
318 E.A. Kalinovskii, V.A. Shustov, V.M. Chaikovskaya and O.L. Prusskaya, Elektrokhimiya, 12 (1976) 1573.
319 V. Plzak, H. Schneider and H. Wendt, Ber. Bunsenges. Phys. Chem., 78 (1974) 1373.
320 T.H. Randle and A.T. Kuhn, in A.T. Kuhn (Ed.), The Electrochemistry of Lead, Academic Press, London, 1979, p. 217.
321 P. Ruetschi, R.T. Angstadt and B.D. Cahan, J. Electrochem. Soc., 106 (1959) 547.
322 G.N. Kokhanov and N.G. Baranova, Elektrokhimiya, 8 (1972) 864.
323 P. Jones, R. Lind and W.F.K. Wynne-Jones, Trans. Faraday Soc., 50 (1954) 972.
324 P. Ruetschi and B.D. Cahan, J. Electrochem. Soc., 105 (1958) 369.
325 A. Fukasawa and M. Ueda, Denki Kagaku, 44 (1976) 646.
326 Ch. Comminellis and E. Plattner, J. Appl. Electrochem., 12 (1982) 399.
327 D.H. Smith, in A.T. Kuhn (Ed.), Industrial Electrochemical Processes, Elsevier, Amsterdam, 1971, p. 127.
328 D. Pletcher, Industrial Electrochemistry, Chapman and Hall, New York, 1982, p. 134.

329 P.W.T. Lu and S. Srinivasan, J. Electrochem. Soc., 125 (1978) 1416.
330 M.H. Miles, G. Kissel, P.W.T. Lu and S. Srinivasan, J. Electrochem. Soc., 123 (1976) 332.
331 B.E. Conway and P.L. Bourgault, Can. J. Chem., 40 (1962) 1690.
332 V.A. Kas'yan, V.V. Sysoeva, A.L. Rotinyan and N.N. Milyutin, Elektrokhimiya, 11 (1975) 635.
333 A.L. Krasil'schikov, Zh. Fiz. Khim., 37 (1963) 273.
334 J.C. Botejue Nadesan and A.C.C. Tseung, J. Electrochem. Soc., 132 (1985) 2957.
335 H.S. Horowitz, J.M. Longo and H.H. Horowitz, J. Electrochem, Soc., 130 (1983) 1851.
336 H.S. Horowitz, J.M. Longo, H.H. Horowitz and J.T. Lewandowski, Am. Chem. Soc. Symp. Ser., 279 (1985) 143.
337 C. Iwakura, T. Edamoto and H. Tamura, Bull. Chem. Soc. Jpn., 59 (1986) 145.
338 A. Bielanski and J. Haber, Catal. Rev. Sci. Eng., 19 (1979) 1.
339 A. Cimino and S. Carra, in S. Trasatti (Ed.), Electrodes of Conductive Metallic Oxides, Part A, Elsevier, New York, 1980, p. 97.
340 S.Trasatti, J. Electroanal. Chem., 111 (1980) 125.
341 V.A. Presnov and A.M. Trunov, Elektrokhimiya, 1 (1975) 71.
342 E. Yeager, Natl. Bur. Stand. (U.S.) Spec. Publ., 455 (1976) 203.
343 J.S. Griffith, Proc. R. Soc. (London) Ser. A, 235 (1956) 23.
344 L. Pauling, Nature (London), 203 (1964) 182.
345 A.M. Trunov and V.A. Presnov, Elektrokhimiya, 11 (1975) 77.
346 A.M. Trunov and V.A. Presnov, Elektrokhimiya, 11 (1975) 290.
347 D. Yohe, A. Riga, R. Greef and E. Yeager, Electrochim. Acta, 13 (1968) 1351.
348 H. Yoneyama, T. Fujimoto and H. Tamura, J. Electroanal. Chem., 58 (1975) 422.
349 A.C.C. Tseung, B.S. Bevan and A.D.S. Tantram, Electrochim. Acta, 15 (1970) 473.
350 E.R.S. Winter, J. Catal., 6 (1966) 35.
351 N.A. Shumilova and V.S. Bagotzky, Electrochim, Acta, 13 (1968) 285.
352 N.A. Shumilova. E.I. Kruscheva and O.V. Moravskaya, Denki Kagaku, 43 (1975) 14.
353 V.S. Bagotsky, N.A. Shumilova and E.I. Kruscheva, Electrochim. Acta, 21 (1976) 919.
354 V.S. Bagotsky, N.A. Shumilova, G.P. Samoilov and E.I. Kruscheva, Electrochim. Acta, 17 (1972) 1625.
355 M. Savy, Electrochim. Acta, 13 (1968) 1359.
356 E.I. Kruscheva, O.V. Mokavokaya, V.V. Paronik, L.V. Erenina, N.A. Shumilova and V.S. Bagotsky, Elektrokhimiya, 11 (1975) 620.
357 E.I. Khruscheva, O.V. Maravskaya, N.A. Shumilova and I.I. Astakhov, Elektrokhimiya, 11 (1975) 1868.
358 E.I. Khruscheva, V.V. Karonik, O.V. Mokavskaya, L.V. Erenina and N.A. Shumilova, Elektrokhimiya, 11 (1975) 836.
359 O.V. Moravskaya, E.I. Khruscheva and N.A. Shumilova, Elektrokhimiya, 13 (1977) 563.
360 Ch. Fabjan, M.R. Kazemi and A. Neckel, Ber. Bunsenges. Phys. Chem., 84 (1980) 1026.
361 A.C.C. Tseung and H.L. Bevan, J. Electroanal. Chem., 45 (1973) 429.
362 K.L.K. Yeung and A.C.C. Tseung, J. Electrochem. Soc., 125 (1978) 878.
363 A.C.C. Tseung, J. Electrochem. Soc., 125 (1978) 1660.
364 F.R. van Buren, G.H.J. Broers, C. Boesveld and A.J. Bouman, J. Electroanal. Chem., 87 (1978) 381.
365 D. Hadjiconstantis, Ph.D. Thesis, University of London, 1978.
366 T. Kudo, H. Obayashi and M. Yoshida, J. Electrochem. Soc., 124 (1977) 321.
367 F.R. van Buren, G.H.J. Broers, A.J. Bouman and C. Boesfeld, J. Electroanal. Chem., 87 (1978) 87.
368 A.G.C. Kobussen, F.R. van Buren and G.H.J. Broers, J. Electroanal. Chem., 91 (1978) 211.
369 Y. Matsumoto and E. Sato, Electrochim. Acta, 25 (1980) 585.
370 D.B. Hibbert and A.C.C. Tseung, J. Electrochem. Soc.,. 125 (1978) 74.
371 Y. Matsumoto, H. Yoneyama and H. Tamura, Chem. Lett. Jpn., (1975) 601.
372 Y. Matsumoto, H. Yoneyama and H. Tamura, J. Electroanal. Chem., 83 (1977) 167.

373 Y. Matsumoto, H. Yoneyama and H. Tamura, J. Electroanal. Chem., 83 (1977) 245.
374 Y. Matsumoto, H. Yoneyama and H. Tamura, Bull. Chem. Soc. Jpn., 51 (1978) 1927.
375 R.R. Dogonadze and Yu. Chizmadzev, Dokl. Akad. Nauk SSSR, 145 (1962) 849.
376 G. Bronoel, J.C. Grenier and J. Reby, Electrochim, Acta, 25 (1980) 1015.
377 Y. Takeda, R. Kanno, T. Kondo, O. Yamamoto, H. Taguchi, M. Shimada and M. Koizumi, J. Appl. Electrochem., 12 (1982) 275.
378 B.C. Wang, J. Molla, W. Aldred and E. Yeager, Proc. 32nd Int. Soc. Electrochem. Mtg., Dubrovnik, Yugoslavia, 1981, p. 130.
379 B.C. Wang and E. Yeager, Proc. 32nd Int. Soc. Electrochem. Mtg., Dubrovnik, Yugoslavia, 1981, p. 125.
380 I.A. Raj, K.V. Rao and V.K. Venkatesan, Prog. Batteries Solar Cells, 5 (1984) 342.
381 L.A. Sazonov, Z.V. Moskvina and E.V. Artamov, Kinet. Catal. (USSR), 15 (1974) 120.
382 G. Karlsson, Electrochim. Acta, 30 (1985) 1555.
383 J. Vondrak and L. Dolezal, Electrochim. Acta, 29 (1984) 477.
384 D.B. Sepa, A. Damjanovic and J.O'M. Bockris, Electrochim. Acta, 12 (1967) 746.
385 J. McHardy and J.O'M. Bockris, J. Electrochem. Soc., 120 (1973) 53.
386 J.P. Randin, Electrochim. Acta, 19 (1974) 87.
387 A.J. Appleby and C. Van Drunen, J. Electrochem. Soc., 123 (1976) 200.
388 B. Broyde, J. Catal., 10 (1968) 13.
389 J.H. Fishman, J.F. Henory and S. Tassore, Electrochim. Acta, 14 (1969) 1314.
390 K. Scott, M.O. Kang and J. Winnick, J. Electrochem. Soc., 130 (1983) 527.
391 H.S. Isaacs and L.J. Olmer, J. Electrochem. Soc., 129 (1982) 436.
392 C.R. Davidson, Ph.d. Dissertation, University of Virginia, 1978.
393 M.R. Tarasevich and B.N. Efremov, in S. Trasatti (Ed.), Electrodes of Conductive Metallic Oxides, Part A, Elsevier, New York, 1980, p. 221.
394 W.J. King and A.C.C. Tseung, Electrochim. Acta, 19 (1974) 485.
395 W.J. King and A.C.C. Tseung, Electrochim. Acta, 19 (1974) 493.
396 A.C.C. Tseung and K.L.K. Yeung, J. Electrochem. Soc., 125 (1978) 1003.
397 A.M. Trunov, A.A. Domnikov, G.L. Reznikov and F.R. Yuppets, Elektrokhimiya, 15 (1979) 783.
398 G.P. Samoilov, E.I. Kruscheva, N.A. Shumilova and V.S. Bagotsky, Elektrokhimiya, 5 (1969) 1082.
399 M.R. Tarasevich, V.S. Vilinskaya, A.M. Khutornoi, R. Kh. Burshtein, F.V. Makordei and Yu. A. Tkach, Elektrokhimiya, 12 (1976) 504.
400 A.N. Frumkin and L.I. Nekrasov, Dokl. Akad. Nauk SSSR, 125 (1959) 115.
401 A. Damjanovic, M.A. Genshaw and J.O'M. Bockris, J. Chem. Phys., 45 (1966) 4057.
402 H.S. Wroblowa, Y.C. Pan and G. Razumney, J. Electroanal. Chem., 69 (1976) 195.
403 R.W. Zurilla. R.K. Sen and E. Yeager, J. Electrochem. Soc., 125 (1978) 1103.
404 V.S. Vilinskaya, N.G. Bulavina, V. Ya. Shepaley and R. Kh. Burshtein, Elektrokhimiya, 15 (1979) 932.
405 N. Nguyen Cong, P. Chartier and J. Brenet, J. Appl. Electrochem., 7 (1977) 383.
406 N. Nguyen Cong, P. Chartier and J. Brenet, J. Appl. Electrochem., 7 (1977) 395.
407 H. Nguyen Cong and J. Brenet, J. Appl. Electrochem., 10 (1980) 433.
408 D.J. Schiffrin, in Specialist Periodical Reports, No. 7, Electrochemistry, The Royal Society of Chemistry, London, 1983, p. 126.
409 V.A. Sadykov and P.G. Tsyrulnikov, Kinet. Katal., 17 (1976) 618.
410 V.A. Sadykov and P.G. Tsyrulnikov, Kinet. Katal., 17 (1976) 626.
411 V.A. Sadykov and P.G. Tsyrulnikov, Kinet. Katal., 18 (1977) 129.
412 V.A. Sadykov and P.G. Tsyrulnikov, Kinet. Katal., 18 (1977) 137.
413 J.R. Goldstein and A. Tseung, J. Phys. Chem., 76 (1972) 3646.
414 M.H. Cota, Nature (London), 203 (1964) 1281.
415 J.R. Goldstein and A.C.C. Tseung, J. Catal., 32 (1974) 452.
416 G.I. Zakharkin, M.R. Tarasevich and A.M. Khutornoi, Elektrokhimiya, 12 (1976) 76.

417 M.R. Tarasevich, A.M. Khutornoi, F.Z. Sabirov, G.I. Zakharkin and V.N. Storoxhenko, Elektrokhimiya, 12 (1976) 256.

418 R.Kh. Burshtein, F.Z. Sabirov and M.R. Tarasevich, Kinet. Katal., 17 (1976) 1333.

419 K. Matsuki, H. Takahashi and E. Yeager, in B. Shumm, Jr., H.M. Joseph and A. Kozawa (Eds.), Manganese Dioxide Symposium, Vol. 2, Cleveland, OH, 1980, p. 76.

420 M.R. Tarasevich, G.I. Zakharkin and R.M. Simirnova, Elektrokhimiya, 8 (1972) 627.

421 F. Haber and J. Weiss, Proc. R. Soc. (London) Ser. A, 147 (1934) 332.

422 P.F. Carcia, R.D. Shannon, P.E. Bierstedt and R.B. Flippen, J. Electrochem. Soc., 127 (1980) 1974.

423 D. Sawyer and E. Seo, Inorg. Chem., 16 (1977) 499.

424 D.T. Sawyer and J.S. Valentine, Acc. Chem. Res., 14 (1981) 393.

425 R.K. Sen, J. Zagal and E. Yeager, Inorg. Chem., 16 (1977) 3379.

426 I. Morcos and E. Yeager, Electrochim. Acta, 15 (1970) 953.

427 E. Yeager, D. Scherson and B. Simic-Glavaski, in J.D.E. McIntyre, M.J. Weaver and E.B. Yeager (Eds.), The Chemistry and Physics of Electrocatalysis, The Electrochem. Society, Pennington, NJ, 1984, p. 247.

428 M.H. Miles, Y.H. Huang and S. Srinivasan, J. Electrochem. Soc., 125 (1978) 1931.

429 B. Parkinson, F. Decker, J.F. Juliao and M. Abramovich, Electrochim. Acta, 25 (1980) 501.

430 P. Clechet, C. Nartelet, J.R. Martin and R. Olier, Electrochim. Acta, 24 (1979) 457.

431 E. Yesodharan, S. Yesodharan and M. Gratzel, Solar Energy Mater., 10 (1984) 287.

432 P. Salvador and C. Gutierrez, Chem. Phys. Lett., 86 (1982) 131.

433 K. Matsuki and H. Kamada, Electrochim. Acta, 31 (1986) 13.

434 K.V. Rao, V.K. Venkatesan and H.V.K. Udupa, J. Electrochem. Soc. India, 31 (1982) 33.

435 E.J. Calvo, Extended Abstracts 164th Meeting of the Electrochemical Society, Washington, DC, 1983, p. 644.

436 M. Stratmann, K. Bohnenkamp and H.J. Engel, Werkst. Korros., 34 (1983) 604.

437 M. Stratmann, K. Bohnenkamp and H.J. Engell, Corros. Sci., 23 (1983) 969.

438 J. Shi Choi and Ki Hyun Yoon, J. Phys. Chem., 74 (1970) 1095.

439 A. Kamalova and N.F. Razina, Izv. Akad. Nauk Kaz. SSR Ser. Khim., 28 (1978) 30.

440 N.S. McAlpine and R.A. Fredlein, Aust. J. Chem., 36 (1983) 11.

441 E.J. Calvo and D.K. Schiffrin, J. Electroanal. Chem., 163 (1984) 257.

442 T. Takei and H.A. Laitinen, Surf. Technol., 16 (1982) 185.

443 T. Takei and H.A. Laitinen, Surf. Technol., 18 (1983) 123.

444 E.J. Calvo and D.J. Schiffrin, in Proc. 32nd Int. Soc. Electrochem. Mtg., Dubrovnik, Yugoslavia, 1981, p. 781.

445 N.F. Razina, Izv. Akad. Nauk Kaz. SSR Ser. Khim., 29 (1980) 75.

446 K. Ogura and Y. Tanaka, Electrochim. Acta, 28 (1983) 1671.

447 W. Fischer, W. Siedlarek, H. Hinuber and R. Lunenshloss, Werkst. Korros., 31 (1980) 774.

448 K. Balakrishnan and V.K. Venkatesan, Electrochim. Acta, 24 (1979) 131.

449 A.T. Kuhn and C.J. Mortimer, J. Electrochem. Soc., 120 (1973) 231.

450 B.V. Tilak, J. Electrochem. Soc., 120 (1979) 1343.

451 T. Arikado, C. Iwakura and H. Tamura, Electrochim. Acta, 22 (1977) 229.

452 B.E. Conway and D.M. Novak, J. Electroanal. Chem., 99 (1979) 133.

453 B.E. Conway and D.M. Novak, J. Chem. Soc., Faraday Trans., 75 (1979) 2454.

454 D.M. Novak, B.V. Tilak and B.E. Conway, in J. O'M.Bockris, B.E. Conway and R.E. White (Eds), Modern Aspects of Electrochemistry, No. 14, Plenum Press, New York, 1982, p. 195.

455 A. Nidola, in S. Trasatti (Ed.), Electrodes of Conductive Metallic Oxides, Part B, Elsevier, New York, 1981, p. 627.

456 I.R. Burrows, D.A. Denton and J.A. Harrison, Electrochim. Acta, 23 (1978) 493.

457 I.R. Burrows, J.H. Entwistle and J.A. Harrison, J. Electroanal. Chem., 77 (1977) 21.

458 R.T. Atanasoski, B.Z. Nikolic, M.M. Jaksie and A.R. Despic, J. Appl. Electrochem., 5 (1975) 155.

459 O. deNora, Chem. Ing. Tech., 42 (1970) 222.
460 D.L. Caldwell, in J.O'M. Bockris, B.E. Conway, E. Yeager and R.E. White (Eds.,), Comprehensive Treatise of Electrochemistry, Vol. 2, Plenum Press, New York, 1981, Ch. 2.
461 R.G. Erenburg, L.I. Krishtalik and I.P. Yaroshevskaya, Elecktrokhimiya, 11 (1975) 1072.
462 L.J.J. Janssen, L.M.C. Starmans, J.G. Visser and E. Barendrecht, Electrochim. Acta, 22 (1977) 1093.
463 R.G. Erenburg, L.I. Krishtalik and V.I. Bystrov, Elektrokhimiya, 8 (1972) 1740.
464 I.A. Ivanter, G.P. Pechnikova and V.L. Bubasov, Elektrokhimiya, 12 (1976) 787.
465 D.A. Denton, J.A. Harrison and R.I. Knowles, Electrochim. Acta, 24 (1979) 521.
466 S. Ardizzone, A. Carugati, G. Lodi and S. Trasatti, J. Electrochem. Soc., 129 (1982) 1692.
467 G.Z. Yrebenik, V.L. Kubasov and V. Zyrebenik, Zh. Prikl. Khim., 51 (1978) 359.
468 I.E. Veselovskaya, E.K. Spasskaya, V.A. Sokolov, V.I. Tkachenko and L.M. Yakimenko, Elektrokhimiya, 10 (1974) 70.
469 M.M. Pecherskii, V.V. Yorodetskii, S.V. Evdokimov and V.V. Losev, Elektrokhimiya, 17 (1981) 1087.
470 S.V. Evdokimov, V.V. Yorodetskii and V.V. Losev, Elektrokhimiya, 21 (1985) 1427.
471 V.V. Losev, Elektrokhimiya, 17 (1981) 733.
472 I.V. Kadija, J. Electrochem. Soc., 127 (1980) 599.
473 V.L. Müller, M. Krenz and M. Zettler, Z. Phys. Chem. (Leipzig), 265 (1984) 729.
474 L. Müller, M. Krenz and R. Landsberg, J. Electroranal. Chem., 180 (1984) 453.
475 G. Faita and G. Fiori, J. Electrochem. Soc., 120 (1973) 1702.
476 T. Arikado, C. Iwakura and H. Tamura, Electrochim. Acta, 23 (1978) 9.
477 R.G. Erenburgh, L.I. Krishtalik and N.P. Rogozhina, Elektrokhimiya, 20 (1984) 1183.
478 R.G. Erenburgh, L.I. Krishtalik and I.P. Yaroskeyskaya, Elektrokhimiya, 11 (1975) 1068.
479 R.G. Erenburgh, L.I. Krishtalik and I.P. Yarosheyskaya, Elektrokhimiya, 11 (1975) 1236.
480 M. Enyo and T. Yokoyama, Electrochim. Acta, 16 (1971) 223.
481 L.D. Burke and J.F. O'Neill, J. Electroanal. Chem., 101 (1979) 341.
482 V.V. Yorodetskii, P.N. Zorin, M.M. Pecherskii, V.B. Busse-Machukas, V.L. Kubasov and Yu.Ya. Tomaskpol'skii, Elektrokhimiya, 17 (1981) 79.
483 M. Inai, C. Iwakura and H. Tamura, Electrochim. Acta. 24 (1979) 993.
484 D.A. Denton, J.A. Harrison and R.I. Knowles, Electrochim. Acta, 25 (1980) 1147.
485 J.A. Harrison, D.L. Caldwell and R.E. White, Electrochim. Acta, 28 (1983) 1561.
486 E.A. Kalinovskii, R.U. Bondar and N.N. Meshkova, Elektrokhimiya, 8 (1972) 1468.
487 K.J. O'Leary and T.J. Navin, in T.C. Jeffrey, P.A. Danna and H.S. Holden (Eds.), Chlorine Bicentennial Symposium, The Electrochemical Society, Princeton, NJ, 1974, p. 174.
488 H. Tamura, C. Iwakura and M. Morita, in B. Schumm, H.M. Joseph and A. Kozawa (Eds.), Proc. 2nd Manganese Dioxide Symp., 1981, p. 349.
489 M. Morita. C. Iwakura and H. Tamura, Electrochim. Acta, 24 (1979) 639.
490 M. Morita, C. Iwakura and H. Tamura, Electrochim. Acta. 25 (1980) 1341.
491 A. Tvarusko, 34th ISE Meeting, Erlangen, Extended Abstr. III-7, 1983.
492 D.M. Shub, M.F. Reznik, V.V. Shalaginov, E.N. Lubnin, N.V. Kozlova and V.N. Lomova, Elektrokhimiya, 19 (1983) 502.
493 V.V. Shalaginov, D.M. Shub, N.V. Kozlova and V.N. Lomova, Elektrokhimiya, 19 (1983) 537.
494 R. Boggio, A. Carugati, G. Lodi and S. Trasatti, J. Appl. Electrochem., 15 (1985) 335.
495 M.B. Konovalov, V.I. Bystrov and V.L. Kubasov, Elektrokhimiya, 11 (1975) 1266.
496 M.B. Konovalov, V.I. Bystrov and V.L. Kubasov, Elektrokhimiya, 11 (1975) 239.
497 R.A. Agapova and G.N. Kokhanov, Elektrokhimiya, 12 (1976) 1649.
498 R. Garavaglia, C.M. Mari, S. Trasatti and C. DeAsmundis, Surf. Technol., 19 (1983) 197.
499 C. Pirovano and S. Trasatti, J. Electroanal. Chem., 180 (1984) 171.
500 R.I. Mostkova, N.F. Nikol'skaya and L.I. Krishtalik, Elektrokhimiya, 19 (1983) 1608.
501 M.J. Hazelrigg and D.L. Caldwell, Electrochem. Soc. Spring Meeting, Extended Abstr. 78–1, 1978, 1139.
502 M. Hayes and A.T. Kuhn, J. Appl. Electrochem., 8 (1978) 327.

503 T. Matsumura, R. Itai, M. Shibuya and G. Ishi, Electrochem. Technol., 6 (1968) 402.
504 T.A. Chertykovtseva, D.M. Shub and V.I. Veselovskii, Elektrokhimiya, 14 (1978) 275.
505 D.M. Shub, T.A. Chertykovtseva and V.I. Veselovksii, Elektrokhimiya, 13 (1977) 415.
506 A.S.W. Johnson and A.C.C. Tseung, J. Appl. Electrochem., 7 (1977) 445.
507 T.A. Chertykovtseva, Z.D. Skuridina, D.M. Shub and V.I. Veselovskii, Elektrokhimiya, 14 (1978) 1412.
508 E.E. Littauer and L.L. Shreir, Electrochim. Acta, 12 (1967) 465.
509 K. Nishibe, Denki Kagaku, 52 (1984) 836; Chem. Abstr., 102 (1985) 193756g.
510 M. Ueda and A. Fukasawa, Nippon Kaisu Yakkaiski, 38 (1984) 233; Chem. Abstr., 102 (1985) 102383u.
511 J. Mozota, M. Vukovic and B.E. Conway, J. Electroanal. Chem., 144 (1980) 153.
512 J. Mozota and B.E. Conway, J. Electrochem. Soc., 128 (1981) 2142.
513 M. Vokovic, H. Angerstein-Kozlowska and B.E. Conway, J. Appl. Electrochem., 12 (1982) 193.
514 E.J. Rudd and B.E. Conway, in B.E. Conway, J.O'M. Bockris, E. Yeager, S.U.M. Khan and R.E. White (Eds.), Comprehensive Treatise of Electrochemistry, Vol. 72, Plenum Press, New York, 1983, p. 641.
515 M.Ya Fioshin and A.P. Tomilov, Elektrokhimiya, 19 (1983) 3.
516 H. Wendt, Electrochim. Acta, 29 (1984) 1513.
517 B.E. Conway, in S. Trasatti (Ed.), Electrodes of Conductive Metallic Oxides, Part A, Elsevier, New York, 1980, p. 433.
518 L.L. Miller, Pure Appl. Chem., 51 (1979) 2125.
519 M.M. Baizer, Pure Appl. Chem., 58 (1986) 889.
520 K. Köster and H. Wendt, in J.O'M. Bockris, B. Conway, E. Yeager and R.E. White (Eds.), Comprehensive Treatise of Electrochemistry, Vol. 2, Plenum Press, New York, 1981, p. 251.
521 M.M. Baizer (Ed.), Organic Electrochemistry, Dekker, New York, 1973.
522 F. Beck, Elektro-Organische Chemie, Verlag Chemie, Weinheim, 1973.
523 S. Swann, Jr., J. Electrochem. Soc., 99 (1952) 125C.
524 H.J. Creighton, J. Electrochem. Soc., 99 (1952) 127C.
525 M.M. Baizer, J. Electrochem. Soc., 124 (1977) 185C.
526 M.J. Allen, Organic Electrode Processes, Reinhold, New York, 1958.
527 C.J. Brockman, Electro-organic Chemistry, Wiley, New York, 1926.
528 W. Vieltich, Fuel Cells, Wiley-Interscience, London, 1970.
529 M. Fleischmann, K. Korinek and D. Pletcher, J. Electroanal. Chem., 31 (1971) 39.
530 J.P Carr and N.A. Hampson, Chem. Rev., 72 (1972) 679.
531 J.P. Pohl and H. Rickert, in S. Trasatti, (Ed.), Electrodes of Conductive Metallic Oxides, Part A, Elsevier, New York, 1980, p. 183.
532 T. Sekine and K. Sugino, J. Electrochem. Soc., 115 (1968) 242.
533 J. Mitzuguchi and S. Matsumoto, J. Pharm. Soc. Jpn., 71 (1951) 737.
534 F. Fichter and K. Grizard, Helv. Chim. Acta, 4 (1921) 928.
535 H. Elbs, Z. Elektrochem., 2 (1896) 552.
536 J. Labhard and K. Zschoche, Z. Elektrochem., 8 (1902) 93.
537 J.S. Clarke, R.E. Ehigamusoe and A.T. Kuhn, J. Electroanal. Chem., 70 (1976) 333.
538 C. Merzbacher and M. Smith, J. Am. Chem. Soc., 22 (1900) 223.
539 C. Michler and J. Pattinson, Chem. Ber., 14 (1891) 2163.
540 M. Frevery, H. Hover and G. Schwarzlose, Chem. Ing. Tech., 46 (1974) 635.
541 J.P. Millington and J. Trotenan, Electrochem. Soc. Spring Mtg., Extended Abstracts 76-1, 1976, p. 700.
542 J.A. Harrison and J.M. Mayne, Electrochim. Acta, 28 (1983) 1223.
543 P. Tissot, H.D. Duc and O. John, J. Appl. Electrochem., 11 (1981) 473.
544 A. Nilsson, A. Ronlan and V.D. Parker, J. Chem. Soc. Perkin Trans. 1, (1973) 2337.
545 M.Ya. Fioshin, G.A. Kokarev, V.A. Koleshnikov, V.I. Bazakin and A.T. Sorokovykh, Elektrokhimiya, 13 (1977) 381.
546 B. Fleszar and J. Ploszynska, Electrochim. Acta, 30 (1985) 31.

547 C.D. Levina, C.M. Kolesova and Yu.B. Vasil'ev, Elektrokhimiya, 13 (1977) 1059.
548 M. Amjad, D. Pletcher and C.Z. Smith, J. Electrochem. Soc., 124 (1977) 203.
549 M. Fleischmann, K. Korinek and D. Pletcher, J. Chem. Soc. Perkin Trans. 2, (1972) 1396.
550 G. Horanyi, G. Vertes and F. Nagy, Acta Chim. Hung., 67 (1971) 357.
551 G. Vertes and F. Nagy, Acta Chim. Hung., 74 (1972) 405.
552 B.S. Remorov, I.A. Avrutskaya and M.Ya. Fioshin, Elektrokhomiya, 17 (1981) 743.
553 P. Robertson, J. Electroanal. Chem., 111 (1980) 97.
554 G. Vertes and G. Horanyi, J. Electroanal. Chem., 52 (1974) 47.
555 K. Manandhar and D. Pletcher, J. Appl. Electrochem., 9 (1979) 707.
556 T.E. Mulina, I.A. Avrutskaya, M.Ya. Fioshin and T.A. Malakhova, Elektrokhimiya, 10 (1974) 481.
557 I.A. Avrutskaya, M.Ya. Fioshin, T.E. Mulina and T.A. Arkhinova, Elektrokhimiya, 11 (1975) 1260.
558 P.M. Robertson, P. Berg, H. Reimann, K. Schluch and P. Seiler, J. Electrochem. Soc., 130 (1983) 591.
559 P.M. Robertson, F. Schwager and N. Ibl, J. Electroanal. Chem., 65 (1975) 883.
560 P.M. Robertson, P. Cetton, D. Matic, F. Schwager, A. Storck and N. Ibl, AIChE Symp. Ser. No. 185, 75 (1979) 115.
561 H.H. Horowitz, H.S. Horowitz and J.M. Longo, in W.E. O'Grady, P.N. Ross, Jr. and F.G. Will (Eds.), Electrocatalysis, The Electrochemical Society, Pennington, NJ, 1982, p. 285.
562 A. Trojanek and R. Kalvoda, Trans. Soc. Adv. Electrochem. Sci. Technol., 12 (1977) 45.
563 E.J.M. O'Sullivan and J.R. White, Electrochem. Soc. Mtg., Boston, 1986, Extended Abstracts 86-1, 1986, p. 713.

Index

P

Q

R

S

T

U

V

W

X